D1069119

Turbulent Flows
in
Engineering

Turbulent Flows in Engineering

A. J. REYNOLDS

Department of Mechanical Engineering,
Brunel University, London

A Wiley–Interscience Publication

JOHN WILEY & SONS

London · New York · Sydney · Toronto

Library of Congress Catalog Card No. 73-8464

ISBN 0 471 71782 7

Printed in Great Britain by
William Clowes & Sons Limited, London, Colchester and Beccles

Preface

'Turbulent Motion. It remains to call attention to the chief outstanding difficulty of our subject.'

Horace Lamb, *Hydrodynamics*, 1932.

Lamb's opinion of turbulence is still valid. He spoke as a mathematician, but the engineer also finds that turbulence provides many of the most challenging problems in fluid mechanics. A turbulent flow comprises motions of widely differing length and time scales, and our understanding varies markedly along the scale of size. The smallest and largest elements are not too mysterious; it is the intermediate scales that pose the really tantalizing questions. The Navier–Stokes momentum equations describe the instantaneous fluid motion and give considerable insight into the smallest scales of turbulence. The traditional approach of engineering fluid mechanics gives integral relationships among the gross attributes of certain technically important flows. The common area of ignorance lies in the middle scales of the flow. From the mathematical point of view, we encounter the problem of closure—averaging of the basic equations gives more unknowns than equations, and further information is needed to define a closed problem. From the experimental point of view, we face the problem of structure—that of isolating simple elements within a turbulent flow which give rise to its characteristic activity.

Here we shall adopt the approach of the engineer, treating first problems which require only a casual knowledge of the smaller scales of motion, and moving progressively to more complex processes and flow geometries. Accordingly, the mathematical analysis is simplest at the beginning, where channel flows are considered, becomes more demanding in the study of developing flows, and attains its greatest complexity in the final chapter. The exception to this pattern is the part of Chapter 2 which treats the analysis of measurements, but it is not necessary to master this material before proceeding to subsequent chapters.

This book is addressed to mechanical, aeronautical and process engineers, either in training or already in practice. An attempt has been made to tie together the applications which arise in these fields, by providing a unified development of the conservation laws and field equations governing turbulent transport processes. The treatment assumes the prior knowledge which might be gained in a single introductory course in engineering fluid mechanics, and is suitable for undergraduate students in their final year. Since related laminar flows are treated in passing, there is no compelling argument against the use of the book by students at an earlier stage in an engineering course, or by readers who have not had this kind of training. The book will also serve as an introduction to the field of turbulence for those intending to undertake more extensive studies, theoretical or experimental.

The primary aim of this book is to present in a rational way the methods of predicting turbulent flows now used in engineering practice, and to indicate how they can be extended. To do this, it is necessary to develop a realistic picture of the processes within turbulent flows, and to introduce appropriate terminology and analytical techniques. Thus the reader is prepared to approach the specialist literature of the subject with some confidence.

By quoting numerical values for many of the quantities which arise in the analysis, I have tried to give a feeling for orders of magnitude and for problems of measurement. Many of these values are sufficiently accurate for use in design calculations, but others are estimates based on limited experiments, and must be treated with some caution. Considering, in particular, the molecular transport properties given in the Appendix, the viscous and thermal properties are fairly trustworthy, while the diffusive characteristics are less well established.

At the end of each chapter will be found a number of analytical and numerical examples, exercises and problems—collectively called 'Examples'. They are designed to amplify the discussion in the body of the text, and to provide the confidence and understanding which can be gained only by working things out for oneself. These 'Examples' are directed at students undergoing a systematic course of study, but other readers will profit by at least reading through them.

The basic pattern of the book is as follows. The first chapter presents the 'conventional wisdom' of the subject, much of it familiar to readers with some previous knowledge of engineering fluid mechanics. Chapter 2, on measurements, departs from the main line of development by dealing explicitly with the statistical aspects of turbulence. Although the detail is not required, some familiarity with these matters is necessary for a full comprehension of the methods used in engineering calculations. The body of the text comprises three chapters on channel flows and three on developing flows. The first chapter of each set (that is, Chapters 3 and 6) provides fundamental results which are used in the chapters devoted to specific applications—friction prediction, heat and mass transfer, free turbulence and boundary layers. The final chapter returns to the statistical point of view: Reynolds momentum equations and allied results are derived, and are used to illustrate some recently developed methods of modelling turbulent flows.

I feel that I should apologize for restricting consideration, for the most part, to a continuous fluid of constant density. It seemed better to deal fairly comprehensively with a few often encountered situations than to try to include the whole range of turbulent activity which arises in engineering. Picking out only a few topics, I regret not being able to discuss particle transport, two-phase flows, atmospheric turbulence, high-speed flows where compressibility is important, and flows in which reactions occur. For those readers whose interests lie in these areas, I can only hope that this book will prove helpful as a basis for further reading.

A. J. Reynolds

References

For the purposes of this book, it seemed appropriate to direct the reader not to the basic literature, but to systematic treatises. At the end of each chapter a number of such works are listed, as 'Further Reading'. Many contain extensive surveys of the specialized literature, and it seemed pointless to repeat them in this book. However, some of the surveys are now a little dated, and many readers will desire an entrée to recent work. Accordingly, at the end of each chapter is provided a second group of 'Specific References', most having appeared within the last decade. There are only half-a-dozen per chapter, but they serve to indicate relevant journals and to identify sources mentioned in the text.

For compactness, some often-mentioned works have been given reference numbers, as indicated below. While most of these are textbooks or specialist treatises, the last four are collections of papers concerning more recent developments.

1. Bird, R. B., W. E. Stewart and E. N. Lightfoot. *Transport Phenomena*, Wiley, New York (1960)
2. Bradshaw, P. *An Introduction to Turbulence and its Measurement*, Pergamon, Oxford (1971)
3. Eckert, E. R. G. and R. M. Drake, Jr. *Analysis of Heat and Mass Transfer*, McGraw-Hill, New York (1972)
4. Goldstein, S. (Editor). *Modern Developments in Fluid Dynamics*, Oxford (1938)
5. Hinze, J. O. *Turbulence: An Introduction to Its Mechanism and Theory*, McGraw-Hill, New York (1959)
6. Launder, B. E. and D. B. Spalding. *Mathematical Models of Turbulence*, Academic Press, London (1972)
7. Lin, C. C. (Editor). *Turbulent Flows and Heat Transfer*, Princeton (1959)
8. Prandtl, L. *The Essentials of Fluid Dynamics*, Blackie, London (1954)
9. Rohsenow, W. M. and H. Y. Choi. *Heat, Mass and Momentum Transfer*, Prentice-Hall, Englewood Cliffs, New Jersey (1961)
10. Rouse, H. (Editor). *Advanced Mechanics of Fluids*, Wiley, New York (1959)
11. Schlichting, H. *Boundary-layer Theory*, McGraw-Hill, New York (1960)
12. Tennekes, H. and J. L. Lumley. *A First Course in Turbulence*, M. I. T. Press, Cambridge, Mass. (1972)
13. Townsend, A. A. *The Structure of Turbulent Shear Flow*, Cambridge (1956)

14. *The Mechanics of Turbulence*, proceedings of a symposium held in Marseilles, edited by A. Favre, Gordon and Breach, New York (1964)
15. *Recent Developments in Boundary Layer Research*, proceedings of a meeting in Naples, AGARDograph 97 (1965)
16. *Computation of Turbulent Boundary Layers*, edited by S. J. Kline, *et al.*, proceedings of a conference organized by the Thermosciences Division, Department of Mechanical Engineering, Stanford University (1969)
17. *Turbulent Shear Flows*, proceedings of a conference hold in London, AGARD-CP-93 (1971)

Acknowledgements

I thank Dr I. S. Gartshore and Dr A. J. Ward Smith for reading parts of the manuscript and offering opinions on it. I also thank my wife, Caroline, for help with the proof reading, and for many other kinds of assistance.

I acknowledge with thanks permission to reproduce wholly or in part the following items; in certain cases a specific source is given at the end of a chapter.

Figure 1.3 with permission of the American Society of Mechanical Engineers; Figures 2.8, 4.6 and 4.7 from *The Journal of Fluid Mechanics* with permission of Cambridge University Press; Figures 3.2, 3.3 and 3.8 by courtesy of the U.S. National Aeronautics and Space Administration; Figure 4.2 from Volume 4 of *Advances in Applied Mechanics* with permission of Academic Press, Inc., and Professor F. H. Clauser; Figures 5.3 and 5.6 from Reference 7 with permission of the U.S. Office of Naval Research; Figures 6.1, 6.2, 6.3 and 6.5 from Reference 13, with permission of Cambridge University Press; Plate I with permission of Professor S. Corrsin and the U.S. Army Ballistics Research Laboratories, Aberdeen Proving Ground; and Plates II and III with permission of Armfield Engineering Ltd.

Notation

Symbols are explained as they are introduced in the text. The notation has been kept simple and conventional, and in consequence some symbols are used in more than one sense. The following conventions have generally been adopted:

(1) Fluctuating quantities. Time-mean values are denoted by capitals, fluctuations by small letters. An overbar ($\overline{\ }$) denotes averaging with respect to time. A prime ($'$) denotes an r.m.s. value.

(2) Fluxes and rates. A dot ($\dot{}$) is used to denote a transfer per unit time. Certain quantities per unit width of a plane flow are denoted by a prime ($'$).

(3) Coordinates. x, x_1, u and U are along a wall or in the direction of the main stream; y, x_2, r, v and V are normal to a wall or in the direction of most rapid change; z, x_3, w and W are normal to the plane of the mean motion; q is the resultant of the velocity fluctuation.

(4) Approximate equality, As a guide, $\simeq$ means 'nearly equal'; $\sim O(x)$ means 'is of the order x'; and $\sim$ means 'something like' or 'behaves like' or, in dimensional analysis, 'depends on'.

(5) Named dimensionless groups are commonly identified by the first two letters of the name, set in roman type.

(6) Correlations and spectra. The special notation is set out in Table 2.1.

(7) Vector quantities are set in bold type. Tensor notation is explained in Section 9.1.1.

Contents

List of Tables

1

The Problems of Turbulence

This chapter asks questions, and takes a few steps towards the answers. It is meant for those who have little knowledge of the subject, and need to be introduced to its terminology, notation, applications, analytical methods and phenomenological basis. This many-faceted introduction is achieved by examining a number of 'simple' flows; in essence, this is a condensed history of the development of our ideas concerning turbulence. The aims of the survey are:

(1) to point out the more obvious features of turbulent flows;
(2) to define terms commonly used in discussing them;
(3) to introduce the notation to be used in describing turbulence and its effects;
(4) to assess methods of analysing turbulent motions; and
(5) to identify the problems which turbulence poses for the engineer.

It may seem odd that we do not begin by defining the object of our investigation. Turbulence will in fact be defined in Section 1.5, when its essential attributes have become clear through a consideration of some specific examples.

1.1 TURBULENT FRICTION AND TRANSITION

Pipe Friction

Figure 1.1 presents experimental results important both for practical calculations and for an understanding of turbulence. The experiment is shown schematically in Figure 1.1(a): a fluid flows steadily through a long, straight, horizontal pipe of uniform diameter d; for simplicity, we take the fluid density ρ and viscosity μ to be constant. Over the length L, well inside the pipe, the fluid pressure drops from p_1 to p_2. Since gravity does not influence the pressure in this horizontal pipe, the fall can be ascribed to friction generated within the flowing fluid.

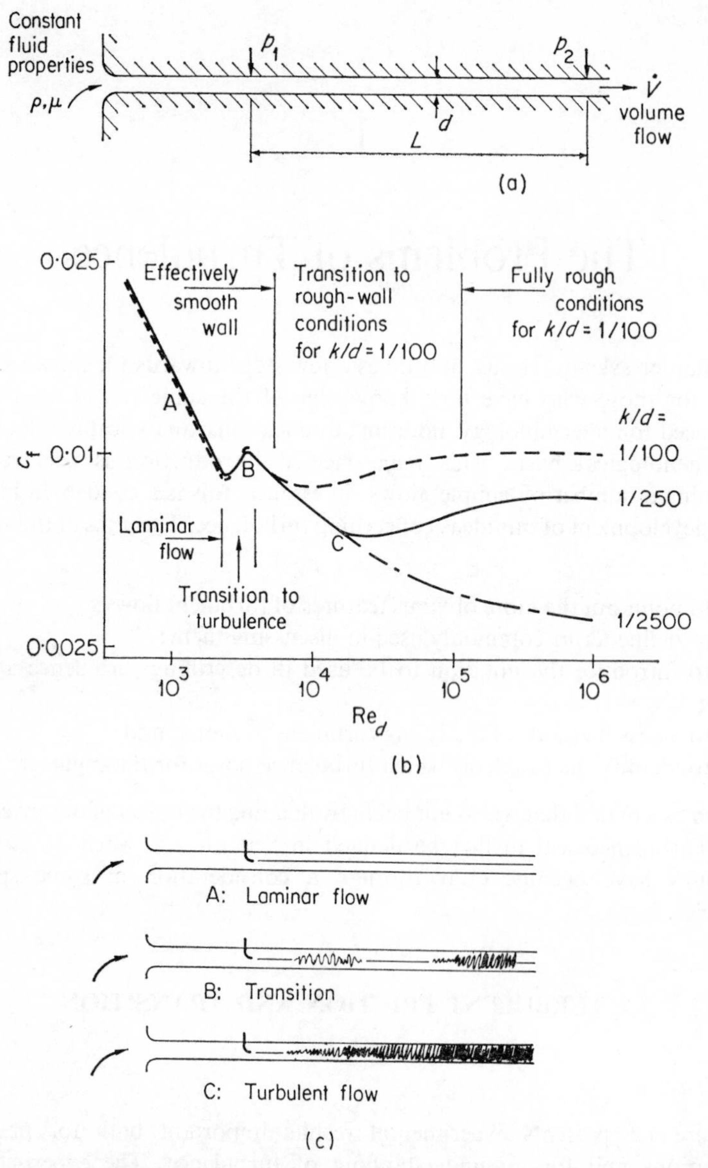

Figure 1.1 Relationship between pipe friction and turbulence, illustrating the roles of transition and roughness. (a) Experimental determination of friction generated by steady flow through a long pipe. (b) Friction characteristics in dimensionless form: friction coefficient $c_f \propto (p_1 - p_2)/\dot{V}^2$ *vs* Reynolds number $\mathrm{Re}_d \propto \dot{V}$, for logarithmic scales. Variations are shown for three levels of wall roughness, measured by the ratio of effective roughness height k to pipe diameter d. (c) Flow visualization corresponding to the flow régimes of (b). A dye filament is introduced along the axis of the pipe

The results of Figure 1.1(b) relate to three kinds of pipe. In two cases the bore has been made rough by the application of uniform roughness elements; this might be done by turning a thread in the pipe wall, or by cementing sand grains to it. The third pipe is fairly smooth, though it will possess a roughness defined by its material and the method of manufacture, and typically with irregularities of various sizes. In the figure, each of the three pipes is identified by its *relative roughness*, the ratio of a typical roughness height k to the pipe diameter d.

In relating the pressure drop and flow rate $\dot{V}$ for these pipes, it is convenient, as in many of the calculations and data correlations of fluid mechanics, to present the results in dimensionless form. We introduce the *Reynolds number* as a dimensionless measure of flow rate:

$$\begin{aligned} \mathrm{Re}_d &= \frac{\rho U_a d}{\mu} = \frac{U_a d}{\nu} \\ &= \frac{4}{\pi}\frac{\dot{V}}{\nu d} \quad \text{for a round pipe} \end{aligned} \tag{1.1.1}$$

with $\nu = \mu/\rho$ the kinematic viscosity of the fluid. The velocity used here

$$\begin{aligned} U_a &= \frac{\dot{V}}{A} \\ &= \frac{4}{\pi}\frac{\dot{V}}{d^2} \quad \text{for a round pipe} \end{aligned} \tag{1.1.2}$$

is the average velocity of the fluid passing along the channel, A being the cross-sectional area. The pressure drop is represented by the dimensionless *friction coefficient*:

$$\begin{aligned} c_f &= \frac{\tau_w}{\frac{1}{2}\rho U_a^2} \\ &= \frac{\pi^2}{32}\frac{d^5}{\rho \dot{V}^2}\frac{p_1 - p_2}{L} \quad \text{for a round pipe} \end{aligned} \tag{1.1.3}$$

Here

$$\begin{aligned} \tau_w &= \frac{(p_1 - p_2)A}{A_w} \\ &= \frac{\frac{1}{4}d(p_1 - p_2)}{L} \quad \text{for a round pipe} \end{aligned} \tag{1.1.4}$$

is the average shear stress on the pipe wall; acting over the wetted area A_w, this just balances the pressure force $(p_1 - p_2)A$. We shall later consider the significance of these dimensionless parameters; for the moment, they are simply devices for presenting empirical data.

The discussion of the results of Figure 1.1(b) falls into two parts, the first relating to the left-hand side of the figure, where the curves lie close together, the second relating to the high-Reynolds-number range, where pipe roughness has a larger role. The roughness-independent results are best understood by considering another simple experiment on pipe flow. Figure 1.1(c) illustrates observations (first reported by Osborne Reynolds in 1883), obtained by introducing a filament of dye on the centre-line of a transparent tube. For sufficiently low Reynolds numbers, the filament stretches undisturbed (save for a slight molecular diffusion of the dye) along the axis; this steady, non-mixing flow is described as *laminar*, since the fluid distortion is akin to that within a pile of thin plates, or laminae, sliding over one another. The laminar flows correspond to region A in Figure 1.1(b), in which the friction coefficient falls steadily, in fact, linearly, in this log–log plot.

At higher Reynolds numbers, the experience of the dye filament is quite different. After progressing undisturbed some way into the pipe, the filament begins to waver from side to side, and finally vanishes in a blur of colour filling the entire cross-section. This unsteady, mixing flow was described by Reynolds as sinuous; nowadays we term it *turbulent*. The turbulent régime corresponds to region C of Figure 1.1(b), where the friction coefficient has a markedly higher value than it would have for laminar flow at the same Reynolds number.

Transition

Our attention is now directed at flows intermediate to those considered above; it is here that the transition from laminar to turbulent flow is most obvious. In this region (B in Figure 1.1) flow visualization reveals patches or flashes of turbulence separated by stretches of nearly laminar flow. The *intermittent* nature of the turbulence, and its sharp demarcation from the adjacent laminar regions, are typical features of the transition between the two modes. Seemingly, the turbulent activity must attain a threshold value in order to sustain itself; if a disturbance does not have the required magnitude, it is damped and rapidly dies away.

Osborne Reynolds found that pipe flow could become turbulent when $\mathrm{Re}_d > 2000$; this provides a useful criterion for assessing many of the channel flows of interest to engineers. However, if the entry to the pipe tapers very gradually and the experimental environment is very still, laminar flow can be maintained at much higher values of the Reynolds number, say, up to $\mathrm{Re}_d = 50{,}000$. Indeed, for the particular case of pipe flow, theoretical arguments suggest that there is no limit to the Reynolds number at which laminar flow is possible, provided that the disturbances are small enough.

The occurrence of transition depends on the *stability* of the basic laminar flow to disturbances imposed by the environment or convected from the inlet.

The flow is unstable if a disturbance gains energy more rapidly than it is lost through viscous dissipation. In the particular case of pipe flow, very weak disturbances cannot extract enough energy from the basic flow; moreover, at sufficiently low Reynolds numbers, even finite perturbations are unable to do so.

An important effect of transition is an increase in the *dissipation* within the pipe, that is, in the rate at which the mechanical energy of the mean flow is reduced by fluid friction. Taking the density to be constant, we calculate the dissipation as the rate at which work is done by the flow against the pressure differential. From equations (1.1.1, 2, 3):

$$\dot{V}(p_1 - p_2) \propto \dot{V}^3 c_f \propto \mathrm{Re}_d^3 c_f$$

for a given pipe and fluid. Figure 1.1(b) indicates that the ratio of the turbulent dissipation to that in laminar flow at the same Reynolds number rises steadily with increasing Reynolds number. In laminar flow, dissipation is linked to the velocity gradients of the mean flow; accordingly, it is greatest near the pipe wall where the velocity changes most rapidly. The flow visualization of Figure 1.1(c) suggests how additional dissipation is achieved within a turbulent flow: the turbulence includes small-scale motions with large velocity gradients and correspondingly large shear stresses; moreover, these fluctuations act to increase the mean velocity gradient near the wall, and its contribution to the dissipation also rises. In order to achieve the observed steady rise in the dissipation rate, the small 'eddies' must become ever smaller as the Reynolds number increases. Thus the range of length scales displayed by the turbulence expands as the Reynolds number rises.

The Wall Layer and Roughness

We turn now to the influence of wall roughness on the friction generated in pipe flow. As indicated in Figure 1.1(b), roughness has a rather small effect in the régimes of laminar flow and transition. (It must be recognized, however, that comparisons of results for smooth and rough walls are bedevilled by the problem of defining an appropriate diameter for a rough pipe.) Even after the flow has become turbulent, the effect of roughness is not large, so long as the Reynolds number is not too high. Only well inside the turbulent régime do different sizes and kinds of roughness produce markedly different friction characteristics. Ultimately, at the highest Reynolds numbers, the friction coefficient depends only on the roughness, being nearly independent of Reynolds number and thus of viscosity. We can distinguish two contrasting patterns of behaviour: at low and moderate Reynolds numbers, even turbulent friction is markedly dependent on viscosity, but little influenced by roughness; at high Reynolds numbers, friction depends primarily on wall roughness and is nearly independent of viscosity.

From these observations we can infer certain features of the *wall layer*, the region nearest the pipe wall, across which the velocity drops rapidly to zero. We noted earlier that the smallest, viscosity-dominated components of the turbulence become steadily smaller as the Reynolds number rises. But immediately next to the wall, the fluid motion is dominated by viscosity, since the turbulent fluctuations are suppressed by the presence of the wall. We may expect that the thickness of this *viscous sublayer* will also decrease as the Reynolds number rises, and that the turbulent activity will extend nearer to the wall.

These ideas provide the basis for an explanation of the trends of Figure 1.1(b). Seemingly, the roughness elements do not participate in friction generation so long as they are buried within the viscous sublayer. However, when the turbulent fluctuations penetrate among the roughness elements, it appears that the stress is transmitted directly from these projections to the turbulent outer flow. The larger the roughness elements, the lower the Reynolds number at which this interaction begins. According to this view, any wall is *hydrodynamically smooth* if its roughness elements do not project into the turbulent part of the flow. Hence smoothness is not an absolute property of a surface, but depends as well on the nature of the flow adjacent to it. A *fully rough* wall is one subjected to a flow whose turbulence is vigorous enough to scour away the viscous layer to expose the roughness elements. For the variable roughness produced by most manufacturing processes, a gradual transition between smooth- and rough-wall conditions is found, as illustrated in the lowest curve of Figure 1.1(b).

These conclusions regarding the wall layer have implications which extend beyond friction generation. When discussing the transfer of any quantity between a solid surface and a turbulent flow—momentum, heat or a dissolved or condensing substance—we must take into account the structure of the wall layer: a viscous sublayer against the wall (if smooth); fully turbulent flow some distance from it; and a blending layer between. Although the time-mean properties of the viscous and turbulent layers merge smoothly, the instantaneous picture is quite different. Highly turbulent fluid penetrates intermittently into the nominally non-turbulent layer, in a manner reminiscent of the flashes of turbulence in pipe flow near transition.

Average Velocities

Finally, we use the example of pipe flow to introduce the problems of describing a turbulent velocity field. In the Reynolds number and friction coefficient we have used an average velocity defined in terms of the flow rate: $U_a = \dot{V}/A$. Figure 1.2 indicates how this is related to the local velocities within the pipe: (a) shows the velocities which might be measured in a cross-section at some instant; (b) shows the variation of the time-mean velocity

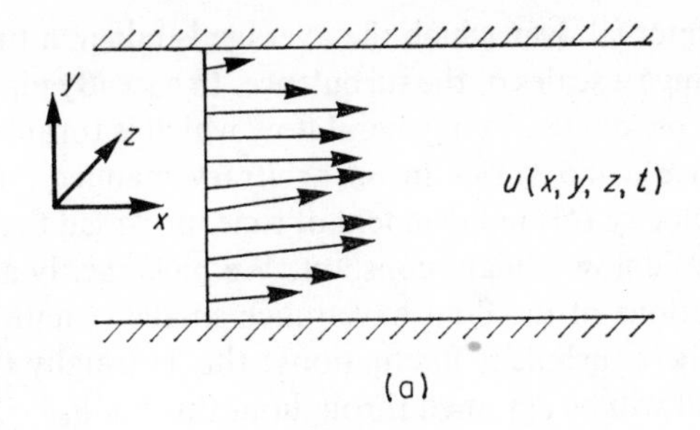

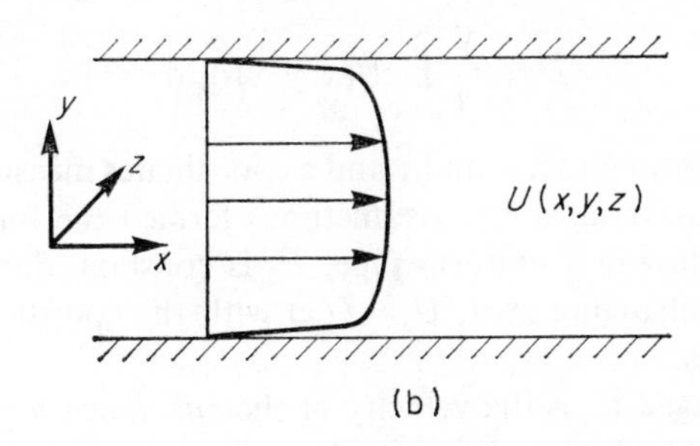

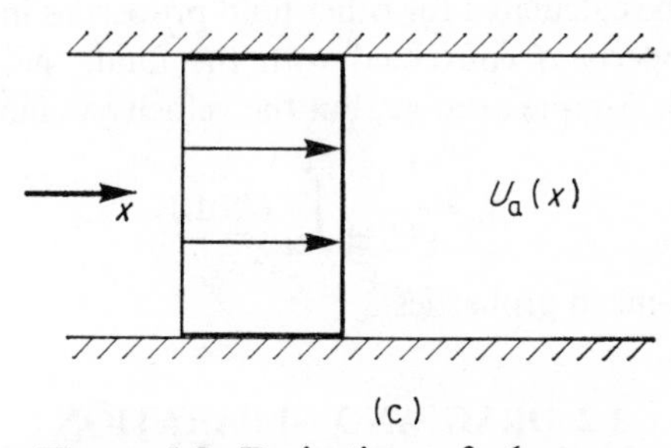

Figure 1.2 Derivation of the one-dimensional approximation to the velocity distribution at a section of a duct. (a) Velocity distribution over a cross-section at an instant in time. (b) The time-mean velocity distribution over the cross-section. (c) The twice-averaged velocity of the one-dimensional model

across the pipe; finally, (c) gives the average of the time-mean velocity over the section.

Analytically, the fluctuating velocity $u(x,y,z,t)$ at a point in the pipe and the time-mean velocity U are related by

$$U = \frac{1}{t_a} \int_{t_0}^{t_0+t_a} u(x,y,z,t)\, dt \tag{1.1.5}$$

with t_a the time interval over which the average is taken, a time substantially longer than the longest scales of the turbulence. In a steady pipe flow, $U=f(r)$, with r the radial coordinate. In a general flow which is turbulent but basically steady, $U=f(x,y,z)$, and varies in an arbitrary manner through space. In either case, the velocity U is independent of time, provided that the controlling features of the basic flow remain constant. We have tacitly assumed that the molecular fluctuations of the fluid have much smaller length and time scales than do the smallest turbulent fluctuations; this is usually the case, save in rarefied gases, and will be assumed throughout this book.

The time-mean value U and the average velocity U_a are related by

$$U_a = \frac{1}{A}\int_A U(x,y,x)\,dy\,dz \tag{1.1.6}$$

with A the cross-sectional area, and y and z coordinates measured in the cross-section. The twice-averaged U_a is sometimes termed the bulk velocity. For constant-density flow in a uniform pipe, U_a is constant along the flow. For flow in a duct of changing area, $U_a=f(x)$ with the coordinate x measured along the duct axis.

The twice-averaged U_a is the velocity of the *one-dimensional model* of duct flow on which much of engineering fluid mechanics is based. Twice-averaged or *bulk values* can be calculated for other fluid properties in a similar manner. For a general property S convected with the fluid, the most appropriate bulk value is not the simple average, but the velocity-weighted value

$$S_b = \frac{1}{U_a A}\int_A US\,dA \tag{1.1.7}$$

with U and S time-mean properties.

1.2 DRAG AND SEPARATION

Drag of a Circular Cylinder

Figure 1.3 illustrates another flow of great practical interest, that around a body immersed in a uniform stream. The particular body considered is a very long circular cylinder of diameter d. The velocity of the undisturbed stream is U_1, and the constant fluid density and viscosity are ρ and μ. Again we interest ourselves in the resistance applied to the flowing fluid, in this case the drag force D. Figure 1.3(a) shows how the drag depends on the flow velocity, the results being given in terms of a dimensionless *drag coefficient*. In general

$$C_D = \frac{D}{\frac{1}{2}\rho U_1^2 A} \tag{1.2.1}$$

with A a representative area of the drag-generating body, and $\frac{1}{2}\rho U_1^2$ the

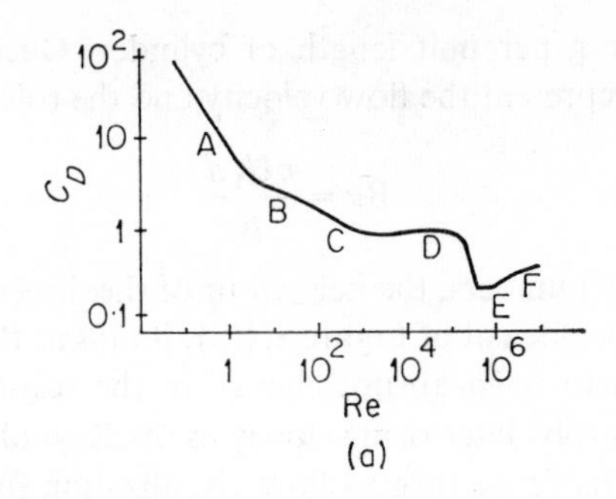

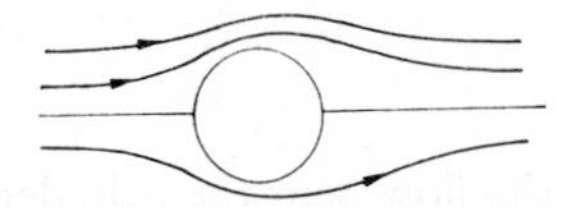

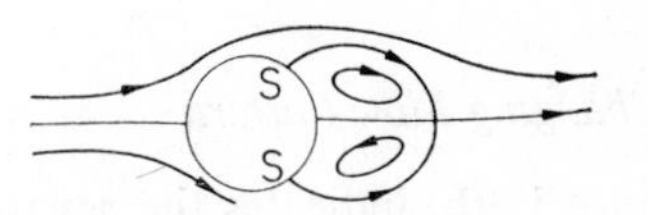

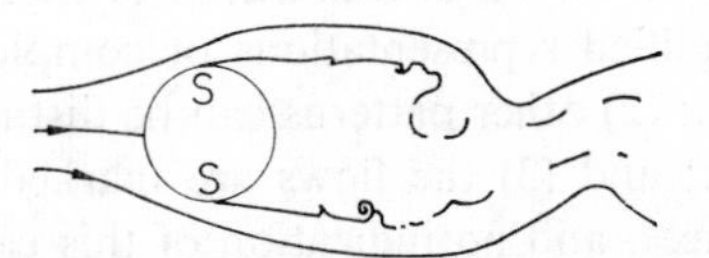

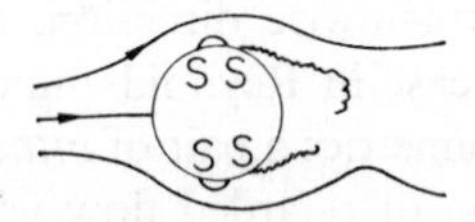

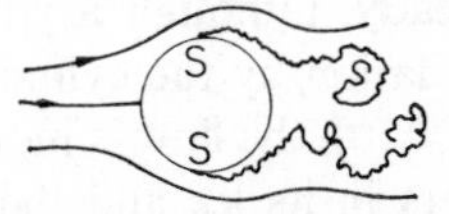

Figure 1.3 Relationship between drag of a circular cylinder and turbulence, illustrating the roles of separation and transition. (a) Drag presented in the form of drag coefficient C_D *vs* Reynolds number Re based on the cylinder diameter. (b) Flow patterns around cylinder for increasing Reynolds numbers corresponding to labels in (a), after Morkovin

dynamic pressure of the stream. For an infinitely long cylinder, we take

$$C_D = \frac{D'}{\frac{1}{2}\rho U_1^2 d} \tag{1.2.2}$$

where D' is the drag per unit length of cylinder. Once again a Reynolds number is used to represent the flow velocity and the role of viscosity; here

$$\mathrm{Re} = \frac{\rho U_1 d}{\mu} \tag{1.2.3}$$

For low Reynolds numbers, the behaviour of the drag coefficient resembles that of the friction coefficient of Figure 1.1(b): it falls as the Reynolds number increases. Again there is an abrupt change in the resistance characteristic; but here it drops sharply, later rising slowly as the Reynolds number increases even further. Once more we turn to flow visualization for an explanation of these trends, in particular, to discover the role played by turbulence.

The Changing Flow Pattern

Figure 1.3(b) indicates the general nature of the flow near the cylinder for several values of the Reynolds number; each sketch shows the instantaneous forms which might be adopted by filaments of dye or smoke convected by the stream. These patterns are based on a survey of cylinder flows made by Morkovin. Three limitations of these sketches should be noted: (1) they are simplified representations of complex patterns, some of which fluctuate in time; (2) other patterns can be distinguished at intermediate Reynolds numbers; and (3) the flows are markedly three-dimensional in certain circumstances, and no indication of this can be given. With these reservations, we can use the sketches to construct a fairly consistent picture of flow development as the Reynolds number increases.

At the lowest Reynolds numbers (region A, $\mathrm{Re} < 1$), the flow around the cylinder is steady, although asymmetric in the streamwise direction, as a result of retardation by the cylinder. With an increase in Reynolds number (region B, $\mathrm{Re} = 50$), the flow is more markedly asymmetric: a pair of *attached vortices* appears in its lee and the *wake*, the region of retarded flow which stretches downstream, begins to waver from side to side. The wake grows narrower as the Reynolds number rises, consistent with the steady fall in drag coefficient.

At still higher Reynolds numbers (region C, $\mathrm{Re} = 250$), the wake contains a double series of evenly spaced vortices. The fluid appears to roll up at the top and bottom of the cylinder, and the shedding of vortices alternately from these two points is the source of the *vortex trail or street* in the wake. The vortex trail is associated with the name of von Kármán, who constructed a theoretical model for it.

In the next stage of the development (region D, $\mathrm{Re} = 10^4$), the flow immediately adjacent to the cylinder is fairly steady. Extending downstream from the top and bottom are two *shear layers*, across which the velocity changes abruptly from the sluggish wake flow to the energetic outer stream.

Such thin shear layers are sometimes referred to as vortex sheets, equivalent to a row of concentrated line vortices lying side by side. The layers roll up to form concentrated vortices as they move downstream, but these are not stable, and break down into the disordered pattern typical of turbulence. The name *free turbulence* is given to that developing within a mass of shearing fluid not directly influenced by solid boundaries. Turbulence generated near a solid boundary, such as a pipe wall or cylinder surface, is termed *wall turbulence*.

The word *vorticity* is used for the rotation of a fluid element, whether in a concentrated vortex (as cases B and C) or in a mass of shearing fluid (as the shear layer of D). For two-dimensional, continuous flows, it can be shown that vorticity or fluid rotation cannot be created within the body of the fluid, but only as a solid boundary. In flow D, for example, the vorticity of the fluid which has passed near the cylinder is convected downstream, first in shear layers, then in concentrated vortices. But when the latter dissolve into a range of turbulent motions, a fundamental change occurs. The disordered motions of turbulence are inherently three-dimensional, and this makes possible the production of vorticity within the body of the flow. In Section 1.4 we shall see how the stretching of vortex lines in three-dimensional flow gives an increase not only in the vorticity of the fluid, but also in the kinetic energy associated with it.

Before we follow the development of these flows to even higher Reynolds numbers, it is helpful to consider the processes near the cylinder surface. A solid body influences fluid flowing near it in two ways: by preventing the passage of the fluid through the solid surface, and by bringing the fluid to rest on the surface. These constraints may be described as the *no-penetration and no-slip conditions*. The former significantly modifies the motion over a distance of a few diameters from the body, even for an ideal, inviscid fluid. The influence of the no-slip condition depends strongly on Reynolds number. For a very low Reynolds number (case A), the diffusive effect of viscosity extends the influence of the fixed surface far into the fluid moving slowly around the cylinder. At a somewhat higher Reynolds number (case B), when the fluid is convected more rapidly past the body, the viscosity-influenced region is thinner, and a well defined wake appears. At even higher Reynolds numbers, the thickness of the viscous region is a small fraction of the diameter. The thin region across which the velocity drops to zero at the surface is termed a *boundary layer*. Beyond it, the velocity distribution is determined by the no-penetration condition, modified by the development of the flow near the body and by the wake.

As indicated above, vorticity is generated where the flowing fluid meets a solid surface; thus the boundary layer on the cylinder is the source of the vorticity which is convected downstream in the wake (cases C and D). The way in which the vortical fluid leaves the surface is of great interest, for it determines the flow over the back of the cylinder and thus the drag on the cylinder. Retarded by surface friction, the fluid of the boundary layer is ultimately unable to advance along the surface against the rising pressure over the back of the cylinder. *Separation* occurs: the boundary-layer fluid ceases to follow the solid profile exactly, and moves downstream a little distance from it. The separation points, at which the layer leaves the surface, are marked in Figure 1.3(b) by the letter s. In some cases, they move in response to the unsteadiness of the flow behind the cylinder.

Beyond the separation point, the direction of the flow right at the surface is reversed, so that the fluid moves towards the point from both directions. Between the separated vortical layer and the surface lies a rather sluggish *recirculating flow* which extends into the wider region behind the cylinder. These eddying flows and the shear layers around them become less steady as the Reynolds number increases, and the region of instability and transition moves progressively closer to the cylinder.

Let us now see how the boundary layer responds when the Reynolds number is increased beyond the values considered above. For case E, with $Re = 10^6$, the role of separation has become more complex: the boundary layer reattaches just downstream of the initial separation; enclosed between the temporarily separated layer and the surface is a laminar *separation bubble*, with a small recirculating flow. In this range of Reynolds numbers, the turbulence of the wake and shear layers extends upstream to the cylinder surface; the boundary layer beyond the separation bubble becomes turbulent, and the final separation is that of a turbulent boundary layer.

It is the occurrence of transition within the boundary layer itself which is responsible for the sharp drop in drag coefficient noted in Figure 1.3(a). Turbulent mixing re-energizes the decelerating boundary layer, and the fluid near the surface is able to proceed further around the cylinder before separating. The consequent narrowing of the wake, and of the low-pressure region just behind the cylinder, corresponds to the fall in drag coefficient. Near the *critical Reynolds number*, at which the coefficient drops sharply, the flow pattern is strongly dependent on the turbulence level of the approaching stream, and on the roughness of the surface. This is not surprising, since these factors must influence the delicate balances which decide when separation and transition will occur.

A further increase in Reynolds number (case F, $Re = 10^7$) eliminates the separation bubbles and extends the turbulent part of the boundary layer forward of the central plane of the cylinder; the drag coefficient slowly rises again. The separating turbulent boundary layers give rise to shear layers which are themselves unstable, and provide further turbulence generation in the wake.

The frequency of vortex shedding from a cylinder is of practical importance, since it defines the excitation applied to cylindrical cables, rods and structures exposed to the wind or to a current of water. The dominant frequency in the wake, and of the fluctuations in drag and lift, is specified by the *Strouhal number*:

$$\mathrm{Sl} = \frac{fd}{U_1} \tag{1.2.4}$$

This depends on the Reynolds number of the flow and on the body shape. For a circular cylinder, $\mathrm{Sl} \simeq 0{\cdot}2$ for $100 < \mathrm{Re} < \mathrm{Re}_c$, and $\mathrm{Sl} \simeq 0{\cdot}3$ for $\mathrm{Re} > \mathrm{Re}_c$, when f is measured in cycles.

Generalizations

The observations detailed above relate to a single class of fluid motions, but from them we can draw more widely applicable conclusions pertaining to instability, transition and separation.

(1) After a period of development, laminar flows—whether in pipes, boundary layers or separated shear layers—become unstable and, subsequently, turbulent. The extent of the turbulent region increases as the Reynolds number of the flow rises.

(2) The *outer flow*, beyond the regions of high shear stress and not directly influenced by viscosity and turbulence, is modified by the growth of boundary layers, shear layers and wakes, and especially by the points at which the boundary layers separate. Transition has an important role in determining these points.

(3) The total drag on a body comprises two elements: *friction drag*, the resultant of the shear stresses on the surface exposed to the flow, and *pressure drag*, the resultant of the pressures acting on the surface. The first element may be expected to increase once transition has occurred. The second is vitally dependent on the location of the separation points and, as in the case of the cylinder, may fall following transition. The net effect of transition depends on the body shape, since friction drag is the major element for a flat, streamlined body, while pressure drag is more important for a bluff body, such as a circular cylinder.

We may consider two extreme cases, between which comes the circular cylinder. The drag characteristic for a flat body aligned with the stream resembles that for pipe flow. On the other hand, for a body with sharp edges (for instance, a cube, or a disc normal to the stream), the separation points will be fixed over a wide range of Reynolds numbers, and the drag coefficient may be expected to be nearly independent of transition and the factors influencing it. This is found to be so in practice.

1.3 SPREADING AND MIXING

Spreading of a Jet

Figure 1.4 shows another kind of flow in which free turbulence can develop—the jet formed by fluid emerging from a round pipe into a large volume of the same fluid. For a generating flow which is laminar but has a sufficiently high Reynolds number, there are two stages in the turbulent development, as indicated in Figure 1.4(a). In the first, the cylindrical shear layer between jet and ambient fluid becomes unstable: the initial well-organized eddies disintegrate into turbulence which spreads both inwards and outwards, and finally absorbs completely the laminar or 'potential' core of the jet. The point of transition moves closer to the nozzle as the Reynolds number increases. In the second stage of development, the turbulence can spread only at the outer boundary, and the jet adopts a pattern of growth consistent with the unchanging interaction between the internal turbulence and the quiescent outer fluid.

Figure 1.4(b) gives velocity histories which might be measured at five points, shown in Figure 1.4(a), within and just outside the turbulent jet. These are instantaneous values, like those of Figure 1.2(a), not averages over time. On the jet axis (point A) the flow has the features typifying well-developed turbulence: irregular fluctuations with a wide range of time scales and peak values. Away from the axis (point B) the character of the record is significantly different, with stretches of fully turbulent activity separated by intervals in which the velocity varies rather slowly. Further out again (point C) the activity is markedly intermittent, with long stretches of gentle variation separating bursts of turbulent activity. Finally, in the outer flow not too far from the jet (points D and E), slow velocity variations are typical, although there are occasional flurries of more intense activity. This intermittent activity is reminiscent of that in pipe flow near transition and in the viscous part of a wall layer.

The intermittency at a point in a basically steady turbulent flow can be specified numerically using the *intermittency factor* γ, defined as the time fraction during which the motion is turbulent, that is, displays rapid velocity fluctuations. Inspection of the velocity history for point C, for example, suggests that $\gamma \simeq 0{\cdot}2$ there. More generally, the intermittency factor can be defined as the probability that the local motion is turbulent.

The significance of these observations is made clearer by flow visualization. The nature of the instantaneous boundary revealed by introducing dense smoke is indicated in Figure 1.4(a). (A photograph displaying these features in another way is presented in Plate I, facing page 46.) Only within the irregularly moving boundary is the fluid actively turbulent, with strong vorticity. Random motions do occur beyond, but they are forced by the distortion of the bound-

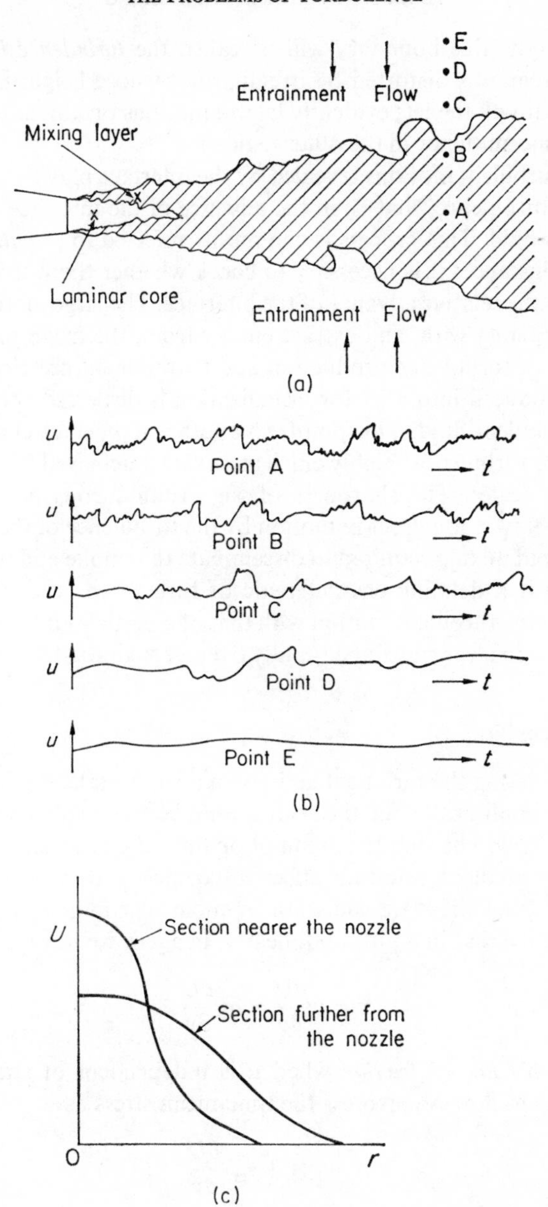

Figure 1.4 Flow in a turbulent jet, illustrating intermittency, entrainment and similarity. (a) Jet emerging from a round nozzle within which the flow is laminar into a quiescent mass of the same fluid. (b) Variations of velocity in time, for points shown in (a) at various distances from the jet centre-line. (c) Variations of time-mean velocity U with distance r from the axis of the jet, for two sections through the jet beyond the laminar core

ary. Accordingly, this boundary will be called the *turbulence interface.* The smoke-filled region is distorted by irregularities whose height is comparable to the half-width of the jet; evidently lateral motions on this scale are responsible for the intermittency in the outer regions.

The large turbulent motions revealed by the interface play a vital part in the spreading of turbulence, that is, in the advance of the interface into the non-vortical outer fluid. This process is commonly referred to as *entrainment*, but in particular instances it is necessary to check whether the word refers to the absolute or to the relative advance of the interface. The large motions are those best able to interact with, and extract energy from, the mean motion. Hence the processes of turbulence production and entrainment are closely linked.

Smoke introduced into a jet for visualization is dispersed very thoroughly within the turbulent fluid. This points to another important characteristic of well-developed turbulence: highly efficient *mixing* is achieved by the profusion of interacting scales. The sharpness of the instantaneous boundary shows that the outer flow, though set in motion by the turbulence of the jet, does not sustain the rapid mixing required to disseminate the smoke and the turbulence itself. Mixing is a defining characteristic of turbulence, as may be seen by comparing a truly turbulent motion with that of a particle embedded in a bowl of jelly situated in a car travelling rapidly over a rough road.

The Eddy Viscosity

While considering the turbulent activity in a jet, we take the further step of looking at its implications for the relationship between stress and strain-rate in a turbulent flow. The simplest assumption links the mean shear stress to the mean velocity gradient using an effective coefficient of viscosity, modelled on that prescribing stress transmission by molecular migration. For parallel laminar flow (like that in a pipe and, nearly, in a jet), we have

$$\tau = \mu \frac{\partial U}{\partial y} = \rho \nu \frac{\partial U}{\partial y} \tag{1.3.1}$$

termed *Newton's law of friction* when μ is independent of strain rate. For parallel turbulent flow, we express the time-mean stress as

$$\tau = \rho(\nu + \epsilon_{\mathrm{m}}) \frac{\partial U}{\partial y} \tag{1.3.2}$$

a step taken by Boussinesq as early as 1877. The kinematic coefficient ϵ_{m}, the turbulent diffusivity of momentum, is called the *apparent or eddy viscosity.* It is not a property of the fluid, as is the molecular viscosity, but depends on the particular flow and on position within it. In fully turbulent conditions, $\epsilon_{\mathrm{m}}/\nu \gg 1$ normally, and the direct contribution of molecular viscosity is negligible.

An obvious step is to extend the parallel between laminar and turbulent flows by assuming ϵ_m to be uniform. In favour of this procedure is the rather simple-minded argument that molecular and macroscopic mixing have the same ultimate effect. On the other hand, the motion-independence of molecular viscosity requires that the scale of the molecular interaction, the mean-free-path, be much less than the length scale of the bulk motion. This condition is not satisfied by the interactions within turbulence (nor in non-Newtonian fluids, even for laminar motion). Note, for example, that the lateral distortions of a jet boundary are only a few times smaller than the half-width across which the change in mean velocity occurs. The hypothesis of constancy of the eddy viscosity appears even less plausible when one remembers the fact of intermittency, which seems to imply a local discontinuity in the stress-transmitting ability of the fluid.

Despite these soundly based reservations, measurements of free-turbulent flows reveal mean-velocity profiles very like those for the corresponding laminar flows. On the other hand, the mean-velocity variations within turbulent wall flows are very poorly represented by a constant eddy viscosity. The mixed success and failure of the constant eddy viscosity suggests two widely applicable conclusions. Some basic features of turbulent flows can be predicted using methods no more difficult than those required for laminar flows, even though the underlying physical processes are more complex. Nevertheless, experimental verification is required when extending the application of a simple model of turbulent activity, even when it has proved satisfactory in other situations.

Townsend has made ingenious use of the hypothesis of a constant eddy viscosity, by applying it only to the smaller scales of turbulence, and imagining the larger motions to develop in this environment. Observations like those of Figure 1.4 provide considerable support for this 'double-structure' postulate.

1.4 STRUCTURE AND SIMILARITY

The Energy Cascade

At this point it is appropriate to gather together some of our deductions, in order to build up a picture of the essential processes within a turbulent flow. We have noted that wide ranges of frequency and length scales are characteristic of well-developed turbulence, and that the scale of the largest motions is not much less than the lateral extent of the turbulent flow. These large motions are presumably rather like the disturbances which first become unstable in the corresponding laminar flow and must, like them, be capable of extracting energy to maintain the turbulence. The apparent source of the energy varies from flow to flow: in a pipe flow, it is the pressure gradient;

in a boundary layer, the outer flow; and in a jet, the initial kinetic energy of the fluid. Whatever the source, the energy extraction can be ascribed to an interaction between the mean flow and large, fairly well-ordered elements of the turbulence. (This will be shown analytically in later chapters.) As we have noted, disturbances below a certain minimum intensity may not be able to sustain this interaction.

When discussing pipe flow, we argued that the smallest motions of the turbulence, with the largest shearing stresses, were responsible for the dissipation of turbulence energy. Thus we come to follow G. I. Taylor in conceiving the energy transfers within a turbulent flow to be a cascade of energy from the largest, energy-extracting scales of motion to the smallest, dissipating scales. This idea explains the wide range of length and time scales within active turbulence: the size of the largest is determined by the mean flow, while the size of the smallest is determined by the fluid viscosity; the intermediate scales interact with the largest and smallest to accomplish the transfer of energy across the range. These ideas are succinctly expressed in lines attributed to E. G. Richardson:

'Big whirls have little whirls that feed on their velocity;
Little whirls have lesser whirls, and so on to viscosity.'

The motions comprising turbulence are frequently referred to as *eddies* of various sizes. This term has much to commend it: it is short, and accurately describes certain swirling motions revealed by flow visualization. Moreover, turbulence is often represented mathematically as a superposition of Fourier components, and each may be interpreted as a cellular pattern of eddies. Thus, if we know the instantaneous variation of the velocity component u along some line, we may use a Fourier cosine transform to write

$$u(r) = \int_0^\infty a(k) \cos kr \, \mathrm{d}k \tag{1.4.1}$$

with the coordinate r measured along the line, and with the inverse transform

$$a(k) = \frac{2}{\pi} \int_0^\infty u(r) \cos kr \, \mathrm{d}r \tag{1.4.2}$$

The amplitude variation $a(k)$ specifies the activity attributable to elements of wavelength $\lambda = 2\pi/k$, which can be identified in a general way with eddies contributing to the overall motion. The transform variable $k = 2\pi/\lambda$ is the *wave-number* of the contribution.

Despite its basis in both observation and theory, the concept of eddies must be used with caution. The dominant motions of most fully turbulent flows are not even approximately cellular, and cannot be adequately represented by a simple superposition of cellular elements. The true complexity of turbulent motion is not readily apparent to the eye, since most visualization techniques provide only a two-dimensional picture, while certain crucial aspects of turbulence are three-dimensional. The large motions which lead to jet spreading, for example, appear as simple swirling motions when a section is taken along the jet. More careful investigations reveal that these boundary-distorting motions are more akin to isolated jets, with balancing flows on either side.

Two other elements have been used to provide a more vivid picture of processes within turbulent flows: fluid lumps and vortex lines. The consideration of a finite lump of fluid helps, for example, in visualizing the stress-transmitting activity of the turbulence. The motion of lumps normal to the mean flow, with some property linked to velocity remaining constant, provides a mechanism for momentum transfer. The distance over which the lump moves before mixing with its new surroundings is termed the *mixing length.* Prandtl took the conserved property to be momentum, and Taylor took it to be vorticity, and both obtained results of practical utility. The name *phenomenological theories* is sometimes applied to such heuristic models which adopt the simplest assumptions consistent with the gross activity of the turbulence.

Vortex Stretching

A consideration of vortex filaments can help us to understand the energy transfer between turbulent motions of differing scales, and between mean flow and turbulence. Consider a cylinder of fluid (length L, diameter d, density ρ), rotating with uniform angular velocity ω. Its mass, kinetic energy and angular momentum have the forms

$$m \propto \rho L d^2, \quad KE \propto (\rho L d^2) d^2 \omega^2, \quad AM \propto (\rho L d^2) d^2 \omega \qquad (1.4.3)$$

Imagine that this vortex element is stretched with its angular momentum remaining constant, as will be the case if diffusion is negligible. During this process, with the mass constant as well:

$$KE \propto \omega \propto d^{-2} \propto L \qquad (1.4.4)$$

Thus stretching will augment the vorticity and energy of the rotating element, the requisite work being done on the ends of the extending cylinder.

Since any turbulent flow can be imagined to be a superposition of vortex lines, we have a simple model of energy transfer from large motions to the smaller scales embedded in them and stretched by the large-scale distortions. Note that the diameter, representing the scale of the velocity variations, is reduced by stretching. These processes can be visualized by dropping a little ink into a container which has just been rapidly filled with water. The filaments of ink behave very much as do vortex lines embedded in the fluid. It is particularly significant that the overall effect of the distortion is a net stretching of the filaments; compression does occasionally occur, but it is inherently less probable than extension.

An important conclusion follows from the fact that the stretching is in the direction normal to the plane of vortex rotation. Clearly this mode of energy transfer requires that the fluid motion be three-dimensional; parallel vortex cells cannot stretch one another. Fluid motion in all directions is a vital feature of turbulence: in its absence, a wide range of scales of motion could not be established nor maintained.

Factors Affecting Large-scale Structure

The problems touched on in the preceding paragraphs—the identification of significant structures and processes within turbulent motions, and their representation in terms of simple elements—are central to the understanding and analysis of turbulence. Tentative progress has been made in identifying the significant features of some simple turbulent flows, but it has not proved

Table 1.1. Variety in the controlling features of plane turbulent flows

Flow species	Number of			
	wall layers	spreading boundaries	convecting outer flows	reversals in mean velocity gradient
Duct flow	2+	0	0	1
Boundary layer	1	1	1	0
Wall jet	1	1	0 or 1	1
Jet	0	2	0 or 1	1
Mixing layer	0	2	1 or 2	0
Wake	0	2	1	1

possible to incorporate these findings into a widely applicable theory. The difficulty of devising such a theory is indicated by Table 1.1, which shows that the controlling features change markedly from one flow species to the next. Only plane or two-dimensional mean flows are considered; other sources of variability are found in axisymmetric flows, for example. Five of the six flow types dealt with in Table 1.1 have been encountered already. The sixth is the *wall jet*, produced by ejecting fluid along a solid surface bounding a fluid at rest or in uniform motion.

The factors considered in Table 1.1 are the numbers of: wall layers bounding the flow, spreading turbulence interfaces, possible outer convecting flows, and reversals in the sign of the mean velocity gradient. (The last element indicates whether turbulence generated in one kind of mean strain field is able to move into a quite different strain field and to interact with turbulence produced there.) Each of these factors must be expected to have a significant influence on the large-scale motions of the turbulence, either allowing or preventing certain modes of development. It is still possible, however, that a degree of universality may exist in the smaller scales.

Self-preservation

We now consider some happier features of turbulent motions. Although the instantaneous interface shown in Figure 1.4(a) is highly convoluted and changes rapidly in time, its mean position has (if the feeding flow is constant) a surprisingly simple shape: once the flow has become fully turbulent, its width varies linearly along the jet. Moreover, the angle of expansion is found to be independent of the species of the fluid comprising jet and surroundings, provided that the complications of compressibility and cavitation are not encountered. Hence the mean motion and the larger scales of the turbulence are independent of density and viscosity, and of the Reynolds number, provided that this is large enough to ensure fully turbulent motion. This characteristic implies that the time-mean shear stresses within the flow are proportional to the time-mean inertia forces, and thus depend on the density rather than on the viscosity. We shall see later how this comes about.

Flows with these features have been described as possessing Reynolds-number similarity; it is more pertinent to say simply that the larger scales of the motion are *Reynolds-number independent*. In such flows, viscosity has a passive function, as in turbulent dissipation; it is dominant in the smallest scales, but plays no direct part in specifying the overall pattern and level of activity. When viscosity does have a direct role in defining the mean flow, for example, in a wall layer, the larger scales will not be independent of Reynolds number.

Another simple feature of fully developed jet flow is illustrated in Figure 1.4(c), which shows cross-stream profiles of time-mean axial velocity, for two sections well away from the nozzle. (Closer to the nozzle the profiles will be flat-topped.) Although they differ in width and height, the profiles have a similar shape. Hence division of the coordinates r and U by suitable scaling functions will reduce the whole family of similar profiles to a single curve representing the velocity variation across all such jets. Profiles and flow patterns with this characteristic are described as *self-similar*, a term applicable to both laminar and turbulent flows. We noted above that the jet width grows linearly, and may expect that the centre-line velocity U_c will decrease according to another power law. Thus

$$\frac{r}{x} \quad \text{and} \quad \frac{U}{x^a} \qquad (1.4.5)$$

are scaled coordinates appropriate to this flow, with a an index not at present known, and x measured along the jet from some *virtual origin*, whose position takes account of the initial, non-similar development. If the profiles are indeed self-similar, the scaled profile can be given as

$$\frac{U}{U_c} = f\left(\frac{r}{x}\right) \quad \text{or} \quad \frac{U}{x^a} = f\left(\frac{r}{x}\right) \tag{1.4.6}$$

In view of the complexity of the turbulent motions within a jet, how can we justify the simplicity of these conclusions? Physically, we can argue that the mean motion and the interacting larger elements of the turbulence must settle into a developing equilibrium in which they can expand without changing their essential character. Such patterns of mean flow and turbulence are said to be *self-preserving*. Now self-preservation implies that a length and a velocity scale of the mean motion can be used to reduce profiles to a common form, as in equations (1.4.6). This is possible only if no other parameter significantly influences the motion. But the Reynolds number formed from the local width and velocity scales will in general vary along the flow. Hence self-preservation is, in many cases, bound up with the Reynolds-number independence of the larger scales of motion.

Dimensional Analysis

In approaching the question of similarity mathematically, we note that the gross features of a jet are independent of the turbulence, being fixed by a few dominant properties of the fluid and the initiating flow. Such a situation is the starting point for dimensional analysis and allied similarity techniques, and we may expect these methods to reveal some simple features of self-similar flow. The dimensionless quantities which are generated are referred to as similarity parameters, since they are associated with the kinematic similarity of families of flows.

Dimensional analysis will be used extensively in later chapters. As examples of its application, we consider here the plane mixing layer, boundary layer and wake. The time-mean velocity U in a turbulent mixing layer depends on position and on the change in velocity across it, ΔU. Taking the flow to be fully turbulent and the density to be uniform, we assume that viscosity and density do not influence the mean motion. Hence

$$U = f(x, y, \Delta U) \tag{1.4.7}$$

The dimensionally homogeneous form is

$$\frac{U}{\Delta U} = f\left(\frac{y}{x}\right) \tag{1.4.7}$$

with the latter function f the same for all mixing layers of the class considered.

Because the defining parameters are few in number, a quite specific result is obtained: the layer grows linearly and is, of course, Reynolds-number independent. Somewhat surprisingly, the mean geometry of a turbulent mixing layer is more simply determined that is that of a laminar layer, in which viscosity plays a part.

The time-mean velocity near a flat plate lying parallel to a uniform stream of velocity U_1 may be expected to have the functional form

$$U = f(x, y, L, U_1, \rho, \mu) \tag{1.4.8}$$

with L the plate length. A dimensionally homogeneous form is

$$\frac{U}{U_1} = f\left(\frac{x}{L}, \frac{y}{L}, \frac{\rho U_1 L}{\mu}\right) \tag{1.4.8}$$

This holds in the boundary layers, in the wake, and in the outer flow. The total drag (per unit width) can be treated in the same way:

$$D' = f(U_1, L, \rho, \mu) \tag{1.4.9}$$

becomes

$$\frac{D'}{\frac{1}{2}\rho U_1^2 L} = f\left(\frac{\rho U_1 L}{\mu}\right) \tag{1.4.9}$$

This is the form adopted in presenting drag data in Figure 1.3(a). The results (1.4.8, 9) suggest that the drag coefficient might replace the Reynolds number as one of the dimensionless quantities defining the velocity field. This conclusion is not strictly correct, and it is instructive to ask where this plausible analysis is at fault.

The Boundary-layer or Thin-flow Approximation

An important simplifying feature of many turbulent and laminar flows has been mentioned already, but deserves emphasis—the thinness of the highly sheared, strongly vortical region. For moderate and large Reynolds numbers, the angle of spread of a jet is small, the thickness of a boundary layer is a small fraction of the dimension of the body, and the spread of a wake is slow. These features have a common cause—the relative slowness of lateral diffusion, molecular or turbulent, in comparison with the basic convection of the mean flow.

Many features of thin flows can be described with sufficient accuracy using simplified equations of motion, the simplification achieved by neglecting terms that are relatively small by virtue of the nearly parallel mean flow. Results obtained in this way are usually referred to as boundary-layer approximations, although the approach is applicable to a wide range of elongated flows. It is interesting to note that the German and French words for a boundary layer are less restrictive, their literal translation being 'limiting layer'.

Almost all self-similar flows are thin in the sense of the boundary-layer approximations, since many thin flows are potentially able to expand in conformity with the profile development. When the flow is both thin and self-similar, we are in an especially good position to predict its main features. The thinness of a flow does not guarantee that it will adopt a self-similar form, however. In particular, turbulent wall flows usually fail to be self-preserving, since different modes of development are required by the turbulence of the wall layer and by that interacting with and spreading into the outer flow.

1.5 DECAY AND RANDOMNESS

Grid Turbulence

Each of the flows we have considered has a source from which it can draw mechanical energy to maintain its turbulence, the source being represented by a gradient of velocity or pressure. Now we consider the development of turbulence in the absence of an energy input; this displays in a more open manner the processes of energy transfer and dissipation which occur, in modified form, in turbulent shear flows.

Figure 1.5 shows a duct within which the time-mean velocity is sensibly uniform, both in time and over each cross-section, except in thin boundary layers. Moreover, the time-mean velocity and pressure are nearly constant along the duct, save in a contracting section, where they change quickly from one set of values to another. Turbulence is introduced by a grid of evenly spaced bars. Their turbulent wakes spread and coalesce, and a little downstream of the grid the characteristics of the turbulence are nearly uniform across the duct, except in the boundary layers. Turbulence produced in this way is called *grid turbulence*.

Once the variation in mean velocity has been erased by turbulent mixing, grid turbulence decays steadily, more and more of its initial stock of energy being transferred to and dissipated in the smallest eddies. But in the contracting section there are gradients of mean velocity, and this provides another opportunity for the turbulence to extract energy from the mean flow. Once the contraction is passed, a new pattern of spontaneous development and decay is established. The behaviour of turbulence which is convected through a large contracting duct is of interest because each part of the fluid experiences much the same mean distortion. Hence energy absorption by the turbulence is displayed without the complication (found in shear flows where the length scales of turbulence and mean flow are comparable) that each fluid element has a unique history of distortion dependent on its wanderings across the mean velocity profile.

Flows similar to that described above occur in wind tunnels and other test channels: before entering the test section the fluid passes through gauzes or through a 'honeycomb' intended to straighten the flow and remove large disturbances. The fluid then accelerates through a contraction and moves at high velocity through the test section. Thus the processes described above have practical importance, as well as casting light on certain fundamental aspects of turbulence.

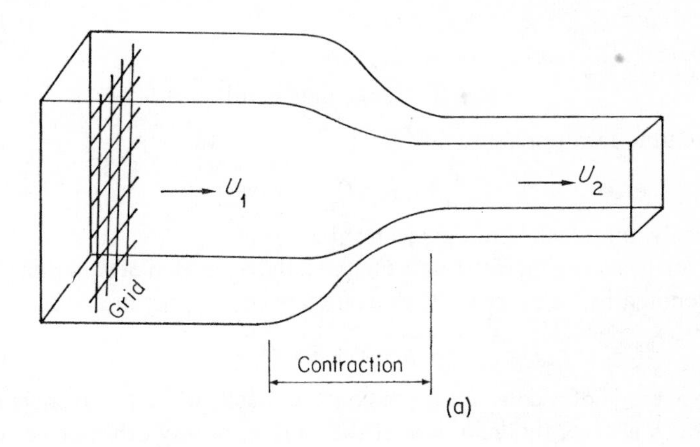

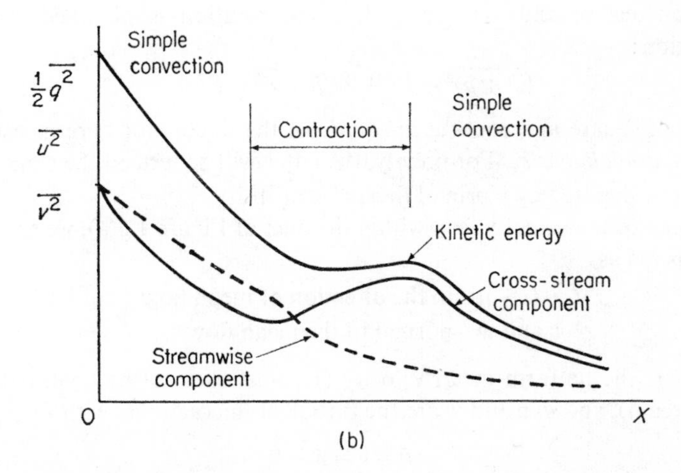

Figure 1.5 Generation, decay and distortion of grid turbulence. (a) A duct containing a turbulence-generating grid and a rapid contraction between two uniform sections. The time-mean velocity is sensibly uniform across most of each cross-section. (b) Variations of turbulence intensity along the duct for streamwise $\overline{u^2}$ and cross-stream $\overline{v^2}$ components, and for turbulence kinetic energy $\frac{1}{2}\overline{q^2}$

Notation for Fluctuating Properties

Figure 1.5(b) indicates how the turbulence behaves during its passage through the central part of the duct, beyond the boundary layers. In order to interpret this information, we must become acquainted with the notation used to describe turbulence and the mean flow in which it is embedded. We take as an example the varying pressure at a point in a basically steady turbulent flow. The instantaneous value is represented as

$$P + p \tag{1.5.1}$$

with

$$P = f(x, y, z) \quad \text{in general}$$

giving the time-mean pressure field, and

$$p = f(x, y, z, t) \quad \text{always}$$

giving the fluctuations about the mean value.

Henceforth, averaging with respect to time, the operation of equation (1.1.5), will be denoted by an overbar. Thus averaging of the pressure gives

$$\overline{P + p} = P + \bar{p} = P \tag{1.5.2}$$

since the average of a constant is the constant itself, while the average of the fluctuation p is zero, by definition. If the square, or any other power of the instantaneous pressure is averaged, a contribution is obtained from the fluctuation:

$$\overline{(P + p)^2} = P^2 + 2P\bar{p} + \overline{p^2} = P^2 + \overline{p^2} \tag{1.5.3}$$

The mean-square and root-mean-square of the fluctuations are measures of the *fluctuation intensity*. For brevity, the latter will be termed the r.m.s. value, and will be denoted by a prime; thus, $p' = (\overline{p^2})^{\frac{1}{2}}$.

Turning to the velocity field within the duct of Figure 1.5(a), we specify the components as

$$\begin{aligned} U + u & \quad \text{in the direction of mean flow} \\ v \text{ and } w & \quad \text{normal to the mean flow} \end{aligned} \tag{1.5.4}$$

Thus U is the uniform mean velocity (U_1 upstream of the contraction, U_2 downstream), and u, v and w are the turbulent fluctuations, with

$$\bar{u} = \bar{v} = \bar{w} = 0 \tag{1.5.5}$$

The intensity of the fluctuations in the three coordinate directions is given by

$$\overline{u^2}, \quad \overline{v^2}, \quad \overline{w^2} \tag{1.5.6}$$

A measure of the overall intensity of the turbulent activity is the mean-square of the resultant q of the fluctuation components:

$$\overline{q^2} = \overline{u^2} + \overline{v^2} + \overline{w^2} \tag{1.5.7}$$

Note that $\frac{1}{2}q^2$ is the instantaneous *kinetic energy* of the turbulence, per unit mass of fluid. Intensity may also be measured by the dimensionless parameter

$$I = \frac{(\frac{1}{3}\overline{q^2})^{\frac{1}{2}}}{U} \tag{1.5.8}$$

variously called relative intensity, turbulence number, turbulence level or simply intensity. As a rough guide, $I \sim 1$ per cent in grid turbulence; $I \sim 10$ per cent near a channel wall; $I > 10$ per cent in a jet or in the part of a wake near the body.

In Figure 1.5(b) the component intensities and turbulence kinetic energy are used to characterize the turbulence. As was anticipated, the intensities decrease steadily, save in the contracting section. Although contraction may increase the turbulence energy, the relative intensity (1.5.8) is reduced, since the mean velocity rises from U_1 to U_2. It is in this sense that the flow in a wind tunnel is made smoother by a contraction.

We see also from Figure 1.5(b) that the relative magnitudes of the streamwise and cross-stream components change, not only as a result of distortion, but also during the two periods of decay. These changes give evidence of selective decay and of interactions among the various scales of turbulent motion.

Homogeneous and Isotropic Turbulence

At a cross-section far enough from the generating grid for the several wakes to have mixed thoroughly, the average history of the fluid elements passing through any point is independent of position in the section, save in the boundary layers. Thus the time-mean properties of the turbulence are uniform across the duct; for example

$$\left.\begin{aligned} \overline{u^2} &= \text{constant} \\ \overline{v^2} &= \text{another constant} \\ \overline{\left(\frac{\partial u}{\partial y}\right)^2} &= \text{constant} \end{aligned}\right\} \text{ in any one section} \tag{1.5.9}$$

Moreover, once the period of rapid initial decay is past, decay proceeds so slowly that the mean values do not change much over short distances measured along the duct, distances comparable to the mesh of the grid and to the scale of the turbulence. Hence, over regions somewhat greater than its length scales, the turbulence is essentially *homogeneous*, having mean properties independent of position.

The restriction to spatial homogeneity greatly simplifies the theoretical treatment of turbulence. Although it is not wholly representative of the more common flows in which mean properties change rapidly through space, homogeneous turbulence has been given a great deal of attention by applied mathematicians; Batchelor's book summarizes the early work in this field. To relate real grid turbulence to the homogeneous model, we must introduce *Taylor's hypothesis*: the statistical properties measured at a fixed point past which slowly decaying turbulence is carried by a uniform flow are identical to those which would be found by averaging over a large volume of homogeneous turbulence.

Further progress can be made in analysing turbulence if it is assumed to be *isotropic* as well as homogeneous, that is, to have mean properties independent of the directions of the axes of reference. For example

$$\left.\begin{aligned} &\overline{u^2} = \overline{v^2} = \overline{w^2} \\ &\overline{\left(\frac{\partial u}{\partial x}\right)^2} = \overline{\left(\frac{\partial v}{\partial y}\right)^2} = \overline{\left(\frac{\partial w}{\partial z}\right)^2} \\ &\overline{\left(\frac{\partial u}{\partial y}\right)^2} = \overline{\left(\frac{\partial v}{\partial z}\right)^2} = \overline{\left(\frac{\partial w}{\partial x}\right)^2} \end{aligned}\right\} \text{at each point within isotropic turbulence} \tag{1.5.10}$$

since each of the equated mean properties is measured in the same way for a different selection of axes. Figure 1.5(b) shows that grid turbulence is not isotropic: the component intensities are not equal. Nevertheless, the departure from isotropy is small enough to allow some useful comparisons to be made between grid turbulence and theoretical results for isotropic turbulence.

Isotropy is not achieved simply by having $\overline{u^2} = \overline{v^2} = \overline{w^2}$; all statistical properties and all scales of the turbulent motion must be independent of direction. The orienting effect of the mean motion prevents these conditions from being satisfied in shear flows. However, the imposed order becomes weaker in the smaller scales of the turbulence, those which do not interact directly with the mean motion, but only with larger elements of the turbulence. At high Reynolds numbers, when there exists a broad range of scales, the fine, energy-dissipating structures are not too far from isotropic. This *local isotropy* of the smallest eddies provides direct practical application of the results for isotropic turbulence, in particular, in representing viscous dissipation. This matter will be considered further in the next chapter.

The Statistical View of Turbulence

In discussing homogeneous and isotropic turbulence, we have approached the boundaries of the *statistical theory* of turbulence, in which it is analysed using the statistical properties of the differential continuity and momentum equations governing the instantaneous random motion. This approach to turbulence will not be adopted until Chapter 9; here we merely take note of some of the concepts on which this body of analysis is founded.

Let us first consider the significance of the term *random,* as applied to a turbulent motion. Strictly interpreted, it would seem to imply that the velocity at any instant is quite independent of that at any other instant, as one result from the rolling of dice is independent of the next. But this interpretation is inappropriate for the turbulent motion of a continuous fluid, since there is a link between the velocities at points that are close in space and time. Hence 'random' means only that the velocities at two points become, on average, less closely related as the separation of the points increases in space or time. Ultimately, it is a matter of judgement whether a process should be described as random or a flow as turbulent. Over a long enough time, the velocity variation in a laminar vortex street will depart from strict periodicity; yet most observers would not describe this flow as turbulent. Over a short time, the large scales in a turbulent wake may be nearly periodic; yet most observers would describe the velocity field as random. In making this interpretation they would perhaps be influenced by the reduction of order at progressively smaller scales of motion.

Now that we have some understanding of the word random, we are in a position to gather our ideas into a *definition of turbulence*: a fluid motion involving random macroscopic mixing, with a large range of length and time scales. The motion must be three-dimensional, vortical and of finite intensity, if the large range of scales is to be achieved. According to this definition, it is not relevant that the larger scales of motion are steady, or nearly steady, or periodic, or nearly periodic, so long as there is an underlying range of random macroscopic activity. Even when this concept is borne in mind, some personal judgement is required in delineating the class of turbulent flows.

Let us next consider the nature of the *statistical properties* of a turbulent flow. Their definition depends on the way in which they change (or, more significantly, do not change) in space and time. Hitherto we have assumed the turbulent fluctuations to be superposed on a basically steady flow. In this case, mean values for a point can be found by integrating over a time much greater than the periods of the fluctuations, as in equations (1.1.5) and (1.5.2). They can also be found using simple experimental techniques, as we shall see in the next chapter. These time-mean values have been assumed to be *stable*, that is, to be repeatable for various starting times and averaging intervals.

Now consider decaying homogeneous turbulence in a mass of fluid in which there is no mean motion. This will not yield stable averages with respect to time, but we can define a stable mean by imagining (it is not usually feasible) that a property is averaged through space at some instant. For example

$$\overline{u^2(t)} = \frac{1}{X}\int_{x_0}^{x_0+X} [u(x,t)]^2\,\mathrm{d}x \tag{1.5.11}$$

with X a length much greater than the lengths typifying the turbulence.

If the average with respect to a coordinate is stable, that is, independent of the (sufficiently long) interval over which it is taken, the varying quantity is said to be a *stationary random function* of that coordinate. Thus, if $\overline{u^2}$ defined in equation (1.5.11) is independent of the point x_0 from which averaging commences, and of the distance X (provided that this is long enough), then u is a stationary random function of x. This is a precise way of stating that the turbulence is homogeneous in the x-direction. Similarly, the requirement that u be a stationary random function of time is a precise way of stating that the mean flow and pattern of turbulence are basically steady.

How can averages be defined for a flow which is neither homogeneous nor steady, for example, the flow through the windpipe into the lungs? Here we must resort to the *ensemble average*, obtained from the values at corresponding points in a number of repetitions of the process. Thus, to find the level of turbulence at a point in the windpipe, we could consider the n instantaneous values u_i registered at the same instant during n successive cycles of breathing. The mean velocity and turbulence intensity are then

$$\frac{1}{n}\sum u_i \quad \text{and} \quad \frac{1}{n}\sum (u_i - U)^2 \qquad (1.5.12)$$

Although velocity fluctuations have been considered for definiteness, these ideas are applicable to any property of a turbulent fluid and, indeed, to any random process.

1.6 DISPERSION AND DIFFUSION

The Turbulent Plume

The final example of this introductory survey is sketched in Figure 1.6: a chimney discharges a hot buoyant jet, or plume, into the turbulent wind. Now we must consider not only the velocity variation, but also variations in temperature and in the concentration of convected substances, for example, sulphur dioxide and ash particles produced by combustion. These variations will influence the overall dynamics of the flow, primarily through their effect on the weight of the plume relative to the surrounding air. The temperature and concentration will, like velocity, exhibit random fluctuations within the turbulent flow, and particularly at its boundaries.

In flows of this kind the dispersion of the convected material is often of primary interest, the turbulent velocities themselves being significant only as agents of dispersion. Hence we must look at turbulence in a different way. No longer do we seek only to describe the flow field and to calculate the forces applied to adjacent bodies. Instead, we ask what happens to individual particles within the turbulent fluid and study the statistical properties of families of particles which are convected by and diffuse through the turbulent flow.

Eulerian and Lagrangian Descriptions

The two ways of looking at a flowing fluid, and in particular at turbulent motion, are called the Eulerian and Lagrangian viewpoints, after two eighteenth-century students of fluid mechanics. The former concentrates on events at a

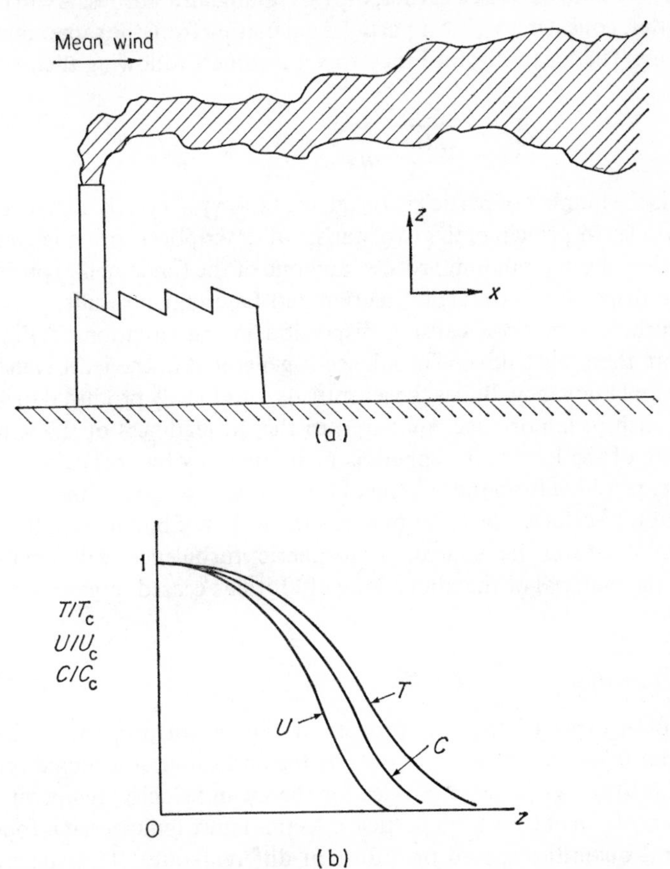

Figure 1.6 Turbulent buoyant jet (or plume) in a turbulent wind, illustrating turbulent dispersion and diffusion. (a) Instantaneous view of plume as indicated by smoke convected by it. (b) Profiles of time-mean temperature, velocity and particle concentration for a fixed value of the coordinate x. The subscript c denotes peak values on the plume centre line

point, specified by functions such as $u(x,y,z,t)$ and $T(x,y,z,t)$, with the coordinates x, y and z selecting the point. The Lagrangian description deals with the history of a particle moving in the flow, given by functions such as $x(x_0,y_0,z_0,t)$ and $T(x_0,y_0,z_0,t)$, with x_0, y_0 and z_0 specifying the particle's position at some initial instant, and thus labelling the particle. The two descriptions have their counterparts in experimental technique: signals from a fixed instrument give an Eulerian description, while flow visualization by convected particles is closer to the Lagrangian picture.

The mean properties of a turbulent flow can be described by Eulerian statistics, the kind discussed earlier, or by Lagrangian statistics. As an example of the latter, consider a series of particles emanating from the same source, say, a chimney. The mean distance they travel in time t following their release is given by

$$\overline{x(t)} = \frac{1}{n} \sum x_i(x_0, t) \tag{1.6.1}$$

where n is the number of particles considered, and $x_i - x_0$ is the distance moved by a particle. In principle, the two statistical descriptions provide equivalent information about a random process, and one of the fundamental problems of turbulent dispersion is to relate Eulerian and Lagrangian results.

The turbulent motions causing dispersion in the situation of Figure 1.6 have more than one source. Turbulence is generated in the jet driven by momentum and buoyancy. But as the jet spreads, this locally produced turbulence will diminish in importance compared to the unsteadiness of the wind. The turbulence of the lower atmosphere is maintained by two processes: thermal instability resulting from the heating of the ground by the sun, and mechanical friction at the surface. The latter includes the wakes of buildings and trees, for example. Whatever its source, atmospheric turbulence will continue to disperse the material of the plume long after it has ceased to generate its own turbulence.

Eddy Diffusivities

It is to be expected that the profiles of time-mean properties will adopt simple and perhaps self-similar forms as the spreading of a plume proceeds. Figure 1.6(b) shows possible profiles for the mean velocity, temperature and the concentration of convected particles. As this figure indicates, it is found that the several quantities spread or diffuse at different rates. This suggests that differing turbulent processes are responsible for conveying momentum, energy and suspended matter across the flow. A simple way of representing this is by introducing a series of effective or eddy diffusivities, possibly one for each attribute of the fluid. Thus, corresponding to each of the molecular transport properties—viscosity, thermal conductivity and mass diffusivity—we have an effective transport property representing an aspect of turbulent mixing.

Earlier we introduced the eddy viscosity or effective diffusivity of momentum ϵ_m through

$$\tau = \rho(\nu + \epsilon_m)\frac{\partial U}{\partial y} \tag{1.3.2}$$

with τ the time-mean shear stress in a nearly parallel flow. In a similar manner we can define a diffusivity to represent heat transfer by the turbulent activity within a nearly parallel mean flow. For molecular conduction, *Fourier's law* expresses the rate of energy transfer per unit area as

$$\dot{q} = -k\frac{\partial T}{\partial y} = -\rho c_p \kappa \frac{\partial T}{\partial y} \tag{1.6.2}$$

with k the thermal conductivity, and $\kappa = k/\rho c_p$ the thermal diffusivity, c_p being the specific heat at constant pressure. Generalized to turbulent flow

$$\dot{q} = -\rho c_p(\kappa + \epsilon_h)\frac{\partial T}{\partial y} \tag{1.6.3}$$

with $\dot{q}$ and $\partial T/\partial y$ the time-mean values, and ϵ_h the *eddy thermal diffusivity.*

The diffusion of a substance through the fluid can be treated in a like manner. For molecular diffusion, *Fick's law* gives the mass flux per unit area

$$N = -\frac{\partial C}{\partial y} \tag{1.6.4}$$

in terms of the mass concentration C and the molecular diffusivity or diffusion coefficient D, different in general for each substance and fluid. Including turbulent diffusion

$$N = -(D + \epsilon_D)\frac{\partial C}{\partial y} \tag{1.6.5}$$

with N and $\partial C/\partial y$ time-mean values, and ϵ_D the *eddy mass diffusivity* of the substance. In a fully turbulent region, it will often be the case that

$$\frac{\epsilon_h}{\kappa} \text{ and } \frac{\epsilon_D}{D} \gg 1 \tag{1.6.6}$$

the effects of turbulence greatly exceeding molecular transport.

The molecular diffusivities of a fluid are related by the *Prandtl and Schmidt numbers*:

$$\mathrm{Pr} = \frac{\nu}{\kappa} = \frac{c_p \mu}{k}$$
$$\mathrm{Sc} = \frac{\nu}{D} = \frac{\mu}{\rho D} \tag{1.6.7}$$

These are properties of the fluid and diffusing substance. Similarly, we introduce *turbulent Prandtl and Schmidt numbers*:

$$\mathrm{Pr}_t = \frac{\epsilon_m}{\epsilon_h} \quad \text{and} \quad \mathrm{Sc}_t = \frac{\epsilon_m}{\epsilon_D} \tag{1.6.8}$$

These are properties of the flow and of position in it, as well as of the fluid and diffusing substance. In fully turbulent regions, they characterize the diffusive activity more closely than do their molecular counterparts. The molecular values (1.6.7) vary over wide ranges, as may be seen by referring to the tables of the Appendix at the end of this book. Moreover, for liquids the Prandtl and Schmidt numbers are strongly dependent on temperature. In a fully turbulent flow, however, the mechanisms of transport are much the same for momentum, energy and suspended substances, namely, convection by the turbulence, admittedly mediated by molecular diffusion. Hence the ratios of the effective diffusivities (1.6.8) differ only a little from unity (say, $2 > \mathrm{Pr_t} > \frac{1}{2}$) for a wide range of flows and fluids.

A much used approach to turbulent diffusion is through the assumptions

$$\mathrm{Pr_t} = \mathrm{Sc_t} = 1 \tag{1.6.9}$$

This simple way of relating diffusive and stress-generating capacities is known as *Reynolds analogy*, after Osborne Reynolds. It will be discussed in some detail in Chapter 5, but we may note two vital criticisms here:

(1) In some circumstances, the various eddy diffusivities are in fact quite different.

(2) In most cases of practical interest, diffusion through an essentially non-turbulent region (perhaps the viscous part of a wall layer) plays an important part in the overall transfer. If the molecular Prandtl or Schmidt number is much different from unity, Reynolds analogy will be seriously in error if applied to the whole process.

It is for these reasons that the profiles of Figure 1.6(b) may have differing widths. Despite these limitations, the basic analogy, and simple modifications of it, provide useful links between the dispersive and stress-transmitting aspects of turbulence. In particular, they make possible the use of a vast body of friction measurements in the prediction of heat and mass transfer rates.

There are special problems associated with the diffusion of heavy particles of a finite size, for example, sand grains in a turbulent stream, or wood-pulp fibres conveyed in a slurry. Such particles respond imperfectly to the turbulent velocity fluctuations: they may be larger than the smaller scales of turbulence, and the time they require to accelerate may be comparable with its shorter time scales.

So far we have considered the effect of turbulence on mean profiles of temperature and concentration and on overall transfer rates. The local effects of turbulent transport are also important, for example, in smoothing out irregularities in concentration, in bringing reagents into contact, and in promoting mass transfer to or from suspended drops, bubbles or particles.

1.7 TURBULENCE AND THE ENGINEER

There are many aspects of turbulence which are not represented in the situations considered in this chapter, for instance, the roles of compressibility and rarefaction in gas flows, and the effects of a free surface, and of boiling and condensation, in liquids and vapours. The phenomena which have been noted in our examples occur in almost all turbulent motions; they may be thought of as the lowest common denominators of turbulence. In the following paragraphs, we shall classify the practical problems associated with turbulence, and isolate some of the fundamental questions which lie behind them. Then we shall survey the available methods of analysis and prediction, and seek a strategy with which the engineer may approach turbulent flow.

Problem Areas

One important group of problems arises from the *dissipation* of energy within turbulence. This manifests itself in a variety of ways—as fluid friction on a wing, drag on a ship's hull, pressure drop in a duct, hydraulic losses in a turbomachine, or loss of head in a hydraulic jump. To calculate most of these, we need to know whether the flow is indeed turbulent, that is, whether transition has occurred. The associated problem of separation from solid boundaries must also be borne in mind, for this can have an important effect on the dissipation, drag and lift generated by the flow. Finally, in treating the linked problems of transition, separation and friction generation, we must allow for their dependence on boundary roughness.

In many situations *forces and pressures* applied by the turbulent fluctuations themselves are important. Critical loads are imposed on buildings and aircraft by gusts, the large turbulent motions in the atmosphere. Here the length scale of the turbulence is comparable to the size of the structure. The high-frequency pressure fluctuations of small-scale turbulence can also be significant, by causing fatigue failure in a surface exposed to a turbulent flow, or by providing excitation near a resonant frequency of a flexible structure. A somewhat similar mechanism plays a part in the generation of water waves: resonant waves are excited by the pressure pattern convected over the surface in the turbulent wind. The pressure fluctuations of noise radiating from turbulent jets are even smaller in magnitude, but they are one of the most widely apparent and most irritating aspects of turbulence.

Associated with the transport of momentum through a turbulent flow are *transfers of heat and matter*. Their prediction is fundamental to the design of heat exchangers, boilers and nuclear reactors, all relying on turbulence to transfer energy between solid surfaces and passing fluids. Similarly, combustion chambers and chemical reactors depend on turbulence to supply or remove process heat, and to bring reagents into contact. Since all of these transfers must be considered in discussing one engineering problem or another, it is advantageous to establish analogies between them, so that information relating to one can be applied to allied problems.

Sometimes there is an *interaction* between the transfer and the turbulent motion by which it is accomplished. In combustion, for example, the turbulence which brings the reagents together is, in part, the result of the violent expansion resulting from the release of thermal energy. A somewhat similar interaction occurs when water flows over an erodible stream bed. The pattern of dunes and ripples on the bed of a natural stream is at once the cause of irregular motions within the flow, and the result of erosive attack by those motions. Thus the prediction of the friction between flowing water and an erodible bed is bound up with the prediction of the boundary which the stream carves for itself. Similar interactions take place between the wind and sand dunes, drifting snow, and water waves.

The turbulent *dispersion* of contaminants—heat, dissolved substances, or finite particles—has several aspects of great practical importance. Perhaps the first to come to mind is the necessity of dispersing man-made wastes—low-grade heat, sewage, smoke and fumes. But dispersion and diffusion are vital in many natural processes; the spread of pollen and spores, and the cycling of the earth's water through the atmosphere are examples. The most straightforward problem of dispersion is the spread of a contaminant through a homogeneous field of turbulence. Diffusion in free-turbulent flows—jets and wakes—is linked to the advance of the interface separating turbulent and ambient fluids. Other difficulties arise when considering dispersion through a flow whose structure is rendered non-uniform by a boundary such as the ground or the sea's surface.

From these examples, it is clear that turbulence cannot be categorized once for all as a nuisance. Often turbulence is deleterious, giving rise to dissipation, unwanted forces, or excessive cooling. On the other hand, the dispersive effects are helpful, indeed vital, to the maintenance of our environment. In some circumstances, good and bad features are inextricably mixed. In a heat exchanger, for instance, turbulence augments the heat transfer, but also increases the power required to force the heat-transfer fluid through the unit. Again, the turbulence which carries water vapour into the atmosphere from the sea's surface also acts to establish troublesome waves there. This serves to remind us that, whatever its economic consequences, turbulent activity is responsible for some of the most attractive and interesting natural phenomena.

A Strategic View

Our first priority in coping with turbulence is to obtain *control* over it: to suppress it when it proves undesirable; to promote it when it is beneficial;

and to obtain the optimum balance when its effects are mixed. This approach has proved fruitful in the design of aircraft wings. By careful design it has been possible to achieve laminar flow over a considerable part of the surface, thus reducing the friction drag. On the other hand, it is sometimes expedient to fit turbulence inducers further back on the wing. These discourage separation, by re-energizing the lower levels of the boundary layer through augmented mixing; thus they delay the rise in drag and fall in lift which occur at low speeds. Another example of control is the use of additives to reduce friction in pipes and on ship's hulls—fibres and long-chain polymers inhibit the turbulence near the wall.

Although some notable successes have been achieved, direct control of turbulence is not possible in most circumstances. It is obvious that control cannot readily be applied to large-scale natural flows in the atmosphere and in rivers and seas. In many 'man-made' flows an attempt to control turbulence entails greater penalties than does the uninhibited turbulence. The tendency of fluid motions to instability is so strong that we have, in many circumstances, no realistic option but to accept it.

This brings us to the second line of attack on turbulence, the attempt to *predict* its effects, in order to put them to use, or, at least, to minimize their harmful aspects. Since no comprehensive theory of turbulence is available, most predictions involve a measure of empiricism. Similarity arguments provide an entrée to many problems of practical interest, reducing the amount of empirical data required, and giving insight into the general features of complex processes. Simple analytical models using fluid lumps or eddies are also helpful, sometimes providing direct numerical predictions, but more commonly giving a skeleton to which measurements can be fitted. For this reason, the label *semi-empirical* is often used for the formulae on which engineering calculations are based. Finally, a few problems have been analysed using the more rational approach which takes explicit account of the statistical properties of the fluctuations. It is possible to predict, for example, certain features of the fine structure of turbulence, and of the response of turbulence on passing through a gauze screen or a rapidly distorting duct. Even here it is necessary to introduce a few bits of experimental data to convert the theoretical structure into working formulae.

From this survey, it will be apparent that any attempt to predict the effects of turbulence, or to construct a working theory of its mechanisms, requires a body of information about real turbulent flows. Hence a prerequisite for the study of turbulence is the ability to *measure* certain of its attributes. This third line of attack on turbulence will be considered in the next chapter.

On commencing the study of turbulence, the engineer or teacher of potential engineers must consider the following questions. Should he seek only the level of understanding needed to use empirical information, or should he try to master the more rigorous methods of the statistical theory of turbulence?

The answers to these questions will, of course, depend on the individual and on the nature of the problems he has to solve. The answers will also depend upon the time at which the questions are asked, for our ability to discuss turbulence rationally (if not succinctly) is steadily being extended.

In assessing the utility of the statistical approach through the basic equations of motion, we must recognize that it is plagued with a fundamental problem which transcends the difficulties of analysis and computation which arise in particular cases. This intractable problem springs from the averaging procedure used to bring order from the chaos of turbulent motion. The difficulty is exemplified in equations (1.5.3):

$$\overline{(P+p)^2} = P^2 + \overline{p^2} \qquad (1.7.1)$$

A single fluctuating quantity, $P+p$, has been eliminated at the price of introducing two statistical properties, P and $\overline{p^2}$. We shall see what this implies for the equations of motion in later chapters, most comprehensively, in Section 9.1.2. The essential point is that we end up with more unknowns than governing equations. Additional information must be provided to establish a closed and solvable mathematical statement; hence this dilemma is called the *problem of closure*. The closing relations must re-introduce certain vital features of turbulence that are eliminated by averaging, and this brings us back to the *problem of structure*, the identification of the dominant elements within turbulent flows. Steady progress is being made, but there is still a gap between soundly based theory and the demands of engineering calculations.

We conclude that the semi-empirical methods developed by engineers will be the basis of prediction for many years to come, at least for the limited situations to which they apply. For many of the complex geometries which arise in engineering equipment, these methods fail to provide sufficient accuracy and surety, and specific experiments will continue to be required in critical cases. However, the growth points in our understanding of turbulence, and in our ability to predict and ultimately to control it, lie at the interfaces between experiment, the proven semi-empirical results, and the more rational theoretical methods derived from the statistical theory.

The approach adopted in this book is consistent with the present position of the engineer with regard to turbulence. In Chapter 2 some statistical ideas will be introduced while considering the analysis of measurements. In Chapter 9 we shall consider the prediction techniques recently made feasible by the availability of high-speed computers. In the intermediate chapters we shall apply the simpler methods to the limited, though important, range of situations for which they are appropriate.

FURTHER READING

Bradshaw, P. Reference 2: Chapter 3
Hinze, J. O. Reference 5: Section 1.1

Kuethe, A. M. and J. D. Schetzer. *Foundations of Aerodynamics*, Wiley, New York (1959): Chapters 14 and 16
Prandtl, L. Reference 8: Sections III.4 and III.6
Schlichting, H. Reference 11: Chapters 1 and 2
Scorer, R. S. *Natural Aerodynamics*, Pergamon Press, London (1958): Chapters 6 and 8
Sutton, O. G. *Atmospheric Turbulence*, Methuen, London (1955): Chapter 1
Tennekes, H. and J. L. Lumley. Reference 12: Chapter 1

SPECIFIC REFERENCES

Batchelor, G. K. *The Theory of Homogeneous Turbulence*, Cambridge U. P. (1957)
Corrsin, S. 'Turbulent flow', *Amer. Scientist*, **49**, pp. 300–325 (1961)
Kline, S. J. 'Observed structure features in turbulent and transitional boundary layers', in *Fluid Mechanics of Internal Flow*, Elsevier, Amsterdam (1967)
Morkovin, M. V. 'Flow around a circular cylinder—a kaleidoscope of challenging fluid phenomena', in *Symposium on Fully Separated Flow*, Amer. Soc. Mech. Eng. (1964)
Tucker, H. J. and A. J. Reynolds. 'The distortion of turbulence by irrotational plane strain', *J. Fluid Mech.*, **32**, pp. 657–673 (1968)

EXAMPLES

1.1 Use dimensional analysis to show that measurements of fluid friction in a long, straight pipe can be correlated as $c_f = f(\mathrm{Re}_d, k/d)$. What special forms apply at high and low Reynolds numbers?

1.2 The laminar boundary layer on a flat plate grows according to the law $\delta = 5(\nu x/U_1)^{\frac{1}{2}}$, with δ the distance at which the velocity is 99 per cent of U_1, the free-stream value. Given the Reynolds number for instability in pipe flow, estimate the Reynolds numbers $\mathrm{Re}_\delta = U_1\delta/\nu$ and $\mathrm{Re}_x = U_1 x/\nu$ at which a laminar boundary layer will become unstable.

1.3 For a laminar pipe flow, where $c_f \mathrm{Re}_d = $ constant, show that the average dissipation per unit volume is proportional to $\mu(U_a/d)^2$, that is, to the square of the characteristic velocity gradient.

1.4 Between the viscosity-dominated sublayer on a smooth wall and the fully turbulent flow lies a part of the wall layer where both viscous and turbulent friction are significant. Experiments show that this region is defined by

$$5 < \frac{y(\tau_w/\rho)^{\frac{1}{2}}}{\nu} < 50$$

(a) For turbulent flow in a pipe $c_f = \tau_w/(\frac{1}{2}\rho U_a^2) = 0{\cdot}079\ \mathrm{Re}_d^{-\frac{1}{4}}$ for a smooth wall and $\mathrm{Re}_d < 100{,}000$. Show that the limiting values of y are proportional to $d\ \mathrm{Re}_d^{-\frac{7}{8}}$.

(b) For a pipe of diameter 10 cm, estimate the thicknesses of viscous and partially viscous layers for the lowest Reynolds number at which turbulence can be maintained, and for $\mathrm{Re}_d = 100{,}000$. For what heights of roughness will the pipe be hydrodynamically smooth under these two flow conditions?

(c) How will the ratio U_a/U_c, with U_c the centre-line velocity, change between the two cases considered in (b)?

1.5 (a) For what value of S does equation (1.1.7) define the average velocity (1.1.6)?

(b) What is the significance of the bulk values obtained by taking $S = U, \frac{1}{2}U^2, T$ and $c_pT + \frac{1}{2}U^2$?

1.6 A rather old-fashioned way of assessing the level of turbulence in a wind tunnel is to measure the drag of a smooth sphere in the test section. The performance is specified in terms of the Reynolds number (based on sphere diameter) at which the drag coefficient attains the value 0·3.

(a) For two tunnels the critical Reynolds number is found to be 100,000 and 400,000. In which is the stream less turbulent?

(b) What is the drag of a sphere of diameter 15 cm at the critical condition in the second tunnel?

(c) Sketch the drag variations with velocity for the sphere considered in (b), and for a similar sphere which has a very rough surface.

1.7 It is proposed to test models in a wind tunnel to estimate the wind loads for two structures: (1) a slab-like building, 60 m high, 25 m by 15 m in plan; and (2) a cylindrical silo with a hemispherical top, 5 m in diameter and 15 m high overall. The design wind-speed is 50 m/sec. The wind tunnel available has a test section 1 m square; not more than 10 per cent of the area is to be filled by a model, in order that excessive blockage be avoided. The maximum tunnel velocity is 25 m/sec.

What relevance will drag coefficients measured in the wind tunnel have for the prototype structures? How might the difficulties which arise for structure (2) be avoided?

1.8 In a turbulent wake, measurements of Townsend give the following variations across the flow; the scales are arbitrary.

Table 1.2

y	$U_1 - U$	τ/ρ	γ	$\overline{u^2}$	$\overline{v^2}$	$\overline{w^2}$
0	2·28	0	1·000	1·50	1·99	1·40
0·1	1·97	0·479	0·996	1·80	1·83	1·50
0·2	1·27	0·675	0·956	1·92	1·42	1·52
0·3	0·48	0·485	0·765	1·25	0·84	0·80
0·4	0·08	0·130	0·519	0·32	0·31	0·22
0·5	–	0·017	0·231	0·05	0·08	0·04

Find the profiles of the turbulence kinetic energy and of the eddy viscosity. How do they vary within the fully turbulent part of the fluid? What support do these results give for the use of a constant eddy viscosity?

1.9 The smallest length scales of turbulence are of the order $\nu^{\frac{3}{4}}\,\varepsilon^{-\frac{1}{4}}$, where ε is the rate at which energy is dissipated per unit mass of fluid. In a gas the mean-free-path is of order ν/a, where a is the sound speed.

(a) Derive a condition for the turbulence scales to be much greater (say, 100 times) than the molecular scales. What does this mean in physical terms?

(b) Show that the mean dissipation rate in a pipe flow is $\varepsilon = 2c_f U_a^3/d$, and hence that the turbulence scale varies as

$$d\mathrm{Re}_d^{-\frac{5}{8}} \quad \text{when} \quad c_f = 0{\cdot}079\,\mathrm{Re}_d^{-\frac{1}{4}}.$$

(c) Using the result (b), estimate the two length scales for atmospheric air and for the flow conditions specified in Example 1.4(b). Under what conditions will it be impossible to distinguish clearly between molecular and turbulent scales?

1.10 (a) Show that the Fourier transform

$$a(k) = \frac{1}{\pi}\left[\frac{\sin\,(k_0 + k)R}{k_0 + k} + \frac{\sin\,(k_0 - k)R}{k_0 - k}\right]$$

is obtained from equation (1.4.2) for $u = \cos(k_0 r)$ for $0 < r < R$, $u = 0$ for $r > R$. Substitute into equation (1.4.1) to check the result. What is the nature of the result when $R \to \infty$? Is this reasonable?

(b) To investigate the effect of the extent of the coherence of the function $u(r)$, consider the ratio $a(k_0)/a(0)$ for $R = \pi n/2k_0$, with $n = 1, 3, \ldots$. Over what distance must the periodicity extend if $a(k_0)$ is to dominate the transform?

1.11 Prepare a table similar to Table 1.1 for axisymmetric flows. For the boundary layer and wall jet, consider developments both within and outside cylinders; add a column indicating intersections of spreading turbulent regions.

1.12 Consider a turbulent jet formed by ejecting fluid from a round pipe into a quiescent mass of the same fluid. The time-mean centre-line velocity U_c and the time-mean effective diameter of the jet b may be supposed to depend upon the momentum flux M (which has dimensions for force), the fluid density ρ, and the distance x measured along the jet from an effective origin.

(a) Use dimensional analysis to show that $b = c_1 x$ and that $U_c = c_2\,(M/\rho)^{\frac{1}{2}}/x$. with c_1 and c_2 universal constants.

(b) Show that the mass flux within the jet varies according to the law $\dot{m} \propto x$.

(c) Show that the mean inflow velocity at the 'edge' of the jet varies as x^{-1}, and that it is a constant fraction of the rate of spread of the turbulent fluid.

(d) Show that the law $U/U_c = f(r/x)$ specifies the velocity variation across the flow, and that it is the same for any jet of the class considered.

1.13 The time-mean streamwise velocity in the turbulent wake formed by a circular cylinder is given by

$$U_1 - c_1 x^{-\frac{1}{2}} \exp\left(-\frac{c_2 y^2}{x}\right)$$

(a) On what defining parameters do the constants c_1 and c_2 depend? How do these enter into the constants?

(b) If the drag of the cylinder is increased by a factor of two, all the other parameters remaining constant, by how much do the velocity deficit $(U_1 - U_c)$ and wake width change? If the velocity of the stream increases by a factor of two, with the drag coefficient remaining constant, by how much do the deficit and width change?

(c) How does the mass-flow deficit vary along the wake? What does this imply for the streamlines beyond the wake and for the apparent rate of spread of the turbulence?

1.14 In a round wake, such as that behind a sphere, it is found that the velocity deficit and width vary as $U_1 - U_c \propto x^{-\frac{2}{3}}$ and $b \propto x^{\frac{1}{3}}$. How does the local Reynolds number of the shear flow vary along the wake? What implications has this for the turbulence of the wake? What conclusions do you reach by applying the same reasoning to the round jet and plane wake (Examples 1.12 and 1.13)?

1.15 Although the eddy viscosity may sometimes be assumed constant across a thin turbulent flow, it does not follow that this quantity may be taken to be constant along the flow. This can be seen as follows.

(a) Assuming the local eddy viscosity to depend on the lateral distance b across which the time-mean velocity drops by ΔU, and on that velocity change, show that $\epsilon_m = K\, b\, \Delta U$, with K a coefficient characteristic of the flow species.

(b) By considering the requirement that the momentum flux be constant along the flow, show that

$$b^{\alpha}(\Delta U)^{\beta} = \text{constant}, \quad \text{with } \alpha, \beta = 0, 1 \text{ or } 2$$

for axisymmetric and plane jets and wakes, and for the plane mixing layer. How does the eddy viscosity vary along each of these flows?

1.16 For the turbulent flow specified by equations (1.5.4), show that the total kinetic energy is $\frac{1}{2}U^2 + \frac{1}{2}q^2$. Taking the total fluid energy to be measured by the total pressure $(p + \frac{1}{2}\rho u^2)$, find the time-mean energy of the mean flow and turbulence.

1.17 Table 1.3 gives instantaneous values of velocity components measured at a point at equal time intervals.

Table 1.3

$U+u$	27	41	58	19	−7	19	−3	18	21	−1	62	26
$V+v$	13	29	6	−1	15	2	−17	−31	−2	13	−9	−18

(a) Calculate U, V, $\overline{u^2}$ and $\overline{v^2}$. If the third mean component W is zero, estimate the turbulence level, assuming that $\overline{w^2} = \frac{1}{2}(\overline{u^2} + \overline{v^2})$.

(b) How do these values differ from those you would expect in isotropic turbulence? If $\overline{u^2}$ and $\overline{v^2}$ were interpreted as the same quantity measured at different points, might the turbulence be homogeneous?

1.18 (a) It is attractive to postulate that homogeneous turbulence left to decay will tend towards isotropy. What evidence is there in Figure 1.5(b) to support this hypothesis?

(b) Explain the trends in the response of turbulence to uniform distortion, as indicated in Figure 1.5(b), by considering three perpendicular vortex filaments, initially identical. Combine the results to see how the components of initially isotropic turbulence would behave.

1.19 Measurements in a turbulent flow give $U = 30$ m/sec and $\overline{u^2} = 0{\cdot}8$ m²/sec². Assuming that the fluctuations follow a normal or Gaussian distribution, find: (1) the standard deviation of the fluctuating signal, and (2) the probabilities that the component $U + u$ will exceed 31 m/sec and will fall short of 27 m/sec.

1.20 State Taylor's hypothesis with reference to stationary random functions.

1.21 (a) What is the Lagrangian description of the flow specified in Eulerian terms by $u = U$, a constant? What is the Eulerian description of the flow specified in Lagrangian terms by $r = r_0$ and $\theta = \theta_0 + \omega_0 t$, using polar coordinates?

(b) Given the Lagrangian mean displacement (1.6.1), how would you calculate the mean convection velocity, if the flow and turbulence are homogeneous and the particles are small and have the same density as the fluid? What difficulties would arise if these conditions were not satisfied?

1.22 Small, neutrally buoyant pellets are introduced into a turbulent pipe flow, and the times taken for them to traverse 50 ft of pipe are measured. The times for twenty of the pellets are: 10·0, 9·2, 9·9, 10·5, 11·0, 8·8, 9·0, 10·1, 10·8, 10·2, 9·4, 9·6, 9·0, 9·8, 9·7, 11·3, 9·3, 9·9, 10·2 and 9·5 sec.

(a) Explain the variability in the transit times, and the 'skewed' nature of the distribution. If the pipe flow were laminar and the pellets were introduced randomly over the cross-section, what distribution would you expect? How does this compare with the distribution for turbulent flow? In what ways does turbulence influence the ability of a pipe flow to disperse particles suspended in it?

(b) Estimate the mean velocity of the flow and the centre-line velocity. How might an estimate be obtained for the positions of the particles at time $t = 8{\cdot}8$ sec after release? What relationship have these calculations to Eulerian and Lagrangian statistics?

1.23 The liquid sodium used to carry heat from the core of a nuclear reactor has a Prandtl number of 0·005. Show that molecular conduction of heat can be significant even in fully turbulent flow.

1.24 (a) Assuming that Reynolds analogy (1.6.9) applies throughout a shear layer across which the velocity, temperature and concentration change by Δ_1, Δ_2 and Δ_3, show that

$$\frac{\tau_w}{\rho \Delta_1^2} = \frac{\dot{q}}{\rho c_p \Delta_1 \Delta_2} = \frac{N}{\Delta_1 \Delta_3}$$

(b) For a class of boundary layers with $c_f = 0{\cdot}046\ \mathrm{Re}_\delta^{-\frac{1}{4}}$, show that $\tau_w \propto U_1^{\frac{7}{4}}$, that $\dot{q} \propto U_1^{\frac{3}{4}}$ and that $N \propto U_1^{\frac{3}{4}}$.

(c) Derive formulae from which $\dot{q}$ and N can be calculated for these boundary layers.

1.25 Consider one-dimensional molecular diffusion of a general substance with concentration S (per unit volume) and diffusivity K.

(a) Assuming simple diffusion, as defined by Fourier's or Fick's laws, apply a conservation argument to an elemental volume to obtain the simple diffusion equation

$$\frac{\partial S}{\partial t} = \frac{\partial}{\partial y}\left(K \frac{\partial S}{\partial y}\right)$$

(b) For constant K, show that this equation implies that a distribution for which there is a fixed scale S_0 is given by

$$\frac{S}{S_0} = f(Kt, y)$$

and hence that

$$\frac{S}{S_0} = f\left(\frac{y}{\sqrt{(Kt)}}\right)$$

(c) Analytically, the simplest problem of this class is that defined by $S = S_0 \cos(\omega t)$ at $y = 0$, and $S \to 0$ for $y \to \infty$. Show that

$$\frac{S}{S_0} = e^{-ay} \cos(\omega t - ay) \quad \text{with} \quad a = \left(\frac{\omega}{2K}\right)^{\frac{1}{2}}$$

is an appropriate solution. What is the thickness of this non-steady layer (to $S/S_0 = 0{\cdot}01$)?

1.26 It is proposed that the flow through a straight pipe can be increased by introducing a number of splitter plates parallel to the axis, to divide the cross-section into passages small enough to retain laminar flow. Decide whether this is feasible, and if so under what conditions. Assume for simplicity that the passages behave like round tubes whose total area equals that of the basic pipe. Take $c_f = 16/Re_d$ for laminar flow, and $c_f = 0{\cdot}08/Re_d^{\frac{1}{4}}$ for turbulent.

1.27 Which of the following terms apply to turbulent flows and which to laminar? If a term is applicable to both, explain any differences in its significance for the two flow types.

Boundary layer, intermittency, dispersion, separation, recirculation, eddy, instability, reattachment, shear layer, transition, mixing, sublayer, spreading, random, eddying, dissipation, isotropic, decay, conduction, similarity, energy cascade, vortex sheet, length scale, Reynolds number, Reynolds analogy, Reynolds-number similarity, friction drag, mean velocity, average velocity, hydrodynamically smooth, diffusivity, statistical theory, Lagrangian description, Taylor's hypothesis, homogeneous, self-preserving, entrainment, pressure drag, no-slip condition, wake, one-dimensional model, Prandtl number, transport properties.

2

Experiments and Their Analysis

There are several reasons for studying turbulence experimentally: one worker seeks empirical data of direct practical application; another wants only a general physical picture of some aspect of turbulence; while a third requires detailed results relating to the fine structure, in order to test theoretical predictions. These varying needs can be satisfied using a variety of experimental techniques: measurements of the overall effects of turbulence; flow visualization to reveal its working; and analysis of the fluctuations themselves. These three kinds of information have been used in the preceding chapter in building up a general picture of turbulent activity and its effects.

This single chapter can provide only an introduction to the basic concepts underlying experimental studies of turbulence; readers who themselves intend to undertake experiments should refer to the specialized works listed at the end of the chapter. The aims of this introductory survey are:

(1) to indicate the range of techniques available, the kinds of information they can provide, and some of their limitations;
(2) to introduce the terminology and symbolism used in presenting experimental results;
(3) to give, through a consideration of the problems and capabilities of measurement, an impression of the variability of turbulence in the four dimensions of space and time; and
(4) to introduce methods of analysing fluctuating signals and, through them, some of the ideas fundamental to the statistical view of turbulence.

The most widely used device for measuring turbulent fluctuations is the *hot-wire anemometer*, a tiny platinum or tungsten wire (say, with diameter 5 μm and length 1 mm), heated by an electrical current which responds to changes in the velocity and temperature of the fluid around the wire. Since so much of our knowledge of turbulence is derived from this device, its operation will be considered in some detail.

Many readers of this book will not have an opportunity to study turbulence using the sophisticated techniques available in specialized laboratories. In these circumstances they may find it difficult to develop a feeling for turbulence.

The simple experiments suggested at a few places in this book should be of some assistance. Also, a good deal can be learned through one's senses: natural streams, the atmosphere, household equipment and industrial installations all generate turbulence, and under some circumstances it can be seen, felt or heard.

2.1 MEASUREMENT OF MEAN PROPERTIES

In the first chapter we were able to infer a good deal about the internal workings of turbulence by examining its overall effects. Thus many tests designed to provide empirical data for engineering design give more fundamental insights into turbulence. Measurements of mean velocities, forces, pressures and shear stresses can tell us something about transition, separation and turbulent friction. Measurements of mean temperature, concentration and transfer rates reveal aspects of turbulent mixing, as do studies of spreading and diffusion in free turbulence. Patterns of erosion, condensation and sedimentation all carry suggestions about the turbulent flows which cause them.

The Effect of Fluctuations

The instruments used to find mean values in highly turbulent flows are often those developed for essentially steady situations: Pitot tubes, manometers, strain gauges, mercury-bulb thermometers, pH meters and so forth. The response of these instruments is (or can be made, using mechanical or electrical filters) slow enough to provide steady readings, independent of the continual fluctuations. In using such instruments for turbulent flows, we assume, often tacitly, that the effect of the fluctuations on the indicated mean value is negligible, either because the fluctuations are small, or because their effects cancel out. For an instrument whose signal, $S+s$, is proportional to some fluctuating property, $F+f$, the latter assumption may be justified, since on averaging

$$\overline{S+s} = K\overline{(F+f)} \quad \text{gives} \quad S = KF$$

Such an instrument will record a simple time average, and calibration can be accomplished using static tests. When the response is non-linear, the value registered will contain a contribution from the fluctuations. Thus for $S+s=K(F+f)^n$ we have

$$\overline{S+s} = S \simeq KF^n\left[1+\tfrac{1}{2}n(n-1)\frac{\overline{f^2}}{F^2}\right] \tag{2.1.1}$$

Here static calibration is insufficient for a determination of the mean of the fluctuating quantity. Either dynamic calibration or a theoretical correction of the static result is required if high precision is sought.

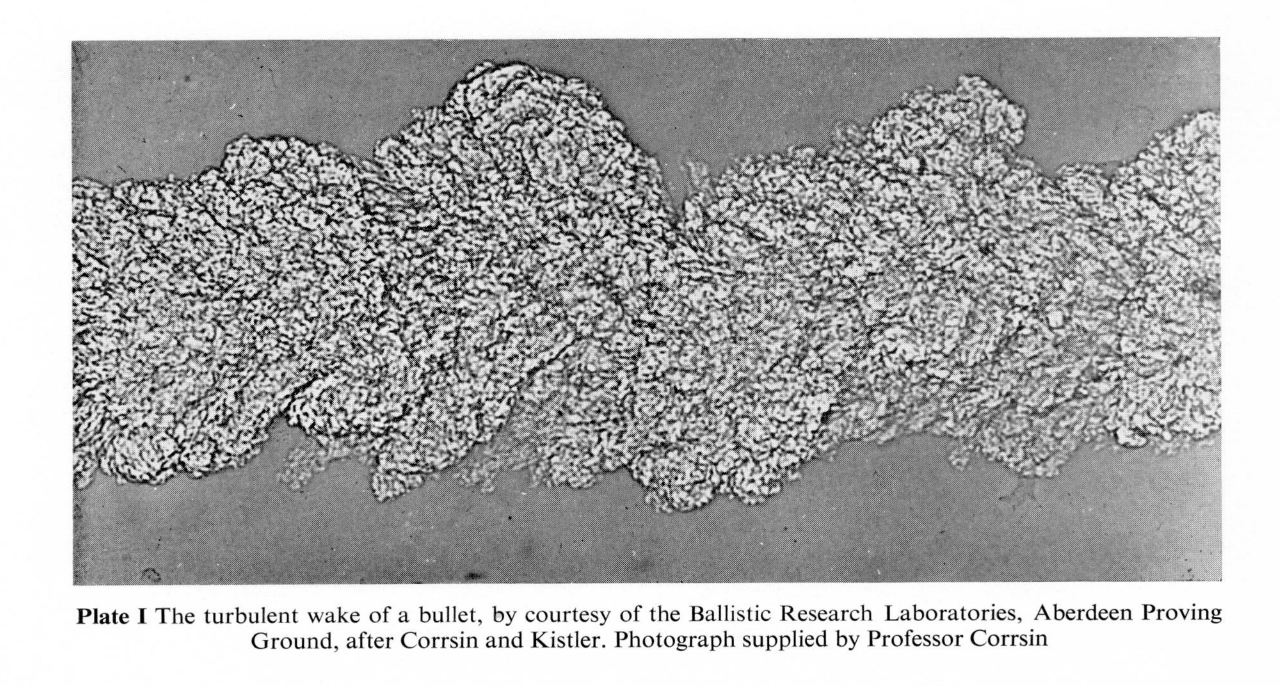

Plate I The turbulent wake of a bullet, by courtesy of the Ballistic Research Laboratories, Aberdeen Proving Ground, after Corrsin and Kistler. Photograph supplied by Professor Corrsin

[*facing page 46*

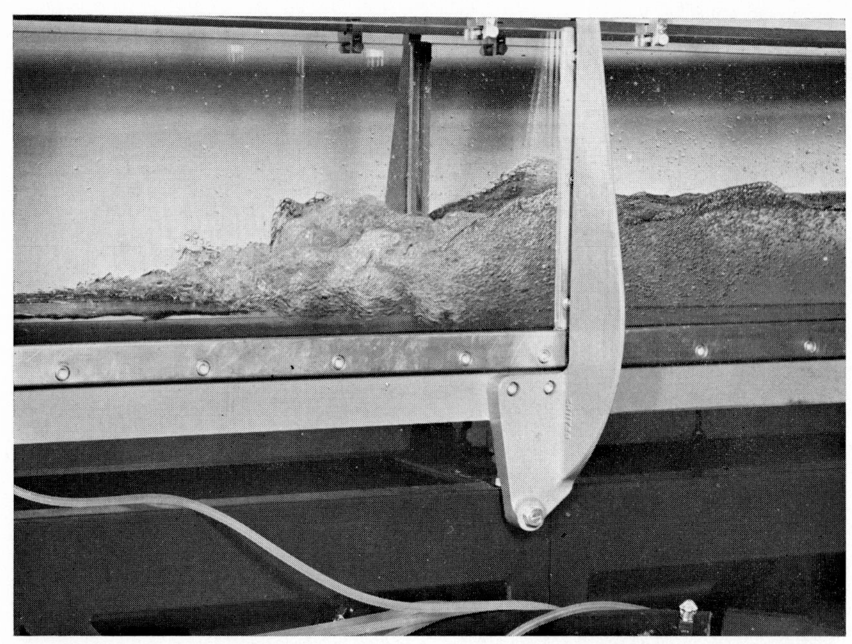

Plate II A hydraulic jump in a glass-sided channel. Photograph supplied by Armfield Engineering Ltd

(a)

(b)

Plate III Use of hydrogen bubbles to indicate separation and flow reversal in an expanding channel. Photographs supplied by Armfield Engineering Ltd

The Pitot-static Tube

Let us see how these ideas apply to the most commonly used velocity-measuring instrument—the Pitot-static tube shown in Figure 2.1. For simplicity, this is taken to be situated in an air flow whose velocity is low enough to render density changes insignificant.

If the flow is steady, and the probe and pressure lines are properly designed and installed, the total pressure p_0 and the static pressure p will be communicated faithfully to the differential pressure meter (perhaps a U-tube manometer) which measures $p_1 - p_2$. Thus

$$p_1 - p_2 = p_0 - p = \tfrac{1}{2}\rho U^2 \tag{2.1.2}$$

and the flow velocity U can be calculated, with an accuracy of perhaps $\frac{1}{4}$ per cent. If the flow is turbulent, the pressures at the sensing points will fluctuate as the local flow direction, velocity and pressure vary. However, if the damping in the pressure lines to the meter is high enough, the recorded differential will

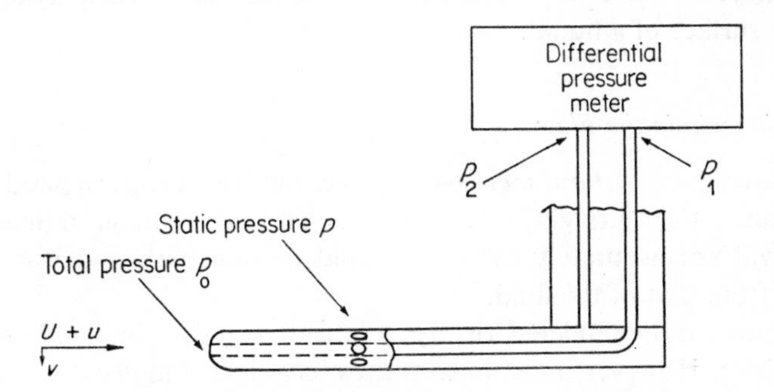

Figure 2.1 Pitot-static tube in a turbulent flow

still be nearly steady. It is not possible to predict the instantaneous pressures just inside the sensing holes for a specified structure of ambient turbulence, but a plausible assessment of the effect of the fluctuations can be obtained by assuming that the fluctuating pressure differential has the form

$$p_0 - p = \tfrac{1}{2}\rho[U^2 + K_1 u^2 + K_2(v^2 + w^2)] \tag{2.1.3}$$

with u, v and w components of the turbulence around the probe, and with the constants K_1 and $K_2 \sim O(1)$, that is, of order of magnitude unity. The steady measured differential will be something like

$$p_1 - p_2 = \overline{p_0 - p} = \tfrac{1}{2}\rho[U^2 + K_1\overline{u^2} + K_2(\overline{v^2} + \overline{w^2})] \tag{2.1.4}$$

with the constants modified to account for distortions introduced within the pressure lines.

Similar difficulties must be anticipated for other slow-responding instruments. Unless you have a cast-iron case, it is unwise to assume that the recorded mean value does not contain a percentage error $\sim O(\overline{f^2}/F^2)$, as indicated in equation (2.1.1).

It should be noted that Pitot-tube readings can deviate from the uniform-flow value for other reasons: the proximity of the tube to a solid boundary, or its location in a strong gradient of mean velocity, either in the direction of the stream or across it.

2.2 FLOW VISUALIZATION

Inevitably, a more detailed knowledge of turbulence is required than is afforded by an analysis of its overall effects. Historically, and for some purposes even now, flow visualization provides the next court of appeal. These techniques can be divided into three broad classes: those giving information about the body of the flow; those indicating what happens on or near a bounding solid surface; and, in some respects intermediate to these, methods displaying events at the free surface of a liquid.

Visualization within the Flow

Some flows carry *natural markers*—smoke, bubbles, drops or sand grains. Here visualization is straightforward, although it must be realized that largish markers will not accurately follow the fluid motion if their density differs markedly from that of the fluid.

Clear flows can be rendered visible by introducing dye, smoke, bubbles or solid particles. However, most intense *dyes* have heavy molecules, and their molecular diffusivities are much smaller than the diffusivity of momentum, the kinematic viscosity. In non-turbulent regions the spread of such dyes will be much slower than the spread of momentum. The increase in weight of the dyed fluid may also be significant. In recent years, much use has been made of the *hydrogen-bubble technique*: at discrete time intervals, myriad tiny bubbles are released by electrolysis from a wire stretched across the flow; the result is a series of patches of bubbles, which are convected and distorted by the flowing fluid. Large, easily followed *particles* can be used, provided that the particle/fluid density difference does not have a crucial effect on the net fluid weight, or on the fidelity with which the particles follow the fluid. Polystyrene has proved popular, as it is nearly neutrally buoyant in water. In the atmosphere, it is possible to use a balloon.

Differences in *density* and *refractive index* occur spontaneously in strongly heated fluids and in gases undergoing large velocity changes, and can also be introduced artificially. These can provide a picture of the flow, provided that suitable optical arrangements are made. Very intense lighting is the basis of the shadowgraph technique; the Schlieren method depends on ray deflection by varying refraction; and optical interference indicates density changes directly. Even without such aids it is often possible to see thermal turbulence, the unsteady convection currents above a heated body.

The pattern of surface elevation in an open-channel flow displays certain features of the flow beneath. The *free surface* can also be utilized by sprinkling powder or bits of paper onto it. These methods are particularly helpful in revealing separation and regions of recirculating flow. However, surface deflections and capillarity can lead to spurious results.

Visualization on a Solid Surface

Our main interest here is finding where transition and separation occur. *Tufts* of thread, wool or paper, fixed by one end to the surface, provide a direct indication of transition (through their fluttering) and separation (through their mean inclination). The dependence of *evaporation* and *sublimation* rates on turbulent activity can also be utilized. The china-clay method, for example, makes use of a substance which is invisible when dissolved in a volatile liquid, but turns white on drying. It provides an indication of transition, since drying proceeds much more slowly beneath a laminar boundary layer.

The *shear stresses* at the surface can be utilized to indicate the flow there, including separation and transition. Oil, which may be coloured to be easily visible, is driven over the surface by the stresses there; it collects along separation lines, and is more thoroughly swept away beyond the point of transition. When there is a strong cross-stream pressure gradient, the direction of the stress and streamlines at the surface can differ greatly from the direction of the flow beyond the boundary layer; hence caution is necessary in interpreting the surface flow pattern.

Examples

Plates I to III illustrate a few of the techniques mentioned above. In Plate I (facing p. 46) the turbulent wake of a bullet is revealed by the shadowgraph method, utilizing density gradients associated with the non-uniform energy and temperature of the strongly shearing fluid. The extreme thinness of the turbulence interface is clear, as are convolutions of various sizes; the interface distortions give evidence of the three-dimensional nature of the turbulence. The apparent uniformity of the internal activity is somewhat deceptive, since the 'shadows' depend on the second derivative of the density, and thus respond most strongly to the smallest structures.

In Plates II and III (facing p. 47) motions in liquids are made visible by bubbles. Plate II shows a hydraulic jump in a laboratory channel; the air bubbles are entrained spontaneously by the violent activity in the roller, and clearly delineate this region of intense dissipation. Its structure is generally similar to that of the recirculating motions which develop when flow separates at a rapid expansion of a closed channel. Again a sharp interface is visible between the active turbulence and the 'outer' flow, in this case, that along the channel bed.

Plate III is not entirely appropriate here, since the expanding free-surface flow is not actively turbulent; but it does indicate how markers, here hydrogen bubbles, can be used to display: (a) separation, and (b) reversed flow or recirculation.

2.3 TRANSDUCERS FOR FLUCTUATING PROPERTIES

A transducer is a device capable of transmitting energy from one system to another; here we are interested in devices which accept energy from some turbulent fluctuation and transmit it to a measuring system. Typically, these fluctuations are rapid, and small in intensity and spatial extent. Hence turbulence transducers must be small, sensitive, and responsive to high-frequency changes. Moreover, they should not distort the quantity being measured. In some cases we must add the requirement of being able to resist a hostile environment, perhaps with large dynamic forces and large ranges of temperature and pressure.

The significance of the words 'small' and 'rapid' varies from flow to flow. The larger elements within the turbulent wind, for instance, have scales measured in minutes and metres; here a robust windmill-like anemometer (that is, wind or velocity meter) will provide meaningful results. In the boundary layer beneath a supersonic air flow, on the other hand, the scales are measured in microseconds and fractions of a millimetre, and a very different instrument is required.

As is suggested by these examples, small time scales are often found in conjunction with small length scales. But this need not be so, for the frequency perceived by a transducer depends on the velocity with which disturbances are convected past it or, more generally, on the mean relative velocity between it and the fluid. Thus the history of the fluctuations measured by the instrument depends both on the changes within the turbulence and on the mean motion of the fluid around the sensor. When the relative convection velocity U is constant and considerably larger than the turbulent fluctuation, and this is often the case, the measured history is very like that generated by the convection of a 'frozen' pattern of homogeneous turbulence past the instrument. Taylor's hypothesis states that the two are exactly equivalent. In such cases, all length and time scales are related by $L = UT$, their ratio directly proportional to the convection velocity.

We shall consider only the more widely used turbulence transducers, looking in turn at those sensing velocity, temperature, concentration, pressure, heat-transfer rate and shear stress. But before embarking on this survey, we take note of a problem generic to measurements of quantities that are defined in terms of activity at a surface—pressure, shear stress and heat transfer. To measure fluctuations in such quantities directly, within the body of the fluid, it is necessary to introduce a sampling surface (albeit a small one) into the turbulent flow. But the activity to be measured is associated with velocity fluctuations normal to the sampling surface; thus the probe fundamentally alters the process which it is meant to measure. We conclude that direct measurements of these fluctuations are often possible only at solid boundaries of the flow. However, if the turbulence scales that are of interest are much larger than the sensor, measurement may be possible by letting the instrument move with the fluid. Moreover, we shall see that there are indirect methods of measurement which can be applied within a turbulent flow: variations in effective shear stresses can be inferred from velocity fluctuations, and variations in effective heat-transfer rate can be obtained from velocity and temperature fluctuations, taken together.

Velocity

Fortunately, the aspect of turbulence which is of widest interest—the velocity fluctuation—is also the easiest to measure. The *hot-wire anemometer* mentioned at the beginning of this chapter has been developed over several decades, and is now a routine (though still very fragile) tool of measurement. Its small size gives it the required features of rapid response, fine resolution and slight disturbance of the flow. In Section 2.3 we shall look in some detail at this very important instrument.

Some disadvantages of the hot wire are:

(1) it is easily broken by mechanical impact or by a stream with a high dynamic pressure, especially if the latter contains particles or drops;

(2) it collects dust and oil droplets, which invalidate the calibration; and

(3) it is not very suitable for use in liquids, especially those with a high electrical conductivity.

The *hot-film probe* alleviates these problems, though with some loss of frequency response and of linearity in the characteristic relating signal and fluctuation. The former deficiency is not too critical for liquid flows, where the mean velocity and the turbulence frequencies linked to it are, typically, at least an order of magnitude below those of gas flows. The hot film is a small area of nickel (perhaps 0·3 mm by 1 mm, and with a thickness measured in Angstrom units), which is deposited by sputtering on an insulating quartz or glass substrate, and is protected by a second quartz layer (perhaps 1 μm thick) sputtered on top. When the film is mounted near the tip of a small wedge-shaped probe, little interference is offered to the turbulence, and dirt is prevented from sticking to the film itself.

Another probe suitable for measurements in liquids is the *fibre probe*. This is a cross between the hot wire and the hot film, obtained by sputtering a nickel film onto a quartz fibre of diameter perhaps 70 μm. Its frequency response, although adequate for most liquid flows, is also inferior to that of a fine, metal hot wire, in that it falls off rapidly above about 200 Hz. *Bead thermistors* (resistance thermometers) have similar characterististics, when used as turbulence transducers.

Hot wires and films do not provide accurate results in very intense turbulence; this is evidenced by the fact that a reversal in the direction of the flow past a hot wire will leave the signal unchanged. A device which copes with this problem is the *pulsed-wire anemometer*. A platinum transmitting wire (perhaps 10 μm in diameter) is located between and normal to two receiving wires (perhaps 5 μm in diameter, and 4 mm apart). Electrical pulsing of the middle wire produces a heat pulse which is convected to one of the receivers. An analysis of many flight times (and directions) gives the statistical properties of the turbulence component normal to the three wires. The frequency response is not as good as that of a simple hot wire and, at the present time, the collection of data is not as easy, but this is nevertheless a very promising instrument.

The anemometers described above make use of the dependence of the wire temperature and electrical resistance on the velocity of the fluid passing the probe. In going directly to these devices, we have neglected some more primitive methods which may prove useful in certain circumstances:

(1) *Direct observation* of suspended particles, bubbles or drops; here we may include the ultra-microscope technique, which determines the velocity of the tiny particles found naturally in water by rotating the objective lens of a microscope through which the brightly illuminated flow is examined.

(2) The *fibre anemometer*: a small fibre fixed normal to the surface indicates the nature of the flow there by its deflections. This and the preceding method are particularly suitable for low velocities. Both are plagued by the very great labour of converting the observations into numerical form.

(3) The *aerofoil probe*: the fluctuating forces on a very small lifting surface indicate the transverse component of turbulence. Similarly, the rotation of a small turbine may be used to sense vorticity.

Optical velocity meters have always seemed desirable because a solid probe need not enter the fluid. Thus the flow is left undisturbed, and measurements can be made in situations where probes might be destroyed by melting, erosion or chemical attack. The basic problems with optical systems have been those of obtaining adequate spatial resolution and, at the same time, providing an output in the form of an easily usable electrical signal. These difficulties are met by a newly developed optical device, the *Doppler-laser anemometer*. At present this is used mostly in situations where the hot-wire anemometer is unsuitable, but it may in time provide far-ranging competition for the hot wire. The instrument utilizes the scattering of a laser beam by particles suspended in the flowing fluid. In some flows scattering particles are already available; in others, smoke or dye can be added without significantly altering the flow. The frequency of the scattered light depends on the velocity (both magnitude and direction) of the particles moving with the fluid; this light is said to experience a Doppler shift. An analogous effect may be noted in the sound emanating from a moving vehicle; the frequency changes when the vehicle passes the listener.

Temperature

The heated-element probes described above—hot wires and films—sense not only velocity fluctuations, but also temperature changes in the fluid. If one wishes to measure only velocities, this fact is an embarrassment. But when temperature fluctuations are required as well, this additional sensitivity is most convenient. The two kinds of fluctuations can be separated either:

(1) by using two adjacent probes, one hot and one only a little above the mean temperature of the fluid, or

(2) by taking readings from a single probe operated at a series of temperatures.

In Section 2.3 the use of hot-wire probes in this manner will be examined in some detail. Thermistors and thermocouples can be used if the temperature does not change too rapidly.

Concentration

The determination of fluctuations in concentration must be based on some readily detectable property of the convected material. Discrete particles (such as those in smoke) or changes in opacity (such as occur in many solutions) can be sensed by measuring the absorption of light passing through the fluid. Conducting solutions can be studied by measuring the fluctuations in the electrical resistance between two electrodes. If the transport properties of the fluid are strongly dependent on concentration, the hot wire can be used, since its heat-transfer characteristic is sensitive to such changes.

Special problems arise in making measurements in strongly inhomogeneous flows, those carrying bubbles, drops or particles. Not only is it hard to measure the concentration of these elements, but they may make it difficult to measure the properties of the convecting fluid.

Pressure

It was noted earlier that locally generated pressure fluctuations cannot be accurately measured within a turbulent flow. However, fluctuations at a solid surface can be determined without flow disturbance by fitting a gauge flush with the surface. *Condenser microphones*, within which changes in capacitance result from the deflection of a flexible diaphragm, are often used in this way, especially when the fluctuations are not too intense. They have a very broad range of flat frequency response, typically from 20 Hz to 40 kHz. For higher intensities, *piezoelectric crystals* (for example, quartz and barium titanate) provide the sensitive element of the gauge; here the pressure change gives rise to an electric charge within the crystal.

Spatial resolution is a major problem with both kinds of transducer. For a sensing-head diameter of 5 mm, about the minimum available commercially, the smallest disturbance which will be correctly recorded has a length scale of about 5 cm. Since the length scales of turbulence are smallest near a solid wall, where these sensors must be installed, the pressures associated with some of the smaller scales of motion will pass virtually unrecognized.

It is possible to circumvent the problem of spatial resolution, by leading the pressure sensed through a small hole to a gauge installed beneath the surface. However, the frequency response of the system will be reduced (to a maximum of perhaps 2 to 5 kHz), since a finite time is required for the pressure at the gauge to adjust to a changed surface value.

Pressure fluctuations can be sensed qualitatively in a more personal manner. By applying a medical stethoscope to the back of a surface exposed to a turbulent flow (this is often possible in a wind tunnel), one can hear the turbulence, and even determine the region of transition. By inserting a hollow tube into a free-turbulent flow, one can hear the turbulence there, and work out a rough picture of the variations of intensity through the flow.

Heat Transfer and Shear Stress

Fluctuations in heat-transfer rate at a solid surface can be measured using a *film resistance thermometer* similar to the hot-film anemometer described earlier. In Section 1.6 we noted that turbulent heat transfer and stress transmission are closely analogous; hence suitable calibration will turn a film thermometer into a fast-responding shear-stress meter.

It was pointed out earlier that direct measurements of shear stress and heat transfer are not usually possible within a turbulent flow. However, they can be determined indirectly from measurements of fluctuations in the velocity components and temperature. For constant density, the fluctuating part of the transfer of x-momentum through unit area perpendicular to the y-direction is $(\rho u)v$, with u and v the x- and y-components of the fluctuating velocity. The effective shear stress generated by turbulent mixing is the time-mean of this instantaneous value:

$$\tau_t = -\rho\overline{uv} \tag{2.3.1}$$

The fluctuating part of the transfer of enthalpy through unit area normal to the y-direction is $(\rho c_p \theta)v$, with θ the temperature fluctuation and c_p the specific heat of the fluid. Hence

$$\dot{q}_t = \rho c_p \overline{v\theta} \tag{2.3.2}$$

gives the effective rate of heat transfer by turbulent mixing.

The results (2.3.1, 2) represent only the activity of the turbulence. Molecular diffusion will also contribute to the mean stress and energy flux, as in equations (1.3.2) and (1.6.3), though its contribution will usually be negligible in fully turbulent parts of the flow. In Section 3.2 the molecular and turbulent contributions to a variety of transport processes will be considered systematically, and the role of mean-flow convection will also be introduced.

The measurements required in equations (2.3.1, 2) appear to make stringent demands on experimental technique, as the fluctuating velocity components and temperature have to be measured at the same time and at essentially the same point. We shall see shortly how a reasonable approximation to these requirements can be obtained using hot-wire anemometers.

2.4 THE HOT-WIRE ANEMOMETER

2.4.1. Probe Construction

The sensing element of the anemometer is a round wire of tungsten, platinum or platinum alloy (10 to 20 per cent rhodium). Tungsten (available in sizes down to 2·5 μm) is stronger and is therefore preferred for high velocities, even though it is more difficult to attach to its supports. Platinum is available in smaller diameters (down to 1 μm), surrounded by a silver sheath of much larger diameter (say, 10 μm). This combination is produced by the Wollaston process of fine-drawing the platinum core inside the enveloping silver. After the Wollaston wire is soldered to the hot-wire supports, the silver is etched away to leave the fine platinum core as the sensing element.

Figure 2.2 shows some of the ways in which hot wires are mounted on probes. The *normal wire* of Figure 2.2(a) is usually placed perpendicular to the mean flow; it is capable of sensing the mean velocity U and the streamwise component of the turbulent fluctuations u and also, under some circumstances, temperature fluctuations in the fluid. The sensitive portion is usually between 0·5 and 3 mm long; a platinum sensor is supported by unetched Wollaston wire; and a tungsten sensor by sleeves of heavy (to a diameter of

30 μm) gold or copper plating. These supports are in turn welded or soldered to steel prongs embedded in the insulating body of the probe.

Figure 2.2(b) shows a *slant wire*; the usual slant angle is $\alpha = 45°$ nominally. Since a hot wire is primarily sensitive to the component of velocity normal to it, slant wires can be used to measure turbulence components normal to the mean flow. However, if only one slant wire is used, the determination of each cross-stream intensity requires that two measurements be made, with the

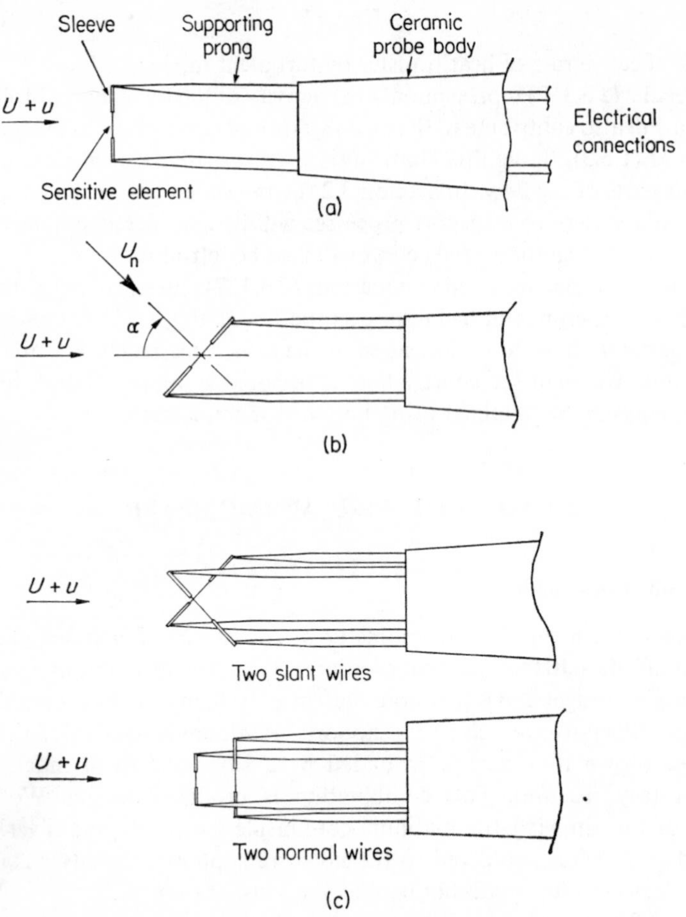

Figure 2.2 Details of hot-wire probes. (a) Normal wire or *u*-probe. (b) Slant wire. (c) X-arrays or X-probes

probe rotated through 180° about its axis between them; in addition, the normal component must be known. The arrangements of Figure 2.2(c) allow a cross-stream component to be measured directly: one form of *X-probe* or X-array probe has two slant wires; the other has two normal wires. About 1 mm must be left between the wires to minimize thermal wake interference. Hopefully, the two wires are identical, and mutually perpendicular, so that their signals can be easily combined to give the turbulence components.

Other varieties of probes are available, including:

(1) probes with two parallel normal wires, one of which can be used to measure temperature fluctuations;

(2) temperature-compensated probes, which maintain a constant calibration in the face of slow (that is, with a period greater than a second or so) changes in fluid temperature; and

(3) three-wire arrays, which can measure all three fluctuating components, provided that the user has enough instruments to handle all the signals.

2.4.2 Heat Transfer from the Wire

Our intention is to see in general terms how a hot wire works and what it can measure. For simplicity, we shall concentrate on operation in air flowing at low velocities.

The rate of heat transfer $\dot{Q}/L$ (per unit length) from a long, fine wire lying normal to a uniform gas flow of velocity U may be taken to depend upon:

d	wire diameter
λ	mean-free-path of the gas molecules
T_w, T_f	absolute temperatures of wire and free-stream fluid
k, ν, κ	fluid thermal conductivity, and diffusivities of momentum and heat at free-stream temperature

In dimensionless form, the heat-transfer law may be written

$$\mathrm{Nu} = f\left(\mathrm{Re}, \mathrm{Pr}, \frac{T_w}{T_f}, \mathrm{Kn}\right) \tag{2.4.1}$$

with $\mathrm{Nu} = \dfrac{\dot{Q}d}{(\pi d L)\, k(T_w - T_f)}$

$= \dfrac{\dot{Q}/L}{\pi k(T_w - T_f)}$ Nusselt number based on wire diameter and surface area

$\mathrm{Re} = \dfrac{Ud}{\nu}$ Reynolds number based on wire diameter

$\mathrm{Pr} = \dfrac{\nu}{\kappa}$ Prandtl number of the fluid

$\dfrac{T_w}{T_f}$ temperature ratio which can be used to take partial account of variations in fluid properties around the wire

$\mathrm{Kn} = \dfrac{\lambda}{d}$ Knudsen number, indicating whether continuum flow is established around the wire

It is not usually necessary to include all of these parameters in the heat-transfer law. The *Prandtl number* is nearly constant for any one gas over the normal range of operating temperatures; it is taken into account by calibrating in the fluid in which measurements are to be made. For gases, continuum flow is established for $\mathrm{Kn} < 0{\cdot}015$; when it is, the role of *Knudsen number* is negligible. Normal operating conditions do in fact lie near this limit, but so long as the gas density does not change much between calibrating and measuring, the calibration will absorb non-continuum effects.

We have neglected entirely the density changes associated with high-speed flow; had this not been done, the Mach number would have appeared in the law (2.4.1). By not including some measure of fluid buoyancy, we have also eliminated the effects of free convection. This is in fact significant only for speeds below about 5 cm/sec. Nor is radiation at all significant at normal wire temperatures.

For air, Collis and Williams have found the remaining parameters of equation (2.4.1) to be related by

$$\begin{aligned} \mathrm{Nu}\left(\frac{T_{\mathrm{f}}}{T_{\mathrm{m}}}\right)^{0\cdot17} &= 0{\cdot}24 + 0{\cdot}56\,\mathrm{Re}^{0\cdot45} \quad \text{for } 0{\cdot}02 < \mathrm{Re} < 44 \\ &= 0{\cdot}48\,\mathrm{Re}^{0\cdot51} \quad \text{for } 44 < \mathrm{Re} < 140 \end{aligned} \tag{2.4.2}$$

The temperature appearing in the temperature loading factor is $T_{\mathrm{m}} = \frac{1}{2}(T_{\mathrm{w}} + T_{\mathrm{f}})$, the mean film temperature. The first of these equations is that most nearly appropriate to hot-wire anemometry.* If the free-stream temperature and temperature-dependent transport properties remain constant, and if in addition the pressure is uniform and the temperature loading factor does not vary significantly, the relationship suggested between heat transfer and velocity is

$$\frac{\dot{Q}}{T_{\mathrm{w}} - T_{\mathrm{f}}} = A + BU^{0\cdot45} \tag{2.4.3}$$

with A and B constants. An earlier, much quoted result—*King's law*—gave $\mathrm{Nu} = A + BU^{\frac{1}{2}}$.

The heat transfer from the sensor of a hot-wire anemometer is influenced by several factors not considered in this derivation. Some of these are:

(1) non-uniformities in temperature and flow pattern near the ends of the sensor;

(2) the accretion of 'dirt' during operation—dust, lint and oil in gases; particles, slime and bubbles in liquids;

(3) 'aging,' the physical and geometric changes occurring during prolonged use at high temperatures and velocities;

* Although this result is widely adopted, the exact form of the heat-transfer law is still a matter for discussion, as some of the references at the end of this chapter show.

(4) rapid variations in the direction and intensity of the turbulence velocity around the wire;

(5) variations in the temperature and other properties of the ambient fluid; and

(6) inclination of the wire to the mean-flow direction, that is, the problem of the slant wire which will be discussed later.

Hence, although the general nature of the heat-transfer characteristic of an actual hot-wire probe is indicated by equation (2.4.3) the exact form is much more difficult to predict than the preceding discussion has suggested.

Many aspects of the first three of these complications can be accommodated by a *calibration* which determines values for the constants, and even the index, of equation (2.4.3) that are appropriate to a particular probe and test environment. A full calibration may not be required for each of a series of nearly identical, commercially produced probes. Re-calibration may be necessitated by aging or by the collection of dirt. The latter problem can also be countered by running the wire briefly at an elevated temperature, by blowing it clear, or by cleaning in an organic solvent or ultrasonic bath.

Wire calibration is usually accomplished by measuring the mean voltage across the wire at a number of mean velocities which span the range of the planned measurements. This procedure avoids certain of the complications (4) by assuming that the quasi-static characteristic applies during dynamic operation in turbulent flow. Recently, methods of dynamic calibration have been developed, and these show that this assumption is not always justified. Even dynamic calibration may not account for the dramatic changes of flow direction which occur in turbulence of very high intensity.

When the free-stream *fluid temperature* is not uniform—this is the fifth of the complications listed above—both the heat-transfer law and the measuring procedure become more complicated. For the first of equations (2.4.2), we have

$$\frac{\dot{Q}/\pi L}{T_w - T_f} = 0{\cdot}24\,k\left(\frac{T_m}{T_f}\right)^{0\cdot17} + 0{\cdot}56\,k\left(\frac{T_m}{T_f}\right)^{0\cdot17}\left(\frac{\rho U d}{\mu}\right)^{0\cdot45} \tag{2.4.4}$$

For a gas at constant pressure, $\rho T_f = \text{constant}$. For air and similar gases, the variations in the transport properties are approximately $\mu, k \propto T_f^{0\cdot76}$. Also

$$\frac{T_m}{T_f} = \tfrac{1}{2}\left(1 + \frac{T_w}{T_f}\right) \propto T_f^{-0\cdot6}$$

approximately, for the typical value $T_w/T_f = 1{\cdot}7$. Hence the 'constants' of equation (2.4.3) vary somewhat as follows:

$$A \propto k\left(\frac{T_{\mathrm{m}}}{T_{\mathrm{f}}}\right)^{0\cdot17} \propto T_{\mathrm{f}}^{0\cdot66}$$
$$B \propto k\left(\frac{T_{\mathrm{m}}}{T_{\mathrm{f}}}\right)^{0\cdot17}\left(\frac{\rho}{\mu}\right)^{0\cdot45} \propto T_{\mathrm{f}}^{-0\cdot13} \qquad (2.4.5)$$

Both quantities, but particularly the intercept A, change with the fluid temperature. This conclusion is important in two ways:

(1) A change in the mean air temperature from the value maintained during calibration will invalidate the calibration, or at least necessitate its correction.
(2) In flows where both velocity and temperature fluctuate, the variations in heat transfer will not be an index of the velocity fluctuations alone.

The first effect can be dealt with by using a temperature-compensated probe, or by arithmetic correction of measured values. The second problem can be minimized by operating the probe at as high a temperature as possible or, more fundamentally, by separating the two contributions using methods which will be derived later.

The mean *wire temperature* maintained by electrical heating usually lies in the range $300\,°\mathrm{C} > T_{\mathrm{w}} > 100\,°\mathrm{C}$. The upper limit is chosen to avoid large changes in material properties and, in the case of tungsten, to avoid oxidation and loss of strength. The lower temperature limit is defined by errors which enter at low overheating ratios, in particular, those produced by fluctuations in the fluid temperature. When a probe is operated in a liquid, its temperature must be even lower, in order that bubbles do not form on it. As was explained earlier, hot-film and fibre probes are the more usual choice for liquid flows.

Since the electrical resistance of the sleeves supporting the fine sensing wire is much lower than that of the sensor, their temperature is little different from that of the fluid. Accordingly, the wire temperature will fall to this value at each end of the sensitive element. Thus the thermal response depends on the wire length, although the effect is small for $L/d > 200$, that is, for $L > 1$ mm with $d = 5\ \mu$m. In any case, this is one of the factors absorbed by calibration of an individual wire.

2.4.3 Modes of Operation

So far we have not considered the source of the heat output from a hot wire—an electrical heating circuit. If, as is easily arranged, the wire resistance is the dominant impedance in the circuit, the energy dissipation within the wire is given simply by

$$\dot{Q} = I^2 R_{\mathrm{w}} = IV \qquad (2.4.6)$$

with I and R_w the instantaneous values of wire current and resistance, and V the instantaneous potential difference across the probe. The temperature differential appearing in the Nusselt number can be expressed as

$$T_w - T_f = \frac{R_w - R_f}{\alpha} \tag{2.4.7}$$

with α the temperature coefficient of the resistance of the wire, and R_f the resistance of the (unheated) wire at the fluid temperature. When the expressions (2.4.6, 7) are introduced into the heat-transfer law (2.4.3), we obtain

$$I^2 R_w = IV = (A + BU^n)(R_w - R_f) \tag{2.4.8}$$

on absorbing the coefficient α into the constants A and B, and introducing the general index n.

Two distinct modes of hot-wire operation are widely used: constant-temperature and constant-current operation. In both, it is the potential difference V which is measured to find the fluctuation in fluid properties. Most turbulence measurements depend implicitly on the existence of a mean convection velocity U which is considerably greater than the fluctuation u, so that the variations in heat-transfer rate and potential difference are relatively small.

Constant-current Operation

Here the heating circuit is designed to maintain the current I sensibly constant; the wire temperature T_w and resistance R_w fluctuate in response to changes in the velocity $U + u$. The relationship between these fluctuations can be found by differentiating equation (2.4.8), holding I constant:

$$\frac{I^2 R_f \, \mathrm{d}R_w}{(R_w - R_f)^2} = \frac{IR_f \, \mathrm{d}V}{(R_w - R_f)^2} = nBU^{n-1} u$$

whence the wire sensitivity is found to be

$$S_I = \left(\frac{\mathrm{d}V}{u}\right)_I = \frac{nB(R_w - R_f)^2}{IR_f \, U^{1-n}} \tag{2.4.9}$$

We have not taken account of the thermal inertia of the wire; its small reservoir of energy ensures that, when the velocity changes rapidly, the temperature lags behind. When this is taken into account, it is found that

$$S_I = \left(\frac{\mathrm{d}V}{u}\right)_I \propto (1 + M^2 \omega^2)^{-\frac{1}{2}} \tag{2.4.10}$$

with ω the frequency of the fluctuation and M a time constant. Typically, the time constant is of the order of 1 msec, and the uncorrected response drops off rapidly above about $\omega = 500$ Hz.

It is often necessary to investigate turbulent fluctuations with much higher frequencies; for example, to pick up a length scale of 5 mm, convected at $U = 100$ m/sec, one must measure up to $\omega = 100/0{\cdot}005 = 20{,}000$ Hz. Frequencies in this range, and indeed beyond 100 kHz, can be retained by passing the output from the constant-current wire through a *compensator*, a circuit whose amplification has the frequency dependence $(1 + M^2\omega^2)^{\frac{1}{2}}$, and is capable of cancelling out the fall in the response of the wire itself. Suitable electrical compensation must be selected for each wire and for each operating condition, by tuning the compensator until its time constant matches that of the wire. Thus constant-current anemometers are less convenient to operate that those using the constant-temperature mode. Nowadays, the latter is the maid-of-all work among anemometers, the constant-current device being used mostly for turbulence of very low intensity (below about 0·1 per cent), and in very high-speed flows.

Constant-temperature Operation

In this mode the wire temperature and resistance are kept nearly constant by an amplifier in a feedback loop. The problem of thermal lag is avoided, and frequencies up to and above 100 kHz can be accommodated. Historically, constant-current anemometers were the first to be widely used, since it proved difficult to design a stable feedback system.

Taking the wire resistance to be constant, we obtain from equation (2.4.8):

$$2R_w I \mathrm{d}I = 2I \mathrm{d}V = nBU^{n-1}(R_w - R_f)u$$

whence the wire sensitivity is found:

$$S_T = \left(\frac{\mathrm{d}V}{u}\right)_T = \frac{nB(R_w - R_f)}{2IU^{1-n}} \tag{2.4.11}$$

The sensitivities (2.4.9, 11) for the two modes depend on the system parameters in much the same way. Their ratio is

$$\frac{S_I}{S_T} = \frac{2(R_w - R_f)}{R_f} \tag{2.4.12}$$

From this it may appear that the constant-current sensitivity is the higher, but it must be borne in mind that a higher overheating ratio can be selected in the constant-temperature mode, since the temperature control system automatically prevents wire burn-out following a large decrease in velocity.

Note that the sensitivities depend on the flow velocity U; this indicates, as indeed do equations (2.4.3, 8), that the velocity *vs* voltage characteristic is non-linear. For low-intensity turbulence, the departure from linearity is not crucial. But for higher intensities (above 10 per cent, or much less if a detailed analysis of the signals is to be made) a non-linear amplifier, usually called a *linearizer*, must be used to make the output signal proportional to the velocity fluctuations. The non-linearity is smaller and more easily corrected in constant-temperature operation, and these are major advantages of the system. For intensities above $u'/U = 20$ per cent, even these measures do not produce signals that can be interpreted confidently.

A good deal has been said about the frequency response of hot wires, but little about their *spatial resolution*. The latter often provides a more stringent limit than does frequency response. Consider a flow near a wall, with $U = 30$ m/sec, and with the smallest scales of turbulence around 1 mm, corresponding to $\omega = 30/0{\cdot}001 = 30{,}000$ Hz. But for a sensor length of 1 mm, the smallest disturbance which will be faithfully recorded has a length scale around 5 mm, corresponding to $\omega = 6000$ Hz.

Temperature Measurements

Equations (2.4.5) indicate that small fluctuations in the fluid temperature will influence the heat transfer from the wire in several ways. We shall not attempt to calculate these terms, but simply express the net result of small temperature and velocity changes as

$$dV = S_u u + S_\theta \theta \tag{2.4.13}$$

with θ the fluctuation in the fluid temperature, $T_f + \theta$. The constants S_u and S_θ can be found by calibration, but we still have to find what parts of the measured signal dV are attributable to the u-variation and to the θ-variation. The solution to this problem utilizes the different ways in which the sensitivities S_u and S_θ depend on wire temperature and heating current.

We consider only constant-current operation. If the velocity-dependent term in the heat-transfer law is a good deal larger than the other, and this will normally be so, equation (2.4.8) becomes

$$I^2 R_w \simeq BU^n(R_w - R_f)$$

and the sensitivity can be written

$$S_u = S_I \simeq \left(\frac{nR_w^2}{BR_f U^{1+n}}\right) I^3 \tag{2.4.14}$$

The sensitivity to temperature fluctuations can easily be found, if we neglect the temperature dependence of the fluid density and transport properties. Then with $U =$ constant, as required in finding $(\partial V/\partial T_f)_u$, equations (2.4.3, 6) give

$$\frac{dR_w}{R_w} = \frac{dV}{V} = \frac{dT_w - \theta}{T_w - T_f} = \alpha \frac{dT_w - \theta}{R_w - R_f}$$

with the use of the relationship (2.4.7). Introducing $dV = \alpha I dT_w$, we obtain

$$S_\theta = \frac{dV}{\theta} = \alpha \frac{R_w}{R_f} I \tag{2.4.15}$$

We now have

$$\left.\begin{array}{l} S_u \propto I^3 \\ S_\theta \propto I \end{array}\right\} \text{ for small currents, for which } R_w \simeq R_f \tag{2.4.16}$$

These results show that the ratio of the two sensitivities varies markedly as T_w and I are changed. For a cold wire, $S_u = S_I \simeq 0$, and the wire operates as a resistance thermometer. Equation (2.4.12) indicates that S_T falls less rapidly as I is decreased; hence the constant-temperature mode does not provide such a satisfactory thermometer. A useful conclusion can be drawn regarding high-temperature operation: when temperature fluctuations cannot be avoided and need not be measured, their contribution can be minimized by operating the hot wire at as high a temperature as possible.

The results obtained above suggest two ways of measuring jointly fluctuating velocities and temperatures:

(a) Adjacent hot and cold wires. The cold wire gives the signal

$$dV = S_L\,\theta \quad \text{with} \quad S_L = \alpha I \frac{R_w}{R_f} \tag{2.4.17}$$

from which θ can be found directly. The hot wire gives $dV = S_u u + S_\theta \theta$. When the signals are combined as follows

$$dV = S_\theta(S_L\,\theta) - S_L(S_u u + S_\theta\,\theta) = -S_L S_u u \tag{2.4.18}$$

the fluctuation u can be obtained.

(b) Same wire operated at different temperatures. Since the measurements are made at different times, they cannot be combined to give signals dependent solely on u or θ. However, our interest is often in the statistical properties appearing in the mean square of the combined signal:

$$\overline{dV^2} = S_u^2 \overline{u^2} + S_\theta^2 \overline{\theta^2} + 2 S_u S_\theta \overline{u\theta} \tag{2.4.19}$$

If $\overline{dV^2}$ is measured for three operating temperatures at which the coefficients S_u and S_θ have been found by calibration, there are available three equations from which $\overline{u^2}$, $\overline{\theta^2}$ and $\overline{u\theta}$ can be calculated. These quantities are presumed to remain the same throughout the measuring procedure.

2.4.4 The Slant Wire

The heat transfer from a fine wire is dependent primarily on the component of velocity normal to it, but the longitudinal component has some influence, and this proves to depend on the wire length and Reynolds number. Tests carried out by Champagne suggest that the effective velocity for the heat-transfer laws (2.4.2, 3, 8) is

$$U_e = U_r(\cos^2 \alpha_r + k^2 \sin^2 \alpha_r)^{\frac{1}{2}} \tag{2.4.20}$$

with U_r the resultant velocity at the probe and α_r the angle between U_r and the plane normal to the wire. For $25° < \alpha_r < 60°$, the factor

$$\begin{aligned} k &\rightarrow 0 \quad \text{for } L/d > 600 \\ &= 0{\cdot}2 \quad \text{for } L/d = 200 \end{aligned} \tag{2.4.21}$$

The latter case is typical of commercially available hot wires. For algebraic simplicity, we shall adopt the cosine law

$$U_e = U_n = U_r \cos \alpha_r \tag{2.4.22}$$

as an approximation to the results (2.4.20, 21), with U_n the velocity component normal to the wire.

Specifying the turbulent flow at the probe by the components $U + u$, v and w, we find the components of the velocity U_n to be

$$(U + u)\cos\alpha + v\sin\alpha \quad \text{in the plane of the wire}$$

and

$$w \quad \text{perpendicular to that plane}$$

with α the angle between the normal and the mean flow. Neglecting the second-order term in w, we have

$$U_e = U_n = (U + u)\cos\alpha + v\sin\alpha$$

For simplicity, King's law will be used to describe the heat transfer:

$$\begin{aligned} \frac{\dot{Q}}{T_w - T_f} &= A + BU_e^{\frac{1}{2}} \\ &\simeq A + B(U\cos\alpha)^{\frac{1}{2}}\left(1 + \tfrac{1}{2}\frac{u}{U} + \tfrac{1}{2}\tan\alpha\frac{v}{U}\right) \end{aligned} \tag{2.4.23}$$

again neglecting second-order terms. It will be appreciated that other second-order terms would be generated by the use of the full result (2.4.20).

The linearized relationship between voltage and velocity fluctuations is seen to be

$$\mathrm{d}V \propto \frac{u}{U} + \tan\alpha\frac{v}{U}$$

The precise form is not important, as the constants for any one wire can be found by calibration. For our purposes it is sufficient to note that the signal from a particular slant wire will be of the form

$$s_1 = au + (b\tan\alpha)v \tag{2.4.24}$$

with $a \simeq b$. When the usual choice of $\alpha = 45°$ is made for the slant angle, the sensitivity is nearly the same for the u- and v-fluctuations.

The question posed in equation (2.4.24) is similar to that of equation (2.4.13): how can we disentangle this mixture of signals? The answer is surprisingly simple. Consider a second slant wire, located near the first, and exactly similar, save that it has the opposite inclination to the flow. Its signal will be

$$s_2 = au - (b \tan \alpha)\, v \tag{2.4.24}$$

since $\tan \alpha = \tan \alpha_1 = -\tan \alpha_2$. The relationships (2.4.24) can be used in two ways:

(a) *X-probe.* The signals from the two wires of the probe can be combined to give

$$s_1 + s_2 = 2au \quad \text{and} \quad s_1 - s_2 = (2b \tan \alpha)\, v \tag{2.4.25}$$

These outputs give u and v individually, and can be analysed to find their statistical properties. The third velocity component w can be found by rotating the probe through 90° about its axis, or by combining the signals from a probe with three slant wires.

(b) *Rotated slant wire.* The use of an X-probe has a number of disadvantages: the spatial resolution is reduced; the two wires must be very nearly alike to give accuracy in the combined signal; and, of course, two sets of instruments are required, together with equipment for combining the signals. If only the simplest statistical properties of the signals are required, and the basic flow is steady, these problems can be avoided by measuring the mean square of the slant-wire signals twice, with a rotation of the probe through 180° about its axis between the two readings. Thus

$$\begin{aligned} \overline{s_1^2} &= a^2 \overline{u^2} + (b \tan \alpha)^2 \overline{v^2} + (2ab \tan \alpha)\, \overline{uv} \\ \overline{s_2^2} &= a^2 \overline{u^2} + (b \tan \alpha)^2 \overline{v^2} - (2ab \tan \alpha)\, \overline{uv} \end{aligned} \tag{2.4.26}$$

lead directly to $\overline{uv}$. The easiest way of finding $\overline{u^2}$ and $\overline{v^2}$ is to determine $\overline{u^2}$ independently using a normal wire.

2.5 THE MEASURING SYSTEM

In Sections 2.3 and 2.4 we considered transducers which can be used to measure turbulent fluctuations, and looked in some detail at one of them—the hot-wire anemometer. But transducers are often only a small part—both physically and functionally—of the system required to obtain a useful description of a rapidly and randomly fluctuating flow. Most of the transducers discussed earlier produce continuous electrical signals representing some aspect of turbulence; accordingly, we shall consider systems which turn continuous signals into meaningful results. Our interest is primarily in what needs to be done and what can be done, rather than in how the results are achieved. Hence we shall only occasionally consider the nature of the electric circuitry involved in signal analysis.

Figure 2.3 is a block diagram of a complete system for the measurement of the characteristics of a turbulent flow. Its first elements are one or more transducers; these may require auxiliary equipment, such as high-frequency oscillators supplying carrier waves for the signals. *Amplifiers* may in fact be

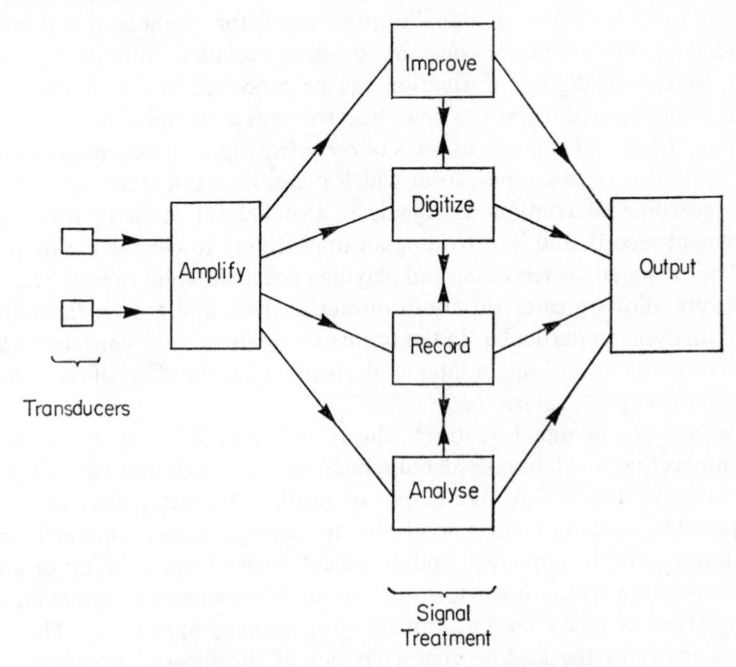

Figure 2.3 Schematic diagram of a system for the measurement of turbulence

introduced at several points in the signal-conversion process, but for convenience they are shown only once in Figure 2.3, immediately following the transducers. Amplification is usually required here since transducer outputs are frequently of very small magnitude. We shall take the amplification to be distortion-free, simply increasing the signal magnitude linearly, without regard to frequency. In practice, amplification is often coupled with signal 'improvement', though in Figure 2.3 the latter is shown as an aspect of signal treatment. Regrettably, some degeneration of the signal usually accompanies amplification, as a result of distortion or the introduction of electronic noise.

Signal Treatment

The amplified signal may be subjected to a variety of operations, in varied order. Accordingly, our diagram shows next a number of alternative paths. The first possibility is that the amplified signals from the transducers are subjected to some form of improvement. We have encountered two examples of signal *improvement*: the frequency compensation of the constant-current hot-wire system, and the linearization of the constant-temperature system.

In recent years, it has become common practice to convert continuous signals to digital form; this is done by sampling at discrete intervals. *Digitizing* opens the way for other forms of signal improvement, for example, the automatic introduction of calibration constants or more complex calibration relationships. Moreover, digital information can be processed in a high-speed computer, using the great variety of analytical routines now available.

There are several convenient ways of *recording* digital data—punched cards or tape, and magnetic tape—from which it can be recalled for analysis. The tape recording of continuous signals is also helpful, both in providing a permanent record, and in providing an opportunity to change the frequency band of the signal, by recording and playing back at different speeds. The latter procedure allows greater variety in output devices, and greater flexibility in signal analysis, in particular, in the frequency analysis to be considered later. Finally, on-line recording for later analysis often has the effect of reducing the load on the experimenter.

The last type of signal treatment shown in Figure 2.3—signal analysis—is so complex that it will be considered separately. Hence we move directly to the final stage of Figure 2.3—the *output* of results. A great variety of outputs are possible, ranging from a single oscilloscope or r.m.s. voltmeter, in the simplest system, to numerical and graphical results from a digital or analog computer. One seldom-used form of output is sometimes enlightening—the loudspeaker. A tape recording can easily be scanned in this way. The sound reveals many of the leading characteristics of turbulence, in particular, its intermittency and random nature. Many listeners are startled by its similarity to the background noise, or 'static', which can be heard on a radio receiver.

It is natural to wonder how much confidence can be placed in an output generated by a long series of electrical transformations, each with its own distortions, drifts and non-linearities. It must be admitted that many aspects of turbulence cannot be measured with a very high degree of accuracy. Indeed, many of the quantities handled with ease by the theoretician cannot be measured even approximately. Even when methods are available which could, in principle, provide a required result, accurate measurements are often impossible because no suitable method of calibration is available. Reasonable accuracy can be obtained, no matter how complex the measuring system, if the entire system can be calibrated as a single unit, with the calibrating signals undergoing exactly the sequence of operations experienced by the measured signals. Thus the search for accuracy reduces to the problem of devising calibration methods that are both:

(1) fully representative of the experimental signals, in magnitude, frequency and general statistical properties; and
(2) interpretable in terms of the properties of turbulence.

Suppose, for example, that we intend to calibrate a velocity probe by vibrating it in still air. A satisfactory calibration would require:

(1) that the imposed vibrations cover representative ranges of amplitude and frequency, and that they contain several superposed scales of motion; and
(2) that we are able to relate the probe movements to velocity fluctuations in fluid convected past a fixed probe.

It is certainly not obvious that these requirements are met by the usual static calibration techniques.

Signal Analysis

Electrical and electronic devices are available to carry out the following operations, many of which we have seen to be required in measuring the instantaneous and statistical properties of a turbulent fluid.

(1) Addition and subtraction. Required in separating velocity and temperature fluctuations and velocity components (equations 2.4.18, 25, 26).

(2) Multiplication and squaring. Required in finding intensities, velocity products and velocity–temperature products (equations 2.4.19, 26).

(3) Averaging. Required in finding mean values in the preceding examples. The time of averaging must be increased as the lowest frequencies of the signal become smaller; hence a variety of instruments and techniques are required.

(4) Filtering. To eliminate low frequencies (a high-pass filter) or high frequencies (a low-pass filter, acting like a transducer with a poor high-frequency response); the combination of the two is a band-pass filter.

(5) Differentiation with respect to time. Useful in distinguishing the fully turbulent parts of intermittent flows, and in measuring dissipation, which depends on velocity gradients.

(6) Probability measurement of the time fraction that a signal exceeds a specified level. This can be used to measure the intermittency factor, and to study the probability distribution of the magnitudes of turbulent fluctuations.

(7) Integration. Required in averaging, and as part of the probability-measurement procedure.

Two further methods of signal analysis:
(8) Correlation and
(9) Spectral analysis

will be considered at length below. They are not, strictly speaking, distinct operations, but are systematic applications of multiplication, filtering and averaging.

These operations must often be carried out sequentially. Consider the mean-square derivative $\overline{(\partial v/\partial x)^2}$, one of the quantities contributing to turbulent dissipation. We assume that two signals are obtained by passing the outputs from an X-probe through a pair of linearizers. The component v is then obtained by subtraction. Next, the combined signal is differentiated with respect to time, Taylor's hypothesis being assumed to connect time and space derivatives:

$$U\frac{\partial v}{\partial x} \simeq \frac{\partial v}{\partial t}$$

Then a low-pass filter removes electronic noise from the signal. Finally, squaring and averaging give the quantity sought. Alternatively, the differentiated signal might be subjected to spectral analysis, to find the contributions to dissipation from the various scales of turbulence.

Conditional Averaging

This more recently developed technique will not be discussed in detail in this book, but the following brief comments will indicate its potential. It is usually referred to as *conditional sampling*, but the term used above is somewhat more descriptive. The average is taken over only those intervals during which a sampling condition is satisfied:

$$\langle S \rangle = \int_0^t cS\,\mathrm{d}t \bigg/ \int_0^t c\,\mathrm{d}t \tag{2.5.1}$$

with the *conditioning function*

$$\begin{aligned} c &= \text{a fixed value} \quad \text{when the condition holds} \\ &= 0 \quad \text{otherwise} \end{aligned} \tag{2.5.2}$$

The technique can be used to investigate distinct elements of a turbulent motion, for example:

(1) the fully turbulent part of an intermittent flow;
(2) the active periods of a flow undergoing transition or of the flow very near a wall; and
(3) elements from one of two intersecting flows.

Appropriate *sampling conditions* for these situations might be:

(1) for intermittent turbulence, the requirement that one or both of $\partial u/\partial t$ and $\partial^2 u/\partial t^2$ be large;

(2) near a wall, the requirement that the product $u(t)v(t)$ be large; and

(3) in intersecting flows with slightly different temperatures, the requirement that the temperature exceed a certain value.

In practice, it is found that most conditional averages are strongly dependent on the selection of the sampling condition and level of discrimination. Nevertheless, the technique is capable of providing the experienced interpreter with considerable insight into turbulent processes. In some ways it is equivalent to spectral analysis and correlation studies, which will be seen later to be essentially equivalent to one another.

2.6 CORRELATIONS AND SCALES

2.6.1 Classification

In equations (2.3.1, 2) we noted that the time-averaged products $\overline{uv}$ and $\overline{v\theta}$ (u, v and θ being velocity and temperature fluctuations) specify net transfers of momentum and energy by turbulent mixing. In equations (2.4.19, 26) we saw that the signals from hot-wire probes could be combined to provide measures of such quantities. These statistical properties are called *correlation functions*, $\overline{uv}$ being a double-velocity correlation function and $\overline{v\theta}$ being a velocity-temperature correlation function. The term *covariance* is sometimes used for such quantities.

The magnitude of $\overline{uv}$ depends on the intensities of u and v, and on the correlation between them, that is, on the degree of interdependence of the two fluctuations. This relationship can be set down as

$$\overline{uv} = Ru'v'$$

with primes denoting r.m.s. values, and R a dimensionless *correlation coefficient* which specifies the interdependence, isolated from the two intensities. A correlation coefficient can be defined in a similar manner for $\overline{v\theta}$; it will, in general, differ from the coefficient for $\overline{uv}$. Such quantities characterize the mean kinematic structure of the turbulence, independent of the magnitude of its fluctuations.

Let us now extend the idea of correlation. The operation denoted by the compact notation $\overline{uv}$ is in full:

$$\overline{uv} = \frac{1}{t_a} \int_{t_0}^{t_0+t_a} u(x_p, t)\, v(x_p, t)\, dt \tag{2.6.1}$$

as in equation (1.1.5); here x_p is a typical coordinate of the point P at which the two velocity components are measured, and t is the time at which the two measurements are made. We shall consider four generalizations of this result:

(1) Point correlations between any two signals (s_1 and s_2) generated simultaneously at the same point P.

(2) Autocorrelations between the values which a signal has at times separated by the constant time interval τ.

(3) Time correlations, a generalization of (2) including two distinct signals emanating from the same point at times separated by τ. These are also called cross correlations.

(4) Spatial correlations between two signals generated simultaneously at two points P_1 and P_2 with spatial separation r. If the same property is measured at the two points, this is called a direct correlation; otherwise, it is a cross correlation.

Other generalizations are possible, but will not be discussed in this chapter:

(5) Space–time correlations, for example, those of signals obtained at points separated by distance r and with time delay $\tau = r/U$, with U an effective convection velocity.

(6) Lagrangian correlations connecting the properties of particles moving through the fluid with their properties at some fixed time.

(7) Higher-order correlations of any of the preceding types, for example, $\overline{v^3}$, $\overline{uv\,\partial u/\partial y}$ and $\overline{v\theta^2}$. Such quantities appear in the energy equations for turbulent motion, to be developed in Chapter 3 and, in more general form, in Chapter 9.

The four generalizations to be considered here are *second-order correlations*, relating only two signals. They are accommodated in the correlation function

$$\overline{s_1 s_2} = \frac{1}{t_a}\int_{t_0}^{t_0+t_a} s_1(x_{p_1}, t_1)\, s_2(x_{p_2}, t_2)\, \mathrm{d}t \tag{2.6.2}$$

with x_{p_1} and x_{p_2} typical coordinates of points P_1 and P_2, and t_1 and t_2 possibly differing by a fixed interval. The general correlation coefficient is

$$R = \frac{\overline{s_1 s_2}}{s_1' s_2'} \tag{2.6.3}$$

with $s' = (\overline{s^2})^{\frac{1}{2}}$ denoting an r.m.s. value. Simple physical arguments (or the use of a mathematical result known as Schwartz's inequality) show that this coefficient must lie in the range

$$-1 \leqslant R \leqslant 1$$

The maximum value $R = 1$ is achieved when $s_1 \equiv s_2$, that is, when $t_1 \to t_2$ and $P_1 \to P_2$, and the two signals measure the same property. The minimum value $R = -1$ might be achieved by correlating a quantity with its negative. The result $R = 0$ is obtained when there is no relationship, even statistical, between the two fluctuations.

In the preceding discussion, we have assumed the signals to be continuous, but the definitions can be applied to discrete items (perhaps generated by digitizing a continuous signal) by introducing summation in the place of integration.

2.6.2 Point Correlations

For these, equation (2.6.2) is restricted by taking $P_1 = P_2$ and $t_1 = t_2$. Such correlations are obtained from simultaneous measurements made, as nearly as feasible, at a single point.

In a general (but statistically steady) turbulent motion, the three correlation functions $\overline{u\theta}$, $\overline{v\theta}$ and $\overline{w\theta}$ will define an effective heat flux, and the three pressure–velocity correlations $\overline{pu}$, $\overline{pv}$ and $\overline{pw}$ (with p the fluctuation in pressure) will give the net rate of working of turbulent pressure and velocity fluctuations. Such correlations between a scalar and the components of a vector can be summarized as

$$\overline{u_i\theta} \quad \text{and} \quad \overline{pu_i} \tag{2.6.4}$$

it being understood that the suffix can take on each of the values $i = 1, 2$ and 3, so that

$$u_1 = u, \quad u_2 = v, \quad u_3 = w \tag{2.6.4}$$

In this *suffix notation* the double-velocity correlations ($\overline{uv}$, $\overline{uw}$, etc.) become

$$\overline{u_i u_j} \quad \text{with} \quad i = 1, 2, 3 \quad \text{and} \quad j = 1, 2, 3 \tag{2.6.5}$$

This group of entities includes correlations related to the shear stresses ($\overline{u_1 u_2} = \overline{uv}$, for example), and the component intensities ($\overline{u_2 u_2} = \overline{v^2}$, for example). The latter can be interpreted, as we shall see later, as effective normal stresses generated by the turbulence. Since we shall often encounter these quantities, we adopt the more compact notation

$$Q_{ij} = \overline{u_i u_j} \tag{2.6.6}$$

The three components $\overline{u_i\theta}$ make up a vector. The nine components $\overline{u_i u_j}$ (actually only six are distinct, since $\overline{u_1 u_2} = \overline{u_2 u_1}$, etc.) make up a *tensor*. This particular tensor is called the double-velocity correlation tensor, or simply the velocity correlation tensor. Tensor quantities are required to describe a three-dimensional stress field since each stress component is associated with two directions, that in which it acts, and that of the normal to the plane on which it acts. Generally similar interpretations can be given for the three-dimensional strains in a solid material, for the rates-of-strain in a fluid, and for the moments-of-inertia of a rigid body. Strictly speaking, these quantities are second-order tensors; in this terminology, vectors are first-order tensors, since their components depend on only one direction.

The suffix notation introduced in equations (2.6.4, 5) has two advantages: it compactly denotes the several components, and also indicates, through the number of distinct suffices, the order of the tensor quantity. Thus, the triple-velocity correlations $\overline{u_i u_j u_k}$ (with each of $i, j, k = 1, 2, 3$) are the twenty-seven components of a third-order tensor. Complex though these multi-component entities may seem, we have, in the hot-wire anemometer and other transducers, the means of measuring many of their components. Tensor notation is treated in more detail in Section 9.1.1.

2.6.3. Time Correlations

We now write equation (2.6.2) in the restricted form

$$\overline{s_1(t)\, s_2(t+\tau)} = \frac{1}{t_\mathrm{a}} \int_{t_0}^{t_0+t_\mathrm{a}} s_1(x_p, t)\, s_2(x_p, t+\tau)\, \mathrm{d}t \tag{2.6.7}$$

with τ the time delay between the two signals. In practice, this delay might be achieved by correlating two tape-recorded signals. The correlation coefficient derived from equation (2.6.7) is denoted

$$R(\tau) = \frac{\overline{s_1 s_2}}{s_1' s_2'} \tag{2.6.8}$$

Autocorrelations are included in equation (2.6.7) as a special case, with $s_1 = s(x_p, t)$ and $s_2 = s(x_p, t+\tau)$. They can be obtained by correlating a tape-recorded signal with a replica delayed by a fixed interval. The autocorrelation coefficient is

$$R(\tau) = \frac{\overline{s_1 s_2}}{\overline{s^2}} = \frac{\overline{s(t)\, s(t+\tau)}}{\overline{s^2}} \tag{2.6.9}$$

since $s_1' = s_2'$ for a stationary random function. In the case of an autocorrelation, but not for time correlations in general

$$R(\tau) \to 1 \quad \text{as } \tau \to 0 \tag{2.6.10}$$

Most of the following discussion relates to the important special case of autocorrelation.

The general character of the dependence of an autocorrelation coefficient on the time delay is indicated in Figure 2.4(a). The curve is shown as 'flat-topped' at $\tau = 0$. This can be justified as follows: the autocorrelation of a stationary random function of time must be an even function of τ, symmetrical about $\tau = 0$; turbulent motions are continuous, with smallest scales of a finite size; hence there is a finite delay before the signal changes at all, and the curve is level at $\tau = 0$. However, in practice, the smallest scales are sometimes comparable with the transducer resolution, and it is not possible to see this levelling out.

Time Scales

We have hitherto applied the term 'time scale' rather loosely. More precise scales can be defined using autocorrelation curves. A scale which measures

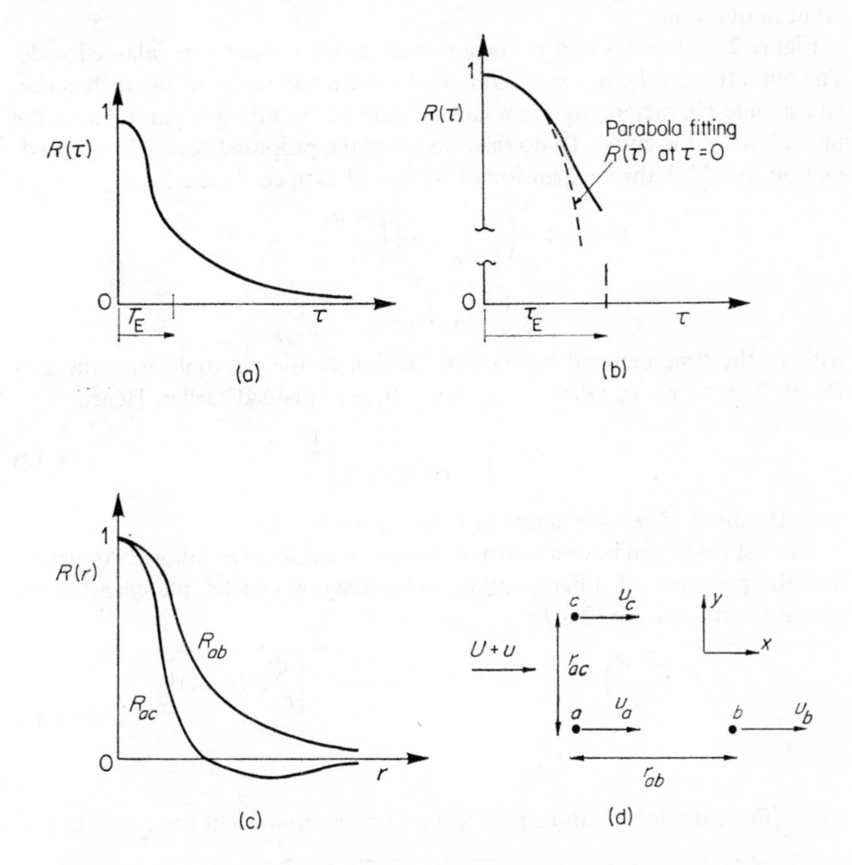

Figure 2.4 Correlation coefficients and associated scales. (a) Dependence of an autocorrelation coefficient on delay time, defining the integral time scale, T_E. (b) Variation of an autocorrelation coefficient for small times, defining the micro time scale, τ_E. Note that the scales are not the same as those used in (a). (c) Dependence of spatial correlation coefficients on the separation between the points shown in (d). (d) Points at which correlated measurements are made: *a–b*, longitudinal correlation; *a–c*, lateral or transverse correlation

the average persistence of turbulent activity at a point is the *integral time scale*

$$T_E = \int_0^\infty R(\tau)\,d\tau \tag{2.6.11}$$

The subscript E marks this as an Eulerian scale, dependent on the velocity history at a fixed point, and distinct from the integral scale which would be obtained from a Lagrangian correlation. Using the integral time scale (2.6.11) we can define the *averaging time* t_a more precisely. Evidently it is necessary that $t_a \gg T_E$ if stable mean values are to be obtained from signals from the point in question.

Figure 2.4(b) shows part of the autocorrelation curve to an enlarged scale. The curvature at the top is determined by the small scales of the turbulence, and a time characterizing them can be defined by fitting a parabola to the upper part of the curve. To do this, we relate the proposed parabolic approximation to a McLaurin expansion of the correlation coefficient:

$$\begin{aligned} R(\tau) &= 1 + \left(\frac{\partial R}{\partial \tau}\right)_0 \tau + \tfrac{1}{2}\left(\frac{\partial^2 R}{\partial \tau^2}\right)_0 \tau^2 + \cdots \\ &\simeq 1 - \left(\frac{\tau}{\tau_E}\right)^2 \end{aligned}$$

with τ_E the time denoted by the intersection of the parabola with the axis $R = 0$. Now at $\tau = 0$, $\partial R/\partial \tau = (\partial R/\partial \tau)_0 = 0$, as explained earlier. Hence

$$\tau_E = \left[-\frac{2}{(\partial^2 R/\partial \tau^2)_0}\right]^{\frac{1}{2}} \tag{2.6.12}$$

gives the *micro time scale*, again in Eulerian terms.

The last result can be cast into a more significant form as follows. Assuming that the processes of differentiation and averaging can be interchanged, we obtain from equation (2.6.9):

$$\begin{aligned} \overline{s^2}\left(\frac{\partial^2 R}{\partial \tau^2}\right)_0 &= \frac{\partial^2}{\partial \tau^2}[\overline{s(t)\,s(t+\tau)}]_0 = \overline{s(t)\left[\frac{\partial^2}{\partial \tau^2}s(t+\tau)\right]_0} \\ &= \overline{s\frac{\partial^2 s}{\partial t^2}} \end{aligned} \tag{2.6.13}$$

since differentiations with respect to t and τ are equivalent for $\tau = 0$. But

$$\frac{d^2\overline{s^2}}{dt^2} = 2\frac{d\overline{(s\,ds/dt)}}{dt} = 2\overline{\left(\frac{ds}{dt}\right)^2} + 2\overline{s\frac{d^2 s}{dt^2}} = 0$$

since $\overline{s^2} =$ constant, for this stationary random process. Combining the last two results, we have

$$\overline{s^2}\left(\frac{\partial^2 R}{\partial \tau^2}\right)_0 = -\overline{\left(\frac{\partial s}{\partial t}\right)^2} \tag{2.6.14}$$

and the micro time scale (2.6.12) can be written

$$\tau_E = \left[\frac{2\overline{s^2}}{\overline{(\partial s/\partial t)^2}} \right]^{\frac{1}{2}} \tag{2.6.15}$$

Thus it is found to have the form

$$\tau_E \propto \frac{s'}{(\partial s/\partial t)'} \tag{2.6.16}$$

simply the ratio of the r.m.s. value of the signal to the r.m.s. value of its derivative.

This analysis can be applied to any fluctuating signal. In practice, however, the time scales of turbulence are most commonly defined using the streamwise velocity fluctuation u, because it is easily measured. Note that time-correlation curves, and the time scales derived from them, depend not only on the structure of the turbulence, but also on the speed with which it is convected past the point(s) of measurement.

2.6.4 Spatial Correlations

We now consider two distinct signals, measured at points distance r apart, and correlated without a time delay:

$$\overline{s_1(x_p)\, s_2(x_p + r)} = \frac{1}{t_a} \int_{t_0}^{t_0 + t_a} s_1(x_p, t)\, s_2(x_p + r, t)\, \mathrm{d}t \tag{2.6.17}$$

with x_p a representative coordinate of the base point P; the separation r can be measured in any direction from P. The correlation coefficient will be denoted

$$R(r) = \frac{\overline{s_1 s_2}}{s_1' s_2'} \tag{2.6.18}$$

For a direct correlation, that between values of the same property measured at different points

$$R(r) \to 1 \quad \text{as} \quad r \to 0 \tag{2.6.19}$$

The variations of direct and cross correlations in space can be obtained, for example, by placing a hot-wire probe at a number of positions along a straight line passing through a fixed probe. These variations are less strongly influenced by the convection velocity of the mean flow than are time correlations, although a difference between the convection velocities at the two measuring points will affect the correlation. Hence the spatial variations of correlation coefficients indicate more directly the characteristics of the turbulence itself.

Figure 2.4(c) shows possible variations of direct correlations. Two complications are found which do not appear in autocorrelations: negative values may occur as the separation between the points is increased; and, though this is not shown, the correlations may be asymmetric with respect to the separation. The latter condition will arise when the character of the turbulence changes significantly over a distance within which the degree of correlation is significant. Consider, for example, turbulent flow near a plane wall, the mean flow $U(y)$ being confined to the x, y-plane: $R(y)$ will be markedly asymmetric; $R(x)$ will be slightly asymmetric, if the flow is still developing; while $R(z)$ should be symmetric. In homogeneous turbulence, symmetry will be found for correlations in any direction; while for isotropic turbulence, a correlation will vary in the same manner, no matter what direction is chosen.

Ways in which correlations like those of Figure 2.4(c) might be obtained are indicated in Figure 2.4(d). Three hot wires sensing the streamwise fluctuation u are shown. The correlation $\overline{u_a u_b} = \overline{u(x)u(x+r)}$, with the separation parallel to the measured components, is termed the *longitudinal correlation*; while $\overline{u_a u_c} = \overline{u(y)u(y+r)}$, with the separation normal to the measured components, is the *transverse correlation*. Both reduce to $\overline{u^2}$ as $r \to 0$.

In general, though not for the very special case of isotropic turbulence, other correlation functions will be required to prescribe the relationship between u-fluctuations, and still others for the other velocity components. Consider the family of correlations which can be derived from the point correlation tensor $Q_{ij} = \overline{u_i u_j}$. These are the twenty-seven* entities

$$Q_{ij}(r_k) = \overline{u_i u_j(r_k)} = u'_i u'_j R_{ij}(r_k) \tag{2.6.20}$$

with $k = 1, 2, 3$ as well as $i, j = 1, 2, 3$. Here R_{ij} is a dimensionless correlation coefficient tensor. It will be appreciated that, taken together, these functions contain a great deal of information about the structure of the turbulence in which they are measured. But all of this data is obtained by averaging with respect to time and, in consequence, cannot reveal certain phase relationships essential to the turbulence.

Length Scales

Using spatial correlations, we can define length scales analogous to the time scales (2.6.11, 12, 15). The *integral length scale* is

$$L = \int_0^\infty R(r)\,\mathrm{d}r \tag{2.6.21}$$

a measure of the average spatial extent or coherence of the fluctuations. For a particular point in a particular flow, its magnitude depends on the quantities correlated and on the direction of separation. In flow near a wall, for

* The fact that there are twenty-seven entities does not indicate that this is a third-order tensor. What we are doing is to describe the variation of the nine components of Q_{ij} throughout three-dimensional space.

example, it is found that the longitudinal scale (based on $Q_{11}(x)$) is considerably greater than the transverse scale (based on $Q_{11}(y)$).

The smaller scales can be measured by the *microscale*

$$\lambda = \left[-\frac{2}{(\partial^2 R/\partial r^2)_0} \right]^{\frac{1}{2}} \tag{2.6.22}$$

(compare with the time scale (2.6.12)), provided that the function $R(r)$ is nearly symmetrical near $r = 0$. For a direct correlation, an analysis like that leading to equation (2.6.15) gives

$$\lambda = \left[\frac{2\overline{s^2}}{\overline{(\partial s/\partial r)^2}} \right]^{\frac{1}{2}} \tag{2.6.23}$$

The most commonly used length scales are those based on u, the streamwise velocity fluctuation. The longitudinal integral scale is

$$L_x = \int_0^\infty R_{11}(r_1)\,\mathrm{d}r_1 \tag{2.6.24}$$

with $R_{11}(r_1) = \overline{u(0)u(r_1)}/\overline{u^2}$ as in equations (2.6.20). The longitudinal microscale is usually taken to be

$$\lambda_x = \frac{1}{\sqrt{2}}\lambda_{11} = \left[\frac{\overline{u^2}}{\overline{(\partial u/\partial x)^2}} \right]^{\frac{1}{2}} \tag{2.6.25}$$

where λ_{11} is derived from R_{11} via equation (2.6.23). This is sometimes called the Taylor microscale, after G. I. Taylor.

In practice, these quantities are usually found, not from $R_{11}(r_1)$, but by applying Taylor's hypothesis to $u(t)$ measured at a fixed point. This gives

$$r_1 \simeq U\tau \quad \text{and} \quad \frac{\partial u}{\partial x} \simeq \frac{1}{U}\frac{\partial u}{\partial t} \tag{2.6.26}$$

The first result can be used to convert an autocorrelation $\overline{u(0)u(\tau)}$ curve into a spatial correlation curve, from which L_x can be found. The second relationship allows λ_x to be calculated directly from $\overline{u^2}$ and $\overline{(\partial u/\partial t)^2}$. In energetic turbulence, the smallest scales are so small that probe interference vitiates an attempt to measure directly the part of $R(r)$ near $r = 0$, on which λ depends.

The microscale is sometimes called the dissipation length scale, but this is not an appropriate name, since the dissipating scales are even smaller than the length λ_x. A more significant interpretation can be found by considering the dissipation within the turbulence. This is most easily done for *isotropic turbulence*, for which it is found* that

$$\varepsilon = -\frac{\mathrm{d}(\frac{1}{2}\overline{q^2})}{\mathrm{d}t} = -\frac{3}{2}\frac{\mathrm{d}\overline{u^2}}{\mathrm{d}t} = 15\nu\overline{\left(\frac{\partial u}{\partial x}\right)^2} \tag{2.6.27}$$

* For example, in Reference 5, p. 179.

gives the rate of dissipation, $\frac{1}{2}\overline{q^2}$ being the turbulence kinetic energy. The last expression may be roughly correct for anisotropic turbulence, by virtue of the nearly isotropic nature (local isotropy) of the small scales which contribute to $\partial u/\partial x$. Combining equations (2.6.25, 27), we have

$$\frac{\mathrm{d}\overline{u^2}}{\mathrm{d}t} = -\frac{\overline{u^2}}{\lambda_x^2/10\nu} \tag{2.6.28}$$

Thus the micro length scale defines a *time scale for decay* of the turbulence energy:

$$\tau_{\mathrm{d}} = \frac{\lambda_x^2}{10\nu} \tag{2.6.29}$$

is a time over which there is a significant change in the energy, provided that there is no energy input.

The length λ_x is intermediate between the large scales which extract and contain most of the energy, and the smallest scales which dissipate it. A Reynolds number based on λ_x is often used to indicate the general level of turbulent activity:

$$\mathrm{Re}_\lambda = \frac{u'\lambda_x}{\nu} \tag{2.6.30}$$

This is called the *turbulence Reynolds number*, although several Reynolds numbers are in fact needed to characterize the various scales which make up turbulence. The decay law can be written

$$\frac{\mathrm{d}(\nu/\overline{u^2})}{\mathrm{d}t} = \frac{10}{\mathrm{Re}_\lambda^2} \tag{2.6.31}$$

showing that Re_λ defines the decay rate for a particular fluid.

2.7 SPECTRAL ANALYSIS

2.7.1 Classification

Correlations represent certain features of the spatial structure and temporal development of turbulence. We shall now examine an alternative method of dissecting turbulence, which breaks it into elements which vary harmonically through space and time. Thus a turbulent motion is represented as a superposition of periodic eddies or waves with different length and time scales, and different orientations in space. The processes of stress generation, diffusion and energy transmission between the various scales of motion can now be described in terms of interactions between these eddies. Since the equations of fluid motion are non-linear, the interactions usually are too. It is these non-linearities, coupled with randomness, which give a peculiarly difficult character to the mathematical analysis of turbulence.

A particular superposition of harmonic components with differing frequencies (or wavelengths) is specified by a function giving the variation of the component intensity with frequency (or wavelength). This is called a *spectrum function* by analogy with the spectrum of visible light. The first types of spectra to be considered are:

(1) *frequency spectra* related to time correlations; those derived from cross correlations are termed cross spectra.

The spatial analogue of the angular frequency ($\omega = 2\pi/T$, inversely proportional to period) is the *wave-number* ($k = 2\pi/\lambda$, inversely proportional to wavelength). Thus we next consider several types of wave-number spectra:

(2) *one-dimensional wave-number spectra* related to spatial correlations in one direction and, through Taylor's hypothesis, to frequency spectra; and

(3) *complex spectra*, which can be used to represent spatial correlations that are not symmetrical or even with respect to the separation distance.

We noted earlier (in equations (2.6.20)) that correlations are functions of the separation vector between the correlated points. In expressing this relationship in terms of wave-numbers, we are led to a spectrum in three-dimensional wave-number space, a function of a wave-number vector. Hence we consider:

(4) *spectra in wave-number space*, related to correlation functions which vary through the three dimensions of physical space; and

(5) *three-dimensional wave-number spectra*, which represent the overall activity associated with a particular length scale, regardless of direction.

The spectra listed above are complementary to the second-order correlations discussed in Section 2.6. It is possible to introduce spectra parallel to other kinds of correlations—space-time, Lagrangian and higher-order—but this will not be done here. We shall concentrate on spectra associated with the velocity correlation tensor of equations (2.6.20).

Since the notation for the several types of correlations and spectra is rather intricate, it is gathered together in Table 2.1.

2.7.2 Frequency Spectra

Wide ranges of time and length scales are defining characteristics of turbulence. In averaging mathematically, or in using an r.m.s. meter, we absorb in a single mean value the contributions from the whole frequency range. But suppose that the complete signal is passed through a set of electrical filters before being measured; then the meter will respond only to those components which the filters have let through. By using a high- and a low-pass filter together to form a band-pass filter, we can, in principle, measure the contributions from a narrow band of frequencies. In practice, the cut-offs will not be abrupt, and the output from a pair of filters will have the bell-shaped form shown as a solid curve in Figure 2.5(a).

Table 2.1. Notation for second-order correlations and spectra.

In addition to having the functional dependence indicated, each quantity will, in general, depend on the base-point position, which may be denoted by the vector $\mathbf{x}_p$ or by the typicalcoordinate x_p

Quantity	General signals s_1, s_2	Velocity components u_i, u_j	Defining equations
Point correlation	$\overline{s_1(t)\, s_2(t)}$	Q_{ij}	(2.6.2, 6)
Autocorrelation	$\overline{s(t)\, s(t+\tau)}$	$Q_{11}(\tau)$	(2.6.9)
Time correlation	$\overline{s_1(t)\, s_2(t+\tau)}$	$Q_{ij}(\tau)$	(2.6.7)
Correlation coefficient	$R(\tau)$	$R_{ij}(\tau)$	(2.6.8, 9)
Frequency spectrum	$F(\omega)$	$E_{ij}(\omega)$	(2.7.1, 3, 8)
Normalized spectrum	$\phi(\omega)$	$E_{ij}/u_i' u_j'$	(2.7.3, 4)
Spatial correlation	$\overline{s_1(x_p)\, s_2(x_p + r)}$	$Q_{ij}(r)$	(2.6.17, 20)
Correlation coefficient	$R(r)$	$R_{ij}(r)$	(2.6.18, 20)
One-dimensional wave-number spectrum	$F(k)$	$E_{ij}(k)$	(2.7.10, 11)
Normalized spectrum	$\phi(k)$	$E_{ij}/u_i' u_j'$	(2.7.9, 10)
Complex spectrum	$F(k)$ or $\phi(k)$	$E_{ij}(k)$	(2.7.12, 13)
Spatial correlation in three-dimensional space	$\overline{s_1(\mathbf{x}_p)\, s_2(\mathbf{x}_p + \mathbf{r})}$	$Q_{ij}(\mathbf{r})$	(2.7.14, 15)
Correlation coefficient	$R(\mathbf{r})$	$R_{ij}(\mathbf{r})$	(2.7.16, 17)
Wave-number spectrum in wave-number space	$F(\mathbf{k})$	$E_{ij}(\mathbf{k})$	(2.7.16)
Normalized spectrum	$\phi(\mathbf{k})$	$E_{ij}/u_i' u_j'$	(2.7.16, 17)
Three-dimensional spectrum	$F_T(k)$	$E(k)$	(2.7.19)

If a consistent range of band-pass filters is available (and this is what commercial spectrum analysers provide), we can, by switching from one to another, find the contributions from every part of the frequency range. At the lower frequencies a longer averaging time will be required, in order that a large enough number of events be included; this requirement can be met by a meter with a range of time-constants. The lower limit ω_L to which the frequency scanning is extended may be determined by the maximum time-constant which is available, or by the intrusion of long-period fluctuations not directly associated with the turbulence. The upper limit ω_U may be decided

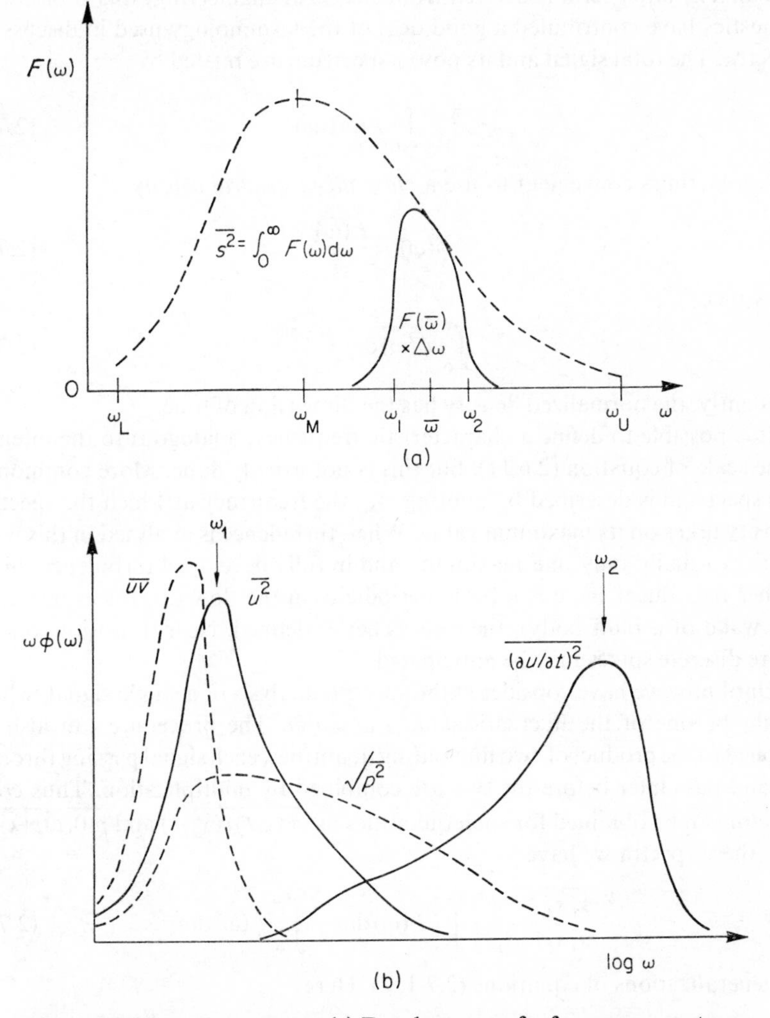

Figure 2.5 Frequency spectra. (a) Development of a frequency spectrum from narrow-band elements: solid curve—contribution for filters with nominal limits ω_1 and ω_2; dashed curve—complete power spectrum. (b) Possible shapes of frequency spectra for effective shear stress ($\overline{uv}$), turbulence energy ($\overline{u^2}$), pressure fluctuations ($\sqrt{\overline{p^2}}$) and dissipation $\overline{(\partial u/\partial t)^2}$

by the appearance of irrelevant electronic noise at high frequencies, or by a limitation in the frequency response of some part of the electrical equipment.

The curve defined by the procedure outlined above, that giving the contribution to $\overline{s^2}$ from each frequency band, is shown dashed in Figure 2.5(a). The function defined is called the *spectral density* or power spectral density of the signal. The latter term is derived from electrical engineering; this subject and acoustics have contributed a good deal of the terminology used in discussing spectra. The total signal and its power spectrum are related by

$$\overline{s^2} = \int_0^\infty F(\omega)\,\mathrm{d}\omega \tag{2.7.1}$$

It is sometimes convenient to use a *normalized spectral density*

$$\phi(\omega) = \frac{F(\omega)}{\overline{s^2}} \tag{2.7.2}$$

for which

$$\int_0^\infty \phi(\omega)\,\mathrm{d}\omega = 1 \tag{2.7.2}$$

Evidently, the normalized density has the dimension of time.

It is possible to define a characteristic frequency, analogous to the integral time scale of equation (2.6.11), but this is not usually done. More commonly, the spectrum is described by quoting ω_M, the frequency at which the spectral density takes on its maximum value. When turbulence is analysed in this way, there is usually only one maximum, and in fully developed turbulence this is rather flat. But if there is a basic periodicity in the flow (as, for example, in the wake of a bluff body), the peak is better defined. Near transition, one or more discrete spikes may be anticipated.

Until now we have considered the spectral analysis of a single signal, which might be one of the fluctuations u, p or $\partial u/\partial t$. The procedure can also be applied to the product of two fluctuating quantities, each signal passing through a band-pass filter before the two are combined by multiplication. Thus *cross spectra* can be obtained for such quantities as $\overline{u(x_p, t)v(x_p, t)}$ and $\overline{p(0, t)p(x, t)}$. For these spectra we have

$$\frac{\overline{s_1 s_2}}{s_1' s_2'} = \frac{1}{s_1' s_2'}\int_0^\infty F(\omega)\,\mathrm{d}\omega = \int_0^\infty \phi(\omega)\,\mathrm{d}\omega \tag{2.7.3}$$

as generalizations of equations (2.7.1, 2). Here

$$\phi(\omega) = \frac{F(\omega)}{s_1' s_2'} \tag{2.7.4}$$

is a normalized spectrum; for a cross spectrum, its integral over all frequencies need not be unity.

The spectra of the various properties of a turbulent flow must be expected to have rather different forms. Figure 2.5(b) suggests the relationships which might be found between the spectral densities of four properties: the turbulence energy $\overline{u^2}$, pressure–fluctuation intensity $\overline{p^2}$, shear stress $\overline{uv}$, and dissipation $\overline{(\partial u/\partial t)^2}$. The spectra have been multiplied by ω in order to give greater prominence to high-frequency elements. Although these are of small magnitude, they extend over very broad frequency bands, a fact masked by the common use (as here) of a logarithmic scale for the ω-axis; hence multiplication by ω merely restores their true importance.

Note that there is a fair degree of segregation in the scales contributing to the four characteristics in Figure 2.5(b). This observation is the starting point for the analysis of certain aspects of turbulence, as we shall see when discussing the scaling of correlations and spectra in Section 2.8.

Consider next the numerical scale which might be given for frequency in Figure 2.5(b). The values of ω_1 and ω_2, at which the energy and dissipation densities have their maxima, can vary over wide ranges. For a pipe flow with $\mathrm{Re}_d = 5 \times 10^5$, corresponding to atmospheric air flowing at $U = 30$ m/sec in a pipe of diameter 25 cm, the values found in the central part of the pipe might be $\omega_1 = 500$ Hz and $\omega_2 = 20{,}000$ Hz. In the atmosphere, with $U = 3$ m/sec, typical values would be $\omega_1 = 0{\cdot}01$ Hz (a period of around one minute) and $\omega_2 = 10$ Hz. The atmospheric energy spectrum also has peaks near 10 μHz (period one day) and near 3 μHz (period about four days); these are associated with daily variations and with longer-period changes in the weather, neither of which would normally be called turbulence. Long-period fluctuations are found in many other situations, and are similarly rejected from the domain of turbulence. This suggests an additional *defining characteristic of turbulence*: it consists of those interacting scales of motion that take part in the local extraction and dissipation of energy.

Apparently a time correlation $R(\tau)$ and spectral density $\phi(\omega)$ give, in different ways, the same information about the development of turbulence. This idea is reinforced by noting that equation (2.7.3) relates the point correlation (that with no time delay) to the spectrum:

$$R(0) = \frac{\overline{s_1 s_2}}{s_1' s_2'} = \int_0^\infty \phi(\omega)\,\mathrm{d}\omega$$

It can be shown,* with a few dozen lines of algebra, that the time correlation and frequency spectrum of the same stationary random fluctuation are related by the *Fourier Transform pair*:

$$R(\tau) = \int_0^\infty \phi(\omega)\cos\omega\tau\,\mathrm{d}\omega$$

and

$$\phi(\omega) = \frac{2}{\pi}\int_0^\infty R(\tau)\cos\omega\tau\,\mathrm{d}\tau \qquad (2.7.5)$$

* For example, in Reference 7, p. 214.

with $R(\tau)=\overline{s_1 s_2}/s_1' s_2'$ and $\phi(\omega)=F(\omega)/s_1' s_2'$. Here ω is the angular frequency, measured not in cycles per second (Hz), but in radians per second (rad/sec).

To indicate the usefulness of these relationships, we consider an immediate application. Using equation (2.6.14), we find the mean-square derivative $\overline{(\partial u/\partial t)^2}$ and the correlation $R_{11}(\tau)=\overline{u(t)u(t+\tau)}/\overline{u^2}$ to be related by

$$\overline{\left(\frac{\partial u}{\partial t}\right)^2} = -\overline{u^2}\left(\frac{\partial^2 R_{11}}{\partial \tau^2}\right)_0$$

From the first of equations (2.7.5) we find

$$\frac{\partial^2 R_{11}}{\partial \tau^2} = -\int_0^\infty \omega^2 \phi_{11}(\omega)\cos\omega\tau\,\mathrm{d}\omega \tag{2.7.6}$$

with $\phi_{11}(\omega)$ the normalized spectral density of $\overline{u^2}$. There follows

$$\overline{\left(\frac{\partial u}{\partial t}\right)^2} = \int_0^\infty \omega^2 E_{11}(\omega)\,\mathrm{d}\omega \tag{2.7.7}$$

with $E_{11}(\omega)$ the spectral density of $\overline{u^2}$. The spectrum of the derivative contains the factor ω^2, which weights it towards the higher frequencies. This explains the very different shapes given to the spectra of $\overline{u^2}$ and $\overline{(\partial u/\partial t)^2}$ in Figure 2.5(b).

The spectrum $E_{11}(\omega)$ is one of the components of a *frequency spectrum tensor* $E_{ij}(\omega)$ related to the velocity correlation tensor of equations (2.6.6) by

$$Q_{ij}=\overline{u_i u_j}=\int_0^\infty E_{ij}(\omega)\,\mathrm{d}\omega \tag{2.7.8}$$

Relationships like (2.7.7) between measurable quantities provide valuable checks on techniques of measurement and data reduction. More generally, the relationships between correlations and spectra assist the experimenter by providing a choice between measuring techniques. Either the time correlation or the frequency spectrum can be found, the other being derived from it, if required.

2.7.3 Wave-number Spectra

The transform relationship between $R(\tau)$ and $\phi(\omega)$ has been seen to be useful. We extend this idea by introducing an analogous spectrum $\phi(k)$ as the transform of the spatial correlation $R(r)$. The wave-number, k, is analogous to frequency, and has the dimension (length)$^{-1}$. For simplicity, we adopt for the present the cosine transform, although it is adequate only when $R(r)$ is an even function of r. Thus

$$R(r) = \int_0^\infty \phi(k) \cos kr \, dk$$

and (2.7.9)

$$\phi(k) = \frac{2}{\pi} \int_0^\infty R(r) \cos kr \, dr$$

The wave-number $k = 2\pi/\lambda$, with λ the wavelength of an harmonic element; as used here, k has the units of rad/m or rad/ft. The normalized spectral density $\phi(k)$ has the dimension of length, and is related to the fully dimensional wave-number spectrum density by

$$\frac{\overline{s_1 s_2}}{s_1' s_2'} = \frac{1}{s_1' s_2'} \int_0^\infty F(k) \, dk = \int_0^\infty \phi(k) \, dk \tag{2.7.10}$$

These statements parallel equations (2.7.3). Since the spectra introduced here are related to spatial correlations along a single direction in space, they are termed *one-dimensional wave-number spectra.*

The wave-number spectrum is a development of the approach suggested in equations (1.4.1, 2). But there the idea was applied to the instantaneous values of a velocity component; the spectrum thus obtained fluctuates from instant to instant, and is of no more use than the velocity fluctuation itself. Here we have applied the technique to a statistical property, and obtain a unique prescription for that aspect of the turbulence. It is still possible to think of the wave-number components as being crudely representative of 'eddies' of various sizes.

The transformations introduced above are often applied to the components of the velocity correlation tensor $Q_{ij} = \overline{u_i u_j(r)}$ of equations (2.6.20). Transforming each of these, we obtain a corresponding set of spectrum components:

$$Q_{ij}(r) = \int_0^\infty E_{ij}(k) \cos kr \, dk \tag{2.7.11}$$

Since the velocity correlation tensor Q_{ij} is often referred to as the energy tensor, the transforms E_{ij} are called the *energy spectrum tensor.*

For simplicity, we have used a cosine transform to generate the wave-number spectrum. For correlations that are not even functions of r, this is inadequate; both odd and even spectral components must be provided, and the range $r < 0$ must be included. It is convenient to use the *complex Fourier transform* in making this generalization. Thus we have the transform pair:

$$R(r) = \int_{-\infty}^\infty \phi(k) e^{ikr} \, dk$$

and (2.7.12)

$$\phi(k) = \frac{1}{2\pi} \int_{-\infty}^\infty R(r) e^{-ikr} \, dr$$

The *complex spectrum* can be separated into real and imaginary parts:

$$\phi(k) = \phi_r(k) + i\phi_i(k) \tag{2.7.13}$$

These are called, respectively, the cospectrum and the quadrature spectrum.

We have not yet taken full account of the three-dimensional character of turbulence, nor of the corresponding three-dimensional character (in wave-number space!) of the spectra used to describe it. This can be done, formally, by transforming with respect to each of the three space coordinates:

$$\overline{s_1(x_1, x_2, x_3)\, s_2(x_1 + r_1, x_2 + r_2, x_3 + r_3)}$$
$$= \int_{-\infty}^{\infty} F(k_1, k_2, k_3) \exp\left[i(k_1 r_1 + k_2 r_2 + k_3 r_3)\right] dk_1\, dk_2\, dk_3 \tag{2.7.14}$$

with x_i coordinates of the base point, r_i separation distances along the co-ordinate axes, and k_i wave-numbers measured along the three axes.

This transformation can be expressed more compactly using *vector notation*:

$$\overline{s_1(\mathbf{x}_p)\, s_2(\mathbf{x}_p + \mathbf{r})} = \int_{-\infty}^{\infty} F(\mathbf{x}_p, \mathbf{k})\, e^{i\mathbf{k}\cdot\mathbf{r}}\, d\mathbf{k} \tag{2.7.15}$$

or in normalized terms:

$$R(\mathbf{x}_p, \mathbf{r}) = \int_{-\infty}^{\infty} \phi(\mathbf{x}_p, \mathbf{k})\, e^{i\mathbf{k}\cdot\mathbf{r}}\, d\mathbf{k} \tag{2.7.16}$$

on the understanding that $\mathbf{k}$ is the wave-number vector with components k_i, $\mathbf{k}\cdot\mathbf{r}$ is the scalar product of $\mathbf{k}$ and $\mathbf{r}$, and $d\mathbf{k} = dk_1\, dk_2\, dk_3$ is a volume element in $\mathbf{k}$-space.

Corresponding to each of the relationships (2.7.16) is an inverse three-dimensional transform:

$$\phi(\mathbf{x}_p, \mathbf{k}) = \frac{1}{(2\pi)^3} \int_{-\infty}^{\infty} R(\mathbf{x}_p, \mathbf{r})\, e^{-i\mathbf{k}\cdot\mathbf{r}}\, d\mathbf{r} \tag{2.7.17}$$

Applying these ideas to the energy tensor Q_{ij} and the associated spectrum tensor E_{ij}, we see that each of these tensors is a function of two vectors: $Q_{ij}(\mathbf{x}_p, \mathbf{r})$ and $E_{ij}(\mathbf{x}_p, \mathbf{k})$. It should be realized, however, that their tensor character does not arise from this dependence on $\mathbf{x}_p$ and $\mathbf{r}$ (or $\mathbf{k}$), but from their relationship to the velocity components $u_i(\mathbf{x}_p)$ and $u_j(\mathbf{x}_p + \mathbf{r})$. Their tensor nature comes not from the way the nine components vary through space, but from the way in which their values at each point are related to the velocity components and, through them, to the directions of the coordinate axes.

The relationship between the complex one-dimensional spectrum (equations 2.7.12, 13) and the complex spectrum in wave-number space (equations 2.7.16, 17) can be established by applying them to the same point correlation:

$$R(0) = \int_{-\infty}^{\infty} \phi(k_1)\, dk_1 = \int_{-\infty}^{\infty} \phi(\mathbf{k})\, d\mathbf{k}$$

with r_1 the direction of separation for the one-dimensional spectrum. Note that, in the symbolism adopted here and summarized in Table 2.1, one-dimensional spectra and space spectra are distinguished only by the nature of the wave-number argument: $\phi(k_1)$ and $\phi(\mathbf{k})$.

The result of the comparison

$$\phi(k_1) = \int_{-\infty}^{\infty}\int_{-\infty}^{\infty} \phi(\mathbf{k})\,\mathrm{d}k_2\,\mathrm{d}k_3 \tag{2.7.18}$$

reveals an awkward feature of the one-dimensional spectrum: it does not give the fluctuation energy associated with a particular scale of disturbances $|\mathbf{k}|$, but the integrated effect of these and certain smaller ones (for which k is larger). This can be seen more easily by considering the analogous frequency spectrum obtained from a fixed point past which turbulence is convected. A short-wavelength element whose direction of variation is nearly across the mean flow will take as long to pass as will a much larger disturbance whose direction of variation is aligned with the mean flow. In wave-number terms, this means that the streamwise components for the two disturbances are equal, and both contribute to the same band on the k_1-scale. This effect is referred to as *aliasing*.

A spectrum which does not have this disadvantage, but is still a function of a single variable, can be derived by integrating $F(\mathbf{k})$ over a spherical shell in $\mathbf{k}$-space. The *three-dimensional spectrum* $F_T(k)$ derived in this way is a function of k, the magnitude of the wave-number vector $\mathbf{k}$, but not of its direction. Since the three- and one-dimensional spectra are both functions of a single variable, they can be related by considering the geometry of wave-number space.

This approach is often applied to the turbulence kinetic energy:

$$\tfrac{1}{2}\overline{q^2} = \int_0^{\infty} E(k)\,\mathrm{d}k \tag{2.7.19}$$

where the three-dimensional spectrum $E(k)$ is called the *energy spectrum function.* Isotropic turbulence provides an example of the relationship* between this kind of spectrum and a one-dimensional spectrum:

$$E(k) = \tfrac{1}{2}k^3 \frac{\mathrm{d}}{\mathrm{d}k}\left[\frac{1}{k}\frac{\mathrm{d}E_{11}}{\mathrm{d}k}\right] \tag{2.7.20}$$

with $E_{11}(k)$ the spectrum of $\overline{u^2} = Q_{11}$.

The result (2.7.20) applies when E_{11} is defined as the spectrum over the range $0 < k < \infty$, as in equation (2.7.11). This is possible because the spectrum is, for isotropic or homogeneous turbulence, an even function of k. If the spectrum were defined over the range $-\infty < k < \infty$, as in equations (2.7.12) for example, the spectral values would be only one-half as large, and the factor $\tfrac{1}{2}$ would not

* Derived, for example, in Reference 7, p. 216.

appear. The values of the components of other even spectra depend in a similar way on the spectral range chosen.

2.7.4 Taylor's Hypothesis

In introducing wave-number spectra, we have been guided by the mathematical analogy with frequency spectra. Now let us return to the main subject of this chapter—measurement. It has not yet proved possible to measure wave-number spectra directly. Indeed, it may not be worth trying to do this, for they can be found indirectly in two ways:

(1) by Fourier transformation of spatial correlation functions, and
(2) by the application of Taylor's hypothesis to frequency spectra.

There is little more to be said about the first method since, once a correlation function has been obtained, routine mathematical procedures can be used to generate its Fourier transform. The second technique will be discussed at some length, however, partly because it is of very great practical importance, and partly because the assumptions underlying Taylor's hypothesis require examination.

Considering turbulence convected by a uniform mean flow, Taylor argued that the sequence of events at a fixed point is nearly equivalent to the movement of an unchanging pattern of turbulence past the point. This idea has already proved useful in several places. In the present context, it is valuable in suggesting that the spectral contribution at frequency ω is generated by a disturbance of wave-number $k = \omega/U$, with U the speed with which the disturbance is convected past the measuring point. Moreover, the time delay τ is equivalent to a separation distance $r = U\tau$ in the direction of convection. These postulates parallel those of equations (2.6.26).

When the relationships of the preceding paragraph are introduced, the frequency spectrum and time correlation function of equations (2.7.5) become

$$R\left(\frac{r}{U}\right) = U \int_0^\infty \phi(kU) \cos kr \, dk$$

and (2.7.21)

$$\phi(kU) = \frac{2}{\pi U} \int_0^\infty R\left(\frac{r}{U}\right) \cos kr \, dr$$

Comparison with equations (2.7.9) shows that the one-dimensional wave-number spectrum for the direction of convection can be estimated as

$$\phi(k) = U\phi(\omega) \quad \text{with} \quad \omega = kU \tag{2.7.22}$$

using a measured frequency spectrum. This result holds for $U = \omega/k$, implying that ω and k have consistent units, that is, both in cycles or both in radians. When this is not the case, a factor must be introduced to account for the contraction or expansion of the frequency scale. Thus, when ω is measured in cycles (say, Hz) and k in radians (say, rad/m), the one-dimensional spectrum is $U\phi(\omega)/2\pi$.

Experimental and theoretical investigations of the validity of Taylor's hypothesis have shown it to apply, not only to weak, nearly homogeneous grid turbulence (for which it was originally proposed), but also through large parts of inhomogeneous, active turbulent flows. The hypothesis has been found to be valid when the intensity is not too high (say, $u'/U < 0{\cdot}1$), and also through the outer 90 to 95 per cent of a boundary layer. These observations suggest that it will be applicable throughout most free-turbulent flows (though not where flow reversal occurs) and in the cores of channel flows (but not too near the walls).

It is apparent that a one-dimensional wave-number spectrum obtained from equations (2.7.21, 22) will have exactly the same shape as the frequency spectrum from which it is derived. Hence the spectra sketched in Figure 2.5(b) can be interpreted directly as one-dimensional spectra. However, from a one-dimensional spectrum we can (in principle) derive a three-dimensional spectrum using a relationship of the kind exemplified in equation (2.7.20). The method of its derivation indicates that in certain respects the three-dimensional spectrum will be quite different from the basic frequency spectrum; indeed, this is the justification for introducing it.

Since wave-number spectra cannot be measured directly, and contain no information which is not implicit in measurable correlations and frequency spectra, are they worthwhile? In one sense, the answer 'yes' is obligatory, for spectra occupy a central place in the literature devoted to turbulence. Some workers—particularly those with a background in physics—have found that wave-number spectra provide powerful tools of analysis and helpful ways of visualizing turbulence. Two specific advantages may be mentioned:

(1) wave-number spectra are much less strongly influenced by mean convection than are frequency spectra, and thus provide more direct insight into the time and length scales of the turbulence; and

(2) three-dimensional spectra are more clearly related to disturbances of a particular size.

It should be recognized, however, that many important contributions to the understanding of turbulent processes have been based directly on correlations.

By considering correlations and spectra, we recover some of the detail lost in the averaging used to bring order into the chaos of turbulence. In principle, everything that has been lost by averaging could be recovered by including more and more correlations (or spectra) in a statistical specification of the turbulence. One might hope that certain aspects of turbulence can be adequately described using only a few statistical properties. This is a possible solution to the *problem of closure* mentioned in Section 1.7. Even if this does not often prove to be the case, it is undoubtedly true that even a superficial penetration into the jungle of statistics provides a much deeper understanding of the overall effects of turbulence.

2.8 THE SCALING OF CORRELATION AND SPECTRUM FUNCTIONS

2.8.1 Survey of the Problem

The problem with which we now concern ourselves is that of correlating pieces of experimental data obtained in turbulent flows that are essentially similar, but have different scales of length, time and velocity. The data might be, for example, frequency spectra measured in grid turbulence for a number of different convection velocities, grid mesh sizes, and distances from the grid.

Three kinds of data correlation will be considered:

(1) scaling of the *largest elements* of the turbulence; here the role of viscosity is assumed to be insignificant, the correlating methods are referred to as *inertial scaling*, and the motions are said to display *Reynolds-number similarity*;

(2) scaling of the *smallest elements* of the turbulence; here the role of viscosity is vital and, coupled with the local rate of energy dissipation, defines a *universal small-scale, dissipating structure*; and

(3) scaling of the *intermediate elements* of the turbulence, those which are not influenced by viscosity, but have none-the-less adopted a universal form defined only by the dissipation rate; this *inertial subrange* of the postulated universal structure will exist only in turbulence which possesses a very wide range of length scales.

The methods of scaling to be developed here apply when there is a substantial difference in size between the larger scales of motion, which contain most of the energy, and the smaller scales, which dissipate it. In spectral terms, this difference in scale manifests itself in a marked separation between the wave-numbers (and frequencies) at which the energy spectrum and the dissipation spectrum take on their peak values; this is illustrated in Figure 2.5(b). The required separation does in fact exist in most turbulent flows since, if the turbulence is to maintain itself, the larger scales cannot be much influenced by viscosity.

Most of the discussion relates to turbulence embedded in a basically steady, but otherwise general turbulent motion, the fluctuations being stationary random functions of time. However, at certain points, decaying homogeneous turbulence will be considered. The results for the smaller scales assume local isotropy, but may be appropriate for turbulence which is either stationary or decaying in the larger scales.

In practice, the proposed scaling procedures may not correlate a body of measured data in the expected manner. Such a failure could result from errors of measurement, or from an inability to control the test environment. More significantly, the failure of a scaling law may point to faults in the postulates on which it is based. Thus scaling of experimental results can provide, not merely a convenient presentation of the results, but also checks on the accuracy of measurement and analysis, and perhaps even a verdict on a theory concerning the structure of turbulence.

2.8.2 Reynolds-number Similarity in the Larger Scales

In Section 1.4 we noted that certain kinematic features of turbulent flows, for example, the angle of spread of a turbulent jet, were independent of the Reynolds number of the mean flow. This led to the conclusion that the larger scales of turbulence are not directly influenced by viscosity—they are said to display Reynolds-number similarity. We saw earlier how this idea, when incorporated into dimensional analysis, could be used to predict certain aspects of the development of the mean flow. Now we investigate the consequences of Reynolds-number similarity for those parts of correlation and spectrum functions which describe the larger, viscosity-independent elements of the turbulence.

Consider first the *spatial correlation* defined in equations (2.6.17, 18). In a particular kind of turbulent flow (for example, grid turbulence or a round jet) we may suppose that the coefficient $R(x_p, r)$ is specified by

$$R \sim x_p, r, L_0, U_0, \nu \tag{2.8.1}$$

where x_p and r are, as before, a representative coordinate of the base point and the separation distance, and L_0 and U_0 are length and velocity scales for the boundary geometry and mean flow. Some possible interpretations for L_0 are: diameter of a wake-producing cylinder, boundary-layer thickness, or mesh size of a generating grid. Possible choices for U_0 are: the free-stream velocity beyond a boundary layer, a convection velocity, the velocity difference across a shear layer, or an initial jet velocity. Sometimes the velocity scale is constructed from more obvious defining parameters; thus the velocity scale for a jet may be defined as $U_0 = (M/\rho L_0^2)^{\frac{1}{2}}$, where M is the jet momentum flux and L_0^2 is a measure of the jet cross-sectional area. Finally, the length and velocity scales of the larger elements of the turbulence are closely linked to those for the mean motion, and it sometimes proves convenient to utilize these turbulence scales rather than mean-flow scales.

Let us now return to the correlation coefficient (2.8.1). In dimensionally homogeneous form

$$R = f\left(\frac{x_p}{L_0}, \frac{r}{L_0}, \frac{U_0 r}{\nu}\right) \tag{2.8.2}$$

For a specified *relative position* of the base point, that is, for x_p/L_0 fixed, Reynolds-number similarity implies that

$$R = f\left(\frac{r}{L_0}\right) \tag{2.8.3}$$

for $U_0 r/\nu >$ a certain value at which viscosity becomes important.

A parallel analysis of the *integral scale* (2.6.21) of the correlation function leads to an alternative statement of the last result. Starting from

$$L \sim x_p, L_0, U_0, \nu \tag{2.8.4}$$

we find

$$\frac{L}{L_0} = f\left(\frac{x_p}{L_0}, \frac{U_0 L_0}{\nu}\right) \tag{2.8.4}$$

Changes in $U_0 L_0/\nu$ alter only a small part of the correlation curve near $r = 0$; hence when the value of x_p/L_0 is fixed

$$\frac{L}{L_0} = \text{constant} \tag{2.8.5}$$

for $U_0 L_0/\nu >$ a certain value. The results (2.8.3, 5) can be combined to give

$$R = f\left(\frac{r}{L}\right) \tag{2.8.6}$$

for $U_0 r/\nu = (U_0 L_0/\nu)(r/L_0) >$ a certain value. This indicates that correlations measured about the same relative position in a particular class of turbulent flows should collapse to a single curve when plotted as R *vs* r/L, save at small values of r/L or $U_0 L_0/\nu$, for which the correlation is influenced by viscosity-dominated motions.

Wave-number spectra can be discussed in a very similar manner, with analogous results. Specifying a normalized spectrum (2.7.9, 16) as

$$\phi \sim x_p, k, L_0, U_0, \nu \tag{2.8.7}$$

we find

$$\frac{\phi}{L_0} = f\left(\frac{x_p}{L_0}, kL_0, \frac{k\nu}{U_0}\right) \tag{2.8.7}$$

For a particular relative position x_p/L_0, Reynolds-number similarity implies that

$$\frac{\phi}{L_0} = f(kL_0) \tag{2.8.8}$$

for $k\nu/U_0 <$ a certain value beyond which small, viscosity-dominated motions are significant. Using the result (2.8.5), we have also

$$\frac{\phi}{L} = f(kL) \tag{2.8.9}$$

for $(\nu/U_0 L_0)(kL_0) <$ a certain value. From equations (2.8.3, 8) we see that, for a particular boundary geometry and relative base-point position, spatial correlations and wave-number spectra are independent of the flow velocity and fluid viscosity, without scaling, except for parts representing scales influenced by viscosity.

For autocorrelations and frequency spectra, the last conclusion does not apply, but in other respects the discussion parallels that above. For an *autocorrelation*, we suppose that

$$R \sim x_p, \tau, L_0, U_0, \nu \tag{2.8.10}$$

and obtain

$$R = f\left(\frac{x_p}{L_0}, \frac{U_0 \tau}{L_0}, \frac{\tau \nu}{L_0^2}\right) \tag{2.8.10}$$

Reynolds-number similarity implies that

$$R = f\left(\frac{U_0 \tau}{L_0}\right) \tag{2.8.11}$$

for x_p/L_0 fixed, and $\tau\nu/L_0^2 >$ a certain value.

For a *frequency spectrum*, we take

$$\phi \sim x_p, \omega, L_0, U_0, \nu \tag{2.8.12}$$

and obtain

$$\frac{U_0 \phi}{L_0} = f\left(\frac{\omega L_0}{U_0}, \frac{x_p}{L_0}, \frac{\omega L_0^2}{\nu}\right) \tag{2.8.12}$$

Reynolds-number similarity implies that

$$\frac{U_0 \phi}{L_0} = f\left(\frac{\omega L_0}{U_0}\right) \tag{2.8.13}$$

for x_p/L_0 fixed, and $\omega L_0^2/\nu <$ a certain value. According to this result, $\phi(\omega)$ will have its maximum at a fixed value of $\omega L_0/U_0$. For a flow with a large-scale element which is nearly periodic, this value is the *Strouhal number* characterizing the flow, as introduced in equation (1.2.4).

Results analogous to (2.8.6, 9) can be obtained for the temporal characteristics of the turbulence. The *integral time scale* (2.6.11) has the form

$$T_E \sim x_p, L_0, U_0, \nu \tag{2.8.14}$$

The dimensionally homogeneous result is

$$\frac{U_0 T_E}{L_0} = f\left(\frac{x_p}{L_0}, \frac{U_0 L_0}{\nu}\right) \tag{2.8.14}$$

giving

$$T_E \propto \frac{L_0}{U_0} \tag{2.8.15}$$

subject to the same conditions as equation (2.8.5). The results (2.8.5, 15) show that, when Reynolds-number similarity exists over most elements of the turbulence, the integral scales vary in proportion to the length and velocity scales of the mean motion.

The relationships (2.8.11, 13) can now be given as

$$R=f\left(\frac{\tau}{T_{\mathrm{E}}}\right) \quad \text{and} \quad \frac{\phi}{T_{\mathrm{E}}}=f(\omega T_{\mathrm{E}}) \tag{2.8.16}$$

for x_p/L_0 fixed, and with $(\nu/U_0L_0)(\tau/T_{\mathrm{E}}) >$ a certain value or $(U_0L_0/\nu)(\omega T_{\mathrm{E}}) <$ a certain value. It is clear that changes in either length or velocity scale will alter the autocorrelation and the frequency spectrum. Looking back to equations (2.7.21, 22), we can see that the transformations suggested by Taylor's hypothesis have the effect of converting the velocity-dependent frequency spectrum into a wave-number spectrum which is independent of the convection velocity, within the range of Reynolds-number similarity. The removal of the dependence on convection leaves a spectrum representing primarily the turbulence. For this reason, experimental results obtained by frequency analysis are often presented as one-dimensional or three-dimensional wave-number spectra. However, the same effect can be achieved by scaling frequencies and spectral values in the manner suggested above.

This line of argument suggests that there is a close relationship between the validity of Taylor's hypothesis and Reynolds-number similarity. Hence we may expect that Reynolds-number similarity will be found in the larger scales of turbulence which is neither very intense nor very inhomogeneous. Experimental results from grid turbulence support this contention, and there is also a considerable body of evidence for Reynolds-number similarity in free-turbulent flows. Hence, as with Taylor's hypothesis, it is only very near a solid boundary that the hypothesis of Reynolds-number similarity fails to indicate correctly the relationships among the larger scales of turbulence.

Scaling procedures which do not use the fluid viscosity are often termed *inertial scaling*. The justification for this name is not apparent in the preceding discussion, since we have considered correlation coefficients and normalized spectra, already scaled using the r.m.s. values s_1' and s_2'. But suppose that we use instead the scales U_0 and L_0, in order to minimize the number of defining parameters. Now some quantity with the dimensions of force or mass must be added, to allow the scaling of correlation or spectrum functions that incorporate these dimensions. The fluid density is the obvious choice.

For an example of explicit inertial scaling, consider the frequency spectrum $E_p(\omega)$ of the pressure-fluctuation intensity $\overline{p^2}$ generated beneath a turbulent boundary layer of thickness δ and outer velocity U_1. We take

$$E_p(\omega) \sim \omega, U_1, \delta, \rho \tag{2.8.17}$$

and obtain

$$\frac{E_p(\omega)}{(\rho U_1^2)^2}\frac{U_1}{\delta}=f\left(\frac{\omega\delta}{U_1}\right) \tag{2.8.18}$$

for the scaled spectrum.

Throughout this discussion we have assumed that the only reason for the failure of inertial scaling is the influence of viscosity on the smaller scales of the turbulence. In practice, scaling often fails in the parts of the length and time spectra corresponding to the largest elements. The reasons are varied: fluctuations in test conditions, inhomogeneity in supposedly uniform upstream conditions, or simply coming to the wall of the test channel or to the end of a taped record of the fluctuations.

2.8.3 Similarity in the Smallest Scales

We turn now to the smallest scales of turbulence. The discussion is based on the observation that, for high mean-flow and turbulence Reynolds numbers, there is a substantial degree of separation between the scales which contain the energy and those which dissipate it. The Russian mathematician Kolmogoroff proposed a *universal equilibrium theory* incorporating the following ideas:

I. At high enough Reynolds numbers, the smallest scales have a universal structure (to be specific, local isotropy) which is statistically independent of the mean flow and the larger scales.

II. This range of small motions dissipates the energy supplied to it by the cascade of energy through the larger scales.

III. Viscosity must play an important role in the smallest, dissipating scales.

Thus it is argued that the kinematic features of the smallest motions are defined by the kinematic viscosity ν and by ε, the rate of dissipation of turbulence kinetic energy per unit mass of fluid, with dimensions of $(\text{length})^2/(\text{time})^3$.

Figure 2.6 attempts to represent these ideas graphically. Three-dimensional wave-number spectra are shown; their shapes are not very different from the modified frequency spectra of Figure 2.5(b). The energy spectrum function $E(k)$ and the dissipation spectrum $k^2 E(k)$ are shown; the latter can be obtained by an argument like that leading to the frequency spectrum of $\overline{(\partial u/\partial t)^2}$ given in equation (2.7.7). The wave-numbers shown correspond roughly to a flow with $U_0 L_0/\nu = 10^5$ and $L_0 = 20$ cm. This might be achieved in air with $U_0 = 7.5$ m/sec. Figure 2.6 will be discussed in more detail later; for the moment we need note only a few points:

(1) The range of dissipating scales R_6 is distinct from the range of energy-containing motions R_1, which are grouped around the wave-number $k_e \simeq 2\pi/L$, with L the integral scale of the turbulence. (For isotropic turbulence, $L_x/L_e = L_x k_e/2\pi \simeq 0{\cdot}75$, where L_x is the longitudinal scale (2.6.24).)

(2) The range of energy extraction R_4 extends from the lowest wave-numbers, those defined by the scale of the mean motion L_0, part-way into the energy-containing range R_1.

(3) There is a flux of energy from low to high wave-numbers—the cascade of energy. Between the ranges of extraction R_4 and dissipation R_6 this is uniform and equal to the dissipation rate ε.

(4) The range of dissipation R_6, and perhaps a broader range R_3, are independent of the special features of the flow in which the turbulence is embedded, unlike the lower wave-number range R_2. This can be justified by arguing that the motions sustained by the energy flux become progressively less ordered, and thus less dependent on the structure of the largest scales, as the energy is transferred to smaller and smaller scales. Ultimately all order is lost and the smallest scales are isotropic.

Applying these ideas to a typical *spatial correlation function*, we take

$$\overline{u(x_p)u(x_p+r)} \sim r, \varepsilon, \nu \tag{2.8.19}$$

for values of r small enough for the correlation to be determined by the scales

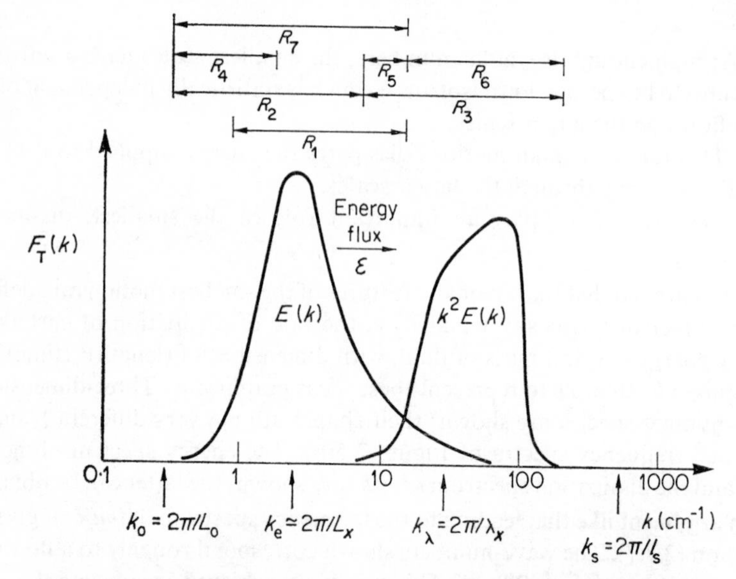

Figure 2.6 Characteristic ranges of a turbulent motion, illustrated by three-dimensional wave-number spectra of energy $E(k)$ and dissipation $k^2E(k)$. The Reynolds number for the case considered, $Re_0 = 10^5$, is about the lowest at which an inertial subrange exists. The wave-numbers correspond roughly to $L_0 = 20$ cm.

R_1 range of energy-containing motions
R_2 range peculiar to the particular flow
R_3 universal equilibrium range
R_4 range of energy extraction
R_5 inertial subrange
R_6 dissipative range
R_7 range of inertial scaling

of motion within the universal range. From ε and ν can be formed the length and velocity scales

$$l_s = \left(\frac{\nu^3}{\varepsilon}\right)^{\frac{1}{4}} \quad \text{and} \quad u_s = (\varepsilon\nu)^{\frac{1}{4}} \tag{2.8.20}$$

The first specifies the size of the smallest scales of the turbulence. It has been found experimentally that dissipation takes place in the band $0{\cdot}1 < kl_s < 1$, corresponding to the length scales $\lambda = 6\,l_s$ to $60\,l_s$. These define the range R_6 of Figure 2.6. Arranged in dimensionally homogeneous form, the parameters (2.8.19) give

$$\overline{u(x_p)\,u(x_p + r)} = u_s^2 f\left(\frac{r}{l_s}\right) = (\varepsilon\nu)^{\frac{1}{2}} f\left[\left(\frac{\varepsilon}{\nu^3}\right)^{\frac{1}{4}} r\right] \tag{2.8.21}$$

This approach can also be applied to wave-number spectra. For the three-dimensional *energy spectrum function*, for example

$$E \sim k, \varepsilon, \nu \tag{2.8.22}$$

giving

$$E = u_s^2\, l_s\, f(l_s k) = (\nu^5\,\varepsilon)^{\frac{1}{4}} f\left[\left(\frac{\nu^3}{\varepsilon}\right)^{\frac{1}{4}} k\right] \tag{2.8.22}$$

for sufficiently large values of k.

Rather similar results can be obtained for the universal equilibrium ranges of autocorrelation functions and frequency spectra, but the discussion is a little more difficult, since the effects of convection must be taken into account.

2.8.4 The Inertial Subrange of Intermediate Scales

Let us add another idea to the three underlying the equilibrium theory:

IV. For sufficiently high Reynolds numbers, there will exist a range of motions which have a universal structure, but are not small enough to be influenced significantly by viscosity.

The kinematic scales of this inertial subrange are determined entirely by ε, the dissipation rate. The range R_5 in Figure 2.6 marks this overlapping of the inertial range R_7 and the universal range R_3. It will be appreciated that, as the Reynolds number is reduced and k_e and k_s move closer together, this region of overlap will vanish.

For the *energy spectrum function* of equations (2.8.22) we have in the inertial subrange:

$$E \sim k, \varepsilon \tag{2.8.23}$$

leading to

$$E = \gamma_1\, \varepsilon^{\frac{2}{3}}\, k^{-\frac{5}{3}} \tag{2.8.23}$$

with γ_1 a universal dimensionless constant. Spectra measured in high-Reynolds-number turbulence display power laws very like this prediction. Slight differences in the index may be attributed, in part, to a relatively minor effect of the larger scales of the turbulence—long-period variations in the mean dissipation rate within the smallest scales.

In treating the *correlation coefficient* (2.8.19) we must modify slightly the procedure used earlier. We have

$$\overline{u(x_p)\,u(x_p+r)} \sim r, \varepsilon \tag{2.8.24}$$

However, it is not the absolute value of the correlation function, but the change in it, which is attributable to the small eddies. Thus we obtain the homogeneous form

$$\overline{u(x_p)\,u(x_p+r)} = \overline{u^2(x_p)} - \gamma_2(\varepsilon r)^{\frac{2}{3}} \tag{2.8.25}$$

with γ_2 another universal constant.

The value of γ_2 depends on the particular correlation considered. For *isotropic turbulence*—and the universal range is assumed to be essentially isotropic—there are only two distinct double-velocity correlations; they are uniquely related to one another and to the energy spectrum function. Detailed calculations give

$$\gamma_2 = 1{\cdot}98\,\gamma_1 \tag{2.8.26}$$

for the longitudinal correlation ($\overline{u_a u_b}$ of Figure 2.4(d)), and

$$\gamma_2 = 2{\cdot}63\,\gamma_1 \tag{2.8.26}$$

for the lateral correlation ($\overline{u_a u_c}$ of Figure 2.4(d)). Experiments to check these relationships in a particular flow provide tests of the hypotheses of the universal equilibrium theory. Determinations of the values of the 'constants' in various flows provide further checks. Values in the range

$$\gamma_1 = 1{\cdot}8 \pm 0{\cdot}5 \tag{2.8.27}$$

have been found; the scatter is moderate, considering the experimental difficulties, and this offers support for the essential elements of the theory.

Part of the variability in the hopefully universal constants can be ascribed to variations in Reynolds number. A diversity of Reynolds numbers can be used to indicate the degree of separation of energy-containing and dissipating scales, in particular:

(1) $\mathrm{Re}_0 = U_0 L_0/\nu$, the mean-flow Reynolds number,
(2) $\mathrm{Re}_L = u' L_x/\nu$, the integral-scale Reynolds number, and
(3) $\mathrm{Re}_\lambda = u' \lambda_x/\nu$, the turbulence Reynolds number introduced in equation (2.6.30).

The last is most widely used for this purpose.

It seems that $\mathrm{Re}_\lambda > 2000$ (corresponding perhaps to $\mathrm{Re}_0 > 3 \times 10^7$) is required to establish a true equilibrium range. Such flows will normally be found only in geophysical or astrophysical applications. However, the universal inertial results do have some application to flows on the scale more usual in engineering equipment. The form of dependence on r and k suggested by the equilibrium theory may be found over a narrow range for any $\mathrm{Re}_\lambda > 100$ (corresponding to perhaps $\mathrm{Re}_0 > 10^5$), although the constants γ_1 and γ_2 will differ somewhat from the universal values for very high Reynolds numbers. Figure 2.6 corresponds roughly to this lower limiting value of Re_λ. Note the intermediate position of $k_\lambda = 2\pi/\lambda_x$ in the range of length scales.

There remains the task of relating the dissipation rate ε to the scales of the mean motion U_0 and L_0. To do this, we introduce a fifth postulate:

V. The rate at which energy is supplied to the energy cascade is determined by the largest elements of the turbulence, those with scales linked to the scaling factors U_0 and L_0 of the mean flow.

The relatively narrow band of energy-supplying motions is shown in Figure 2.6 as range R_4. We have now

$$\varepsilon \sim U_0, L_0, x_p \tag{2.8.28}$$

giving

$$\varepsilon = \frac{U_0^3}{L_0} f\left(\frac{x_p}{L_0}\right) \tag{2.8.28}$$

It may be concluded that

$$\varepsilon = \text{constant} \times \frac{U_0^3}{L_0} \tag{2.8.29}$$

for a specified relative position in a particular kind of flow.

For nearly homogeneous grid turbulence, it is found that

$$\varepsilon \simeq 1{\cdot}6 \frac{u'^3}{L_x} \tag{2.8.30}$$

with L_x the longitudinal integral scale (2.6.24). The rate of dissipation in actively generating turbulence may be roughly estimated by taking

$$\frac{u'}{U_0} = \frac{L_x}{L_0} = \frac{1}{8} \tag{2.8.31}$$

to relate the energy-containing scales to those of the local mean flow. These estimates are broadly consistent with experimental observations.

Using the result (2.8.29) we can write equations (2.8.21, 22, 23, 25) in terms of mean-flow scales. In particular, for the inertial subrange

$$E = \text{constant} \times U_0^2 L_0 (kL_0)^{-\frac{5}{3}}$$

and

$$\Delta\,[\overline{u(x_p)\,u(x_p + r)}] = \text{constant} \times U_0^2 \left(\frac{r}{L_0}\right)^{\frac{2}{3}} \tag{2.8.32}$$

As would be expected, these results are consistent with equations (2.8.3, 8), which give the general pattern of inertial scaling.

The result (2.8.29) shows also that the length and velocity scales (2.8.20) of the dissipating elements are related to the mean-flow scales by

$$\frac{l_s}{L_0} = \text{constant} \times \left(\frac{\nu}{U_0 L_0}\right)^{\frac{3}{4}} \propto \mathrm{Re}_0^{-\frac{3}{4}}$$

and (2.8.33)

$$\frac{u_s}{U_0} = \text{constant} \times \left(\frac{\nu}{U_0 L_0}\right)^{\frac{1}{4}} \propto \mathrm{Re}_0^{-\frac{1}{4}}$$

Note that the range of length scales extends rapidly as the mean-flow Reynolds number is increased. The nature of the change in the dissipation rate resulting from the joint variation (2.8.33) can be seen from equations (2.6.27):

$$\varepsilon \propto \nu \overline{\left(\frac{\partial u}{\partial x}\right)^2} \propto \nu \left(\frac{u_s}{l_s}\right)^2 \propto \frac{U_0^3}{L_0} \tag{2.8.34}$$

This is, of course, consistent with equation (2.8.29).

The dependence of the length-scale ratios L_0/l_s and L_x/l_s on the turbulence Reynolds number can be found as follows. Using equations (2.6.27, 28), we obtain

$$\varepsilon \propto \nu \left(\frac{u'}{\lambda_x}\right)^2 \propto \frac{U_0^3}{L_0}$$

whence

$$\mathrm{Re}_\lambda^2 = \left(\frac{u' \lambda_x}{\nu}\right)^2 \propto \frac{u'^4 L_0}{\nu U_0^3}$$

Reynolds-number similarity for the larger scales suggests that $u' \propto U_0$ and $L_x \propto L_0$, as in equations (2.8.31). Hence

$$\mathrm{Re}_\lambda^2 \propto \mathrm{Re}_0 \tag{2.8.35}$$

and the first of equations (2.8.33) indicates that

$$\frac{L_0}{l_s}, \frac{L_x}{l_s}, \frac{k_s}{k_0}, \frac{k_s}{k_e} \propto \mathrm{Re}_\lambda^{\frac{3}{2}} \tag{2.8.36}$$

Thus the turbulence Reynolds number does, as was anticipated, provide an index of the range of scales within a particular turbulent motion. However, it is clear that the Reynolds numbers Re_0 and Re_L might also be used for this purpose.

FURTHER READING

Batchelor, G. K. *The Theory of Homogeneous Turbulence*, Cambridge U. P. (1953): Chapters 2, 3 and 6

Boothroyd, R. G. *Flowing Gas-solids Suspensions*, Chapman and Hall, London (1971): Section 3.5 and Chapter 4

Bradshaw, P. *Experimental Fluid Mechanics*, Pergamon, Oxford (1970): Chapters 3, 4 and 6

Bradshaw, P. Reference 2: Chapters 2 and 4 to 8
Cockrell, D. J. (Editor). *Fluid Dynamic Measurements in the Industrial and Medical Environments*, Leicester U. P. (1972)
Corrsin, S. *Turbulence: Experimental Methods*, Volume VIII, 2 of *Hand buchder Physik*, Springer, Berlin (1963)
Goldstein, S. Reference 4: Chapter VI, Sections II and III
Hinze, J. O. Reference 5: Chapters 1 to 3
Kovasznay, L. S. G. *Turbulence Measurements*, Section F of *Physical Measurements in Gas Dynamics and Combustion*, Princeton U. P. (1954)
Lin, C. C. Reference 7: Section C, Chapters 1 to 3
Lumley, J. L. and H. A. Panofsky. *The Structure of Atmospheric Turbulence*, Interscience, New York (1964): Part I
Pankhurst, R. C. and D. W. Holder. *Wind-tunnel Technique*, Pitman, London (1952): Chapter 3
Townsend, A. A. Reference 13: Chapters 1 and 3
Tranter, C. J. *Integral Transforms in Mathematical Physics*, Methuen, London (1956): Chapters I and III

SPECIFIC REFERENCES

Bradbury, L. J. S. and I. P. Castro. 'A pulsed-wire technique for velocity measurement in highly turbulent flows', *J. Fluid Mech.*, **49**, pp. 657–691 (1971) and 'Some comments on heat-transfer laws for fine wires', *J. Fluid Mech.*, **51**, pp. 487–495 (1972)
Brüel and Kjaer A/s, Naerum, Denmark: Catalogs and Technical Review
Corrsin, S. and A. L. Kistler, 'The free-stream boundaries of turbulent flows', *Rep. 1244, U. S. Nat. Adv. Com. Aero.* (1955)
Dansk Industri Syndikat A/s, Herlev, Denmark: Manuals, Catalogs and DISA Bulletin
Grant, H. L. 'The large eddies of turbulent motion', *J. Fluid Mech.*, **4**, pp. 149–190 (1958)
Koch, F. A. and I. S. Gartshore. 'Temperature effects on hot wire anemometer calibrations', *J. Physics*, Section E, **5**, pp. 58–61 (1972)
Raichlen, F. 'Some turbulence measurements in water', *Proc. Amer. Soc. Civ. Eng.*, **93**, *J. Eng. Mech. Div.*, pp. 73–97 (1967)

EXAMPLES

2.1 At a point near the wall of a duct the turbulence components are given by $\frac{1}{3}\overline{u^2}/U^2 = \frac{4}{3}\overline{v^2}/U^2 = \frac{2}{3}\overline{w^2}/U^2 = c_f$, with c_f the friction coefficient at the wall and U the local mean velocity, Taking $K_1 = K_2 = 1$ in equations (2.1.4), estimate the error in the velocity found using a Pitot-static tube for a smooth wall with $c_f = 0{\cdot}003$, and for a rough wall with $c_f = 0{\cdot}01$.

2.2 A Pitot-static tube is found to indicate the velocity $U(1 + A\cos^2\theta)^{\frac{1}{2}}/(1 + A)^{\frac{1}{2}}$, with A a constant, when placed in a steady flow of velocity U with angle θ between tube axis and flow direction.

(a) Use this result to show that $[U_1^2 + \overline{u^2} + (\overline{v^2} + \overline{w^2})/(1 + A)]^{\frac{1}{2}}$ provides an estimate of the reading given by the tube when it is placed in a turbulent stream whose mean velocity U_1 coincides with the tube axis.

(b) What result is obtained if the two do not coincide?

2.3 Experiments on the diffusive action of a turbulent jet involve the release of a contaminant:

(1) inside the turbulent fluid of the jet, and
(2) in the essentially non-turbulent fluid outside.

In each case, it is found that the concentration drops abruptly at the instantaneous boundary of the jet. Why?

2.4 The distances between the transmitting wire of a pulsed-wire anemometer and the two receiving wires are 2·0 mm and 1·75 mm. The instrument is located in a turbulent wake, with its axis parallel to the free stream. The following instantaneous velocities are inferred from flight times recorded at intervals of 10 msec (negative values correspond to pulses arriving at the second receiver):
1·6, 6·4, 4·7, 2·4, 5·5, 7·0, 10·2, 8·1, 8·3, 6·1, 7·9, 2·3, 4·0, 3·8, 0·7, –0·6, 1·5, –2·0, –5·4, –3·0, –3·2, 1·6, –0·6, 2·0, 5·1, 3·1, 3·9, 7·0, 9·0, 8·2, 9·8, 8·6, 3·9, 1·5, 4·1, 5·1, 2·3, 0·2, –1·5, –5·0, –4·7, –2·5, –3·0, –3·1, 0·7, 0·9 m/sec.

(a) What is the general character of the motion? What range of flight times has to be accommodated?

(c) Estimate the mean velocity, turbulence intensity, and relative intensity of the turbulence.

(c) In what way would the results be different had they been obtained from a normal hot wire? (Remember the cross-stream fluctuations.) Assuming that the hot-wire output had been sampled at the same instants in time, estimate the mean velocity, turbulence intensity and relative intensity which might have been inferred from its signals.

2.5 Figure 2.7 shows schematically a pressure-measuring system in which the surface pressure p is transmitted through a small opening to a cavity of volume V where a gauge senses the pressure p_i.

(a) Assuming that the density changes by only a small fraction of its mean value, show that the mass flow into the cavity is related to the internal pressure by

$$m(K + \kappa)\frac{dp_i}{dt} = \dot{m}$$

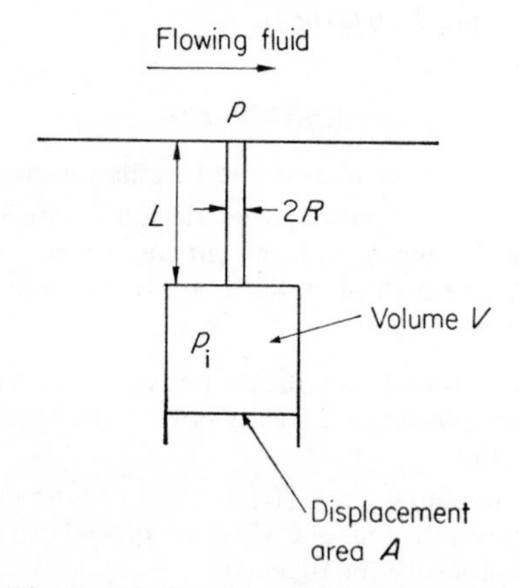

Figure 2.7 See Examples 2.5 and 2.6

with $m = \rho V$, $K = (1/V)\,\mathrm{d}V/\mathrm{d}p_i$ and $\kappa = (1/\rho)\,\mathrm{d}\rho/\mathrm{d}p_i$. What do the terms involving K and κ represent? In what circumstances would one of them be negligible?

(b) Taking the flow resistance to be the result of an expansion loss, so that $\dot{m} = \rho C[2(p - p_i)/\rho]^{\frac{1}{2}}$, show that the time taken for the internal pressure to rise from p_1 to the constant surface value p_2 is $[m(K+\kappa)/C][2(p_2 - p_1)/\rho]^{\frac{1}{2}}$.

(c) Taking the flow resistance to be the result of laminar friction, so that $\dot{m} = r(p - p_i)$, show that the time constant of the system is $m(K+\kappa)/r$.

(d) Consider the particular case of a gas flowing through a circular tube (length L, radius R) to a liquid manometer (bore area A, effective density ρ_m). Assuming the compression to be adiabatic, show that the time constant is $(8\nu L/\pi R^4)(V/a^2 + \rho A/\rho_m g)$, with a the sound speed of the gas.

(e) What additional effect would be found if the inertia of the manometer fluid were taken into account?

2.6 In addition to the time lags discussed in Example 2.5, which arise from flow resistance in leads, pressure-measuring systems experience delays owing to the finite speed of sound in the pressure-transmitting fluid.

(a) Neglecting the effects considered in Example 2.5, derive a formula giving a limiting frequency which can be accurately sensed through a tube of length L, which contains a fluid of sound speed a.

(b) Evaluate for air and for water, taking the tube length to be 2 cm. Is this source of delay likely to provide a serious limitation? Under what circumstances will the elasticity of the tube walls be significant?

2.7 (a) Express the eddy diffusivities of equations (1.3.2) and (1.6.3) in terms of the mean products (2.3.1, 2).

(b) By analogy, write down a parallel relationship between the effective mass flux and the eddy mass diffusivity associated with turbulent mixing. What is the relationship analogous to equations (2.3.1, 2)?

2.8 For air, the kinetic theory of gases gives the mean-free-path of the molecules as $\lambda = (2 \times 10^{-5})\,T/p$ metres, with T in °K and p in N/m^2. Determine whether pressure changes in the range $p = 0{\cdot}5$ to $2{\cdot}0$ atm will produce significant changes in the Knudsen number, for hot wires with diameters 2, 5 and 70 μm.

2.9 (a) For what range of velocities is the first of equations (2.4.2) valid for a 5 μm hot wire operating in atmospheric air? What effect do pressure changes have on these limits?

(b) Estimate Re and Nu for a 5 μm wire operating at 250 °C in atmospheric air at 20 °C.

(c) Why does the heat-transfer law change at Re = 44? At high velocities, vortex shedding from a hot wire introduces spurious signals. At what velocity might this be expected to begin for a 5 μm wire operating in atmospheric air, and for a 70 μm fibre probe operating in water?

2.10 The calibration of a normal hot wire operating at constant temperature in a uniform stream of air gives the following results.

Table 2.2

U(ft/sec)	9·5	18·5	30·5	46	63·5	72	82	94·5
V(volts)	6·95	7·5	8·0	8·45	8·9	9·05	9·3	9·4

(a) Derive an expression of the form $V^2 = A + BU^n$ for this wire.

(b) What is the sensitivity of the wire when operated under the test conditions with the mean velocity $U = 70$ ft/sec?

(c) During a series of tests, the fluid temperature is $T_f = 30$ °C, rather than the value 20 °C at which the calibration was carried out. However, the wire is operated at the same overheating ratio T_w/T_f and resistance ratio R_w/R_f as obtained during calibration. Find the modified calibration curve and the sensitivity at $U = 70$ ft/sec. Why is it convenient to maintain the overheating ratio constant?

2.11 The resistance of tungsten is given by

$$R = R_0[1 + a_1(T - T_0) + a_2(T - T_0)^2]$$

with $R_0 = R(0\text{ °C})$, $a_1 = 4 \times 10^{-3}$, $a_2 = 8 \times 10^{-7}$, and the temperature difference measured in C deg. Is the non-linearity of this characteristic likely to be significant for a hot wire made of this material?

2.12 (a) Show that the thermal inertia of a hot wire may be taken into account by adding a term $C\,dR_w/dt$ to the right-hand side of equation (2.4.8), with $C = c_w m_w/\alpha$, a measure of the thermal capacity of the wire.

(b) Show that the time constant of the wire is $M = R_w C/R_f(A + BU^n)$ where R_w and U have their mean values.

(c) By what proportion is the response reduced when the frequency is $\omega = 1/M$? What is the phase shift of the signal in this case?

2.13 Consider two similar slant wires which give the signals

$$s_1 = a(u + v) = \sqrt{2}\,au_1$$

$$s_2 = a(u - v) = \sqrt{2}\,au_2$$

where u and v are streamwise and transverse components in the plane of the wires, and u_1 and u_2 are their resultant components at 45° to the u-axis.

(a) For isotropic turbulence, show that: $\overline{s_1 s_2} = 0$, $\overline{uv} = 0$ and $\overline{u^2} = \overline{s_1^2}/2a^2$.

(b) Show that, in general, the correlation between u and v is given by $\frac{1}{2}(\overline{u_1^2} - \overline{u_2^2})$.

2.14 The quantity $\overline{u^2 v}$ is to be determined using the signals from two similar slant wires, which give $s_{1,2} = au \pm bv$. What statistical properties of s_1 and s_2 must be measured, and how must they be combined?

2.15 The results obtained from an X-wire probe are very sensitive to small differences in the two wires. To see this, consider a pair of wires that are similar in every respect, save in their inclination to the mean flow. The outputs are $s_1 = au + bv$ and $s_2 = au - rbv$, with $r = -\tan\alpha_2/\tan\alpha_1$.

(a) Calculate $\overline{u^2}$, $\overline{v^2}$ and $\overline{uv}$ in terms of $\overline{s_1^2}$, $\overline{s_2^2}$, $\overline{s_1 s_2}$ and r.

(b) Consider a set of measurements obtained near a wall: $\overline{s_1^2} = 1$, $\overline{s_2^2} = 2$ and $\overline{s_1 s_2} = 0{\cdot}5$. Find $\overline{u^2}$, $\overline{v^2}$ and $\overline{uv}$ for

(1) $\alpha_1 = -\alpha_2 = 45°$, and
(2) $\alpha_1 = 45°$, $\alpha_2 = -40°$.

(c) What percentage errors will be produced in $\overline{u^2}$, $\overline{v^2}$, $\overline{uv}$ and R if the values (1) are used, when in fact the angles are those specified in (2)?

2.16 Draw block diagrams showing the equipment necessary to measure

(a) the mean-square derivative $\overline{(\partial v/\partial x)^2}$,

(b) the velocity-temperature point correlation $\overline{v\theta}$,
(c) the double-velocity correlation coefficient $R_{12}(y)$,
(d) the autocorrelation $\overline{p(t)p(t+\tau)}$,
(e) the microscale λ_x,
(f) the frequency spectrum $E_{12}(\omega)$, and
(g) the wave-number spectrum of $\partial u/\partial t$.

2.17 Distinguish between the precise meanings of the words 'random' and 'uncorrelated'. What is the significance of the former term in everyday speech?

2.18 It is sometimes possible to remove contaminating noise from a signal by making use of the lack of correlation between the two.

(a) If each of two signals $S_1 + s_1$ and $S_2 + s_2$ contains an element of noise (s_1, s_2) which is uncorrelated with both S_1 and S_2, show that $\overline{(S_1 + s_1)^2} = \overline{S_1^2} + \overline{s_1^2}$ and $\overline{(S_1 + s_1)(S_2 + s_2)} = \overline{S_1 S_2} + \overline{s_1 s_2}$.

(b) A normal hot wire placed in a certain duct senses not only the local turbulence, but also fluid motions associated with intense plane sound waves travelling along the duct. Explain how $\overline{u^2}$ can be found for the local turbulence, by considering the signals from two similar hot wires placed side-by-side in the duct, but some distance apart.

(c) Explain how the magnitude of the signals associated with electronic noise or probe vibration can be evaluated by considering the signals from two similar hot wires placed close together in the flow.

2.19 Find the correlation coefficient relating the fluctuating part of the signals defined in Table 1.3.

2.20 Using the data of Table 1.2, and interpreting τ/ρ as $-\overline{uv}$:

(a) Find the variation across the wake in the correlation coefficient relating u and v.
(b) Determine its value within the fully turbulent region.
(c) Explain why the correlation is negative.

2.21 (a) Find the autocorrelations of $\sin(at)$ and e^{-at}.

(b) An autocorrelation curve has the form $\exp(-\tau^2/T^2)$. Show that the corresponding integral time scale is $(\sqrt{\pi}/2)T$.

2.22 Will a second-order correlation necessarily be a second-order tensor? What is the relationship between these two kinds of ordering?

2.23 The following table gives lateral correlation coefficients of the u-fluctuations generated by a grid of mesh size 3 in by 3 in, for a convection speed of $U = 15$ ft/sec.

Table 2.3.

y (in)	0·035	0·050	0·070	0·1085	0·197		0·284
$R_{11}(y)$	0·981	0·962	0·928	0·851	0·716		0·565
y (in)	0·512	1·00	2·00	3·00	4·00	6·00	8·00
$R_{11}(y)$	0·370	0·180	0·036	−0·022	−0·026	−0·015	0·00

(a) Find L_x and λ_x from the correlation curve. What is the decay time scale τ_d? How far does the turbulence travel while its intensity decreases by one-half? What relevance has this for Taylor's hypothesis?

(b) Plot:
(1) $R = (1 - y/L_x)\exp(-y/L_x)$, and
(2) $R = (1 - y^2/\lambda_x^2)\exp(-y^2/\lambda_x^2)$

together with the experimental results. Does this suggest any other way of finding L_x and λ_x?

2.24 Use continuity arguments to explain why, in Figure 2.4(c), R_{ac} is negative at intermediate values of r, while R_{ab} is positive throughout. These correlations are defined in Figure 2.4(d).

2.25 Consider the highly restricted case of isotropic turbulence.

(a) What values do the nine double-velocity correlation coefficients $R_{ij}(0)$ take on at each point?

(b) Show that

$$Q_{11}(r_2) = Q_{22}(r_1) = Q_{33}(r_1) = Q_{11}(r_3)$$

(c) Show that the longitudinal and transverse correlations are the only distinct functions required to define the twenty-seven correlation functions $Q_{ij}(r_k)$.

2.26 Figure 2.8 shows variations of modified spatial correlations defined by $R_{ij} = \overline{u_i(x_p)\,u_j(x_p+r)}/\overline{u_i(x_p)\,u_j(x_p)}$. They are based on measurements made by Grant in a boundary layer; the solid curves relate to $y/\delta \simeq 0{\cdot}035$, and the dashed curves to $y/\delta \simeq 0{\cdot}45$.

What can you deduce about boundary-layer turbulence from these results?

2.27 The result (2.6.22) applies only when $R(r)$ is symmetrical near $r = 0$. Why? Is it likely that this condition will be satisfied?

2.28 (a) Use Taylor's hypothesis to show that the integral and microscales are related by

$$L_x = U\tau_{\mathrm{E}} \quad \text{and} \quad \lambda_x = \frac{U\tau_{\mathrm{E}}}{\sqrt{2}}$$

(b) Show that the time scale of decay and the micro time scale are related by $20\nu\tau_{\mathrm{d}} = U^2\tau_{\mathrm{E}}^2$.

(c) Find τ_{E} and τ_{d} for the data of Table 2.3.

2.29 (a) Explain how the effective speed of convection of some element of turbulence can be determined from time correlations between the signal from one point and signals from several downstream points.

(b) Suppose that this procedure is applied to the signals from pressure transducers mounted on the wall beneath a turbulent boundary layer, and that the signals are first passed through identical band-pass filters. What changes in the convection velocity and correlations might be expected as the frequency band is changed?

2.30 (a) Show that the intensity of decaying grid turbulence is given by

$$\frac{\mathrm{d}(U/u')^2}{\mathrm{d}(Ux/\nu)} = 10\,\mathrm{Re}_\lambda^{-2}$$

(b) Find the decay law for $\mathrm{Re}_\lambda = $ constant; this result appears to apply during an initial period of decay defined approximately by $x/M < 200$, with M the mesh size.

2.31 Show that the normalized density of a frequency spectrum is related to the integral time scale of the signal by $\phi(0) = (2/\pi)\,T_{\mathrm{E}}$.

2.32 Generalize equations (2.7.8) to give the relationships between $Q_{ij}(\tau)$ and $E_{ij}(\omega)$.

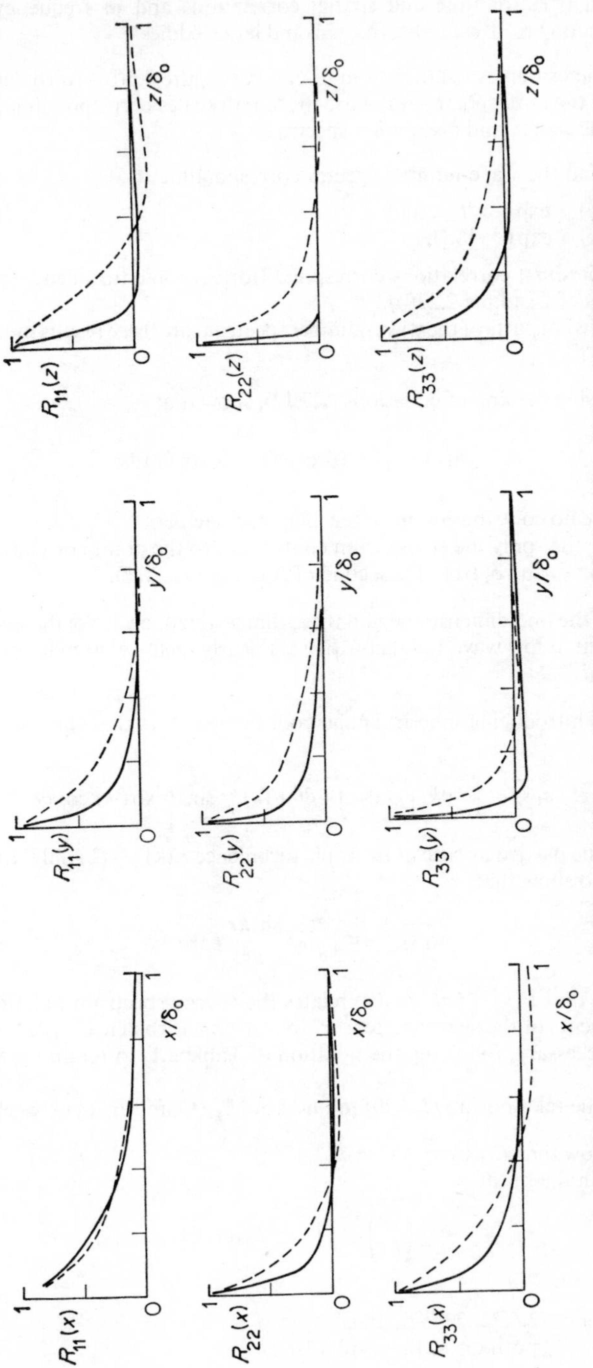

Figure 2.8 See Example 2.26. The scale δ_0 is about $0 \cdot 7\delta$, where δ is the total thickness of the boundary layer

2.33 Which parts of time and spatial correlations and of frequency and wave-number spectra are attributable to small and large 'eddies'?

2.34 For the examples mentioned in discussing Figure 2.5(b)—turbulence in a pipe flow and in the atmosphere—what are the length scales corresponding to maximum values of the energy and dissipation spectra?

2.35 (a) Find the wave-number spectra corresponding to

(1) $R_{11}(x) = \exp(-x/L_x)$, and
(2) $R_{11}(x) = \exp(-x^2/\lambda_x^2)$.

These longitudinal correlations correspond (for isotropic turbulence) to the lateral correlations of Example 2.23(b).

(b) Over what parts of the wave-number spectrum are these results most likely to be meaningful?

2.36 (a) Using the first of equations (2.7.12), show that

$$R(r) = \int_{-\infty}^{\infty} (\phi_r \cos kr - \phi_i \sin kr)\,dk$$

Why is there no contribution from the imaginary element?

(b) Show that only the cospectrum contributes to the point correlation $R(0)$.
(c) Obtain ϕ_r and ϕ_i from the second of equations (2.7.12).

2.37 While the one-dimensional and three-dimensional spectra of the same signal are very different at low wave-numbers, they are nearly identical at high wave-numbers. Why is this?

2.38 (a) By introducing spherical polar coordinates, show that equation (2.7.14) can be written

$$\overline{s_1 s_2} = \int_0^{\infty} dk \int_0^{2\pi} d\phi \int_0^{\pi} d\theta\, F(\mathbf{k})\, k^2 \sin\theta \exp(ikr\cos\theta)$$

(b) For the particular case of isotropic turbulence $F(\mathbf{k}) = f(k)$ only. Integrate over the sphere to show that

$$\overline{s_1 s_2} = 4\pi \int_0^{\infty} k^2 \frac{\sin kr}{kr} F(\mathbf{k})\,dk$$

(c) Show that $E(k) = 6\pi k^2 E_{11}(\mathbf{k})$ relates the energy spectrum function to one of the components of the spectrum tensor. In isotropic turbulence $E_{11}(\mathbf{k}) = f(k)$ alone. Why is it necessary, following the notation of Table 2.1, to retain the vector argument?

(d) Use the relationship (2.7.20) to find how $E_{11}(\mathbf{k})$ and $E_{11}(k)$ are related.

2.39 (a) Show that $E_{11}(0)/\overline{u^2} = (2/\pi)L_x$

(b) Obtain the result

$$\overline{\left(\frac{\partial u}{\partial x}\right)^2} = \int_0^{\infty} k^2 E_{11}(k)\,dk$$

by applying

(1) equations (2.6.22, 23, 57), and
(2) Taylor's hypothesis to the result (2.7.7).

2.40 A frequency analysis of the u-fluctuations of grid turbulence gives the following results; the convection velocity is $U = 15$ m/sec.

Table 2.4.

ω	1	1·5	3	5	10	15	30	50
$1000\phi_{11}$	6·5	8·8	6·6	7·9	6·8	9·2	8·1	5·0
ω	100	150	300	500	1000	1500		
$1000\phi_{11}$	2·8	1·71	0·64	0·21	0·042	0·0135		

with ω in Hz and $\phi_{11}(\omega) = E_{11}(\omega)/\overline{u^2}$ in sec.

(a) Plot as a one-dimensional wave-number spectrum using log–log scales.

(b) Compare with the result $\phi(k) = (2/\pi)L_x/(1 + k^2L_x^2)$ to find the integral scale L_x, and plot this formula with the experimental results. This wave-number spectrum corresponds to the correlation (1) of Example 2.35(a).

(c) Find the autocorrelation and streamwise correlation functions.

(d) Find the normalized energy spectrum function $E(k)/\overline{u^2}$ on the assumption that the turbulence is isotropic.

(e) Use the result of Example 2.38(c) to find the component $E_{11}(\mathbf{k})$.

2.41 (a) Why is Taylor's hypothesis less accurate for high-intensity, inhomogeneous turbulence?

(b) An autocorrelation function and a frequency spectrum are converted into a spatial correlation and a wave-number spectrum by applying Taylor's hypothesis. Which parts of the resulting curves are most likely to be in error?

(c) How can Taylor's hypothesis be checked experimentally?

2.42 For decaying grid turbulence, it may be assumed that the energy spectrum function depends on the parameters defining the flow as follows:

$$E(k, t) \sim U, M, k, t, \nu$$

with M the mesh size.

(a) Show that $E/U^2M = f(kM, UM/\nu, Ut/M)$

(b) Develop similar expressions for the turbulence kinetic energy $\frac{1}{2}\overline{q^2}$ and integral scale L.

(c) What additional restriction is implied by the adoption of the self-preserving form $E/\overline{q^2}L = f(kL, UM/\nu)$?

2.43 (a) Why does the spectrum of $\partial u/\partial t$ not display Reynolds-number similarity?

(b) What is the value of the Reynolds number $u_s l_s/\nu$ characterizing the smallest scales of the turbulence? What does this imply?

2.44 (a) For isotropic turbulence, show that $E(k) \propto k^n$ implies that $E_{11}(k) \propto k^n$ as well.

(b) What is the form of the three-dimensional spectrum of dissipation in a region where $E_{11}(k) \propto k^n$?

(c) If $E_{11}(k)$ has the form (2.8.23) in the inertial subrange, what value would you expect the constant to have for high-Reynolds-number turbulence? What is the form of the associated frequency spectrum? Show that these results can be used to find the dissipation rate, from measurements of a single turbulence component.

2.45 Pressure fluctuations are measured at the wall of a water channel of square section. The measurements are made beneath the region of recirculating flow behind a ridge of height $h = 0·1$ m which spans one wall of the channel. Tests are carried out at two mean-flow velocities, $U_1 = 5$ m/sec and $U_2 = 10$ m/sec. The power spectrum densities of the two sets of fluctuations are given in the following table.

Table 2.5.

ω	2·5	5	10	25	50	100	200	400
E_1	18,000	22,000	6300	1200	63	6·3	0·84	0·078
E_2	220,000	325,000	225,000	24,000	3250	450	27·5	2·0

with ω in Hz and $E_p(\omega)$ in $(\mathrm{N/m^2})^2$ sec.

(a) Use inertial scaling to reduce these results to dimensionless form. Plot the results on log–log paper and comment on the cohesion of the data.

(b) What is the Strouhal number of this non-steady flow? Does this have the order-of-magnitude which you would expect?

(c) By considering the associated wave-number spectrum, show that $E_p(\omega) \propto \omega^{-\frac{7}{3}}$ in the inertial subrange. What power-law variations are found in the experimental data?

2.46 Show that $L_0/\lambda_x \propto \mathrm{Re}_0^{\frac{1}{2}} \propto \mathrm{Re}_\lambda$. How do the separations between the scales L_0, L_x, λ_x and l_s change as the Reynolds number is increased?

2.47 The mean-free-path of a gas is given by the kinetic theory as about $\lambda = 1{\cdot}5\nu/a$, with a the sound speed.

(a) Show that $\lambda/l_s = 1{\cdot}5\, u_s/a$ follows from this result.

(b) Show also that $\lambda/l_s \propto M_0 \mathrm{Re}_0^{-\frac{1}{4}}$, with $M_0 = U_0/a$ the mean-flow Mach number. Under what conditions could the mean-free-path be comparable with the smallest scales of the turbulence?

2.48 (a) Note that equation (2.8.29) can be written as

$$\varepsilon \propto \frac{U_0^2}{L_0/U_0} = U_0 L_0 \left(\frac{U_0}{L_0}\right)^2$$

Devise physical interpretations for these results, given that $u' \propto U_0$ and $L_x \propto L_0$.

(b) Introduce the energy scale (2.8.29) into equations (2.8.21, 22) and write them in meaningful form.

2.49 (a) Subject to the conditions (2.8.30, 31) show that the various Reynolds numbers characterizing turbulence are related by

$$\mathrm{Re}_0 = 64\,\mathrm{Re}_L = 6{\cdot}8\,\mathrm{Re}_\lambda^2$$

Note that $\mathrm{Re}_s = 1$ in every case.

(b) Show also that the various length scales are related by

$$0{\cdot}26\left(\frac{\lambda_x}{l_s}\right)^2 = 9{\cdot}4\frac{L_x}{\lambda_x} = \mathrm{Re}_\lambda$$

Note that $L_0/L_x = 8$ has been assumed.

(c) Find Re_L, Re_λ, λ_x/L_0 and l_s/L_0 for $\mathrm{Re}_0 = 10^8$ and for $\mathrm{Re}_0 = 10^5$.

(d) Taking $L_0 = 20$ cm and $\mathrm{Re}_0 = 10^5$, estimate the wave-numbers marked in on Figure 2.6.

3

Channel Flows I: Fundamentals

This and the following two chapters deal mostly with flows in channels of uniform section. These are of great importance, since it is often expedient to build fluid-transfer systems with long uniform elements. Thus in a heat exchanger, the required interface between two fluids is provided by passing one fluid through a long tube surrounded by the second. Again, to simplify construction and maintenance, an irrigation canal will be given the same cross-section over many miles, unless some environmental feature dictates a change.

Flows in uniform channels are important in another way: in some respects they are the simplest turbulent motions. Following a period of development, the mean flow and turbulence remain the same at successive sections; the flow is then said to be *fully developed.* We shall find that the conservation laws governing the motion are greatly simplified when the flow is statistically uniform in one direction. Moreover, since fully developed turbulence occupies the entire channel, we need not consider the mechanism of its spreading into non-turbulent fluid. These simplifications are partially annulled by the complexity of the *wall layer*, the region close to the channel wall where the velocity rises abruptly and the nature of the turbulence changes rapidly.

Most of the results in this chapter relate to two kinds of fully developed flow:

(1) *plane parallel flow*, with $U=f(y)$, as in the central part of a very broad, flat channel, and

(2) *axisymmetric parallel flow*, with $U=f(r)$, as in a round pipe or annulus.

However, some consideration will be given to the less restricted motions from which these parallel flows develop:

(3) *plane or two-dimensional flow*, with U, $V=f(x,y)$, as in the entry to a broad channel or in a plane wake, and

(4) *axisymmetric flow without swirl*, with U, $V=f(x,r)$, as in the entry to a round pipe or in a round jet.

These more general results will prove useful in Chapter 6, when we establish the equations governing developing flows.

Chapters 4 and 5 will deal with two classes of practical problems: the prediction of friction, and the calculation of transfers of heat and mass. In the present chapter we provide the foundation for these studies:

(1) using experimental results to build up a picture of the turbulent activity in channel flows, and in wall layers generally,

(2) developing methods of describing the turbulent transport of mass, momentum and energy,

(3) deriving conservation laws for mass, momentum and mechanical and thermal energy, and

(4) introducing some of the simple theoretical models which have been used to represent turbulent activity—the eddy diffusivity, Reynolds mass flux and mixing length.

These simple models will serve us through most of this book, but in Chapter 9 we shall make a more realistic attack on the problem of representing turbulent transport processes.

3.1 CHANNEL TURBULENCE

To provide a realistic background for the analysis of channel flows, we shall examine some experimental results relating to an important motion of this class—pipe flow. But before doing this, we consider the general features to be expected in turbulence near a fixed wall.

The constraints applied by the wall influence the flow in opposing ways. The necessity that the tangential velocity drop to zero, coupled with the scouring action of the turbulence, generates a very high velocity gradient. This provides a large direct dissipation, and leads to further dissipation through the generation of turbulence. However, the requirement that the velocity vanish at the wall has a contrary effect: by suppressing turbulent activity very near the surface, it limits the extraction of energy from the mean flow.

As a consequence of the mixed effects of the wall constraints, the turbulence structure varies profoundly across the channel. The most rapid changes occur very near the wall: the intensity reaches a peak and then falls away, while the integral length scales increase linearly with distance from the wall, although they are ultimately limited by the size of the channel or the thickness of the boundary layer. Such changes in length scales are evident in the boundary-layer results displayed in Figure 2.8.

Figure 3.1 indicates the several regions which can be distinguished in a channel flow. The particular case considered is flow in a round pipe, where the major changes are confined to the outer fifteen per cent of the pipe radius. An exaggerated scale has been used for the regions very near the wall. In reality, Regions I and II occupy only a very small fraction of the radius: for $\mathrm{Re}_d = 10^5$, for example, $y_1/R \simeq 0{\cdot}001$ and $y_2/R \simeq 0{\cdot}01$. Despite their small extent, these inner layers influence the entire flow, since a significant fraction of the velocity variation occurs within them. Figure 3.1 is idealized in representing the wall as a featureless plane; realistic surfaces have protrusions whose height is comparable with the thickness of the inner layers.

Let us consider in turn the regions indicated in Figure 3.1:

I. The viscous sublayer. Here the mixing stress, $\rho\overline{uv}$ of equation (2.3.1), is negligible, since $v \simeq 0$. The variation in the mean velocity is determined by the molecular viscosity, and is nearly linear, as in laminar flow. This region is also

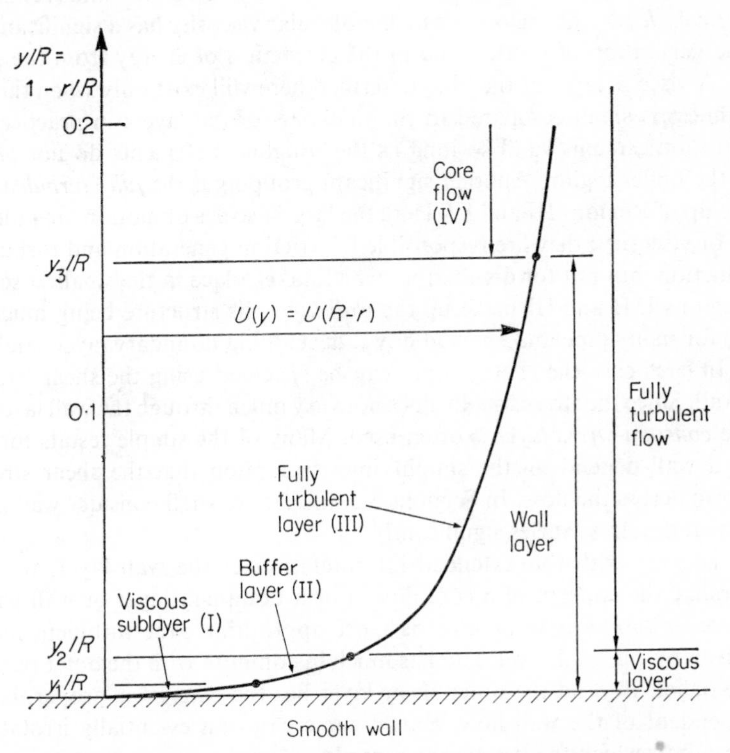

Figure 3.1 Mean-velocity variation near the smooth wall of a pipe of radius R, illustrating the wall layer and its dissection. The scale of the regions near the wall has been exaggerated

called the *linear sublayer* or the *laminar sublayer*. The latter term is incorrect, since the tangential components, u and w, maintain significant magnitudes in the sublayer; indeed, the relative intensity u'/U takes on its maximum value there.

II. The buffer layer. In this region the viscous and mixing stresses are of comparable magnitude. It is sometimes called the transition region, an appropriate description of its function, but somewhat misleading in suggesting a true change from laminar to turbulent flow.

III. The fully turbulent layer. Here the flow is still dominated by the wall, but the turbulence develops sufficiently to render the viscous stress negligible. We shall see later that the mean velocity varies nearly logarithmically in this region; hence it is often called the *logarithmic layer.*

IV. The turbulent core. In fully developed flow this is wholly turbulent but, unlike the logarithmic layer, it is influenced by the constraints of the entire periphery of the pipe.

These four regions can be grouped in several ways. Regions I and II comprise the *viscous layer*, the region in which molecular viscosity has a significant role in the generation of friction and in the extraction of energy from the mean flow. A viscous layer of the kind described here will exist only when the wall roughness is small compared to the thickness of the layer; in practice, our description remains valid so long as the roughness elements do not project into the buffer region. Another significant grouping is the *fully turbulent flow* made up of Regions III and IV. Here the largest scales of motion are independent of viscosity; they are responsible for friction generation and turbulence production, but not for dissipation, which takes place in the smallest scales.

Regions I, II and III make up the *wall layer*, its structure being much the same for many pipe and channel flows, developing boundary layers and wall jets. In each case the energy input can be specified using the shear stress at the wall. Since the stress usually does not vary much through the wall layer, the name *constant-stress layer* is often used. Many of the simple results for flow near a wall depend on the simplifying assumption that the shear stress is uniform across the flow. In Sections 3·5 and 5·5 we shall consider wall layers in which the stress varies significantly.

In keeping with this extended interpretation of the wall layer, we must generalize the concept of a core flow. For a boundary layer or wall jet, the term *outer turbulent shear layer* is more appropriate; the turbulent motion is driven by the wall layer, and has much in common with the outer part of a wake or jet. Beyond the outer shear layer lies the *outer flow* proper, largely independent of the wall flow, and in some cases an essentially irrotational motion into which the turbulence spreads.

The structure of the core or outer flow must be expected to differ from one flow species to the next, since the determining factors are highly varied, as is made clear in Table 1.1. However, three often encountered flows—those in a pipe, plane channel and constant-pressure boundary layer—have one vital feature in common: the velocity variation is small in the core or outer flow. Hence fairly accurate predictions of friction and transfer rates can be obtained using only a crude model of the flow beyond the wall layer.

Having deduced some of the features to be expected, let us turn to the experimental results for pipe flow, based on measurements made by Laufer. Figure

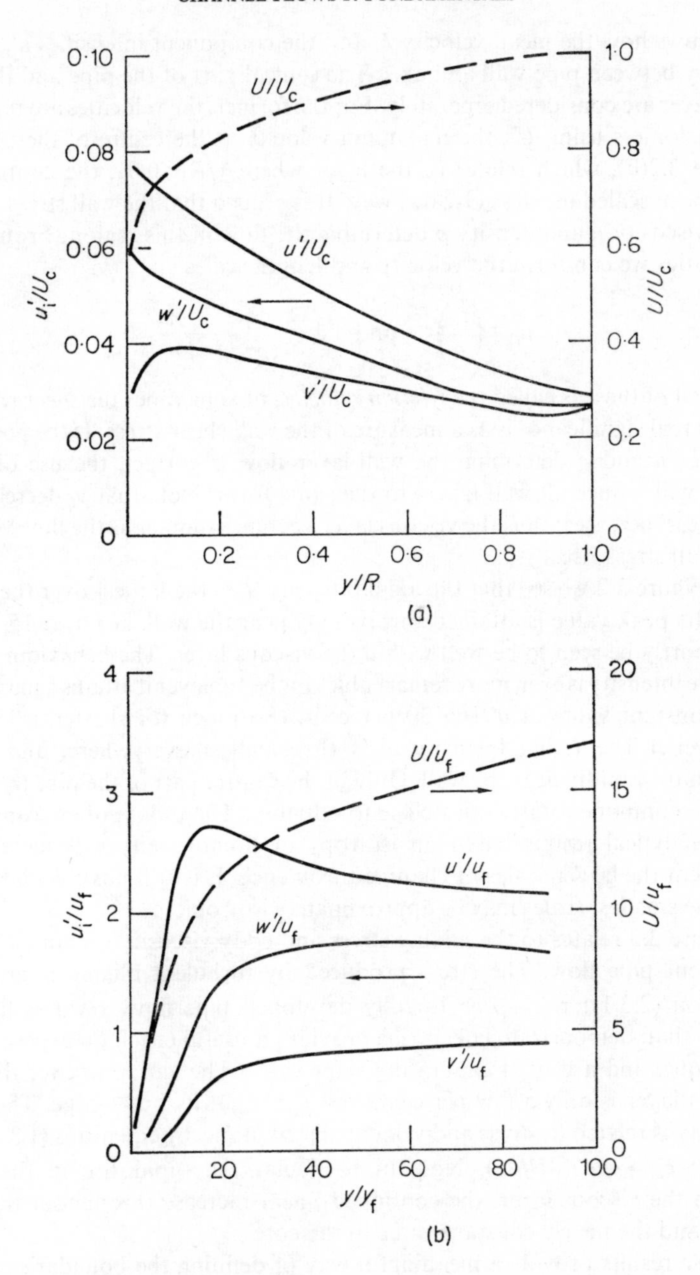

Figure 3.2 Variations of component intensities and mean velocity in pipe flow, based on the measurements of Laufer for $Re_d = 5 \times 10^5$. (a) Fully turbulent region. (b) Viscous and wall layers

3.2 shows how the mean velocity U and the component intensities u', v' and w' vary between pipe wall and axis. The central part of the pipe and the thin wall layer are considered separately. For the former, the velocities are rendered dimensionless using U_c, the maximum velocity at the centre of the pipe. In Figure 3.2(b), which relates to the layer where $y/R < 0{\cdot}01$, the coordinates have been scaled in a less obvious way. It is argued that the wall stress τ_w and fluid viscosity μ and density ρ determine the flow in this region. From these quantities we can form the velocity and length scales

$$u_f = \left(\frac{\tau_w}{\rho}\right)^{\frac{1}{2}} \quad \text{and} \quad y_f = \frac{\mu}{(\rho\tau_w)^{\frac{1}{2}}} = \frac{\nu}{u_f} \tag{3.1.1}$$

The first of these is called the *friction velocity*, or sometimes the *shear velocity*, but its real significance is as a measure of the wall shear stress. If the postulate that τ_w, μ and ρ determine the wall-layer flow is correct, the use of these scales will reduce all wall layers to the same form. Note that y_f decreases as the stress increases; thus the viscous layer becomes thinner as the flow velocity and wall stress rise.

In Figure 3.2 we see that the axial intensity u' is the largest over the entire flow. Its peak value is attained surprisingly near the wall, at $y/y_f \simeq 15$, which will shortly be seen to be well within the viscous layer. The behaviour of the relative intensity is even more remarkable: in the sublayer it attains a maximum and constant value of $u'/U \simeq 30$ per cent. (So much for the term 'laminar sublayer'.) The radial intensity v' is the smallest everywhere, and drops away most rapidly near the wall. Only in the central part of the pipe ($r/R < \frac{1}{3}$) are the component intensities close to equality. These data offer scant hope that analytical results based on isotropy or homogeneity will successfully represent the larger scales of channel turbulence. It is still possible, however, that the smallest scales may be approximately isotropic.

Figure 3.3 relates to the mixing stress and eddy viscosity generated within turbulent pipe flow. The stress produced by turbulent mixing is given by equation (2.3.1): $\tau_t = -\rho\overline{uv}$. In fully developed pipe flow, τ varies linearly (as we shall demonstrate later); this provides a useful check on experimental technique, and a way of calibrating slant wires. The variation over the thin viscous layer is only a few per cent, and $\tau_t \simeq \tau_w$ at its outer edge. The eddy viscosity is related to stress and velocity distributions by equations (1.3.2) and (2.3.1): $\epsilon_m = -\overline{uv}/(dU/dy)$. Note in particular the rapid rise in the ratio ϵ_m/ν in the viscous layer, the continued linear increase throughout the wall layer, and the nearly constant value in the core.

These results provide a meaningful way of defining the boundaries of the sublayer and complete viscous layer. We take the limit of the sublayer to be

$$\frac{y_1}{y_f} = 5 \quad \text{where} \quad \frac{\tau_t}{\tau_w} \simeq 10\% \tag{3.1.2}$$

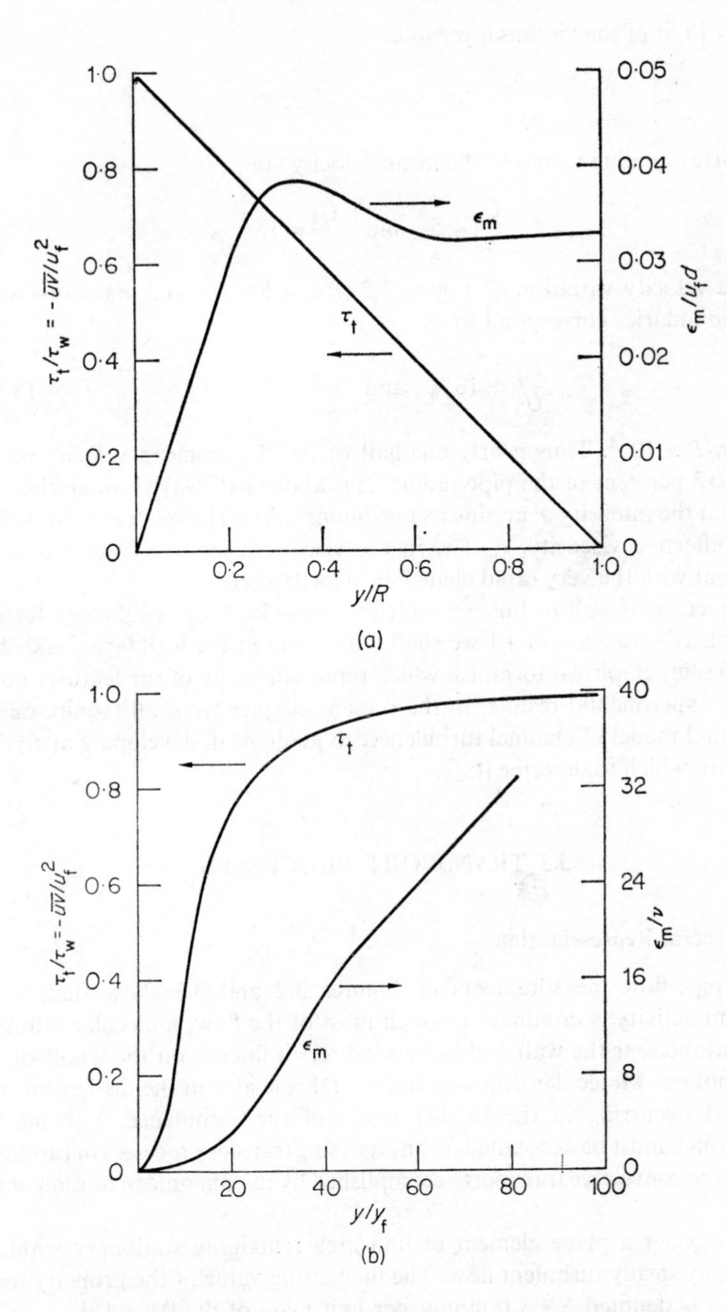

Figure 3.3 Variation across a pipe flow of the turbulent mixing stress $\tau_t = -\rho\overline{uv}$ and of the corresponding eddy viscosity ϵ_m. (a) Fully turbulent region. (b) Viscous and wall layers

and the limit of the viscous layer to be

$$\frac{y_2}{y_f} = 30 \quad \text{where} \quad \frac{\tau_t}{\tau_w} \simeq 90\% \tag{3.1.2}$$

The corresponding values of the mean velocity are

$$\frac{U_1}{u_f} = 5 \quad \text{and} \quad \frac{U_2}{u_f} = 13 \tag{3.1.3}$$

For the velocity variation of Figure 3.2 ($\mathrm{Re}_d = 5 \times 10^5$ and a smooth wall), these boundaries correspond to

$$\frac{U_1}{U_c} \simeq 18\% \quad \text{and} \quad \frac{U_2}{U_c} \simeq 47\% \tag{3.1.4}$$

while $y_f/R \simeq 10^{-4}$. Thus nearly one-half of the rise in mean velocity occurs within 0·3 per cent of the pipe radius. It is about half-way through this thin layer that the intensity u' attains its maximum value. The relatively low values of the effective viscosity, $\epsilon_m + \nu$, in the viscous layer (see Figure 3.3) are consistent with the very rapid change in velocity there.

The picture of wall turbulence which we have built up is sufficient for our present needs. In Section 4.1 we shall look again at the wall layer, and shall develop semi-empirical formulae which represent many of the features noted in these experimental results. In the present chapter we shall require only a conceptual model of channel turbulence, to guide us in developing analytical tools with which to describe it.

3.2 TRANSPORT PROCESSES

3.2.1 General Representation

The pipe-flow measurements of Figures 3.2 and 3.3 show that, while turbulent activity is dominant through most of the flow, molecular diffusion is important near the wall and has an indirect influence on the whole of the mean motion. Molecular diffusion has a vital role also in the dissipation and transport occurring in the smaller scales of the turbulence. Yet another factor which must be accounted for in analysing transfer processes in turbulent flows is the convective transport accomplished by the time-mean motion of the fluid.

We consider a plane element of unit area remaining stationary within a statistically steady turbulent flow. The fluctuating value of the property to be discussed is denoted $S + s$ (amount per unit mass of fluid), and the varying flux of this property through the elemental area is $J + j$ (amount per unit area per unit time). The fluctuating component of velocity in the direction y normal to the plane is denoted $V+v$, as indicated in Figure 3.4. Finally, the fluid density, assumed uniform, is denoted by ρ, and the molecular diffusivity of the property S is denoted by K (with dimensions of area per unit time). In terms of these quantities, the instantaneous flux of S is

$$J+j=-\rho K\frac{\partial(S+s)}{\partial y}+\rho(S+s)(V+v)$$

This is the sum of the fluctuating convection and simple gradient diffusion, like that of Fourier's and Fick's laws (1.6.2, 4).

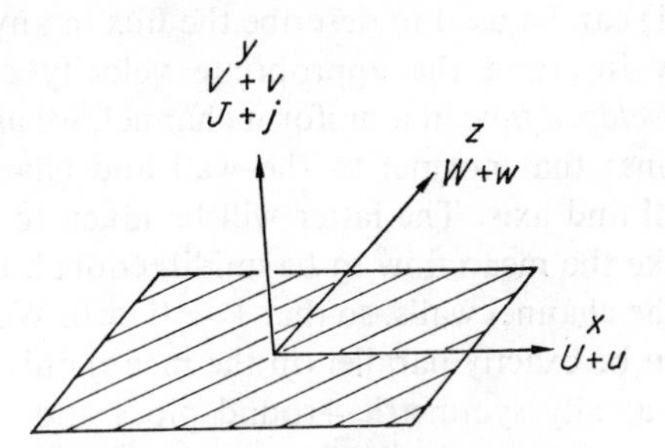

Figure 3.4 Plane element of area considered in analysing transport mechanisms, showing coordinate axes, fluctuating velocity components and flux $J+j$ in the positive y-direction

The time-mean value of the flux of S is then found to be

$$J=-\rho K\frac{\partial S}{\partial y}+\rho SV+\rho\,\overline{sv} \tag{3.2.1}$$

It has been assumed that K is locally constant, that is, does not fluctuate significantly. This result shows that the net flux is the result of:

(1) molecular diffusion down the mean concentration gradient,
(2) bulk convection associated with the mean concentration and the mean velocity normal to the unit plane, and
(3) the average transfer effected by turbulent mixing, which will depend on the degree of correlation between s and v, and on their intensities.

This last element may be formally converted to a diffusion term by introducing an *eddy diffusivity*:

$$\overline{sv}=-\epsilon_s\frac{\partial S}{\partial y} \tag{3.2.2}$$

This has already been done in equations (1.3.2) and (1.6.3, 5), which introduce the eddy viscosity and eddy diffusivities of heat and mass. Alternatively, the effect of turbulent mixing can be represented by an *effective lateral convection velocity* V_s or by an *effective mass flux* or *Reynolds flux* G_s:

$$\overline{sv} = \rho V_s S = G_s \Delta S \tag{3.2.2}$$

with ΔS the change in S across some finite interval Δy. Whether or not the transformations (3.2.2) prove useful depends on whether the hypothetical quantities ϵ_s, V_s and G_s can be specified realistically. We shall look further into this matter in Section 3.4.

The result (3.2.1) can be used to describe the flux in any direction within a turbulent flow by inserting the appropriate velocity components. When discussing *fully developed flow* in a uniform channel, we are interested in two particular directions: that normal to the wall and channel axis, and that parallel to the wall and axis. The latter will be taken to be the x-direction. In addition, we take the mean flow to be unidirectional, that is, to be everywhere parallel to the channel walls, so that $V = W \equiv 0$. We shall see later that turbulent flows can be exactly parallel (in the mean) only in channels whose cross-sections are axially symmetric—round pipes and annuli. However, a reasonable approximation to parallel flow can be obtained in the central part of a very broad flat channel, which may be thought of as part of an annulus with nearly equal inner and outer radii.

To reduce the axial and normal flux components to their simplest possible forms, we require also that the variation of the mean value S is

(1) the same across each section of the channel, and
(2) at worst, linear along the channel.

Thus the mean value of S may vary uniformly along the channel (as does the pressure with a uniform wall friction, or the temperature with a uniform heat input), or remain constant (as does the time-mean kinetic energy). Such a fully developed, parallel flow with mean properties varying linearly along the channel (if at all), we shall term a *linearly varying flow*. It is specified by the conditions:

$$V = W \equiv 0, \qquad U = f(y) \quad \text{or} \quad f(r) \tag{3.2.3}$$

for the mean flow;

$$\overline{u^2}, \overline{uv}, \text{etc.} = f(y) \quad \text{or} \quad f(r) \tag{3.2.3}$$

for functions of the turbulent fluctuations; and

$$S = S_i(y) + \frac{\partial S}{\partial x}(x - x_i) \tag{3.2.4}$$

with

$$\frac{\partial S}{\partial x} = \text{constant throughout the flow} \tag{3.2.4}$$

for properties such as the mean pressure, temperature and concentration. The subscript i denotes some virtual origin.

In a linearly varying flow, the axial or streamwise component of the flux has the form

$$J_x = -\rho K \frac{\partial S}{\partial x} + \rho S U + \rho \overline{su} = \rho S U + f(y) \tag{3.2.5}$$

on the basis of the following arguments:

(1) The variation in K may be expected to be small and, in any case, the streamwise molecular transfer is normally negligible in comparison with the other terms.

(2) The fluctuations s are statistically independent of the mean level of S, being determined by the velocity fluctuations and by the gradients $\partial S/\partial x$ and $\partial S/\partial y$, both independent of x; thus $\overline{su} = f(y)$.

These arguments indicate that the lateral or cross-stream component of the flux has the form

$$J_y = -\rho K \frac{\partial S}{\partial y} + \rho \overline{sv} = f(y)$$

or (3.2.6)

$$J_r = -\rho K \frac{\partial S}{\partial r} + \rho \overline{sv} = f(r)$$

Our interest is primarily in this component, since it represents the transfer between the wall and the body of the fluid. The several ways in which this component will be interpreted in the following sections are outlined in Table 3.1.

3.2.2 Mass Transfer

We convert the general flux component (3.2.1) by introducing the symbols used to describe mass transfer in equations (1.6.4, 5). These changes are indicated in the first line of Table 3.1:

$J \rightarrow N$, the mass flux (mass* of substance transferred per unit area per unit time),

$\rho S \rightarrow C$, the mass concentration (mass* of the substance per unit volume), and

$K \rightarrow D$, the mass diffusivity of the substance in the particular fluid (area per unit time).

Thus the resultant mass flux is given by

$$N = -D \frac{\partial C}{\partial y} + VC + \overline{cv} \tag{3.2.7}$$

* Often measured in moles of the transferred substance.

Note that the diffusivity D may be strongly dependent on the concentration C, and that the mean motion and turbulence will be affected when C is large enough.

Table 3.1. Interpretations of the general cross-stream flux (3.2.6)

Type of flux (and equation)	Symbol (J_y)	Amount per unit mass (S)	Fluctuation (s)	Diffusivity (K)
Mass (3.2.7)	N	C/ρ	c/ρ	D
Cross-stream velocity → Pressure (3.2.8)	$-P$	—	v	0
Streamwise velocity → Shear stress (3.2.10)	$-\tau$	U	u	ν
Pressure → Work (3.2.13)	$\dot{W}_p$	—	p	0
Turbulence kinetic energy (3.2.14)	J_q	$\frac{1}{2}\overline{q^2}$	$\frac{1}{2}(q^2-\overline{q^2})$	ν
Internal energy (3.2.15)	$\dot{q}_e$	E	$e=c_v\theta$	$k/\rho c_v$
Enthalpy (3.2.16)	$\dot{q}_h$	H	$h=c_p\theta$	$\kappa=k/\rho c_p$

A finite mean-motion contribution CV will arise when the result is applied in one of the following ways:

(1) in any but the normal direction of a parallel flow, as in equation (3.2.5),

(2) in the direction normal to the wall in a developing plane flow, where $U, V=f(x,y)$,

(3) in the direction normal to the wall in a fully developed, but non-parallel channel flow, where $U, V, W=f(y,z)$,

(4) in the direction normal to the wall when the rate of injection of contaminant or basic fluid is significant; a small lateral flow may be important in the viscous layer even though it is insignificant in the outer flow.

The steam condenser provides an interesting example in which a variety of transfer mechanisms interact. Inevitably the steam contains a small proportion of permanent, non-condensing gases, usually air. These are convected towards the condensing surface by the steam (CV), and in steady operation this flow is balanced by an opposed diffusion ($\overline{cv}-D\,\partial C/\partial y$), supplemented by mean-flow convection (CU). Near the surface the permanent gases form a blanket through which the water vapour must be transferred. Since molecular diffusion is very slow, effective operation of the condenser requires that the gases be scavenged from the surface by a combination of turbulent mixing ($\overline{cv}$) and convection parallel to the surface (CU).

3.2.3 Stress Generation

Let us now see what the transfer relationship (3.2.6) tells us about the stresses within turbulent fluid. We consider first the implications for the *pressure* within a parallel flow. For $s \to v$, we have $\rho\overline{sv} \to \rho\overline{v^2}$, giving the net transfer of y-momentum in the y-direction. Considering a cubical element between the wall and the unit plane of Figure 3.4, we obtain the force balance

$$P + \rho\overline{v^2} = P_w \tag{3.2.8}$$

with P the time-mean pressure within the fluid, and P_w the pressure at the wall. It appears that the flux $J_y = -P$ in this application, as indicated in Table 3.1.

For the fully developed parallel flow under discussion

$$P_w = P_{w_i} + \frac{dP_w}{dx}(x - x_i)$$

and (3.2.9)

$$\frac{dP_w}{dx} = \text{constant}$$

These are particular forms of equations (3.2.4). But $\overline{v^2} = f(y)$ in fully developed flow, giving

$$\frac{\partial P}{\partial x} = \frac{dP_w}{dx} = \text{constant} \tag{3.2.9}$$

throughout the channel.

Here we have found that variations of a direct mixing-stress component correspond to variations in the mean pressure within the turbulent fluid. These pressure changes are of small magnitude, but they have important consequences when they are not self-equilibrating as a result of the symmetry of the flow. Unbalanced direct stresses acting across the channel will give rise to *secondary flows* normal to the channel walls. The magnitude of the secondary flow is determined by the requirement that it generate cross-stream shear stresses which balance the variations in direct stresses. Only when the flow has axial symmetry, or is flat and very broad, can the normal mixing stresses be balanced without the mean flow departing from the axial direction. Thus an analysis based on the restricted forms (3.2.5, 6) for the flux components, which assume parallel flow with $V = W \equiv 0$, applies rigorously to only a few geometries:

(1) a round pipe,
(2) an axisymmetric annulus, and
(3) a very wide flat channel, though this must have sides somewhere, unless it is a thin-gap annulus.

We next consider the *shear stress* acting in the x-direction on a unit plane $y =$ constant. We make the substitutions set out in Table 3.1:

$J_y \to -\tau$, with τ the effective shear stress on the upper (y increasing) side of the unit plane,
$S \to U$, the velocity parallel to the plane, and
$K \to \nu$, the kinematic viscosity of the fluid.

The effective stress is then, for a parallel flow

$$\begin{aligned}\tau &= \mu\frac{\partial U}{\partial y} - \rho\overline{uv} \\ &= \tau_v + \tau_t\end{aligned} \tag{3.2.10}$$

the sum of the mean viscous component τ_v and the apparent stress τ_t generated by the net transfer of x-momentum. The latter component was introduced in terms of an eddy viscosity in equation (1.3.2) and in explicit form in equation (2.3.1).

In fully developed channel flow the two contributions, τ_v and τ_t, have the same sign everywhere, and

$$\epsilon_m = \frac{\tau_t}{\rho\, dU/dy} = \nu\frac{\tau_t}{\tau_v} > 0$$

as shown in Figure 3.3. This is true in many other turbulent shear flows, since the fluid coming from a region of higher velocity ($dU > 0$) will more frequently exhibit a positive fluctuation ($u > 0$). Thus for $dU/dy > 0$, $u \gtrless 0$ will typically correspond to $v \lessgtr 0$, so that $\tau_t = -\rho\overline{uv} > 0$. However, this argument need not apply if there is an abrupt reversal in the sign of the velocity gradient dU/dy, coupled with an asymmetric distribution of the turbulence properties on the two sides of the maximum or minimum in mean velocity. Thus in a wall jet, or in an annulus flow, there is a region near the maximum in mean velocity in which $\epsilon_m < 0$.

Through most of a turbulent channel flow

$$\frac{\tau_t}{\tau_v} = \frac{\epsilon_m}{\nu} \gg 1$$

However, Figure 3.3 shows that $\tau_t/\tau_v < 1$ for $y/y_f < 10$, while $\tau_v \gg \tau_t$ within the sublayer, $y/y_f < 5$. Right at the channel wall

$$\tau_w = \tau_v = \mu\left(\frac{dU}{dy}\right)_0 \tag{3.2.11}$$

for both laminar and turbulent outer flows.

The normal stress $\rho\overline{v^2}$ and shear stress $\rho\overline{uv}$ are members of a family of momentum transfers related to the *double-velocity correlation tensor* introduced in equations (2.6.5, 6):

$$\tau_{ij} = -\rho\overline{u_i u_j} = -\rho Q_{ij} \quad \text{with} \quad i, j = 1, 2, 3 \tag{3.2.12}$$

Since $\overline{u_1 u_2} = \overline{u_2 u_1}$, with two similar results, there are only six distinct stress components; this fact is consistent with the balancing of the torque acting on an infinitesimal element of fluid. Three of the stress components are normal or direct stresses, with $i = j$, the stress being normal to the unit plane specified by i and j. The remaining three of the distinct components are shear stresses, with $i \neq j$, the stress being tangential to the plane considered. These stresses are known by a variety of names: *mixing, apparent or Reynolds stresses.* Momentum equations in which these stresses appear are called *Reynolds equations.*

3.2.4 Energy Flows

In order to calculate the several kinds of energy flux normal to the parallel mean streamlines of a turbulent flow, we introduce into the general result (3.2.6) the fluctuations in pressure, kinetic energy and thermal energy, as set out in Table 3.1.

Considering the pressure fluctuation p, we obtain

$$\dot{W}_p = \overline{pv} \tag{3.2.13}$$

giving the *work flux*, the mean rate at which the turbulent pressure fluctuations do work on the fluid above ($y > 0$) the unit plane of Figure 3.4. The molecular diffusion term has no apparent significance, and has been omitted.

We denote the turbulence kinetic energy by $\frac{1}{2}q^2$, as in equations (1.5.7, 8). The substitutions of Table 3.1 give the *kinetic-energy flux*:

$$J_q = -\tfrac{1}{2}\mu\frac{\partial\overline{q^2}}{\partial y} + \tfrac{1}{2}\rho\,\overline{vq^2} \tag{3.2.14}$$

The molecular term is retained, since there is a kinetic-energy transfer associated with the diffusion of momentum.

Finally, we consider the transfers associated with two fluctuating measures of thermal energy—the specific internal energy (denoted $E + e$) and the specific enthalpy (denoted $H+h$). These are associated with the fluctuating temperature $T+\theta$. The molecular and mixing terms must be approached in different ways. For the former, we take $S \rightarrow T$ and $\rho K \rightarrow k$, the thermal conductivity. For the latter, we take $s \rightarrow e$ or h. Thus for the *internal-energy flux*:

$$\dot{q}_e = -k\frac{\partial T}{\partial y} + \rho\,\overline{ve} \qquad (3.2.15)$$

When $E = E(T)$ only, so that $dE = c_v dT$ and $e = c_v\theta$, with c_v the specific heat at constant volume for the fluid, we have the alternative forms

$$\begin{aligned} \dot{q}_e &= -k\frac{\partial T}{\partial y} + \rho c_v \overline{v\theta} \\ &= -\frac{k}{c_v}\frac{\partial E}{\partial y} + \rho\,\overline{ve} \end{aligned} \qquad (3.2.15)$$

The latter corresponds to the substitutions indicated in Table 3.1.

In a parallel manner, we obtain for the *enthalpy flux*:

$$\begin{aligned} \dot{q}_h &= -k\frac{\partial T}{\partial y} + \rho\,\overline{vh} \\ &= -k\frac{\partial T}{\partial y} + \rho c_p \overline{v\theta} \\ &= -\frac{k}{c_p}\frac{\partial H}{\partial y} + \rho\,\overline{vh} \end{aligned} \qquad (3.2.16)$$

The last two are applicable for $H = H(T)$ only, with c_p the specific heat at constant pressure.* These results parallel equations (1.6.3), since $\kappa = k/(\rho c_p)$ is the thermal diffusivity. The heat transfer associated with the temperature-velocity correlation function $\overline{v\theta}$ has already been identified in equation (2.3.2).

The fluxes of internal energy and enthalpy have a common molecular-conduction term, but differ in the terms representing turbulent mixing. However, the fluctuations are related by

$$h = e + \frac{p}{\rho}$$

for constant density, giving

$$\overline{vh} = \overline{ve} + \frac{\overline{pv}}{\rho} \qquad (3.2.17)$$

Comparison of equations (3.2.15, 16) indicates that

$$\dot{q}_h = \dot{q}_e + \overline{pv} = \dot{q}_e + \dot{W}_p \qquad (3.2.17)$$

with $\dot{W}_p$ the work flux of equation (3.2.13).

* For the constant-density fluid considered here, there is no distinction between c_p and c_v. The two symbols are retained so that the results are, in some respects, applicable to fluids of varying density.

3.3 CONSERVATION LAWS

3.3.1 General Results

We shall shortly derive statements expressing the principles of conservation of mass, momentum and energy in forms appropriate to a fluid in turbulent motion, but with parallel mean streamlines. We begin by considering a general conservation law for two-dimensional mean flow. This includes the parallel-flow results as special cases, and reveals the limitations imposed by the restriction to parallel mean flow.

For the fixed cubical element shown in Figure 3.5, we require that the net

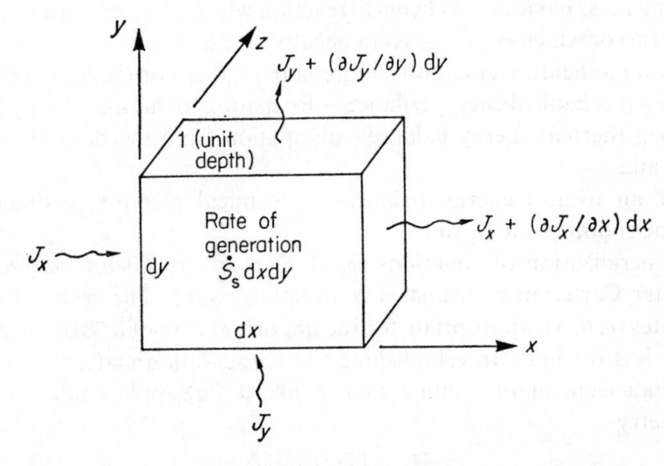

Figure 3.5 Cubical volume element considered in developing a general conservation law. Transfers in two dimensions only are shown: J_x and J_y are components of the flux of a quantity S, and $\dot{S}_s\, dxdy$ is the rate at which S is generated within the element

outflow of some quantity S balance the rate at which it is created within the element. Thus the flux components of S, J_x and J_y, are related by

$$\frac{\partial J_x}{\partial x}\mathrm{d}x\,\mathrm{d}y + \frac{\partial J_y}{\partial y}\mathrm{d}y\,\mathrm{d}x = \dot{S}_s\,\mathrm{d}x\,\mathrm{d}y$$

or

$$\frac{\partial J_x}{\partial x} + \frac{\partial J_y}{\partial y} = \dot{S}_s \tag{3.3.1}$$

where $\dot{S}_s$ is the source strength (or, for $\dot{S}_s < 0$, a sink) of the substance S, measured as mass of S emitted per unit volume per unit time.

To obtain results appropriate to turbulent flows, we introduce into equation (3.3.1) the flux components

$$J_x = -\rho K\frac{\partial S}{\partial x} + \rho SU + \rho\,\overline{su}$$

and (3.3.2)

$$J_y = -\rho K\frac{\partial S}{\partial y} + \rho SV + \rho\,\overline{sv}$$

Some interpretations of the source term of equation (3.3.1) are:

(1) in general—a variation of the S-distribution in time, with $\dot{S}_s = -\partial S/\partial t$; for turbulent flow, the change of S must be slow enough to be separable from the fluctuations s;

(2) for a mass balance—a chemical reaction which gives rise to S (a positive, source term) or which eliminates it (a negative, sink term);

(3) for a momentum equation—a pressure gradient or the fluid weight;

(4) for a mechanical energy balance—dissipation to thermal energy (a sink);

(5) for a thermal energy balance—dissipation from mechanical energy (a source); and

(6) for an overall energy balance—a chemical reaction, exothermic (a source) or endothermic (a sink).

The generalization of equations (3.3.1, 2) to *three-dimensional flow* using rectangular Cartesian coordinates is straightforward. The results for polar coordinates (r, θ, x), appropriate for the important case of *axisymmetric mean flow*, are less obvious. In establishing the conservation statement, we must consider an element of volume shaped like a pineapple chunk. Thus for axisymmetry

$$\frac{\partial J_x}{\partial x} + \frac{1}{r}\frac{\partial (rJ_r)}{\partial r} = \dot{S}_s \qquad (3.3.3)$$

with the radial flux given simply by

$$J_r = -\rho K\frac{\partial S}{\partial r} + \rho SV + \rho\,\overline{sv} \qquad (3.3.3)$$

where $V+v$ is the radial component of velocity. Note that the gradient derivative transforms simply as

$$\frac{\partial}{\partial y} \rightarrow \frac{\partial}{\partial r} \qquad (3.3.4)$$

while the derivative of a flux component transforms as

$$\frac{\partial}{\partial y} \rightarrow \frac{1}{r}\frac{\partial (r\ )}{\partial r} = \frac{\partial^2}{\partial r^2} + \frac{1}{r}\frac{\partial}{\partial r} \qquad (3.3.4)$$

Hence the source of each differentiation must be borne in mind when a Cartesian result is converted to polar coordinates.

To obtain the specialized conservation laws appropriate to parallel, *linearly developing flow*, we introduce the restricted flux components (3.2.5, 6):

$$J_x = \rho S U + f(y)$$
$$J_y = -\rho K \frac{\partial S}{\partial y} + \rho \overline{sv}$$

and similar results for axisymmetric flow. Equations (3.3.1, 3) give

$$U\frac{\partial S}{\partial x} = \frac{\partial}{\partial y}\left[K\frac{\partial S}{\partial y} - \overline{sv}\right] + \frac{\dot{S}_s}{\rho} \quad \text{for plane parallel flow} \tag{3.3.5}$$

$$= \frac{1}{r}\frac{\partial}{\partial r}\left[r\left(K\frac{\partial S}{\partial r} - \overline{sv}\right)\right] + \frac{\dot{S}_s}{\rho} \quad \text{for axisymmetric parallel flow} \tag{3.3.6}$$

Note that the molecular diffusion term $K\partial S/\partial x$ has been omitted from J_x, with the result that a term $\partial(K\,\partial S/\partial x)/\partial x$ is neglected in equations (3.3.5, 6). Even in laminar flow the neglected direct molecular diffusion is normally very small compared with an indirect streamwise diffusion, often termed *accelerated streamwise diffusion*. This consists of lateral diffusion to a region of higher velocity ($\Delta y > 0$, say), followed by downstream convection, and finally lateral diffusion in the opposite direction ($\Delta y < 0$).

The parallel-flow results (3.3.5, 6) are more widely applicable than may at first appear, since near a plane wall the flow may be nearly parallel even though the outer flow is developing in the streamwise direction. Moreover, since the wall layer occupies only a small fraction of the channel width, it is often possible to use the simpler plane-flow result (3.3.5) in analysing wall flows in pipes and annuli.

3.3.2 Mass Transfer and Continuity

There are two aspects of the conservation of mass to be examined: the conservation of the basic fluid, taken here to have uniform density, and the conservation of a substance transferred within this basic fluid, with local mass concentration $C + c$. In the following discussion the concentration C will be taken to be small, so that the properties of the basic fluid and its flow are not significantly altered by the presence of the contaminant.

To obtain the *constant-density continuity equation*, the result expressing the fact that the fluid has uniform density and does not contain expanding or contracting cavities, we take the source term $\dot{S}_s = 0$, and set $J_x = U$ and J_y, $J_r = V$ in equations (3.3.1, 3). Thus

$$\frac{\partial U}{\partial x} + \frac{\partial V}{\partial y} = 0 \quad \text{for plane flow} \tag{3.3.7}$$

and

$$\frac{\partial U}{\partial x} + \frac{1}{r}\frac{\partial (rV)}{\partial r} = 0 \quad \text{for axisymmetric flow} \tag{3.3.8}$$

These continuity results for the basic flow allow us to express the mean-flow contributions to the general conservation statements (3.3.1, 3) in simpler forms. For plane flow

$$\frac{\partial(\rho US)}{\partial x}+\frac{\partial(\rho VS)}{\partial y}=\rho S\left(\frac{\partial U}{\partial x}+\frac{\partial V}{\partial y}\right)+\rho\left(U\frac{\partial S}{\partial x}+V\frac{\partial S}{\partial y}\right)=\rho\frac{\mathrm{D}S}{\mathrm{D}t} \quad (3.3.9)$$

Similarly, for axisymmetric flow

$$\frac{\partial(\rho US)}{\partial x}+\frac{1}{r}\frac{\partial(r\rho VS)}{\partial r}=\rho\left(U\frac{\partial S}{\partial x}+V\frac{\partial S}{\partial r}\right)=\rho\frac{\mathrm{D}S}{\mathrm{D}t} \quad (3.3.9)$$

Here $\mathrm{D}S/\mathrm{D}t$ is the derivative following the mean motion. Such a derivative following a moving element of the fluid is called an *Eulerian, convective or substantive derivative.* Note that the left-hand side of equations (3.3.5, 6) is a restricted form of the convective derivative. The generalization to three-dimensional motion is obvious.

Now consider the conservation law for a substance transferred through the basic fluid. For simplicity, we again take $\dot{S}_s = 0$. Introducing flux components of the form (3.2.7) into equation (3.3.1), and using the simplification obtained in equations (3.3.9), we obtain

$$\frac{\mathrm{D}C}{\mathrm{D}t}=\frac{\partial(D\partial C/\partial x)}{\partial x}+\frac{\partial(D\partial C/\partial y)}{\partial y}-\frac{\partial(\overline{cu})}{\partial x}-\frac{\partial(\overline{cv})}{\partial y} \quad (3.3.10)$$

Here the rate of change of the mean concentration is represented as the sum of molecular-diffusion and turbulent-mixing terms. This result generalizes the diffusion equation of classical physics, whose one-dimensional form is discussed in Example 1.25. The two-dimensional form is

$$\frac{\partial C}{\partial t}=D\left(\frac{\partial^2 C}{\partial x^2}+\frac{\partial^2 C}{\partial y^2}\right) \quad (3.3.11)$$

specifying the development of the concentration distribution in a stationary medium with constant diffusivity D.

The result (3.3.10) can be written more compactly by introducing *eddy diffusivities* to represent the turbulent mixing, as explained in equations (3.2.2). However, for most turbulent flows it is not realistic to assume that these coefficients are constant, nor that the same coefficient will represent the transfers in different directions at a particular point in the turbulent fluid.

Generalized diffusion equations for axisymmetric flow and parallel flows with linearly varying concentration can be obtained from the general results (3.3.3, 5, 6) using the mass-transfer substitutions of Table 3.1. In principle, we might proceed in the same way to obtain momentum and energy equations for these species of turbulent flow. This programme is not easy to put into practice, however, for the precise nature of the source term is not always readily apparent. Accordingly, we shall approach each of these balances afresh, only later drawing parallels with the general results.

3.3.3 The Balance of Forces

In fully developed, parallel, constant-density flow, the mean velocity distribution does not change along the flow. Hence the shear stresses acting on a fluid element exactly balance the net pressure force on it. We consider in turn the cases of plane and axisymmetric parallel flow. Referring to Figure 3.6(a), we obtain for the former:

$$\frac{\mathrm{d}\tau}{\mathrm{d}y} = \frac{\partial P}{\partial x}$$

Using equations (3.2.9), which sum up the earlier discussion of pressure variations in fully developed parallel flow, we write this force balance as

$$\frac{\mathrm{d}\tau}{\mathrm{d}y} = \frac{\mathrm{d}P_{\mathrm{w}}}{\mathrm{d}x} = \text{constant} \tag{3.3.12}$$

throughout the flow. Here we have taken the mean pressure to be linearly varying in the sense of equation (3.2.4).

For simplicity, we have neglected the weight of the fluid. However, this can be introduced by the transformation

$$p_{\mathrm{w}} \to p_{\mathrm{w}} + \rho g z \tag{3.3.13}$$

Equation (3.3.12) then becomes

$$\frac{\mathrm{d}\tau}{\mathrm{d}y} = \frac{\mathrm{d}P_{\mathrm{w}}}{\mathrm{d}x} + \rho g \sin\theta \tag{3.3.13}$$

with $\sin\theta = \mathrm{d}z/\mathrm{d}x$, the inclination of the channel to the horizontal.

Integration of equation (3.3.12) gives a linear cross-stream variation of the effective shear stress, $\tau = \tau_{\mathrm{v}} + \tau_{\mathrm{t}}$, whether the flow is laminar or turbulent. Three species of *plane parallel flow* can be distinguished:

Type A. Pressure or Poiseuille Flow

Here the stresses applied *to* the fluid at the two walls are equal and in the same direction. The stresses *within* the fluid are related by $\tau_1 = -\tau_2 = \tau_{\mathrm{w}}$, and the force balance for the entire channel gives

$$2\tau_{\mathrm{w}} = -b\frac{\mathrm{d}P_{\mathrm{w}}}{\mathrm{d}x} \tag{3.3.14}$$

Hence

$$\tau = \tau_{\mathrm{w}}\left(1 - \frac{2y}{b}\right) = \frac{\mathrm{d}P_{\mathrm{w}}}{\mathrm{d}x}(y - \tfrac{1}{2}b) \tag{3.3.14}$$

gives the stress variation across the channel.

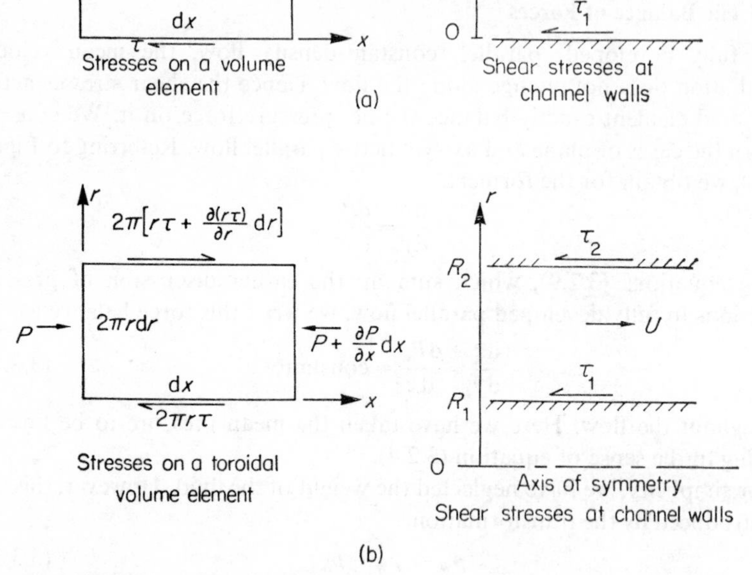

Figure 3.6 Force balances for parallel mean flow. (a) Plane flow in a channel. (b) Axisymmetric flow in a pipe or annulus

Type B. Shearing or Couette Flow

Here the stresses applied at the walls are equal but have opposite directions. The stresses *within* the fluid are given by $\tau_1 = \tau_2 = \tau_w$, and the overall force balance gives

$$\frac{dP_w}{dx} = 0 \tag{3.3.15}$$

Hence the stress is constant across the channel:

$$\tau = \tau_w \tag{3.3.15}$$

In order that this kind of flow be maintained, it is necessary that one wall of the channel move relative to the other.

Type C. Flow under a Combination of Pressure and Shear—Combined Poiseuille–Couette Flow

Here the wall stresses are in general unequal: $\tau_1 \neq \pm\tau_2$. The overall force balance is

$$\tau_2 - \tau_1 = b\frac{dP_w}{dx} \tag{3.3.16}$$

and the stress distribution is

$$\tau = \tau_1 + (\tau_2 - \tau_1)\frac{y}{b}$$
$$= \tau_1 + \frac{dP_w}{dx}y \qquad (3.3.16)$$

This kind of flow will occur when:

(1) the roughnesses of two stationary walls differ,

(2) one wall moves parallel to the other, as in a hydrodynamic bearing film,

(3) the fluid considered is a gas bounded by a (perhaps wavy) liquid surface, or

(4) the fluid considered is a liquid with a (perhaps wavy) free surface; the shear stress there will be very small.

Now let us turn to *axisymmetric parallel flow*. Figure 3.6(b) shows a cross-section through an annular element. Once again using equations (3.2.9) to rewrite the pressure gradient, we obtain

$$\frac{1}{r}\frac{d(r\tau)}{dr} = \frac{dP_w}{dx} = \text{constant} \qquad (3.3.17)$$

throughout the flow. Two species of axisymmetric parallel flow have to be considered:

Type D. Pipe or Poiseuille Flow

For a uniform wall stress, the overall force balance is

$$2\tau_w = -R\frac{dP_w}{dx} \qquad (3.3.18)$$

Integration of equation (3.3.17) again gives a linear stress variation:

$$\tau = \tau_w\frac{r}{R} = -\frac{1}{2}\frac{dP_w}{dx}r \qquad (3.3.18)$$

on requiring that the solution be finite on the axis $r = 0$.

Type E. Annulus Flow

The stress is taken to be uniform over each wall, but in general the two stresses are unequal: $\tau_1 \neq \pm\tau_2$. The overall force balance gives

$$2(R_2\tau_2 - R_1\tau_1) = -(R_2^2 - R_1^2)\frac{dP_w}{dx} \qquad (3.3.19)$$

and integration of equation (3.3.17) gives the distribution

$$\tau = \frac{\tau_1 R_1}{r} - \frac{1}{2}\frac{dP_w}{dx}\left(r - \frac{R_1^2}{r}\right) \qquad (3.3.19)$$

The stress variation is non-linear unless a rather artificial condition, $\tau_1 R_2 = \tau_2 R_1$, is satisfied. However, when $R_2/R_1 \simeq 1$, the variation will be nearly linear. It is possible to envisage flows of this kind in all of the circumstances listed following equations (3.3.16); moreover, unequal stresses will arise even when the roughness is the same for the two walls.

This discussion has provided one fixed point for our analysis of fully developed channel flows: whether the flow is laminar or turbulent, the variation of the mean shear stress is known. For *laminar flows*, the introduction of

$$\tau = \mu \frac{\partial U}{\partial y} \quad \text{or} \quad \mu \frac{\partial U}{\partial r}$$

leads, through further integrations, to the velocity variation, flow rate, and friction laws. For reference, these results are given in Table 3.2, where $\dot{v}$ is the

Table 3.2. Characteristics of fully developed laminar flows of a Newtonian fluid. For compactness, $\beta = \tau_w/\mu$

	Type A Plane pressure flow	*Type B* Plane shearing flow	*Type D* Pipe flow
Velocity U	$\beta y(1 - y/b)$	βy or $U_w y/b$	$\frac{1}{2}\beta(R - r^2/R)$
Flow rate	$\dot{v} = \beta b^2/6$	$\dot{v} = \frac{1}{2}\beta b^2$ or $\frac{1}{2}U_w b$	$\dot{V} = \frac{1}{4}\beta\pi R^3$
Velocity ratio	$U_a/U_c = \frac{2}{3}$	$U_a/U_w = \frac{1}{2}$	$U_a/U_c = \frac{1}{2}$
Friction coefficient	$12\nu/U_a b$	$8\nu/U_w b$	$16\nu/U_a d$

flow per unit breadth and $\dot{V}$ the total flow, and U_a is the average velocity, $\dot{V}/A$. Since the basic equations are linear, the velocity, flow rate and stress distribution for combined Poiseuille–Couette flow (Type C) can be obtained by adding the results for Types A and B. The results for annulus flow (Type E) have a generally similar structure, but are more complicated algebraically.

For *turbulent flows* it is not possible to make such rapid progress. Introducing the effective stress (3.2.10) into equation (3.3.12), we have

$$\frac{d(\mu\, dU/dy - \rho\, \overline{uv})}{dy} = \frac{dP_w}{dx} \tag{3.3.20}$$

Integration to find the velocity variation is possible if we are able to say how the mixing stress varies across the flow. One obvious way of doing this is to take the eddy viscosity to be constant. This assumption provides useful results for two parts of a channel flow:

I. Viscous Sublayer: $\epsilon_m = 0$

This layer is so thin that we may, without significant error:

(1) assume that the shear stress is uniform, and

(2) for axisymmetric flow, neglect the transverse curvature of the wall. Thus

$$\mu \frac{dU}{dy} = \tau_w + \frac{dP_w}{dx} y \simeq \tau_w$$

and

$$U = \frac{\tau_w}{\mu} y = \frac{\tau_w}{\mu}(R - r) \tag{3.3.21}$$

as for laminar Couette flow.

II. Fully Turbulent Core: $\epsilon_m = \epsilon_c$, a constant

Since the velocity variation in the core is a smallish fraction of the total change in velocity, its exact form need not be known for many purposes. Thus useful results follow from the adoption of a constant eddy viscosity, even though it does in fact vary considerably in the core, as shown in Figure 3.3(a). Only changes in mean velocity can be found in this way: the absolute values depend on the structure of the wall layer, where the rapidly varying eddy viscosity has much lower values than in the core. Adapting the results of Table 3.2, we obtain those set out in dimensionless form in Table 3.3.

The parameters used in this table are

$$R_f = \frac{u_f b}{\epsilon_c} \quad \text{or} \quad \frac{u_f d}{\epsilon_c}$$

and (3.3.22)

$$R_0 = \frac{\Delta U \Delta y}{\epsilon_c} = \frac{(U_{max} - U_{min})\Delta y}{\epsilon_c}$$

They may be interpreted either as Reynolds numbers incorporating the eddy viscosity, or as dimensionless forms of the eddy viscosity. Such *flow constants* will be used extensively in Chapter 7 to characterize free-turbulent flows.

In Table 3.3 experimental values of R_f are given for three kinds of channel flow. For each case a range of values is found, since R_f rises slightly as the mean-flow Reynolds number increases, and since an element of judgement is required in selecting a coefficient to represent the results. The values of R_f for the two pressure flows (Types A and D) are nearly the same, but that for shearing flow (Type B) is much lower. However, this comparison is somewhat inappropriate since, for the last case, the mean velocity increases steadily across the entire channel. The ranges of the velocity deficit (calculated as though the wall layers did not exist) prove to be nearly the same for the three flows. Note that this range is only a little greater than the velocity change across the viscous sublayer, which is $\Delta U/u_f = 5$, according to equations (3.1.3). Finally in Table 3.3, we consider the values of R_0 derived from the values of R_f given above. These too vary considerably from flow to flow.

Table 3.3. Characteristics of fully turbulent core flows, on the basis of a constant eddy viscosity

	Type A Plane pressure flow	*Type B* Plane shearing flow	*Type D* Pipe flow
Velocity deficit $(U_c - U)/u_f$	$R_f(y/b - \frac{1}{2})^2$	$R_f(y/b - \frac{1}{2})$	$\frac{1}{4}R_f(r/R)^2$
R_f ($\pm 15\%$)	26	7·4	27
Range of deficit $(\Delta U/u_f)$	6·5	7·4	6·8
Δy	$\frac{1}{2}b$	b	R
R_0	85	55	91

Note: Results *A* apply to a broad open channel, with $\frac{1}{2}b$ interpreted as the depth.

Although the parabolic velocity distributions of Table 3.3 are adequate for some purposes, the variation is in fact more like

$$\frac{U_c - U}{u_f} \propto \left(1 - \frac{y}{b}\right)^n \quad \text{or} \quad \left(\frac{r}{R}\right)^n$$

with (3.3.23)

$$\begin{aligned} n &= 1{\cdot}9 \pm 0{\cdot}2 \quad \text{for channel flow} \\ &= 1{\cdot}5 \pm 0{\cdot}2 \quad \text{for pipe flow} \end{aligned}$$

We now have adequate descriptions of the mean velocity in the turbulent core and in the viscous sublayer. The discussion of the velocity variation between these regions will be deferred until Chapter 4, since it does not make direct use of the force balances which are of primary interest here.

3.3.4 The Mechanical Energy Equation

For the constant-density fluid considered here, it is possible to distinguish a group of energy contributions which are independent of the temperature and internal energy of the fluid. We consider the energy balance for these mechanical-energy elements alone, before examining the overall energy balance which includes thermal terms. The mechanical elements are kinetic energy, displacement (or pressure) energy and potential energy. For simplicity, gravitational effects will be omitted; the results can easily be generalized to include them, as indicated in equations (3.3.13).

Figure 3.7 shows the time-mean flows of energy and work which act to change the mechanical energy within a cubical element of a plane parallel flow. The convective contributions in the direction of the mean flow cancel one another. The net flux of all forms of mechanical energy into this element can differ from zero only as a result of dissipation, that is, the conversion of mechanical to internal energy. Requiring that the net outwards flux equal the negative of the internal dissipation, we obtain the energy balance

$$\frac{\partial(UP)}{\partial x} - \frac{\mathrm{d}(U\tau)}{\mathrm{d}y} + \frac{\mathrm{d}\dot{W}_p}{\mathrm{d}y} + \frac{\mathrm{d}J_q}{\mathrm{d}y} = -\rho\varepsilon \tag{3.3.24}$$

The rate of dissipation is $\rho\varepsilon$ per unit volume, and ε per unit mass.

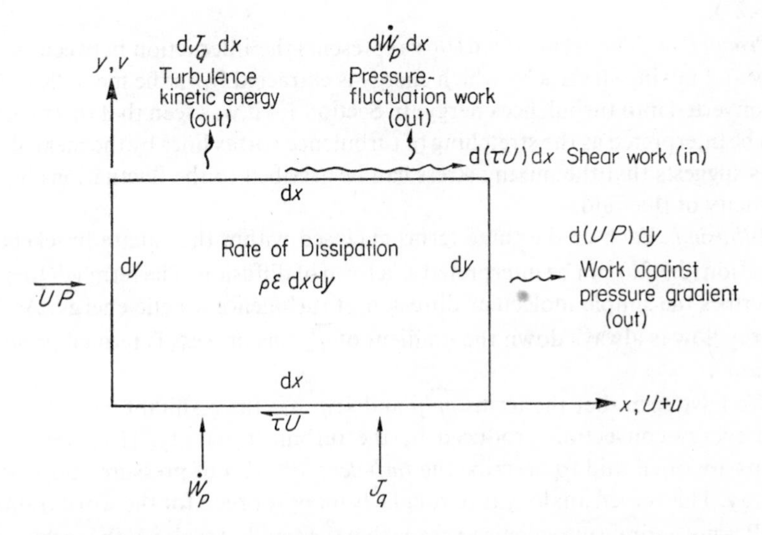

Figure 3.7 Net flows of mechanical energy from an element of volume (of unit depth in the z-direction) for plane parallel mean flow $U(y)$. The rate of dissipation of mechanical energy within the element is $\rho\varepsilon\,\mathrm{d}x\,\mathrm{d}y$

Noting that $U = f(y)$ and $\mathrm{d}\tau/\mathrm{d}y = \mathrm{d}P_w/\mathrm{d}x = \partial P/\partial x$ in plane parallel flow, we reduce the energy balance to

$$\tau\frac{\mathrm{d}U}{\mathrm{d}y} - \frac{\mathrm{d}(\dot{W}_p + J_q)}{\mathrm{d}y} = \rho\varepsilon \tag{3.3.25}$$

Introducing the shear stress and energy fluxes given in equations (3.2.10, 13, 14), we have finally

$$\nu\left(\frac{\mathrm{d}U}{\mathrm{d}y}\right)^2 - \overline{uv}\frac{\mathrm{d}U}{\mathrm{d}y} - \frac{\mathrm{d}}{\mathrm{d}y}\left[\overline{v\left(\frac{p}{\rho} + \tfrac{1}{2}q^2\right)} - \nu\frac{\mathrm{d}(\tfrac{1}{2}\overline{q^2})}{\mathrm{d}y}\right] = \varepsilon \tag{3.3.26}$$

The interpretation of the several terms follows.

Dissipation. The term

$$\varepsilon_v = \nu\left(\frac{dU}{dy}\right)^2 \tag{3.3.27}$$

gives the dissipation associated with the parallel mean motion. Thus the total dissipation is

$$\varepsilon = \varepsilon_v + \varepsilon_t \tag{3.3.27}$$

with ε_t the additional dissipation within the turbulence. In the fully turbulent part of the flow, $\varepsilon_t \gg \varepsilon_v$. For homogeneous, isotropic turbulence, there can be no mean velocity gradient; then $\varepsilon = \varepsilon_t$ can be calculated using equations (2.6.27).

Production. The term $-\overline{uv}\, dU/dy$ represents the interaction between mean flow and mixing stresses by which energy is extracted from the mean flow and is converted into turbulence energy. In Section 1.4 it was seen that this process can be interpreted as the stretching of turbulence vortex lines by the mean flow. This suggests that the mixing stress can be ascribed to the fluctuations in the vorticity of the fluid.

Diffusion. Each of the three terms enclosed within the square brackets of equation (3.3.26) can be interpreted as a form of diffusion. The term $\nu\, d(\frac{1}{2}\overline{q^2})/dy$ describes the simple molecular diffusion of turbulence kinetic energy. As the energy flow is always down the gradient of $\overline{q^2}$, this process is termed *gradient diffusion.*

We have seen that the terms $\overline{pv}/\rho$ and $\frac{1}{2}\overline{vq^2}$ are the resultant work transfer and energy convection produced by the turbulent activity. However, these terms are often said to describe the *turbulent diffusion* of pressure and kinetic energy. This forced analogy is particularly inappropriate for the work transfer $\overline{pv}$. It is not entirely unrealistic to argue that the smallest scales of the turbulence will produce something close to gradient diffusion, and to represent their activity by an eddy diffusivity. But this interpretation cannot be correct for the larger scales; these convey fluid over distances comparable to the scale of the mean motion; their effect cannot be specified by the gradient at any one point. Moreover, even for the smallest scales, the resultant transfer must depend on the link between the fluctuations in the convecting velocity and in the convected quantity. In particular, we may expect that the 'diffusive' effect of turbulent mixing will be fundamentally different for pressure and kinetic-energy fluctuations, and different again for fluctuations in temperature and contaminant concentration.

The three 'diffusion' terms of equation (3.3.26) have one feature in common: none contributes to the energy balance for the entire cross-section of the channel. At the channel walls, each of the terms in square brackets is zero, since $v=0$ and $\mathrm{d}q^2/\mathrm{d}y=2q\,\mathrm{d}q/\mathrm{d}y=0$ at a rigid boundary where no slip occurs. Hence integration of equation (3.3.26) over the section gives simply

$$-\int_A \overline{uv}\frac{\mathrm{d}U}{\mathrm{d}y}\mathrm{d}A=\int_A \varepsilon_t\,\mathrm{d}A \tag{3.3.28}$$

for this fully developed flow.

Although it was derived for a statistically steady flow, the last result bears on the question of the *instability* which leads to turbulence. Evidently a disturbance will develop when the left-hand side exceeds the right. We can see in general terms what this implies about the flow and the growing disturbance by the following simple analysis. We take the disturbance to have scales l' and u', the channel width and mean velocity being b and U. Instability may be expected for

$$\frac{u'^2\,U/b}{\nu(u'/l')^2} > \text{a certain number}$$

the form of the dissipation being inferred from the expressions for it which we encountered earlier. This criterion may be rewritten as

$$\frac{Ub}{\nu}\left(\frac{l'}{b}\right)^2 > \text{a certain number} \tag{3.3.29}$$

We conclude that:

(1) the least stable disturbances are the largest, for which $l' \to b$, and
(2) instability will occur when $Ub/\nu >$ a certain value.

This shows why the mean-flow Reynolds number is an index of sensitivity to instability in a laminar flow.

Another integral energy balance can be obtained from equation (3.3.24):

$$-\dot{v}\frac{\mathrm{d}P_w}{\mathrm{d}x}+U_2\tau_2=\rho\int_0^b \varepsilon\,\mathrm{d}y \tag{3.3.30}$$

Here

$$\dot{v}=\int_0^b U\,\mathrm{d}y$$

is the mean flow per unit breadth of the channel, and

$$\tau_2=\mu\left(\frac{\mathrm{d}U}{\mathrm{d}y}\right)_2$$

is the shear stress within the fluid at $y=b$, where the wall is moving with velocity $U=U_2$. The result (3.3.30) shows that the total dissipation balances the flow work of the pressure gradient and the shear work at the moving wall. While the sum of these two terms is necessarily positive, one of them can be negative.

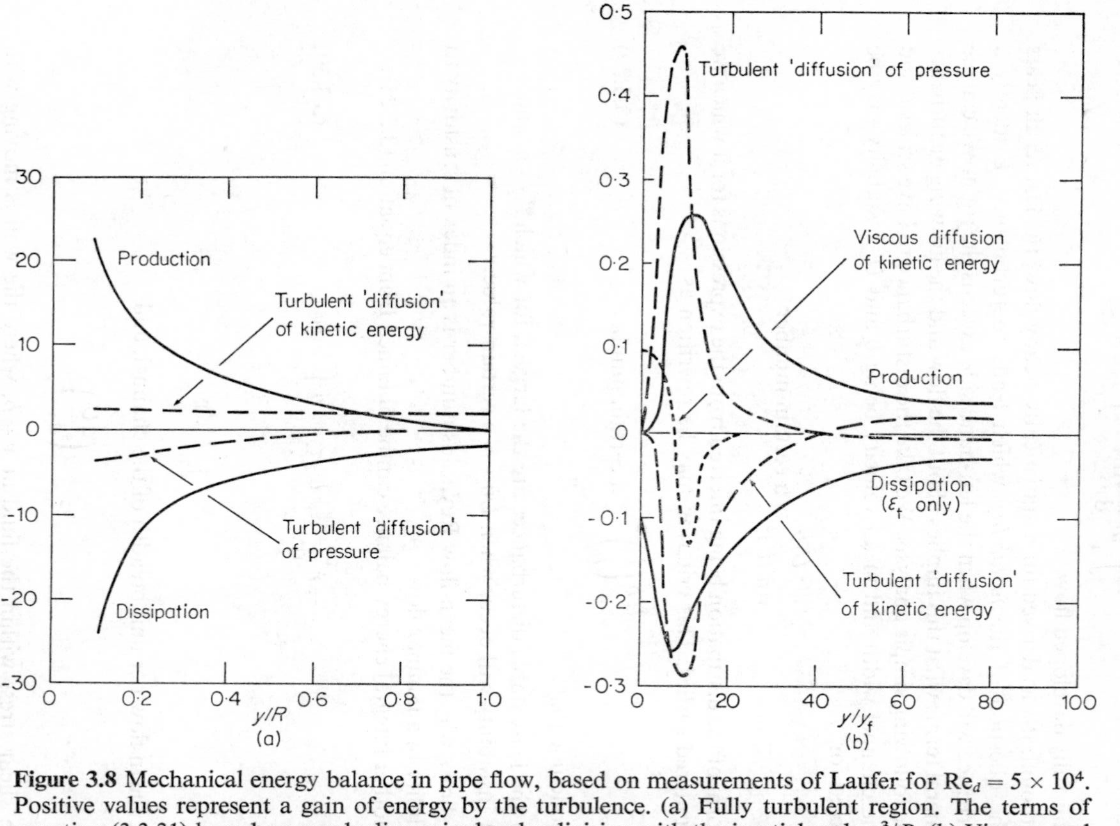

Figure 3.8 Mechanical energy balance in pipe flow, based on measurements of Laufer for $\mathrm{Re}_d = 5 \times 10^4$. Positive values represent a gain of energy by the turbulence. (a) Fully turbulent region. The terms of equation (3.3.31) have been made dimensionless by division with the inertial scale u_τ^3/R. (b) Viscous and wall layers. The terms of equation (3.3.26) have been made dimensionless by division with the viscous scale u_τ^4/ν

We return now to the local balance of the mechanical-energy components for fully developed turbulent channel flow. We consider the particular case of pipe flow, using equation (3.3.26) to describe the wall layer, which is so thin that the transverse curvature is negligible, and using

$$-\overline{uv}\frac{\mathrm{d}U}{\mathrm{d}r}-\frac{1}{r}\frac{\mathrm{d}}{\mathrm{d}r}\left[\overline{rv\left(\frac{p}{\rho}+\tfrac{1}{2}q^2\right)}\right]=\varepsilon_t \tag{3.3.31}$$

for the fully turbulent flow. Here the polar transformations (3.3.4) have been introduced, and the molecular-diffusion term is omitted, since it is significant only in the viscous layer.

The trends shown in Figure 3.8 are based on measurements made by Laufer. He measured all of the terms of the energy balances (3.3.26, 31) except:

(1) the dissipation, which was calculated from limited measurements by assuming that the dissipating scales had some of the characteristics of isotropic turbulence, and

(2) the so-called pressure-diffusion term involving $\overline{pv}$, which was found as the balancing term of the equation.

This procedure is dubious, since the dissipation term is normally one of the largest in the energy balance, and a small error in it will lead to a grossly incorrect prediction of the $\overline{pv}$ term. There is a school of thought which holds that a more realistic assessment of the energy balance can be obtained by taking $\overline{pv}=0$ through much of the flow, the dissipation being found as the balancing item. For the purpose of this discussion, we shall accept Laufer's energy balance, while bearing in mind the doubtful aspects of its derivation.

Note first that the dissipation is everywhere negative, according to the sign convention adopted here, while the production term is everywhere positive. The signs of the other terms change across the flow, as they must if their integrals are zero. Referring to Figure 3.8(a), which describes the fully turbulent part of the flow, we note that the production and dissipation are in rough balance, except in the centre of the pipe, where the turbulence is maintained by the 'diffusion' of kinetic energy. All of the contributions change rapidly as the wall layer is approached.

The more complicated activity within the wall layer is shown in Figure 3.8(b). This pattern is representative of the whole class of wall layers within which the stress is nearly constant. We note that:

(1) The two turbulent 'diffusion' terms are large, but have opposite signs, so that their combined effect on the energy balance is comparatively small. The process of kinetic-energy transfer extracts energy from the viscous layer and carries it to the fully turbulent fluid. The process of pressure-diffusion acts in the opposite direction, and appears to be particularly important in

maintaining the turbulence in the viscous layer. Remember, however, that its true variation is uncertain.

(2) Molecular diffusion carries energy towards the wall down the steep kinetic-energy gradient within $y/y_f < 20$; compare Figure 3.2(b).

(3) The turbulent dissipation ε_t and production remain roughly in balance over most of the viscous layer, although within the sublayer only the former retains its significance. In this region the mean-flow dissipation ε_v is very large, though it does not enter into this energy balance.

(4) A consideration of the overall balance suggests that, for this flow, the total dissipation is composed of three roughly equal elements: ε_v for the viscous layer, ε_t for the wall layer, and ε_t for the core region.

These results provide a picture of turbulent pipe flow which is at once too simple and too complex. Each term simplifies by adding contributions from successive turbulent motions of different sizes and forms. However, at a particular point and instant in time the local flow pattern is usually uncomplicated, and the local, instantaneous energy balance involves only a few of the processes identified in the time-averaged equations.

3.3.5 Thermal Energy Equations

The most convincing way of deriving the equations governing the flow of heat through a turbulent fluid is to obtain first a complete energy balance, comprising both thermal and mechanical contributions. Figure 3.9 shows the fluxes of internal energy which must be considered in conjunction with the mechanical contributions of Figure 3.7. For this parallel, statistically steady flow, the *overall energy balance* is obtained by requiring that the resultant outwards flux of energy be zero:

$$\frac{\partial(\rho UE)}{\partial x}+\frac{\partial \dot{q}_e}{\partial y}+\frac{\partial(UP)}{\partial x}-\frac{\mathrm{d}(\tau U)}{\mathrm{d}y}+\frac{\mathrm{d}\dot{W}_p}{\mathrm{d}y}+\frac{\mathrm{d}J_q}{\mathrm{d}y}=0 \tag{3.3.32}$$

Since the fluid density is taken to be constant, no term representing work done in changing its volume has been included. Note also that the internal-energy fluxes have been calculated as in equations (3.2.5, 6), which apply to linearly varying flow (where $\partial E/\partial x=$ constant) and omit a small term representing streamwise diffusion, as was noted in connection with equations (3.3.5, 6).

Using the requirement of parallel flow, $U=f(y)$, and the mechanical energy balance (3.3.24) we can reduce the overall balance to

$$\rho U\frac{\partial E}{\partial x}+\frac{\partial \dot{q}_e}{\partial y}=\rho\varepsilon \tag{3.3.33}$$

This may be termed the *thermal energy equation.* The dissipation does not appear in the overall equation (3.3.32), but does appear in the equations governing the mechanical and thermal components alone—in the former as a sink, in the latter as a source. Introducing the flux (3.2.15), we obtain an equation for the temperature variation:

$$U\frac{\partial(c_v T)}{\partial x}+\frac{\partial}{\partial y}\left[-\frac{k}{\rho}\frac{\partial T}{\partial y}+c_v\overline{v\theta}\right]=\varepsilon \tag{3.3.33}$$

As in the mass and momentum balances (3.3.10, 20), the equation for the mean value (here T) contains a mixing term involving the fluctuations (here θ). Thus each of these simple equations poses the problem of closure, to be discussed in Section 3.4.1.

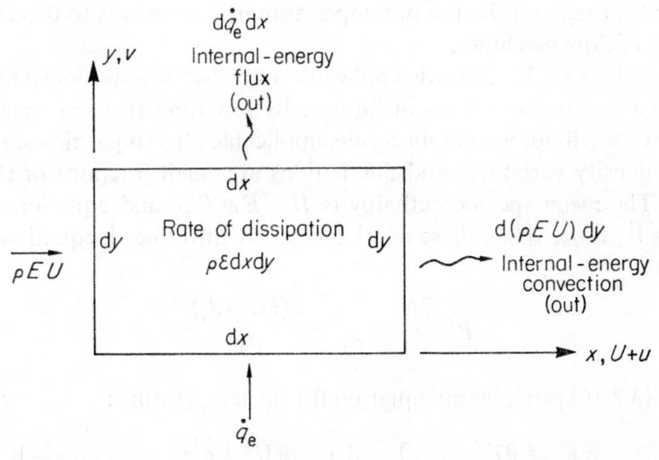

Figure 3.9 Net flows of thermal energy from an element of volume for plane parallel mean flow. The corresponding flows of mechanical energy are shown in Figure 3.7

The equations (3.3.33), and those to be derived from them, can be applied to:

(1) a linearly varying field, with

$$\frac{\partial E}{\partial x}=\text{constant} \quad \text{and} \quad \frac{\partial \dot{q}_e}{\partial y}=\frac{d\dot{q}_e}{dy} \tag{3.3.34}$$

corresponding to a *uniform heat flux* at the channel walls, or

(2) other forms of development, with $\partial E/\partial x \neq$ constant, for example, that corresponding to *uniform wall temperature.*

The equations omit the streamwise diffusion term $\partial\dot{q}_e/\partial x$. For case (1), this comprises only a molecular contribution; but for case (2), there is a non-zero mixing term $\partial\overline{ue}/\partial x$ as well (see equations (3.2.5)).

It is implicit in the assumptions of parallel flow that the *dissipation* $\varepsilon = f(y)$ only. In practice, it will vary slowly along the channel, as a result of changes in wall friction engendered by the temperature dependence of the fluid viscosity. This reminds us that, strictly speaking, a fully developed flow is possible only when the temperature is nearly constant along the channel.

The dissipation is usually negligible in comparison with the other terms of equations (3.3.33). It will be of significance only when the flow is nearly adiabatic, or when the velocity is very high, that is, comparable with the speed of sound, and thus beyond the range of the present discussion. We conclude that the dissipation term may be omitted when analysing heat-transfer equipment, but must be retained in certain other applications, for example, aerodynamic heating, and the use of temperature measurements to determine the efficiency of flow machines.

The results (3.3.32, 33) are applicable only for strictly constant-density flows, that is, for most flows of liquids. By rewriting them in terms of the *enthalpy* of the fluid, we obtain results applicable also to gas flows, provided that the density variations and fluctuations are small fractions of the mean density. The mean specific enthalpy is $H = E + P/\rho$, and equations (3.2.17) give $\dot{q}_e + \dot{W}_p = \dot{q}_h$. When these relationships are introduced, equation (3.3.32) becomes

$$\rho U \frac{\partial H}{\partial x} + \frac{\partial \dot{q}_h}{\partial y} = \frac{d(\tau U - J_q)}{dy} \tag{3.3.35}$$

The flux (3.2.16) provides an equation for the temperature:

$$U\frac{\partial(c_p T)}{\partial x} + \frac{\partial}{\partial y}\left[-\frac{k}{\rho}\frac{\partial T}{\partial y} + c_p\overline{v\theta}\right] = \frac{d}{dy}\left[\tfrac{1}{2}\nu\frac{d(U^2 + \overline{q^2})}{dy} - \overline{uv}\,U - \tfrac{1}{2}\overline{vq^2}\right] \tag{3.3.35}$$

The turbulence-dependent terms on the right-hand sides of equations (3.3.35) are of the same order as the dissipation appearing in equations (3.3.33), and are, like them, negligible in many circumstances.

Most of the turbulence terms can be absorbed into the basic energy equation by introducing the *total enthalpy* of the turbulent fluid:

$$H_0 + h_0 = H + h + \tfrac{1}{2}(U + u)^2 + \tfrac{1}{2}v^2 + \tfrac{1}{2}w^2 \tag{3.3.36}$$

Averaging, we find the mean value:

$$H_0 = H + \tfrac{1}{2}U^2 + \tfrac{1}{2}\overline{q^2} \tag{3.3.36}$$

Then the fluctuation is

$$h_0 = h + Uu + \tfrac{1}{2}q^2 - \tfrac{1}{2}\overline{q^2} \tag{3.3.36}$$

giving

$$\overline{vh_0} = \overline{vh} + U\overline{uv} + \tfrac{1}{2}\overline{vq^2} \tag{3.3.36}$$

The enthalpy equation (3.3.35) can now be regrouped as

$$
\begin{aligned}
U\frac{\partial H_0}{\partial x} &= \frac{\partial}{\partial y}\left(\kappa\frac{\partial H}{\partial y} - \overline{vh_0}\right) + \frac{\mathrm{d}}{\mathrm{d}y}\left[\tfrac{1}{2}\nu\frac{\mathrm{d}(U^2+\overline{q^2})}{\mathrm{d}y}\right] \\
&= \frac{\partial}{\partial y}\left(\kappa\frac{\partial H_0}{\partial y} - \overline{vh_0}\right) + \frac{\mathrm{d}}{\mathrm{d}y}\left[\tfrac{1}{2}\nu\left(1-\frac{1}{\mathrm{Pr}}\right)\frac{\mathrm{d}(U^2+\overline{q^2})}{\mathrm{d}y}\right]
\end{aligned}
\tag{3.3.37}
$$

Here $\mathrm{Pr} = \nu/\kappa = \mu c_p/k$ is the *Prandtl number* of the fluid; for $\mathrm{Pr} = 1$, the turbulence-dependent terms vanish. An even simpler form can be derived from equations (3.3.37) by introducing

$$
\dot{q}_{h_0} = -\rho\left(\kappa\frac{\partial H_0}{\partial y} - \overline{vh_0}\right) \tag{3.3.38}
$$

Energy balances for the entire cross-section can be obtained by integrating the elemental results obtained above. Integrating the first of equations (3.3.33) from wall to wall of a plane channel, we have

$$
\begin{aligned}
\rho\int_0^b U\frac{\partial E}{\partial x}\,\mathrm{d}y &= \rho\frac{\partial}{\partial x}\int_0^b UE\,\mathrm{d}y \\
&= \dot{q}_1 + \dot{q}_2 + \rho\int_0^b \varepsilon\,\mathrm{d}y \\
&= \dot{q}_1 + \dot{q}_2 - \dot{v}\frac{\mathrm{d}P_\mathrm{w}}{\mathrm{d}x} + U_2\tau_2
\end{aligned}
\tag{3.3.39}
$$

on using the result (3.3.32). Here

$$
\begin{aligned}
\dot{q}_1 &= -k_1\left(\frac{\partial T}{\partial y}\right)_1 \quad \text{for } y=0 \\
\dot{q}_2 &= +k_2\left(\frac{\partial T}{\partial y}\right)_2 \quad \text{for } y=b
\end{aligned}
\tag{3.3.40}
$$

are the rates of heat transfer *into* the fluid.

The results (3.3.39) show that changes in the flux of internal energy along the channel result from:

(1) heat transfer at the walls, which may be either inwards or outwards, and

(2) internal dissipation, necessarily positive, equalling the work done on the fluid by the pressure gradient and the shear stress at the moving wall. As was pointed out earlier, this contribution is often negligible.

A similar treatment of the enthalpy equation (3.3.35) gives

$$
\rho\frac{\partial}{\partial x}\int_0^b UH\,\mathrm{d}y = \dot{q}_1 + \dot{q}_2 + U_2\tau_2 \tag{3.3.41}
$$

Note that

$$
\dot{q}_\mathrm{e} = \dot{q}_\mathrm{h} = \dot{q}_{h_0} = \dot{q}_\mathrm{w}
$$

at a rigid wall, where $v = 0$. Finally, we integrate the equation (3.3.37) for the total enthalpy, and find

$$\rho \frac{\partial}{\partial x} \int_0^b U H_0 \mathrm{d}y = \dot{q}_1 + \dot{q}_2 + \left(1 - \frac{1}{\mathrm{Pr}}\right) U_2 \tau_2 \tag{3.3.42}$$

For $\mathrm{Pr} = 1$, the change in the flux of total enthalpy is exactly equal to the total heat transfer. We shall see later that energy equations usually take on simpler forms when $\mathrm{Pr} = 1$. These simpler results often provide acceptable approximations for flows of gases.

Integrated results for pipe flow can be written down by analogy. For simplicity, we consider the case of a linearly varying field, where, for example, $\partial E/\partial x = $ constant throughout the flow. Hence the flux integral simplifies to

$$\rho \frac{\partial}{\partial x} \int_A U E \, \mathrm{d}A = \rho \int_A U \frac{\partial E}{\partial x} \mathrm{d}A = \dot{m} \frac{\partial E}{\partial x} \tag{3.3.43}$$

with $\dot{m}$ the mass flux through the channel. The integral results for pipe flow are then:

$$\dot{m} \frac{\partial E}{\partial x} = C \dot{q}_{\mathrm{w}} - \dot{V} \frac{\mathrm{d}P_{\mathrm{w}}}{\mathrm{d}x} \tag{3.3.44}$$

$$\dot{m} \frac{\partial H}{\partial x} = C \dot{q}_{\mathrm{w}} \tag{3.3.45}$$

$$\dot{m} \frac{\partial H_0}{\partial x} = C \dot{q}_{\mathrm{w}} \tag{3.3.46}$$

Here $C = \pi d$ is the pipe circumference, and $\dot{q}_{\mathrm{w}} = k(\partial T/\partial r)_{\mathrm{w}}$ is the energy flux *into* the fluid, assumed uniform around the periphery. We need not allow for motion of the boundaries in this case, though this might occur in annulus flow.

The applicability of the preceding results is as follows:

(a) equations (3.3.33, 39, 44): parallel flow with strictly constant density;

(b) equations (3.3.35, 41, 45): parallel variable-density flow, but with changes in density small fractions of the mean value; and

(c) equations (3.3.37, 42, 46): parallel flows with significant variations in mean density and velocity in the streamwise direction, but with the fluctuations in density remaining small fractions of the local mean value.

The extended applicability of equations (b) and (c) is not apparent from their derivation, which assumed constancy of density, but identical results do in fact emerge when this restriction is relaxed.

It would be possible to parallel the preceding discussion of the momentum equation by applying these energy relationships to some laminar flows, and to the limited parts of turbulent flows—the sublayer and core flow—where the effective diffusivities are nearly constant. We shall not do this immediately, but defer the discussion until Chapter 5 when a more detailed picture of the wall layer will be available, and a more comprehensive analysis is possible.

3.4 SIMPLE THEORETICAL MODELS

3.4.1 The Problem of Closure

The earlier part of this chapter provides two of the pre-requisites for the solution of practical problems involving turbulent channel flows:

(1) a realistic picture of the turbulent motion, based on time-averaged measurements, and

(2) rational equations expressing the time-average requirements for the conservation of mass, momentum and energy.

But these equations are not sufficient to define even the simplest features of the turbulence or of the associated mean motion. Consider, for example, the fully turbulent region of a plane parallel flow, for which

$$\frac{\mathrm{d}\overline{uv}}{\mathrm{d}y} = -\frac{1}{\rho}\frac{\mathrm{d}P_{\mathrm{w}}}{\mathrm{d}x}, \quad \text{a constant} \tag{3.4.1}$$

is the momentum equation derived from equation (3.3.20), and

$$-\overline{uv}\frac{\mathrm{d}U}{\mathrm{d}y} - \frac{\mathrm{d}}{\mathrm{d}y}\left[\overline{v\left(\frac{p}{\rho} + \tfrac{1}{2}q^2\right)}\right] = \varepsilon_{\mathrm{t}} \tag{3.4.2}$$

is the energy equation derived from equation (3.3.26). This pair of equations relates four characteristics of the flow: U, $\overline{uv}$, $\overline{v(p/\rho + \frac{1}{2}q^2)}$ and ε_{t}. While further relationships among these quantities can be developed from the basic conservation laws, these additional constraints inevitably introduce still more characteristics of the turbulence. Thus we face the problem of closure: how is it possible to cut short this infinite regression, to obtain a closed mathematical problem?

Some possible lines of attack are:

(1) the mathematical approach: arbitrary simplifications of the existing equations by dropping terms or by expressing them in convenient forms, perhaps suggested by analogy with laminar flow;

(2) the physical approach: use of a specific model of turbulent activity to establish new relationships, that is, relationships not implicit in the basic physical principles already introduced; and

(3) the combined approach: interaction of physical arguments with simplification and manipulation of mathematical results.

To obtain numerical predictions from any of these methods of analysis, we must supply some empirical information, usually in the form of 'universal' (hopefully) constants or functions.

The most recent attempts to predict turbulent motions follow the third of these paths, in combining mathematical and physical arguments. The relatively sophisticated methods which have been developed for boundary layers and other complex flows will be introduced in Chapters 8 and 9. For the present we shall be concerned with the results obtained by the earlier students of turbulence who adopted the first two of the approaches described above by introducing eddy diffusivities, mixing lengths, and similar artifices. These early analyses are often referred to as *phenomenological or semi-empirical theories,* since they combine heuristic arguments and empirical data to predict the gross phenomena arising from turbulent activity.

In some respects the primitive analyses are misleading, while their useful consequences can mostly be obtained by more realistic arguments—often from dimensional analysis alone. Nevertheless, it is necessary to have some knowledge of the simple models, for they permeate the literature of turbulence. Moreover, they give correct results, perhaps fortuitously, for a limited number of situations of practical importance, and thus provide the basis for many present-day calculations relating to turbulent flows.

Models of turbulence have been introduced here as ways of coping with the problem of closure. They can be looked at in another way—and undoubtedly were by those who developed them. The simple models introduce fictitious entities—eddy diffusivities, mixing lengths, and mass flows—that are easier to visualize than the fluctuations of the actual motion. Viewed in this way, they are seen to be tentative solutions to the *problem of structure*, the physical counterpart of the mathematical problem of closure.

3.4.2 Eddy Diffusion and Convection

In equations (3.2.2) we noted that the difficult mixing component $\rho\overline{sv}$ of the total flux J could be represented, formally, either as gradient diffusion or as steady lateral convection. Thus the general result (3.2.1) becomes

$$J = -\rho(K + \epsilon_s)\frac{\partial S}{\partial y} + \rho SV \tag{3.4.3}$$

or

$$J = -\rho K\frac{\partial S}{\partial y} + \rho S(V + V_s) \tag{3.4.4}$$

with ϵ_s the eddy diffusivity and V_s the effective velocity of lateral convection. Meteorologists often call the eddy diffusivity an *exchange coefficient*, or follow German usage in the term *Austausch coefficient*. The fluid density is sometimes incorporated, to give a 'dynamic' rather than a 'kinematic' diffusivity.

Although the formulation (3.4.4) is adequate for some purposes, the magnitude of the fluctuations s contributing to $\overline{sv}$ is often linked to the overall change in the mean value across some part of the flow. This change is shown in Figure 3.10 as ΔS. In such circumstances, it is appropriate to write

$$J = -\rho K \frac{\partial S}{\partial y} + \rho S V + G_s \Delta S \tag{3.4.5}$$

with G_s the *Reynolds flux*, the effective lateral transfer rate across the range ΔS, expressed as a mass flow of the basic fluid per unit area.

The hypothetical measures of turbulent activity are related to each other and to the basic mixing component by

$$\overline{sv} = G_s \frac{\Delta S}{\rho} = V_s S = -\epsilon_s \frac{\partial S}{\partial y} \tag{3.4.6}$$

Formally, any mixing process can be described in any of these three ways. However, for a specific situation one of the representations may be considerably simpler and more meaningful. In particular, the appearance of a negative

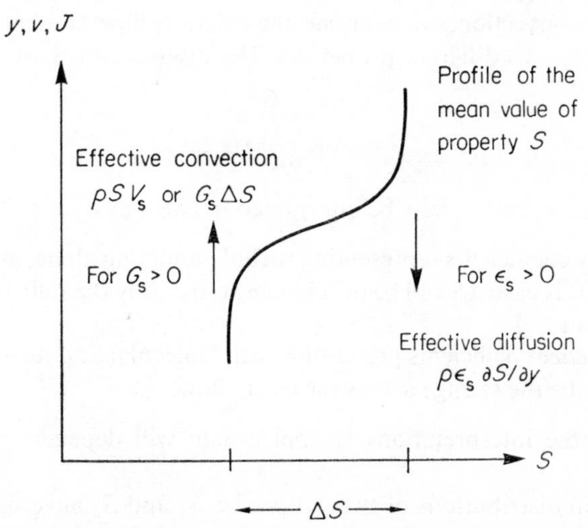

Figure 3.10 Representation of the flux across a gradient in the mean property S by an effective diffusivity ϵ_s and by an effective mass flux or Reynolds flux G_s. The latter is defined in terms of the change ΔS across the part of the flow of interest

value for the eddy diffusivity may be taken to indicate that the process is fundamentally different from the microscopic mixing which provides the pattern for gradient diffusion. The effective convection does not suffer from this conventional restriction, and in this sense has a wider range of application.

A more general representation of turbulent mixing is achieved by admitting both diffusion and convection:

$$\overline{sv} - \epsilon_s \frac{\partial S}{\partial y} = \overline{sv_s} = V_s S = \frac{G_s}{\rho} \Delta S \tag{3.4.7}$$

Here v_s, V_s and G_s represent a residual flux which cannot be realistically described as gradient diffusion. The residual convection is generally conceived as representing the larger scales of the turbulence, while the gradient diffusion is ascribed to the smaller scales.

We are now faced with the problem of deciding, for each individual property S, how the net effect $\overline{sv}$ should be split between diffusion and convection. This problem has already been noted in connection with those terms of the mechanical energy equation (3.3.26) which represent turbulent 'diffusion'. For any one property, the division between diffusion and convection must be expected to vary from one flow to the next (for example, from wake to pipe flow) and from one part of a flow to another (for example, from core flow to wall layer).

Returning now to the representation of the mixing transfer entirely as diffusion or convection, we examine the relationships between the transfer coefficients for two different properties. The discussion is based on the relations

$$J = -\rho \epsilon_s \frac{\partial S}{\partial y} = G_s \Delta S \tag{3.4.8}$$

The quantities ϵ_s and G_s may be interpreted in two ways:

(1) as *eddy* coefficients representing turbulent mixing alone, as in equations (3.4.3, 5); in this case ΔS will be the change across only the fully turbulent part of the flow; and

(2) as *effective* coefficients presenting both molecular and turbulent mixing; here ΔS may be the change across the entire flow.

Which of these interpretations is appropriate will depend on the specific application.

If the mean distributions of two properties S_1 and S_2 have similar shapes, so that

$$\frac{1}{\Delta S_1} \frac{\partial S_1}{\partial y} = \frac{1}{\Delta S_2} \frac{\partial S_2}{\partial y} \tag{3.4.9}$$

we find from the relations (3.4.8) that

$$\frac{\epsilon_{s_1}}{\epsilon_{s_2}} = \frac{G_{s_1}}{G_{s_2}} = \frac{J_1/\Delta S_1}{J_2/\Delta S_2} \tag{3.4.10}$$

These results are the kernel of the much-used analogies between transfers of momentum, heat and mass. But for the moment we are more interested in what they imply about the relationship between eddy diffusivities and Reynolds fluxes. For similar distributions of S_1 and S_2, the diffusivities of the two properties are proportional to the Reynolds fluxes. When the two distributions are not similar, the relationships (3.4.10) will not apply, but we may still expect that differing eddy diffusivities will correspond to differing Reynolds fluxes. Non-similar distributions may be expected, in particular, when the two transfer processes are distributed in different proportions between diffusion and convection.

Before looking at the eddy diffusivity and Reynolds flux separately, let us consider the circumstances in which one or the other may be expected to be useful, that is, to provide a simple, yet fairly accurate description of a turbulent mixing process. Figure 3.11 suggests the ways in which these two quantities may be expected to vary across a turbulent channel flow. For the Reynolds flux, it is important to distinguish between the two interpretations noted above,

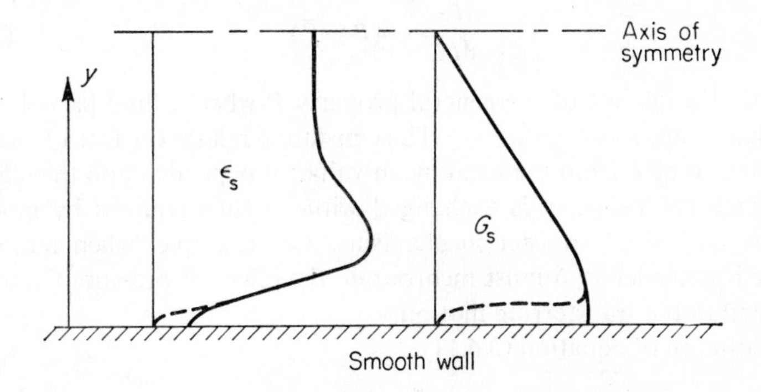

Figure 3.11 Typical variations of eddy and effective diffusivity and Reynolds flux within a turbulent channel flow. Full lines represent variations of 'effective' measures of transfer, which incorporate both molecular and turbulent transport. Dashed lines represent 'eddy' properties, which describe only the turbulent activity. For clarity, the thickness of the viscous layer has been exaggerated, as has the difference between the effective and eddy diffusivities in that layer

following equations (3.4.8); only for the second does G_s remain nearly constant across the viscous layer, as shown in Figure 3.11. We see that the eddy diffusivity and Reynolds flux are complementary: one is constant (or slowly varying) when the other changes rapidly. In the core flow ϵ_s is fairly constant; this fact has already been used in the calculations leading to Table 3.3. In the wall layer, on the other hand, G_s is fairly constant, since the variation is nearly linear across the channel (exactly so when the shear stress is represented, as will be seen later).

We conclude that eddy diffusivities find their most fruitful applications away from solid walls, in free turbulent flows, and in the cores of channel flows. The Reynolds flux provides a more useful (that is, simpler) description of the transfers in the wall layer. It is also useful in specifying the transfer across the whole of a channel flow, comprising wall layer and core or outer flow, since this transfer is normally controlled by the flow near the wall. We saw above that the changes in eddy diffusivity and Reynolds flux from property to property are of the same order. Hence neither possesses an advantage from this point of view.

3.4.3 The Eddy Diffusivity

We shall now develop a simple mathematical model for the diffusive action of turbulent mixing across the mean streamlines of a parallel flow. Adopting the Lagrangian viewpoint, we take

$$\frac{\mathrm{d}P}{\mathrm{d}t} = -\delta(P - \bar{P}) \tag{3.4.11}$$

to specify the change of the general property P when a fluid particle moves through a mean distribution $\bar{P}(y)$. The constant δ relates the rate of change to the deviation of P from the local mean value; it depends upon the efficiency with which the property is exchanged with its surroundings by molecular diffusion and other transfer mechanisms. For example, when momentum transfer is considered, δ must incorporate the effect of pressure fluctuations correlated to the transferring motion.

The solution of equation (3.4.11) is

$$P(t) = \delta \int_0^\infty \bar{P}(t-\tau)\,\mathrm{e}^{-\delta\tau}\,\mathrm{d}\tau \tag{3.4.12}$$

as may be demonstrated by substituting into that equation, and then integrating by parts. This solution represents the instantaneous value in terms of the mean values in the vicinity of the particle at earlier times; the factor $\mathrm{e}^{-\delta\tau}$ has the effect of progressively reducing the importance of earlier contacts.

Assuming the particle displacements to be small, and the gradient of $\bar{P}$ to be fairly uniform, we calculate the historical mean-profile value as

$$\begin{aligned} \bar{P}(t-\tau) &= \bar{P}(t) + \frac{\partial \bar{P}(t)}{\partial y}\Delta y(t-\tau) \\ &= \bar{P}(t) + \frac{\partial \bar{P}(t)}{\partial y}\int_t^{t-\tau} v(\tau')\,\mathrm{d}\tau' \end{aligned} \tag{3.4.13}$$

Substitution in equation (3.4.12) gives

$$P(t) = \bar{P}(t) - \frac{\partial \bar{P}}{\partial y}\int_0^\infty v(t-\tau)\,\mathrm{e}^{-\delta\tau}\,\mathrm{d}\tau \tag{3.4.14}$$

following changes in the variables and in the order of integration. Finally, assuming that the field is statistically stationary, we obtain

$$\overline{P(t)v(t)} = -\frac{\partial \bar{P}}{\partial y}\int_0^\infty Q_{22}(\tau)\,\mathrm{e}^{-\delta\tau}\,\mathrm{d}\tau \tag{3.4.15}$$

where, in the notation of Table 2.1

$$Q_{22}(\tau) = \overline{v(t)\,v(t-\tau)} = \overline{v^2}\,R_{22}(\tau) \tag{3.4.16}$$

This is the autocorrelation function of v, the lateral velocity fluctuation. Note that this is a *Lagrangian correlation*, relating the velocities of particles passing the point considered to the velocities they had at times τ earlier.

Finally, we replace $P(y,t)$ by $S(y)+s(y,t)$, our earlier notation, and obtain from equation (3.4.15):

$$\epsilon_s = \int_0^\infty Q_{22}(\tau)\,\mathrm{e}^{-\delta_s\tau}\,\mathrm{d}\tau \tag{3.4.17}$$

This indicates that the eddy diffusivity depends on the basic turbulent activity, through the correlation function $Q_{22}(\tau)$, and on the particular substance transferred (and other aspects of the turbulence), through the transfer coefficient δ_s.

The derivation of this expression for the diffusivity indicates the conditions in which gradient diffusion may be expected to be dominant. The crucial step (3.4.13) requires that the scale of the mixing be small compared to the scale of changes in the mean distribution $\bar{P}(y)$. When this condition is not satisfied, for example, near the rather sharp maximum in the mean velocity of a wall jet, we must expect the gradient-diffusion model to be inadequate.

A very obvious failure of the diffusion model is in representing the work transfer associated with pressure fluctuations, the so-called pressure-diffusion term of equation (3.3.26). Neither of the basic statements (3.4.11, 13) can be meaningfully applied to pressure fluctuations. These remarks are consistent with the anomalous activity of the pressure diffusion within a wall layer. In Figure 3.8(b) it is seen to produce a flow of energy *towards* the region of highest turbulence energy. However, as noted in Section 3.3.4, the derivation of this contribution is suspect.

The general nature of the transfer patterns in turbulent flows in which the diffusivity varies only moderately can be discovered by studying laminar flows of a Newtonian fluid. Such information can be applied only with great caution to the generality of turbulent motions, for the eddy diffusivity must be expected to depend on:

(1) the flow species;

(2) the length and velocity scales, L_0 and U_0, specifying a flow of that species, and the Reynolds number formed from them;

(3) position in the flow, specified by the coordinates x and y;
(4) the entity S which is transferred, at least, through its diffusivity K.

Thus

$$\frac{\epsilon_s}{U_0 L_0} = f\left(\frac{x}{L_0}, \frac{y}{L_0}, \frac{U_0 L_0}{\nu}, \frac{\nu}{K}, \text{flow species}\right) \tag{3.4.18}$$

This line of argument has been used in equations (3.3.22), where the parameters R_f and R_0 are defined using mean-flow length and velocity scales: the channel breadth b and the distance Δy over which the velocity change occurs, and the friction velocity u_f and the velocity change across the flow ΔU.

Attempts have been made, notably by von Kármán, to lessen the dependence of the dimensionless characteristic (3.4.18) on the flow species and relative position within the flow. This is done by adopting scales associated with the local turbulence:

$$\epsilon_s \propto \frac{\overline{v^2}\,\lambda_x^2}{\nu} \quad \text{and} \quad \frac{\overline{q^2}\,\lambda_x^2}{\nu} \tag{3.4.19}$$

with λ_x the longitudinal microscale of the turbulence. None of these scaling procedures has produced a widely applicable expression for the eddy diffusivity. Moreover, if this kind of analysis is to be undertaken, it is better applied directly to readily interpreted quantities such as $\overline{uv}$, rather than to the hypothetical ϵ_s.

3.4.4 The Reynolds Flux

The Reynolds flux is not obtained by analogy with molecular diffusion, and is not often used to describe transfer processes in laminar flows. However, the examples to be considered below encompass both laminar and turbulent motions, since no assumption is made about the mechanism of cross-stream transport. It will be seen that certain parameters commonly used to specify momentum, heat and mass transfers—friction coefficients and Stanton numbers—can be interpreted as scaled forms of the Reynolds flux. Thus

$$\frac{G_s}{\rho U_0} = \text{dimensionless transfer coefficient} \tag{3.4.20}$$

with ρ the uniform density of the basic fluid, and U_0 a characteristic mean velocity, perhaps the average velocity of a channel flow, the free-stream velocity beyond a boundary layer, or the velocity change across a mixing layer. In general, the dimensionless characteristic (3.4.20) will exhibit the functional dependence indicated in equation (3.4.18) for the eddy diffusivity.

(*a*) *Heat Transfer*. Here $\Delta S = \Delta H$, an enthalpy difference, and equations (3.4.8) give the enthalpy flux:

$$\dot{q}_h = G_h \Delta H \tag{3.4.21}$$

with G_h the Reynolds mass flow for enthalpy. In terms of the *Stanton number*, we have

$$\mathrm{St} = \frac{\dot{q}_h}{\rho U_0 \Delta H} = \frac{G_h}{\rho U_0} \tag{3.4.22}$$

These results hold for any part of the flow, but are most often applied between a solid wall and the body of the passing fluid. For a channel, with $U_0 = U_a$, the Stanton number can be interpreted as the ratio of the energy flux from the wall $\dot{q}_w$ to a convective energy flux along the channel or, more simply, as the ratio of the Reynolds flux (for thermal energy) to the channel mass flow. If the Reynolds flux is given the second of the interpretations which follow equations (3.4.8), this applies for both laminar and turbulent flow.

(b) *Mass Transfer.* Here $\Delta S = \Delta C/\rho$ with ΔC a change in mean concentration. Equations (3.4.8) give the mass flux of the substance whose concentration is C:

$$N = \frac{G_c}{\rho} \Delta C \tag{3.4.23}$$

with G_c/ρ a mass-transfer coefficient with dimensions of velocity. In terms of a dimensionless parameter analogous to the Stanton number, we have

$$\mathrm{St}_c = \frac{N}{U_0 \Delta C} = \frac{G_c}{\rho U_0} \tag{3.4.24}$$

The close parallel with equation (3.4.22) will be apparent.

(c) *Momentum Transfer: Shear Stress.* In this case, we usually have $S = \Delta S = U_0$, the mean-velocity scale, and

$$\tau = G_m U_0 = \rho V_m U_0 \tag{3.4.25}$$

with G_m the Reynolds flux for momentum, and V_m the lateral convection velocity of equation (3.4.4.). Note that the measures of the hypothetical lateral flow vary across the flow in the same way as the shear stress, their variation being linear in fully developed channel flow. For the *friction coefficient*, we have

$$c_f = \frac{\tau}{\frac{1}{2}\rho U_0^2} = 2\frac{G_m}{\rho U_0} = 2\frac{V_m}{U_0} \tag{3.4.26}$$

giving a simple kinematic relationship between the coefficient and the hypothetical velocity V_m.

(d) *Lateral Flow and Friction.* There is a momentum transfer associated with the convection velocity V, and the effective stress is

$$\tau_e = \rho(-VU + V_m U_0) \tag{3.4.27}$$

with U the local mean velocity. The effective stress exceeds the mixing stress when $V < 0$, as when the actual flow is towards an adjacent wall. Applied at the edge of a wall layer, with $U_0 = U = U_a$, the average velocity in the channel, the last result becomes

$$\tau_e = \rho U_a(-V + V_m) = U_a(-\rho V + G_m) \tag{3.4.28}$$

showing how the actual and hypothetical flows combine to give the total momentum transfer or effective shear stress. This formulation is often used in discussing the flow near a wall at which condensation, transpiration or suction takes place.

The relationships (3.4.22, 24, 26) lead to the much-used analogies between the transfers of momentum, heat and mass. Combined, they give

$$2\frac{G_m}{c_f} = \frac{G_h}{\mathrm{St}} = \frac{G_c}{\mathrm{St}_c} = \rho U_0 \tag{3.4.29}$$

These results are particular interpretations of the right-hand pair of equations (3.4.10). If we assume that the Reynolds fluxes are the same for the three transfers, we obtain

$$\tfrac{1}{2}c_f = \mathrm{St} = \mathrm{St}_c = \frac{G_m}{\rho U_0} \tag{3.4.30}$$

or

$$\frac{\tau}{U_0} = \frac{\dot{q}_h}{\Delta H} = \frac{\rho N}{\Delta C} = G_m \tag{3.4.30}$$

This series of results is commonly referred to as *Reynolds analogy*, the several fluxes usually being evaluated at a wall. It is possible that two of the Reynolds fluxes will be equal, but not equal to the third; most plausibly

$$G_h = G_c \neq G_m$$

In this case a partial analogy will hold. The validity of Reynolds analogy, and improvements to it, will be discussed in detail in Chapter 5.

Provided that the mean distributions of U, H and C are similar, in the sense of equation (3.4.9), equations (3.4.10) indicate that the three diffusivities are equal when the three Reynolds fluxes are. If the diffusivities are interpreted as *eddy diffusivities*, describing only the turbulent mixing, this implies that the turbulent Prandtl and Schmidt numbers are unity:

$$\mathrm{Pr}_t = \mathrm{Sc}_t = 1 \tag{3.4.31}$$

However, if they are interpreted as *effective diffusivities*, describing the combined molecular and turbulent transfers, the molecular Prandtl and Schmidt numbers are also unity:

$$\mathrm{Pr} = \mathrm{Sc} = \mathrm{Pr}_t = \mathrm{Sc}_t = 1 \tag{3.4.32}$$

Reynolds analogy is often derived from these requirements, but the derivation based on equality of the Reynolds fluxes, that given above, is more clear cut.

3.4.5 The Mixing Length

For many years the most widely used way of reducing the equations of turbulent motion to a closed, solvable form was to follow Prandtl in introducing the mixing length. This was conceived as the distance which a particle travelled across the flow before mixing with its new environment. When the variation of the mixing length was specified in an appropriate way, the velocity distribution and other characteristics of the mean flow could be predicted.

The original mixing-length argument took turbulent mixing to be analogous to molecular transport in a gas, where, between collisions, the properties of the molecules are unchanged. We shall not adopt this restricted view, but instead develop the mixing-length concept as an alternative formulation of the gradient-diffusion model of turbulent mixing.

The result (3.4.14), which gives the fluctuations resulting from small random motions across the mean gradient $\partial\bar{P}/\partial y$, can be written more simply as

$$P(t) = \bar{P}(t) - \frac{\partial \bar{P}}{\partial y} L \tag{3.4.33}$$

where

$$L = \int_0^\infty v(t-\tau)\,\mathrm{e}^{-\delta\tau}\,\mathrm{d}\tau \tag{3.4.33}$$

is the effective distance travelled by the particle currently at the point of interest. The distance L fluctuates randomly, and is a function of δ, the parameter defining the exchange of the property P with its historical environment. Introducing $S(y) + s(y,t)$ in place of $P(y,t)$, we calculate the mixing correlation function and eddy diffusivity:

$$\overline{sv} = -\overline{vL_s}\frac{\partial S}{\partial y} \quad \text{and} \quad \epsilon_s = \overline{vL_s} \tag{3.4.34}$$

with L_s the length appropriate to a particular entity S.

So far we have merely interpreted the gradient-diffusion results. Now a new element is added, an explicit way of calculating the transferring fluctuation v in terms of the mean velocity gradient $\mathrm{d}U/\mathrm{d}y$. When applied to the velocity $U+u$, equations (3.4.33) give

$$U + u = U - \frac{\mathrm{d}U}{\mathrm{d}y} L_\mathrm{m}$$

We argue that v depends on the mean-velocity gradient in much the same way as does u, and introduce

$$v = \left|\frac{\mathrm{d}U}{\mathrm{d}y}\right| L_v \tag{3.4.35}$$

with $L_v(\sim L_\mathrm{m})$ a randomly varying length characterizing the v-fluctuations.

The absolute value bars ensure that $v \gtrless 0$ always corresponds to $L_v, L_m \gtrless 0$; thus $\overline{uv} < 0$ for $\mathrm{d}U/\mathrm{d}y > 0$.

Equations (3.4.34) now become

$$\overline{sv} = -\overline{L_v L_s}\left|\frac{\mathrm{d}U}{\mathrm{d}y}\right|\frac{\partial S}{\partial y} \quad \text{and} \quad \epsilon_s = \overline{L_v L_s}\left|\frac{\mathrm{d}U}{\mathrm{d}y}\right| \tag{3.4.36}$$

Finally, we introduce the mixing length l_s, through

$$\overline{sv} = -l_s^2\left|\frac{\mathrm{d}U}{\mathrm{d}y}\right|\frac{\partial S}{\partial y} \quad \text{and} \quad \epsilon_s = l_s^2\left|\frac{\mathrm{d}U}{\mathrm{d}y}\right| \tag{3.4.37}$$

This step can be justified by arguing that $L_v \sim L_s$, so that a single length suffices. However, l_s^2 may be thought of simply as a compact notation for $\overline{L_v L_s}$, although this notation assumes a positive correlation between L_v and L_s, and thus implies that $\epsilon_s \geqslant 0$.

Like the eddy-diffusivity and Reynolds-flux formulations, the mixing-length model is helpful only when it leads easily to a reasonably accurate description of some aspect of a turbulent flow. Fortunately, it does this in some circumstances, although some would say unfortunately, feeling that these successes have led to undue emphasis on a model of limited applicability. Below, we consider some of the assumptions which have been made regarding the variation of the mixing length within turbulent fluid. We shall see later that most of the correct results which follow from them can be obtained from dimensional analysis of the basic problem, without any reference to the mixing length.

The proposed expressions for the mixing length contain empirical coefficients which must be selected to match the observed behaviour of turbulent flows. When the appropriate factors are introduced, it often turns out that the magnitude of the mixing length is inconsistent with the arguments which led to it, that is, the mixing length is comparable to the flow width, rather than very much smaller. This observation suggests that both the mixing-length and gradient-diffusion hypotheses are seriously in error in their representation of the physical processes of mixing, and that they cannot provide a suitable basis for the prediction of more varied and more complex turbulent motions.

Some characteristics which have been assumed for mixing lengths are indicated below. In each case a formula for the eddy diffusivity follows from equations (3.4.37); some of the proposals were made originally in terms of the eddy diffusivity, the mixing length being derived as a consequence.

(a) The mixing lengths are the same for all quantities transferred:

$$l_m = l_h = l_D \tag{3.4.38}$$

Subject to this restriction, the eddy diffusivities are also equal. Provided that the mean-property distributions are similar, the Reynolds flux is the same for every property, and Reynolds analogy (3.4.30) applies. It should be realized that the assumptions (3.4.38) are made for convenience, and are not inherent in the mixing-length model; some inconsistencies in these assumptions will become apparent shortly.

(b) For a *free shear layer* of width b:

$$\frac{l_m}{b} = \text{an empirical constant} \tag{3.4.39}$$

for a particular species of flow. This assignment can be justified by arguing that the mixing across the layer is unrestricted, while its scale increases with the width of the flow. Values of the constant (~0·1) are indicated in Table 7.2.

(c) In *fully turbulent flow near a wall*:

$$\frac{l_m}{y} = K, \quad \text{an empirical constant} \tag{3.4.40}$$

The constant, often termed von Kármán's constant, is found to be around 0·4. This assumption can be justified by arguing that the size of the dominant 'eddies' is defined by the distance to the wall. Introduced into equations (3.4.37) and applied to a *constant-stress layer*, it gives

$$K^2 y^2 \left|\frac{dU}{dy}\right| \frac{dU}{dy} = -\overline{uv} = \frac{\tau_w}{\rho} = u_f^2$$

with u_f the friction velocity. Re-arranging, we have

$$\frac{dU}{dy} = \frac{u_f}{Ky} \tag{3.4.41}$$

Integration gives

$$\frac{U}{u_f} = \frac{1}{K} \ln y + \text{constant} \tag{3.4.42}$$

This *logarithmic law of the wall* will later be derived using dimensional arguments alone.

(d) For an *open channel* of depth d:

$$\frac{l_m}{y} = K\left(1 - \frac{y}{d}\right)^{\frac{1}{2}} \tag{3.4.43}$$

This gives somewhat more consistent results in the body of the flow than does the simpler assumption (3.4.40). By replacing d by R or $\frac{1}{2}b$ we obtain results for a pipe of radius R or a closed channel of width b.

(e) For the outer part of a turbulent *boundary layer* of thickness δ:

$$\frac{l_m}{y} = f\left(\frac{y}{\delta}\right) \tag{3.4.44}$$

with f supposed to be a universal function, sometimes chosen so that l_m is constant for $y/\delta > 0{\cdot}2$, say.

(f) Using dimensional arguments, von Kármán defined the mixing length in terms of the local mean flow:

$$l_m = K \frac{dU/dy}{d^2 U/dy^2} \tag{3.4.45}$$

When this result is applied to a constant-stress layer, the results (3.4.41, 42) are recovered; hence the constant K is the same as that appearing above.

A significant development of the mixing-length argument was made by Taylor. He reasoned that it was unrealistic to treat the transfers of momentum and heat in the same way, since the velocity of a moving fluid element will be influenced by the associated pressure fluctuations. This led him to consider the *transfer of vorticity*, for in a two-dimensional motion this quantity behaves like heat, being conserved, save for the effects of small-scale diffusion. See the discussion centering on equations (1.4.3).

The role of the vorticity fluctuations can be demonstrated by rewriting the force balance (3.3.12) for plane parallel flow. For the fully turbulent part of the flow

$$\frac{1}{\rho}\frac{dP_w}{dx} = \frac{1}{\rho}\frac{d\tau}{dy} = -\frac{d\overline{uv}}{dy} = -\overline{u\frac{\partial v}{\partial y}} - \overline{v\frac{\partial u}{\partial y}}$$

If the turbulent motion is assumed to be confined to the same plane as the mean motion, it is governed by a continuity equation which has the form of the mean-motion result (3.3.7). Hence

$$\frac{\partial u}{\partial x} + \frac{\partial v}{\partial y} = 0 \tag{3.4.46}$$

and one of the terms of the momentum equation is seen to be zero:

$$\overline{u\frac{\partial v}{\partial y}} = -\frac{1}{2}\frac{\partial \overline{u^2}}{\partial x} = 0$$

in this flow. By adding in the term

$$\overline{v\frac{\partial v}{\partial x}} = \frac{1}{2}\frac{\partial \overline{v^2}}{\partial x} = 0$$

we convert the momentum equation to the form

$$\frac{1}{\rho}\frac{d\tau}{dy} = \overline{v\left(\frac{\partial v}{\partial x} - \frac{\partial u}{\partial y}\right)} = \overline{v\omega} \tag{3.4.47}$$

The fluctuation in the vorticity or rotation of the fluid is

$$\omega = \frac{\partial v}{\partial x} - \frac{\partial u}{\partial y} \tag{3.4.48}$$

Its mean value is simply $-dU/dy$ for a parallel flow.

The mixing-length results can be used to calculate the correlation function $\overline{v\omega}$. Replacing $S+s$ by $-\mathrm{d}U/\mathrm{d}y+\omega$ in equations (3.4.37), we obtain from this discussion of vorticity transport:

$$\frac{1}{\rho}\frac{\mathrm{d}\tau}{\mathrm{d}y}=\overline{v\omega}=l_\omega^2\left|\frac{\mathrm{d}U}{\mathrm{d}y}\right|\frac{\mathrm{d}^2U}{\mathrm{d}y^2}$$

and (3.4.49)

$$\epsilon_\omega=l_\omega^2\left|\frac{\mathrm{d}U}{\mathrm{d}y}\right|$$

This relationship between shear stress and mean velocity differs from that obtained directly from $\overline{uv}$, namely

$$\frac{1}{\rho}\frac{\mathrm{d}\tau}{\mathrm{d}y}=-\frac{\mathrm{d}\overline{uv}}{\mathrm{d}y}=\frac{\mathrm{d}}{\mathrm{d}y}\left(l_\mathrm{m}^2\left|\frac{\mathrm{d}U}{\mathrm{d}y}\right|\frac{\mathrm{d}U}{\mathrm{d}y}\right)$$

and (3.4.50)

$$\epsilon_\mathrm{m}=l_\mathrm{m}^2\left|\frac{\mathrm{d}U}{\mathrm{d}y}\right|$$

In this context, Prandtl's original analysis is often referred to as the *momentum-transport* model, to distinguish it from Taylor's *vorticity-transport* theory.

The most important lesson to extract from this comparison of momentum-transport and vorticity-transport results is the warning that both are unrealistic. Each represents a form of gradient diffusion, at best only a rough approximation to turbulent mixing, as is made apparent by the inconsistencies of eddy-diffusivity and mixing-length results.

With this caveat in mind, let us see if anything more positive can be learned from the contradictory results of Prandtl and Taylor. For simplicity, we assume that

$$l_\mathrm{m},\ l_\omega=\text{constants across the flow} \tag{3.4.51}$$

This restricts the discussion to free turbulence and core flows, where the mean velocity is predicted reasonably well on this basis. It is in such flows, if anywhere, that the vorticity-transport model may apply, for significant two-dimensional motions are observed there, for example, the eddies shed into the wake of a cylinder lying normal to the stream. Equations (3.4.49, 50) predict the same stress distribution if

$$2l_\mathrm{m}^2=l_\omega^2=\text{constant} \tag{3.4.52}$$

This result casts doubt on the assumption (3.4.38).

Suppose that we now seek to develop an analogy between heat transfer and friction. Which of the choices

$$\epsilon_\mathrm{h}=\epsilon_\mathrm{m} \quad\text{and}\quad \epsilon_\mathrm{h}=\epsilon_\omega \tag{3.4.53}$$

is the more realistic? For free turbulence, where the mechanisms of heat and vorticity transport are most likely to be similar, the second is found to give better results. For wall turbulence, which in any case is not governed by the relations (3.4.52), the former choice is more successful. We conclude that the restrictions (3.4.38) are not justified in general.

3.4.6 The Eddy-viscosity and Mixing-length Hypotheses

We have developed the mixing-length model as a particular expression of gradient diffusion. It can also be considered as an extension of the kinetic theory of gases, following Prandtl's original thinking. The result

$$\epsilon_s = \overline{vL_s} \tag{3.4.54}$$

provides the starting point. In a gas at low pressure, the molecules may be assumed rigid and non-attracting, and the particle velocity and the distance travelled are unrelated. Moreover, since these scales are very small compared to those of the bulk motion, they are not influenced by it. Hence the molecular viscosity is given by

$$\nu \propto \lambda V' \tag{3.4.54}$$

with λ the mean-free-path (dependent on density) and V' the r.m.s. velocity of the molecules (dependent on temperature). Detailed calculations give $\frac{1}{3}$ for the constant of proportionality appropriate to the kinematic viscosity, and constants of the same order for other diffusivities.

In a turbulent transfer process, the elements v and L_s are correlated. The forms obtained for the eddy diffusivity take account of this:

$$\epsilon_s = \overline{v^2} \int_0^\infty R_{22}(\tau) \mathrm{e}^{-\delta_s \tau} \, \mathrm{d}\tau$$

from equations (3.4.16, 17), and

$$\epsilon_s = l_s^2 \left| \frac{\mathrm{d}U}{\mathrm{d}y} \right|$$

from equations (3.4.37). The latter uses the velocity gradient $\mathrm{d}U/\mathrm{d}y$ to express the correlation.

Considering now only the eddy viscosity, we can extend the parallel with kinetic theory. The results

$$-\overline{uv} = \frac{\tau}{\rho} = l_m^2 \left(\frac{\mathrm{d}U}{\mathrm{d}y} \right)^2 = \epsilon_m \frac{\mathrm{d}U}{\mathrm{d}y} \tag{3.4.55}$$

combine to give

$$\epsilon_m = l_m \left(\left| \frac{\tau}{\rho} \right| \right)^{\frac{1}{2}} = l_m \left(l_m \frac{\mathrm{d}U}{\mathrm{d}y} \right) \tag{3.4.56}$$

For a constant-stress wall layer with $\tau_w/\rho = u_f^2$, we obtain

$$\epsilon_m = l_m u_f = K u_f y \tag{3.4.57}$$

on taking $l_m = Ky$, as in equation (3.4.40). (These last results can in fact be obtained by dimensional arguments, on requiring that l_m and ϵ_m depend only on u_f and y.) The parallels with the molecular diffusivity (3.4.54) are evident: the mean-free-path is replaced by the mixing length, and the molecular velocity by the friction velocity or its equivalent, $l_m \mathrm{d}U/\mathrm{d}y$.

It is also of interest to express the eddy viscosity in terms of the *effective velocity of lateral convection,* which is related to the Reynolds flux of momentum by $V_m = G_m/\rho$. Using the results (3.4.25), we obtain

$$\epsilon_m = \left(\frac{U_0}{\mathrm{d}U/\mathrm{d}y}\right) V_m \tag{3.4.58}$$

with U_0 a mean-velocity scale. Here V_m appears as the 'mixing velocity', and it is the length scale which is expressed in terms of the mean velocity distribution. This length scale does not appear in the Reynolds-flux model, but enters because the eddy viscosity has been expressed in terms of that model.

The results (3.4.55) to (3.4.58) define the empirical or conceptual inputs required by the simple theoretical models of turbulent activity:

(1) the mixing-length model requires the variation of a length scale of the turbulence;

(2) the Reynolds-flux model requires the variation of a velocity scale of the turbulence; and

(3) the eddy-viscosity model requires the variation of the product of length and velocity scales.

As was pointed out earlier, the ease with which these requirements can be met varies from flow to flow and from one part of a flow to another.

The analytical results obtained above reveal the nature and limitations of the eddy-viscosity and mixing-length hypotheses. The *eddy-viscosity hypothesis* is simply that

$$\epsilon_m = \frac{\tau}{\rho} \Big/ \frac{\mathrm{d}U}{\mathrm{d}y} \tag{3.4.59}$$

be finite and positive. A variety of distributions $\epsilon_m(y)$ may be introduced to represent varied flows, but in every case the points y_m, where $\mathrm{d}U/\mathrm{d}y = 0$, must coincide with the point y_0, where $\tau = 0$:

$$y_m = y_0 \tag{3.4.59}$$

The *mixing-length hypothesis** is that

$$\epsilon_m = l_m^2 \left|\frac{\mathrm{d}U}{\mathrm{d}y}\right| = l_m \left(\left|\frac{\tau}{\rho}\right|\right)^{\frac{1}{2}} \tag{3.4.60}$$

* This term is sometimes applied to the results (3.4.57), applicable to a constant-stress wall layer. This usage is unfortunate in giving an excessively restricted impression both of the mixing-length model and of the wall-layer results.

with l_m finite and positive. This adds the restriction that

$$\epsilon_m = 0 \quad \text{where} \quad y = y_m = y_0 \tag{3.4.60}$$

Any function $l_m(y)$ which remains finite at this point will give this characteristic.

The conditions (3.4.59, 60) restrict the use of the eddy-viscosity and mixing-length models. Here we consider only fully developed channel flows, but analogous limitations apply in other situations.

(1) In calculating the flow rate and friction for a symmetrical duct—a pipe, for example—little difficulty arises. From symmetry, $y_0 = y_m$, and the vanishing of ϵ_m at the axis introduces only a small error into the results obtained from the mixing-length model.

(2) To find the flow rate and friction in an asymmetric duct—for example, one with walls of differing roughness—we must remove the possible singularity by assuming $y_0 = y_m$, contrary to experimental evidence. This need not introduce an important error, as the detail of the core flow is not usually very significant in this calculation.

(3) The calculation of heat transfer from wall to wall poses a more serious problem. The vanishing of ϵ_m implies that the thermal diffusivity ϵ_h also vanishes, and that very high temperature gradients exist near $y_0 = y_m$. Thus the mixing-length hypothesis is inappropriate for this calculation. However, no fundamental difficulty arises with the eddy-viscosity model, since a finite value of ϵ_m can be assigned once it is assumed that $y_0 = y_m$.

3.5 STRESS-SCALING OF THE ENERGY EQUATION

The limitations of the eddy viscosity and mixing length necessitate the development of more realistic models of turbulent diffusion. Here we shall take two steps in this direction. A property of the turbulence itself—the local mixing stress—will be used as an index of turbulent activity, in the place of the mean velocity gradient. And an additional source of information will be introduced—the energy equation (3.4.2) for parallel flow. Although the results go only a small way towards removing the limitations of the simplest models, they serve as a transition to the more complex models which will be considered in Chapter 9. We also obtain a somewhat more consistent picture of the wall layer, and see why the mixing length is reasonably successful there.

Since the mixing stress and velocity gradient are closely linked in many flows, the use of the stress as a scale of turbulent activity cannot provide a comprehensive solution to the problems of gradient-diffusion models. Moreover, in this discussion of parallel flows, typified by developed pipe flow, we shall not take account of mean-flow convection. These two fundamental weaknesses will be removed in the final chapters of this book.

Equilibrium Layers

Our discussion of models of turbulence started with the momentum and energy equations (3.4.1, 2) for parallel flow. In seeking to predict mean-velocity variations, we have so far attacked the momentum equation or the effective stress which appears in it. Now we turn to the energy equation and, using the local shear stress as a scale of turbulent activity, see what this equation reveals about turbulent wall layers. In doing this, we introduce the concept of an equilibrium layer:

(1) Strictly, this is a region in which the production and dissipation of turbulence energy are locally in balance. For exact energy equilibrium, the energy equation gives

$$\overline{uv}\frac{\mathrm{d}U}{\mathrm{d}y} + \varepsilon_t = 0 \tag{3.5.1}$$

Figure 3.8 shows that this is nearly true in the fully turbulent part of the wall layer in a pipe. It is also true for the inner part of a developing boundary layer, where both mean-flow convection and energy diffusion are unimportant.

(2) More loosely, the term 'equilibrium' denotes a region where lateral energy diffusion may be significant, but is, like dissipation and production, determined by local properties of the turbulence. Such a region is better called a *locally determined layer*; in it, the complete energy equation (3.4.2) must be used.

(3) Even more loosely, the name 'equilibrium' is applied to any layer whose scaled form remains constant while the layer develops, in other words, to a *self-preserving or self-similar layer.*

In this chapter we shall consider only equilibrium layers of types (1) and (2); in Chapter 8, we shall treat self-preserving or equilibrium boundary layers.

We consider first a *constant-stress layer*, for which

$$-\overline{uv} = \frac{\tau_w}{\rho} \tag{3.5.2}$$

in the fully turbulent region. Arguing that the dissipation is maintained by energy extraction in motions whose scale is determined by the constraint of the wall, we take

$$\varepsilon_t \sim \frac{\tau_w}{\rho}, y \tag{3.5.3}$$

whence

$$\varepsilon_t = \frac{A}{y}\left(\frac{\tau_w}{\rho}\right)^{\frac{3}{2}} \tag{3.5.3}$$

with A a dimensionless constant. Substitution in the requirement of energy equilibrium (3.5.1) gives

$$\frac{dU}{dy} = A\frac{u_f}{y} \tag{3.5.4}$$

and integration leads to a logarithmic mean-velocity variation. These results are equivalent to those obtained from Prandtl's hypothesis (3.4.40): $l_m = Ky$ for the wall layer, with $K = 1/A$.

Generalizing this argument to include the part of a boundary layer where the stress varies significantly, we take

$$\varepsilon_t = \frac{(\tau/\rho)^{\frac{3}{2}}}{L_\varepsilon} = \frac{(-\overline{uv})^{\frac{3}{2}}}{L_\varepsilon} \tag{3.5.5}$$

For this part of the flow we cannot easily guess the characteristic length relating stress and dissipation; instead we invert the procedure, and use these quantities to define L_ε, the *dissipation length scale*. For a constant-stress wall layer, we recover

$$L_\varepsilon = l_m = Ky$$

showing that near a wall this scale reduces to the mixing length. When the stress varies, and the two lengths differ, we are still able to calculate the scale L_ε characteristic of the local turbulence, independent of the velocity gradient.

In order to correlate the outer flows in a variety of boundary layers, Bradshaw adopted

$$\frac{L_\varepsilon}{\delta} = f\left(\frac{y}{\delta}\right) \tag{3.5.6}$$

with δ the boundary-layer thickness, as an alternative to the mixing-length proposal (3.4.44). The data correlation (3.5.6) does in fact provide a somewhat better representation of boundary layers (that is, the function $f(y/\delta)$ defined in this way varies less from flow to flow), and is thus a better basis for boundary-layer calculations.

Locally Determined Layers

Turning now to less rigidly defined equilibrium layers, the word now having sense (2), we make use of the full equation (3.4.2). Considering a wall layer where the local length scale is y, we follow Townsend in specifying all of the turbulence properties in terms of the local shear stress. This leads to the dimensionally homogeneous forms:

$$-\overline{uv} = \frac{\tau}{\rho} \qquad \varepsilon_t = \frac{A}{y}\left(\frac{\tau}{\rho}\right)^{\frac{3}{2}}$$

and

$$\overline{v\left(\frac{p}{\rho} + \tfrac{1}{2}q^2\right)} = B_1\left(\frac{\tau}{\rho}\right)^{\frac{3}{2}} \tag{3.5.7}$$

with B_1 another empirical constant. The first two are generalizations of the constant-stress results (3.5.2, 3). The energy equation (3.4.2) now gives

$$\frac{dU}{dy} = \frac{A}{y}\left(\frac{\tau}{\rho}\right)^{\frac{1}{2}}\left(1 + \frac{3B_1}{2A}\frac{y}{\tau}\frac{d\tau}{dy}\right) \tag{3.5.8}$$

with the second term in square brackets representing the diffusion of energy.

For fully developed pipe and channel flows, we have seen in equations (3.3.16, 18) that the stress varies linearly:

$$\tau = \tau_w + \alpha y \tag{3.5.9}$$

This provides a reasonable approximation for certain other near-wall flows, notably, that in the constant-pressure boundary layer. For a linear stress variation, equation (3.5.8) can be integrated, giving a generalization of the logarithmic law found for constant stress; see Example 3.28. For values of y that are small, but not small enough to fall within the viscous layer, the result approaches the logarithmic form. Moreover, there is a more extensive region where the equation

$$\frac{dU}{dy} = \frac{A}{y}\left(\frac{\tau}{\rho}\right)^{\frac{1}{2}} \tag{3.5.10}$$

with the energy-diffusion term omitted, gives results indistinguishable from those of the full equation (3.5.8). This simpler equation is in fact the result obtained from Prandtl's wall-layer assumption $l_m = Ky$, with the stress allowed to vary.

In deriving equation (3.5.8) we have assumed that the wall layer is not influenced by the opposite wall, nor by the core or outer flow. As the distance from the wall increases, this hypothesis becomes untenable. The question arises: does the simpler result (3.5.10) cease to apply because energy diffusion becomes important, or because the outer flow begins to have an influence? Experiments suggest that the constant $3B_1/2A \simeq -0{\cdot}2$. This small value indicates that, for wall layers across which the stress does not vary much, penetration of the outer flow will cause equation (3.5.10) to fail before the diffusive effect becomes significant. Hence for pipe and plane channel flows and for the constant-pressure boundary layer, it seems unlikely that equation (3.5.8) will ever be appropriate.

When the stress varies by a considerable factor within the wall layer, the opposite conclusion can be reached: since the diffusion-specifying factor $\alpha y/\tau$ is much larger, the full equation (3.5.8) becomes relevant. This is the case in boundary layers subjected to strong pressure gradients, and in some of the asymmetric channel flows governed by the stress variation (3.3.16). The family of equilibrium layers defined by local values of the stress and stress gradient has as its extreme members:

(1) the *constant-stress layer*, determined solely by the wall stress, and

(2) the *zero-stress layer*, where the wall stress is very small, and the flow is determined by the stress gradient. In this case the turbulence near the wall is maintained not by local production, but by diffusion of momentum and energy from the outer turbulence.

As was pointed out earlier, these results based on the use of the shear stress as a scale of turbulent activity retain many of the limitations of the mixing-length hypothesis. But they have a more certain physical and analytical basis, and illustrate an approach which has proved capable of far-reaching development.

FURTHER READING

Bird, R. B., W. E. Stewart and E. N. Lightfoot. Reference 1: Chapters 1, 2, 3, 5, 8, 10, 16 and 20
Bradshaw, P. Reference 2: Chapters 1 and 3
Goldstein, S. Reference 4: Chapter V, Sections 68 to 85
Hinze, J. O. Reference 5: Sections 5-1 to 5-3 and 7-8 to 7-10
Prandtl, L. Reference 8: Chapter III, Sections 4 and 5
Schlichting, H. Reference 11: Chapters 18 and 19
Schubauer, G. B. and C. M. Tchen. Reference 7: Section B, Chapters 1 to 3
Spalding, D. B. *Convective Mass Transfer*, Arnold, London (1963): Section 2.2
Townsend, A. A. Reference 13: Chapters 2 and 9

SPECIFIC REFERENCES

Drummond, G. 'Steamside pressure gradients in surface condensers', *Proc. I. Mech. E.*, **186**, pp. 117–124 (1972)
Hinze, J. O. 'Turbulent pipe flow', in Reference 14
Laufer, J. 'The structure of turbulence in fully developed pipe flow', *Rep. 1174, U.S. Nat. Adv. Com. Aero.* (1954)
Reynolds, A. J. 'Wall layers with non-uniform shear stress', *J. Fluid Mech.*, **22**, pp. 443–448 (1965)
Townsend, A. A. 'Equilibrium layers and wall turbulence', *J. Fluid Mech.*, **11**, pp. 97–120 (1961)
Wallis, G. B. and R. S. Silver. 'Studies in pressure drop with lateral mass extraction', *Proc. I. Mech. E.*, **180**, pp. 27–42 (1965–66)

EXAMPLES

3.1 The turbulent motion in the viscous sublayer is governed by:

(1) the continuity equation for the velocity fluctuations

$$\frac{\partial u}{\partial x}+\frac{\partial v}{\partial y}+\frac{\partial w}{\partial z}=0$$

or, in tensor notation

$$\frac{\partial u_1}{\partial x_1} + \frac{\partial u_2}{\partial x_2} + \frac{\partial u_3}{\partial x_3} = 0$$

a result obtained in Example 3.6, and

(2) the equations of motion

$$\mu \frac{\partial^2 u_i}{\partial y^2} = \frac{\partial p}{\partial x_i} \quad \text{with } i = 1, 2, 3$$

which assume that the pressure gradient is balanced by the viscous stresses, the fluid acceleration and mixing stresses being negligible.

(a) Show that at a fixed solid wall:

$$u = v = \frac{\mathrm{d}v}{\mathrm{d}y} = 0$$

(b) Using the results (a), and taking the shear stress to be constant, show that at the wall:

$$\nu \frac{\mathrm{d}U}{\mathrm{d}y} = \frac{\tau_w}{\rho}$$

$$\nu \frac{\mathrm{d}^2 U}{\mathrm{d}y^2} = \frac{\mathrm{d}(\overline{uv})}{\mathrm{d}y} = 0$$

$$\nu \frac{\mathrm{d}^3 U}{\mathrm{d}y^3} = \frac{\mathrm{d}^2(\overline{uv})}{\mathrm{d}y^2} = 0$$

$$\nu \frac{\mathrm{d}^4 U}{\mathrm{d}y^4} = \frac{\mathrm{d}^3(\overline{uv})}{\mathrm{d}y^3} = 3\overline{\frac{\mathrm{d}u}{\mathrm{d}y}\frac{\mathrm{d}^2 v}{\mathrm{d}y^2}}$$

all these quantities being evaluated at the wall.

Express $U(y)$ as a power series in y. Why does the mean velocity variation depart suddenly from the linear form?

(c) Show that in the sublayer

$$\overline{u^2} = \overline{u_1^2} = \overline{\left(\frac{\mathrm{d}u_1}{\mathrm{d}y}\right)^2} y^2 + \frac{1}{\mu}\overline{\frac{\mathrm{d}u_1}{\mathrm{d}y}\frac{\partial p}{\partial x_1}} y^3 + \frac{1}{4\mu^2}\overline{\left(\frac{\partial p}{\partial x_1}\right)^2} y^4$$

the derivatives being evaluated at the wall. How do $\overline{v^2}$ and $\overline{w^2}$ vary with y near the wall?

(d) Find an expression for the variation of $\overline{uv}$ similar to that given for $\overline{u^2}$ in (c). How does this quantity vary with y? Use the data of Figure 3.3 to find a numerical value for the coefficient of y^3.

3.2 (a) Using the results of Example 3.1, show that the relative intensity u'/U may be expected to remain constant across the sublayer, $U(y)$ being the local mean velocity in the x, y-plane.

(b) How do v'/U and w'/U vary in the sublayer?

3.3 Using the data of Figures 3.2 and 3.3, and the fact that the shear stress varies linearly across a pipe, find the variation of the correlation coefficient relating the fluctuating components u and v.

3.4 Air at atmospheric pressure flows through a smooth-walled pipe with a mean velocity of 100 m/sec. Estimate the variation of the mean pressure within a section of the pipe. Are pressure differences of this magnitude measurable? What experimental problems may they pose? Do they provide a feasible way of measuring the intensity $\overline{v^2}$?

3.5 Consider a turbulent shear flow in which the mean motion is two-dimensional, with $W\,(=U_3)=0$, and nearly parallel, with

$$\frac{\partial U}{\partial y}\left(=\frac{\partial U_1}{\partial x_2}\right) \gg \frac{\partial U}{\partial x}\left(=\frac{\partial U_1}{\partial x_1}\right)$$

(a) Write down the nine apparent-stress components τ_{ij} in the form of a three-by-three matrix. Do the same for the nine mean rates-of-strain $\partial U_i/\partial x_j + \partial U_j/\partial x_i$.

(b) Repeat (a) for exactly parallel mean flow, and show how the stresses act on, and the strains deform, the fluid within a cubical element of volume.

(c) Use the data describing a plane wake which are given in Table 1.2 to find the principal axes of the Reynolds stresses in the wake flow. These axes are defined by $\tan\alpha = 2\tau_{12}/(\tau_{11}-\tau_{22})$. What is their physical significance?

(d) Find also the principal axes of mean strain, taking the mean flow to be exactly parallel. In what way does the relationship between stress and strain differ from that in laminar flow? Why?

3.6 (a) Generalize equation (3.3.1) to three dimensions using:

(1) Cartesian coordinates x, y, z (velocities U, V, W), and
(2) polar coordinates x, r, θ (velocities U, V, W).

Note that the results apply to fluctuating components $J_i + j_i$ as well as to the mean values J_i.

(b) Obtain the continuity equations for constant-density flow for mean flows that are:

(1) plane $(U, V, 0)$, and
(2) axisymmetric without swirl $(U, V, 0)$.

Do the components U and V need to be constant in time?

(c) Using the mean-flow equations (3.3.7, 8), find the equations governing the fluctuations in velocity about the mean values U, V.

3.7 (a) What form will the flux (3.2.1) take on when the density fluctuates? Take the fluctuating value to be $R+\rho$, and assume that K is constant when calculating the effective flux.

(b) Show that the flux of the basic fluid is given by

$$J_y = RV + \overline{\rho v}$$

What is the direction of the mean velocity V just above a heated, impermeable surface along which a fluid flows uniformly? For U and $\mathrm{d}U/\mathrm{d}y > 0$, what is the sign to be expected for $\overline{\rho u}$?

(c) Obtain the generalization of equation (3.2.10) including density fluctuations. For the situation discussed in (b), what signs do you expect for the several terms?

(d) Generalize the continuity equation (3.3.7) to include flows in which:

(1) the mean density varies slowly in time, and
(2) the density fluctuates rapidly.

3.8 (a) Reduce the generalized diffusion equation (3.3.10) to the form of the simple diffusion equation (3.3.11) by following equations (3.2.2) in introducing

(1) eddy diffusivities ϵ_x and ϵ_y, and
(2) effective convection velocities V_x and V_y.

(b) What interpretations may be given to the terms of the equations produced in this way?

3.9 (a) Find the variation of velocity in the laminar flow of a Newtonian fluid through an annulus, in terms of the stresses at the two walls, and in terms of the velocities there.

(b) Taking the walls to be fixed, find the flow rate through the annulus as a function of the pressure gradient.

3.10 Use the results of Example 3.9 to discuss the following laminar flows under gravity:

(1) flow vertically downwards through an annulus with fixed walls,
(2) flow through an annulus with fixed walls and with axis inclined at the angle θ to the horizontal,
(3) flow downwards on the outside of a round rod, and
(4) flow downwards on the inside of a circular cylinder.

(a) What boundary conditions are appropriate?

(b) How are the pressure gradient and wall stresses related, and what force balances should replace equation (3.3.17)?

(c) Examine the limits $R_1 \to 0$ and $R_1 \to R_2$. In the former case, note particularly the velocity variation and shear stress near the inner cylinder.

3.11 Two immiscible fluids of differing densities and velocities move in uniform, laminar, plane flow along a very broad horizontal channel which has fixed walls.

(a) How is the pressure drop along the channel related to the flow rates and depths of the two fluids?

(b) Suppose that the two fluids are air and water. State in general terms how the velocity varies across the channel when:

(1) the depths are equal,
(2) the water is a thin film on the bottom, and
(3) the air is a thin film at the top.

For a given pressure drop, how do the flow rates (2) and (3) compare with those in the absence of the thin films?

3.12 In the Bingham model of a non-Newtonian fluid, which represents some suspensions and pastes reasonably well, the velocity gradient and shear stress are related (for τ, $\mathrm{d}U/\mathrm{d}y > 0$) by

$$\tau = \mu_0 \frac{\mathrm{d}U}{\mathrm{d}y} + \tau_0 \quad \text{for } \tau > \tau_0$$

and

$$\frac{\mathrm{d}U}{\mathrm{d}y} = 0 \quad \text{for } \tau < \tau_0$$

with μ_0 and τ_0 positive constants.

(a) What is the physical significance of these relationships?

(b) Obtain results paralleling those of Table 3.2 for a Bingham plastic undergoing
(1) plane shearing flow, and
(2) plane pressure flow.

(c) What similarities and dissimilarities are there between the behaviour of this material and a Newtonian fluid capable of developing turbulent motion?

3.13 (a) After multiplying by U, the local mean velocity, interpret equation (3.3.13) as an energy equation for the mean motion.
(b) What process is represented by each of the terms of equation (3.3.24)?

3.14 Equation (3.3.28) relates the dissipation within the turbulence to the production of turbulence energy.

(a) Use equation (3.3.25) to obtain a similar relationship for the total dissipation within a channel with fixed walls:

$$\int_A \varepsilon \, \mathrm{d}A = \int_A \tau \frac{\mathrm{d}U}{\mathrm{d}y} \mathrm{d}A = -\frac{\mathrm{d}\tau}{\mathrm{d}y} \int_A U \mathrm{d}A = -\dot{V}\frac{\mathrm{d}P_\mathrm{w}}{\mathrm{d}x}$$

State the physical significance of these results.
(b) What is the corresponding result for a channel with one moving wall?

3.15 Interpret the following conservation laws in terms of the general result (3.3.5):
(a) the force balance (3.3.12),
(b) the thermal energy balance (3.3.33, 35, 37)
(c) the mechanical energy equations (3.3.24, 25), and
(d) the overall energy equation (3.3.32).

3.16 (a) State the physical requirements which must be satisfied in order to set up turbulent pipe flows with the following special features:
(1) fully developed flow,
(2) feature (1) plus a linearly varying mean temperature,
(3) features (1) and (2) plus a linearly varying mean concentration of: (i) helium, and (ii) water vapour.

(b) Are these conditions easy to satisfy in practice? If not, are the equations of linearly varying flow still likely to provide useful approximations?

3.17 The temperature distribution within a circular tube is given by

$$\frac{1}{rU}\frac{\partial}{\partial r}\left(r\frac{\partial T}{\partial r}\right) = \frac{1}{\kappa}\frac{\partial T}{\partial x}, \text{ a constant}$$

(a) Derive this result from equation (3.3.3). What assumptions and restrictions have to be made?
(b) Introducing the appropriate velocity distribution $U(r)$ from Table 3.2, show that

$$T = \frac{U_\mathrm{c}}{4\kappa}\frac{\partial T}{\partial x} r^2 \left(1 - \frac{r^2}{4R^2}\right) + B\log r + \frac{\partial T}{\partial x} x + C$$

with B and C constants of integration. By applying appropriate boundary conditions, show that

$$T - T_\mathrm{c} = \frac{U_\mathrm{c}}{4\kappa}\frac{\partial T}{\partial x} r^2 \left(1 - \frac{r^2}{4R^2}\right)$$

where T_c is the temperature on the axis.

(c) Show that the average temperature of the fluid is

$$T_{\mathrm{a}} = \frac{2\pi}{A}\int_0^R Tr\,\mathrm{d}r = T_{\mathrm{c}} + \frac{5}{48}\frac{U_{\mathrm{c}}R^2}{\kappa}\frac{\partial T}{\partial x}$$

and that the bulk temperature (see equation 1.1.7) is

$$T_{\mathrm{b}} = \frac{2\pi}{\dot{V}}\int_0^R uTr\,\mathrm{d}r = T_{\mathrm{c}} + \frac{7}{96}\frac{U_{\mathrm{c}}R^2}{\kappa}\frac{\partial T}{\partial x}$$

(d) What physical significance for the bulk temperature is suggested by equations (3.3.43) to (3.3.46)? It is also termed the mixing-cup temperature; why?

3.18 (a) For the flow discussed in Example 3.17, show that

$$\dot{q}_{\mathrm{w}} = \tfrac{1}{4}\rho c_p U_{\mathrm{c}} R \frac{\partial T}{\partial x}$$

by considering:

(1) the temperature gradient obtained from the results given there, and
(2) the total energy input $\rho \dot{V} c_p \partial T/\partial x$ per unit length of the tube.

(b) Show that the heat transfer is specified by

$$\mathrm{Nu} = \frac{\dot{q}_{\mathrm{w}} d}{k(T_{\mathrm{w}} - T_{\mathrm{f}})} = \tfrac{8}{3}, 6, \tfrac{48}{11}$$

when the fluid temperature is defined in the following ways:

$T_{\mathrm{f}} = T_{\mathrm{c}}$, the temperature on the axis,
$= T_{\mathrm{a}}$, the average temperature, and
$= T_{\mathrm{b}}$, the bulk temperature.

(c) Using the friction coefficient given in Table 3.2, show that

$$\frac{\mathrm{St}}{\frac{1}{2}c_{\mathrm{f}}} = \frac{\mathrm{Nu}}{8\,\mathrm{Pr}} = \frac{3}{4\,\mathrm{Pr}}$$

when Nu is based on T_{a}. For what kinds of fluids is Reynolds analogy most appropriate for this laminar flow?

3.19 The calculations of Examples 3.17 and 3.18 are based on the assumption that the dissipation within the fluid is negligible compared with the transfer of energy by conduction.

(a) To see when this is valid, show that the ratio of the dissipation in unit length to the energy flow through the wall is

$$\frac{\int_A \varepsilon\,\mathrm{d}A}{\int_C \dot{q}_{\mathrm{w}}\,\mathrm{d}C} = \frac{\mathrm{d}P}{\mathrm{d}x}\Big/\rho c_p\frac{\partial T}{\partial x} = \frac{U_{\mathrm{a}}\tau_{\mathrm{w}}}{\dot{q}_{\mathrm{w}}} = \frac{8\,\mathrm{Pr}}{3}\,\frac{U_{\mathrm{a}}^2}{2c_p(T_{\mathrm{w}} - T_{\mathrm{a}})}$$

the last result applying only to laminar pipe flow.

(b) In what kinds of fluids and in what sorts of flows is dissipation likely to be important?

(c) Explain in general terms how to find the temperature distribution and Nusselt number when dissipation is significant.

3.20 (a) Write down the equation governing the distribution of concentration in a round tube through whose walls there is uniform mass transfer into a fluid in laminar motion.

(b) Compare with the energy equation of Example 3.17. Why can a direct analogy be drawn between these heat and mass transfers, but not between them and the momentum transfer defined by equation (3.3.17) with $\tau = \mu \mathrm{d}U/\mathrm{d}r$? In what way is the analogy between heat and mass transfer not exact?

(c) Using the parallels drawn in Table 3.1, convert the results of Example 3.18 into expressions describing the analogous mass transfer.

3.21 (a) Taking the Lagrangian correlation coefficient of equations (3.4.16) to be $R_{22}(\tau) = \exp(-\tau/T_L)$, with T_L the Lagrangian integral time scale, show that

$$\epsilon_s = \frac{\overline{v^2}\,T_L}{1 + \delta_s T_L}$$

gives the eddy diffusivity.

(b) When δ_s is large, which 'eddies' cease to influence the transfer? What is the effect on the eddy diffusivity?

(c) Considering the transfer of two properties, show that

$$\frac{\epsilon_1}{\epsilon_2} = \frac{1 + T_L \delta_2}{1 + T_L \delta_1}$$

What simple results are obtained when δ_2 and δ_1 are both small and both large? If $\delta_2 \gg \delta_1$, what can be said about the ratio ϵ_1/ϵ_2? What is the effect of increasing the Reynolds number of the flow, so that T_L becomes smaller?

(d) For the particular case of momentum and heat transfer, show that the turbulent Prandtl number is given by

$$\mathrm{Pr_t} = \frac{1 + T_L \delta_h}{1 + T_L \delta_m}$$

What values of $\mathrm{Pr_t}$ would you expect for $\mathrm{Pr} \gg 1$ (as for oils) and for $\mathrm{Pr} \ll 1$ (as for liquid metals)?

3.22 Consider the flow of a condensing vapour through a tube of constant cross-sectional area A and circumference C.

(a) Neglecting changes in fluid density, show that the pressure gradient is given by

$$\frac{\mathrm{d}P_w}{\mathrm{d}x} = -\frac{C}{A}\tau_w - \frac{\mathrm{d}(\rho U_a^2)}{\mathrm{d}x} = -\frac{C}{A} U_a (G_m - 2G_L)$$

where U_a is the average velocity across the tube, $G_m = \tau_w/U_a$ is the Reynolds flux characterizing the wall stress, and $G_L = -(1/C)\,\mathrm{d}(\rho U_a A)/\mathrm{d}x = -(\rho A/C)\,\mathrm{d}U_a/\mathrm{d}x$ is the lateral flow (per unit area) generated by condensation at the tube wall.

(b) The Reynolds flux of momentum does not have the value G_0 which is found in non-condensing flow of the vapour under the same conditions. The relationship between the two may be estimated by arguing that, in non-condensing flow, $\tau_w = \tau_0$ is generated by momentum transfer through two equal flows $\frac{1}{2}G_0$, one towards and one away from the wall; and that, in condensing flow, the flow towards the wall is $\frac{1}{2}(G_0 + G_L)$ and that away from it is still $\frac{1}{2}G_0$. Show that

$$G_m = G_0 + \tfrac{1}{2}G_L$$

according to this model. What does it imply about the structure of the turbulence?

(c) Using the preceding results, show that

$$\frac{dP_w}{dx} = -\frac{C}{A}U_a\left(G_0 - \frac{3G_L}{2}\right) = -\frac{2}{d}c_{f_0}\rho U_a^2 - \frac{3}{2}\rho U_a\frac{dU_a}{dx}$$

where $d = 4A/C$ is the tube diameter, and $c_{f_0} = \tau_0/(\frac{1}{2}\rho U_a^2)$ is the friction coefficient for non-condensing flow under the same conditions. When is it possible for the pressure to rise along the tube?

(d) Considering a uniform rate of condensation over a length L of the tube, show that

$$U_a = U_1 - (U_1 - U_2)\frac{x}{L}$$

with U_1 and U_2 the initial and final velocities. Find the variation in pressure within the length L, assuming that d, ρ and c_{f_0} are constant.

3.23 (a) What distributions of eddy viscosity are implied by the mean-velocity variations (3.3.23)?

(b) What velocity-deficit variation is predicted for the cores of pipe and channel flows, if the mixing length is assumed to be constant?

(c) Which of the assumptions, constancy of mixing length and constancy of eddy viscosity, is more useful in describing these flows?

(d) What velocity variation within the core is implied by the assumption (3.4.43)?

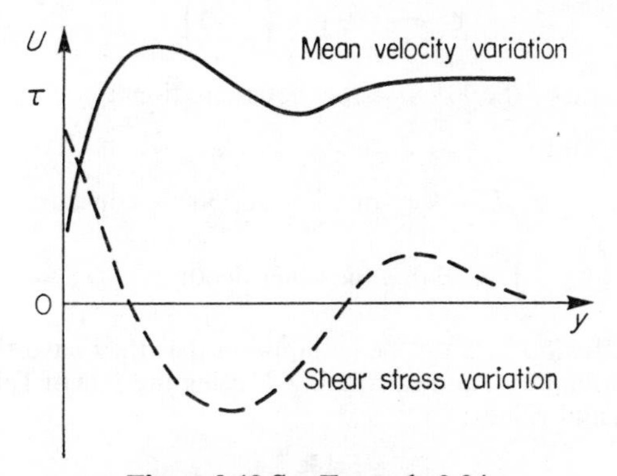

Figure 3.12 See Example 3.24

3.24 Figure 3.12 shows distributions of mean velocity and shear stress in a nearly parallel flow near a fixed wall.

(a) How might these distributions be produced?

(b) Sketch the corresponding distributions of the eddy viscosity and mixing length. What can you infer about the applicability of these devices for representing turbulent transfer of momentum?

3.25 (a) As an improvement to his original mixing-length model, Prandtl suggested taking

$$\epsilon_m = l_m^2\left[\left(\frac{dU}{dy}\right)^2 + l_1^2\left(\frac{d^2U}{dy^2}\right)^2\right]^{\frac{1}{2}}$$

with l_1 an auxiliary length. Why is this result more likely to provide an adequate description of the velocity variations within turbulent flows? What are its disadvantages?

(b) What is the eddy viscosity corresponding to von Kármán's assumption for the mixing length (3.4.45)? What unrealistic features has this result? Show that, for a constant-stress layer, it gives a logarithmic distribution of mean velocity.

3.26 The distribution of sediment particles in the fully turbulent part of the flow in a wide channel with an erodible bed is defined by

$$J_y = \overline{c_s v} - C_s V_s = 0$$

for an equilibrium distribution, in which the tendency to settle under gravity is everywhere balanced (statistically) by the upward 'diffusion' associated with turbulent mixing. Here $C_s + c_s$ is the fluctuating mass concentration of sediment, and V_s (>0) is the settling velocity of the particles when the fluid is still. All of the particles are assumed to be identical.

(a) Show that the distribution is determined by

$$\frac{dC_s}{dy} = -\left(\text{Sc}\frac{V_s}{\epsilon_m}\right) C_s$$

with Sc the Schmidt number for particle diffusion.

(b) Assuming that Sc is uniform, show that

$$C_s \propto e^{-\beta y},\ y^{-\gamma},\ \left(\frac{d}{y} - 1\right)^{\gamma}$$

with $\beta = \text{Sc}\ V_s/\epsilon_m$ and $\gamma = \text{Sc}\ V_s/(Ku_f)$, for the assumptions:

(1) $\epsilon_m = \text{constant}$

(2) $\epsilon_m = Ku_f y$ or $l_m = Ky$ or $\dfrac{U}{u_f} = K^{-1} \ln y + \text{constant}$

(3) $l_m = Ky\left(1 - \dfrac{y}{d}\right)^{\frac{1}{2}}$, with d the water depth

(c) Plot these distributions on the assumption that they have the same value at $y = a = 0{\cdot}05\ d$, taking $\gamma = 1$, and calculating βd using the data of Table 3.3. Compare with the experimental values:

Table 3.4

$(y-a)/(d-a)$	0·1	0·3	0·5	0·7	0·9
$C(y)/C(a)$	0·30	0·13	0·06	0·03	0·006

(d) Which of the assumptions of (b) seems most plausible? What is the significance of the parameters β and y? Do the predicted distributions depend upon the parameters defining the flow in the way you would expect?

3.27 The arbitrary nature of the postulates of the mixing-length theory is made clear when an attempt is made to extend it to axisymmetric, plane, swirling flow, where the axial and radial components of the mean velocity are $U = V = 0$. The mean vorticity and rate-of-strain are $dW/dr + W/r$ and $dW/dr - W/r$, with W the mean swirl or θ-component. The term W/r accounts for the difference between the actual rotation and distortion of the fluid, and those which appear to take place in this curvilinear coordinate system.

In fully developed flow in an annulus between concentric rotating cylinders, the torque is constant across the flow; thus $r^2\tau = \text{constant}$. Moreover, when the inner cylinder rotates, with the outer cylinder fixed, the mean angular momentum is found to be uniform, to a good approximation, within the core flow; thus, $rW = \text{constant}$ in the core.

Consider how it is possible to modify the parallel-flow results

$$\frac{\tau}{\rho} = \epsilon_m \frac{dU}{dy} = l_m^2 \left(\frac{dU}{dy}\right)^2$$

to deal with this situation. The following ways of replacing one or both of the derivatives dU/dy may be postulated, in conjunction with either $\epsilon_m = \text{constant}$ or $l_m = \text{constant}$:

(1) $\dfrac{dU}{dy} \to \dfrac{dW}{dr}$

(2) $\dfrac{dU}{dy} \to \dfrac{dW}{dr} - \dfrac{W}{r} = r\dfrac{d(W/r)}{dr}$

(3) $\dfrac{dU}{dy} \to \dfrac{dW}{dr} + \dfrac{W}{r} = \dfrac{1}{r}\dfrac{d(rW)}{dr}$

(4) $U \to rW$ or $\dfrac{r\tau}{\rho} = -\overline{vrw} = \epsilon_m \dfrac{d(rW)}{dr}$

(a) Comment on the physical significance of these substitutions, and on their plausibility, and see what they imply about the velocity distribution and torque.

(b) Show that postulate (2), in conjunction with $\epsilon_m = \text{constant}$, provides the most consistent predictions.

3.28 (a) For the stress distribution

$$\tau = \tau_w + \alpha y \quad \text{with } \alpha > 0 \text{, a constant}$$

show that equation (3.5.8) implies that

$$U = Au_f \ln\left[\frac{(\tau/\rho)^{\frac{1}{2}} - u_f}{(\tau/\rho)^{\frac{1}{2}} + u_f}\right] + (2A + 3B_1)\left(\frac{\tau}{\rho}\right)^{\frac{1}{2}} + C$$

with C a constant of integration.

(b) Show that this result tends to

$$U = \frac{u_f}{K} \ln y + \text{constant} \quad \text{for } \alpha y/\tau_w \ll 1$$

in keeping with the result (3.4.42) which is given by a number of simple analyses of the constant-stress layer. This simple result will achieve reality only if the flow remains fully turbulent until the logarithmic limit is approached. What condition must be satisfied if this is to be so?

(c) Show that the result of (a) tends to

$$U = (2A + 3B_1)\left(\alpha\frac{y}{\rho}\right)^{\frac{1}{2}} + C \quad \text{for } \alpha y/\tau_w \gg 1$$

This gives the velocity distribution in a zero-stress layer, found close to the surface in a boundary layer near separation.

3.29 For flow in a plane channel with equal wall stresses at the two fixed walls, the simple result (3.5.10) is in error by about one-tenth when $y/b = 0{\cdot}1$, that is, its predictions differ by one-tenth from experimental results Use this information, and the value $3B_1/2A \simeq -0{\cdot}2$, to show that $\tau_2/\tau_1 > 5$ must relate the dissimilar wall stresses in an asymmetric channel flow, if the full equation (3.5.8) is to be required anywhere in the flow.

3.30 The core of a channel flow is sometimes represented by assuming the mixing length to have a constant value $l_m = l_c$ there. The mixing lengths appropriate to the three core flows of Table 3.3 can be determined by requiring that the range of each velocity deficit match that given by the eddy-viscosity model. Show that:

$$\frac{l_c}{\frac{1}{2}b} = \frac{2}{3}\frac{u_f}{\Delta U} = 0{\cdot}103 \text{ for plane pressure flow}$$

$$\frac{l_c}{b} = \frac{u_f}{\Delta U} = 0{\cdot}135 \text{ for plane shearing flow}$$

$$\frac{l_c}{R} = \frac{2}{3}\frac{u_f}{\Delta U} = 0{\cdot}098 \text{ for pipe flow}$$

4

Channel Flows II: Friction and Flow Rate

In this chapter the methods developed in Chapter 3 will be used to study the relationship between the rate at which a fluid passes through a channel and the friction resisting the motion. This relationship is the key to the questions which engineers most often ask about flowing fluids: what is the pressure drop; what pumping power is required; how much fluid is delivered? These must be answered for a wider variety of flows than were considered in the preceding chapter, which dealt mostly with parallel mean flow in a smooth-walled channel. Hence it will be necessary to extend our picture of turbulent channel flows in a number of ways. Breaking down the generalization into stages, we shall:

(1) develop an empirical description of the constant-stress wall layer, guided by dimensional analysis, and use it to study friction in smooth-walled pipes;

(2) examine the effects of wall roughness on friction and on the flow near a wall, and introduce the methods used to specify them;

(3) extend the analysis to asymmetric flows, those in channels where the wall stresses differ, including annuli and channels with walls that move or have differing roughness;

(4) consider secondary flows in channels with more varied cross-sections, including open channels;

(5) introduce the special methods used to specify friction in open channels; and

(6) look briefly at methods of predicting the flows and losses in systems and networks of channels.

4.1 THE WALL LAYER

4.1.1 Results of Dimensional Analysis

As in Section 3.1, our study of the wall layer is linked to the particular case of a round pipe: we concentrate on this one geometry for ease of visualization;

we determine empirical constants using data obtained for the most part from pipe flows; and our first application of the results will be in predicting pipe friction. However, the flow in any wall layer across which the shear stress varies only slightly—a constant-stress layer—may be expected to be essentially like that in a pipe.

In Section 3.1 we saw that dimensional analysis allowed experimental results for the wall layer to be presented in a compact form; the quantities

$$u_f = \left(\frac{\tau_w}{\rho}\right)^{\frac{1}{2}} \quad \text{and} \quad y_f = \frac{\mu}{(\rho\tau_w)^{\frac{1}{2}}} = \frac{\nu}{u_f} \tag{3.1.1}$$

were used as scales. We now use dimensional analysis as a predictive technique. Table 4.1 indicates the parameters which determine the mean velocity variation $U(y)$ within fully developed flow in a smooth-walled pipe. For analytical

Table 4.1. Specification of the velocity variation in a smooth-walled pipe

Region (as in Figures 3.1, 4.1)	Related parameters	Results
Viscous sublayer ($\mathrm{I}, 0 < y < y_1$)	$U \sim \tau_w, \mu, y$	$(\mu/\tau_w)\, U/y = (U/u_f)/(y/y_f) = 1$
Buffer layer ($\mathrm{II}, y_1 < y < y_2$)	$U \sim \tau_w, \mu, \rho, y$	$U/u_f = f(y/y_f)$
Turbulent constant-stress layer ($\mathrm{III}-, y_2 < y < y_{3-}$)	$\mathrm{d}U/\mathrm{d}y \sim \tau_w, \rho, y$	$(y/u_f)\,\mathrm{d}U/\mathrm{d}y = A$ or $\Delta U/u_f = A \ln y$
Turbulent equilibrium layer ($\mathrm{III}+, y_2 < y < y_{3+}$)	$\mathrm{d}U/\mathrm{d}y \sim \tau, \rho, y$	$[y/\sqrt{(\tau/\rho)}]\,\mathrm{d}U/\mathrm{d}y = A$
Core flow ($\mathrm{IV}, y_3 < y < R$)	$\Delta U \sim \tau_w, \rho, R, y$	$\Delta U/u_f = f(y/R)$
Constant-stress layer ($0 < y < y_{3-}$)	$U \sim \tau_w, \mu, \rho, y$	$U/u_f = f(y/y_f)$
Overall ($0 < y < R$)	$U \sim \tau_w, \mu, \rho, R, y$	$U/u_f = f(y/y_f, y_f/R)$

convenience, it is the velocity gradient which is considered in the fully turbulent part of the wall layer, and the velocity defect in the core. The parameters are selected from:

τ_w, the time-mean stress at the wall, assumed uniform over it,
μ and ρ, the viscosity and density, assumed uniform throughout the flow, (4.1.1)
$R(=\frac{1}{2}d)$, the pipe radius, and
$y(=R-r)$, the distance from the wall to the point considered.

The regions considered in Table 4.1 are those distinguished in Figure 3.1. The choice of parameters for each region expresses in symbols the verbal definitions given in Section 3.1. In particular, since the turbulent mixing stress depends on the density ($\tau_t = -\rho\overline{uv}$), this property will influence the mean velocity in regions of turbulent flow. For the fully turbulent regions, we recognize the *Reynolds-number similarity* of the larger scales of the motion, by dropping the viscosity μ from the defining parameters. Note that the quantity y_f/R, which plays the role of a mean-flow Reynolds number, appears only in the overall velocity variation. Thus the scaling of Table 4.1 presents both wall-layer and core velocity variations in forms independent of Reynolds number.

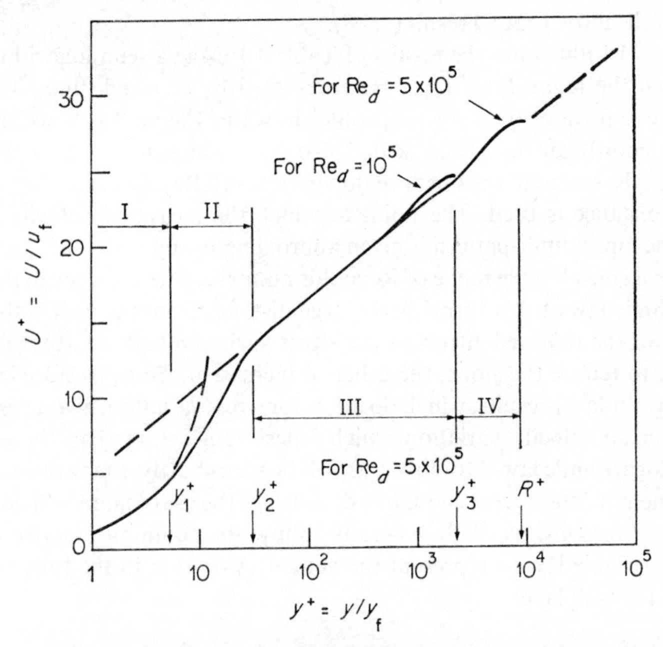

Figure 4.1 Semi-logarithmic representation of the mean-velocity variation in a pipe, illustrating the 'universal law of the wall'. Labels as in Figure 3.1 and Tables 4.1 and 4.3:
I—viscous or linear sublayer
II—buffer layer
III—fully turbulent, logarithmic layer
IV—core flow

The discussion of equilibrium layers in Section 3.5 suggests some alternative ways of selecting the parameters defining the wall layer. The stress gradient

$$\alpha = \left(\frac{d\tau}{dy}\right)_0 \quad \text{in general}$$
$$= -\frac{\tau_w}{R} \quad \text{in a pipe} \tag{4.1.2}$$

might be used instead of R; but the apparent increase in generality is largely illusory, at least for wall layers with relatively large wall stress. This possible substitution does demonstrate, however, that the inclusion of R has the effect of allowing for the stress variation. A more fruitful modification of the list (4.1.1) is the replacement of both τ_w and R by the local stress $\tau(y)$. In Table 4.1 the equilibrium or locally determined layer defined in this way is labelled Region III+, to distinguish it from the apparently less extensive constant-stress layer, here labelled Region III–. The result given by dimensional analysis is equivalent to equation (3.5.10), which lacks the energy-diffusion term of the more general result (3.5.8).

Figure 4.1 illustrates the results of Table 4.1 using a semi-logarithmic plot, in which the logarithmic layer is represented by a straight line. This is an alternative form of the velocity profile shown in Figure 3.1. Since the cross-stream coordinate has been scaled using y_f, which depends on Reynolds number, the extent of the linear region varies with Reynolds number when the scaled distance is used. The point at which the measured velocity deviates from the logarithmic pattern is given approximately by $y/R = 0{\cdot}15$, and varies little for a considerable range of Reynolds numbers. The accuracy of the simple logarithmic law at this surprisingly large distance from the wall is the consequence of the opposed effects of the stress variation and the outer flow, one tending to reduce the slope, the other to increase it. Thus, paradoxically, the equilibrium-layer results, which do allow for stress variation, give a description of the mean velocity variation which is less satisfactory than the constant-stress, logarithmic law. Moreover, as will be seen shortly, the variability of the experimental data is great enough to discourage the use of more refined models. For these reasons, we shall generally follow the common practice of using the logarithmic law to represent the velocity variation in the fully turbulent part of the wall layer.

4.1.2 The Logarithmic Layer

The result for the turbulent constant-stress layer which is given in Table 4.1 (and was obtained earlier from mixing-length and energy-equilibrium arguments) can be rewritten in the variables appropriate to the wall layer:

$$\frac{y}{y_f}\frac{d(U/u_f)}{d(y/y_f)} = A \tag{4.1.3}$$

where A is a dimensionless constant, expected to be nearly the same for all such layers. Integration gives

$$\frac{U}{u_f} = A \ln\left(\frac{y}{y_f}\right) + B \tag{4.1.3}$$

For compactness, we shall use the *scaled mean velocity and lateral distance*

$$U^+ = \frac{U}{u_f} \quad \text{and} \quad y^+ = \frac{y}{y_f} \tag{4.1.4}$$

the logarithmic layer becoming

$$U^+ = A \ln y^+ + B \tag{4.1.3}$$

The dimensionless constant B specifies the 'slip' within the viscous layer interposed between the *smooth* wall and the fully turbulent flow. Different constants will be found for flows near rough walls.

The constant A, which appears elsewhere in the form $1/K$, can be interpreted in terms of the eddy viscosity near the wall:

$$\epsilon_m = \frac{\tau_w}{\rho} \Big/ \frac{dU}{dy} = \frac{1}{A} u_f y = K u_f y \tag{4.1.5}$$

or

$$\frac{\epsilon_m}{\nu} = K y^+ \tag{4.1.5}$$

Thus the constants A and K specify the rate at which turbulent mixing develops at points progressively further from the wall.

The empirical constants A and B can be determined by plotting a measured velocity distribution in the manner of Figure 4.1, and then finding the slope and intercept of the linear region. Since many of the numerical predictions relating to channel flows and boundary layers depend upon the values assigned to A and B, we shall look in some detail at the results obtained in this way. The analysis also proves instructive by revealing some problems characteristic of turbulence measurements. The two questions to be answered are:

(1) Is there a universal wall layer in whose outer part the velocity varies logarithmically?

(2) What values should be assigned to the constants A and B?

The values listed in Table 4.2 are drawn mostly from an analysis by Hinze of measurements in pipe and channel flows; some results for channel flows and boundary layers have been added. Values for the 'standard' roughness comprised of sand grains glued to the wall have been given, in addition to those for effectively smooth walls. The variability in the supposed constants A and B can be ascribed in part to 'real' effects:

(1) differences in channel geometry and in the outer flow;

(2) a residual influence of Reynolds number, not accounted for in the scaling adopted; and

(3) effects of wall roughness.

We must also admit the existence of a number of 'artificial' effects:

(4) deviations of the flow from the ideal, through failure to achieve symmetry, full development, or uniformity of roughness;

(5) instrument errors resulting from velocity fluctuations or proximity to the wall; and

(6) difficulties of interpretation, in particular, (a) in fitting constants to a fairly narrow range of data (typically, $1{\cdot}8 < \log_{10} y^+ < 3$), (b) in deciding how far from the wall to apply the logarithmic law, and (c) for rough walls, in selecting the origin of y.

On surveying parts I to III of Table 4.2, we conclude that each of the 'real' effects (1–3) does influence the values of the constants A and B, although the precise nature of these influences is difficult to assess, owing to the scatter attributable to the 'artificial' factors (4–6). Nevertheless, the variability in A and B is small enough to allow the definition of a *universal law of the wall* broadly representative of wall layers in which the stress varies only moderately.

Parts IV and V of Table 4.2 give the values of the constants A, K, B and B_s which appear to be most probable, together with an indication of the variability in each. The uncertainty in the velocity profile is not as great as that in the individual constants, since a high value of A is often found in conjunction with a low value of B. Hence the predictions of velocity given by the most probable values are within perhaps 5 per cent of the true value (within the nearly logarithmic range) for the range of conditions surveyed.

Parts IV and V also give values of the constants inferred from the pipe-friction laws to be discussed later. These values differ from the near-wall results, as they are effective values characterizing the entire flow. Although these effective constants can be determined with greater precision than the near-wall values, they need not provide an accurate indication of the velocity variation in the constant-stress layer near the wall. The fact that the effective values are little different from the near-wall values indicates that the logarithmic distribution does provide a fairly accurate description of the mean velocity throughout the core. This point will be investigated more fully later.

4.1.3 The Buffer Layer

For the region between the linear sublayer and the logarithmic layer we have from Table 4.1:

$$\frac{U}{u_f} = f\left(\frac{y}{y_f}\right) \quad \text{or} \quad U^+ = f(y^+) \tag{4.1.6}$$

To specify this part of the velocity profile, we must supply not merely a pair of empirical constants, but the functional form of the variation. A number of workers have done this, using a combination of dimensional analysis, curve

Table 4.2. Constants for the logarithmic law of the wall

I. Smooth-walled pipes and plane channels

Nature of results	Re_d or Re_b	A	B
Most probable	5×10^3 to 5×10^6	2·7	4 (pipes)
			5 (channels)
	10^4	2·6	5
	10^6	2·7	3
Range	5×10^4	2·9	2·5
		2·7	6
		2·5	5

II. Boundary layers on smooth walls with small pressure gradients

	A	B
Clauser's correlation	2·49	4·9
Cole's correlation	2·5	5·1
Patel's correlation	2·39	5·45

III. Pipes roughened with uniform sand grains of size k_s

Nature of results	Re_d	A	B_s
Most probable	10^4 to 10^6	2·37	8·8
	10^4	2·15	9·5
	10^5	2·4	9·0
	10^6	2·33	8·7
Range	10^5	2·35	9·7
		2·4	7·7
		2·43	9·0
Most probable	10^5 to 10^6	2·5	8·7 ($R/k_s = 500$)
		2·33	9·2 ($R/k_s = 125$)

IV. Composite results for smooth walls

Nature of results	A	K	B
Most probable for smooth walls only	2·6 (±10%)	0·385 (±10%)	4·5 (±25%)
Most probable universal law	2·5 (±15%)	0·4 (±15%)	5 (±25%)
Effective values from pipe-friction law	2·46	0·406	5·67

V. Composite results for sand-grain roughness

Nature of results	A	K	B_s
Most probable for rough walls only	2·35 (±7%)	0·425 (±7%)	9 (±15%)
Most probable universal law	2·5 (±12%)	0·4 (±15%)	8 (±15%)
Effective values from pipe-friction law	2·46	0·406	8·61

fitting, and knowledge of the sublayer and logarithmic layer. Although some of these models are very crude, it is necessary to know of their existence since they are the foundations of formulae that are still used to predict heat and mass transfer.

We shall see later that the details of the buffer layer are not usually important for the prediction of friction. But this is not always so for heat- and mass-transfer calculations. In a fluid with a high Prandtl number ($\mathrm{Pr} = \nu/\kappa \gg 1$), for example, a lubricating oil, the temperature drop takes place almost entirely in the viscous layer, and a realistic prediction of heat transfer requires that the diffusion there be accurately specified.

With such applications in mind, most workers have expressed their buffer-layer models in terms of the eddy viscosity, which in the buffer layer has the form

$$\frac{\epsilon_m}{\nu} = f(y^+) \tag{4.1.7}$$

For more complex models, this takes the equivalent form (see equations (4.1.6)):

$$\frac{\epsilon_m}{\nu} = f(y^+, U^+) \tag{4.1.7}$$

From equations (4.1.5) we obtain the requirement for the smooth merging of the buffer and fully turbulent layers:

$$\frac{\epsilon_m}{\nu} \to Ky^+ \text{ for } y^+ > 50\text{, say} \tag{4.1.8}$$

Finally, the reasoning of Example 3.1 leads to the conclusion that, near a smooth wall

$$\frac{\epsilon_m}{\nu} \propto y^{+3} \text{ (or a higher power)} \quad \text{for } y^+ < 5\text{, say} \tag{4.1.9}$$

since $\overline{uv} \propto y^3$ or a higher power near the wall.

Some of the more commonly encountered ways of specifying the buffer layer are set out below, in order of increasing sophistication. Some are comprehensive, incorporating the linear and logarithmic layers, while others specify only the buffer or viscous regions.

(1) The simplest approach is to ignore the buffer layer entirely, by assuming that the linear and logarithmic profiles meet at the point $y_i^+ = y_i/y_f$ defined by

$$y_i^+ = A \ln y_i^+ + B \tag{4.1.10}$$

Inserting constants from Table 4.2, IV, we find

$$\begin{aligned} y_i^+ &= 10{\cdot}7 \quad \text{for } A = 2{\cdot}6,\ B = 4{\cdot}5 \\ &= 11 \quad\ \ \text{for } A = 2{\cdot}5,\ B = 5 \end{aligned} \tag{4.1.10}$$

(2) von Kármán represented the buffer layer by a modified logarithmic law:

$$U^+ = 5 \ln y^+ - 3{\cdot}05 \quad \text{for } 5 < y^+ < 30 \tag{4.1.11}$$

with the usual linear and logarithmic formulae applying outside these limits.

(3) Reichardt proposed

$$\frac{\epsilon_m}{\nu} = K\left[y^+ - y_s^+ \tanh\left(\frac{y^+}{y_s^+}\right)\right] \tag{4.1.12}$$

where y_s^+ corresponds to the nominal sublayer boundary.

(4) Rannie adopted the simpler form

$$\frac{\epsilon_m}{\nu} = \sinh^2\left(\frac{y^+}{y_s^+}\right) \quad \text{for } y^+ < 27{\cdot}5 \tag{4.1.13}$$

with the scale length $y_s^+ = 14{\cdot}5$.

(5) Deissler took, first

$$\frac{\epsilon_m}{\nu} = n^2 y^+ U^+ \quad \text{for } y^+ < 26, \text{ with } n = 0{\cdot}109 \tag{4.1.14}$$

and later

$$\frac{\epsilon_m}{\nu} = n^2 y^+ U^+ (1 - e^{-n^2 y^+ U^+}) \quad \text{for } y^+ < 26, \text{ with } n = 0{\cdot}124 \tag{4.1.15}$$

(6) A form considered by Hama

$$\frac{\epsilon_m}{\nu} = n^4 y^{+4} \frac{dU^+}{dy^+} \tag{4.1.16}$$

is identical to Deissler's (4.1.15) for small y^+.

(7) Finally, van Driest introduced a 'damping factor' into the mixing length (3.4.40), the corresponding eddy viscosity (3.4.37) becoming

$$\frac{\epsilon_m}{\nu} = K^2 y^{+2} (1 - e^{-y^+/y_s^+})^2 \frac{dU^+}{dy^+} \tag{4.1.17}$$

with the scale length $y_s^+ = 26$. This result has been widely used, the argument y^+/y_s^+ being modified in various ways to account for local variations in stress and for mass transfer.

Each of the eddy viscosities (4.1.12) to (4.1.17) is associated with a particular velocity variation in the fully turbulent region, the constants y_s^+ and n being chosen to ensure that the buffer model tends to or matches the logarithmic law selected. The velocity variation can be calculated from

$$\frac{dU}{dy} = \frac{\tau_w/\rho}{\nu + \epsilon_m} \tag{4.1.18}$$

or

$$\frac{dU^+}{dy^+} = \left(1 + \frac{\epsilon_m}{\nu}\right)^{-1} \tag{4.1.19}$$

If $\epsilon_m/\nu = f(y^+)$ or $f(y^+, \mathrm{d}U^+/\mathrm{d}y^+)$, an explicit integral can sometimes be found for $U^+(y^+)$. For $\epsilon_m/\nu = f(y^+, U^+)$, it will usually be necessary to integrate a differential equation numerically to determine $U^+(y^+)$. In Section 5.5.4 we shall encounter this problem again, in a more general form.

4.2 THE SMOOTH-WALLED PIPE

4.2.1 Variation of Mean Velocity

The analysis of this particular flow is useful not only in giving predictions for a case of great practical importance, but also in suggesting relationships and techniques of wider applicability.

We have available sufficiently accurate models of the velocity variation in the several segments of this flow: the sublayer in equation (3.3.21), the core flow in Table 3.3 and equations (3.3.23), the logarithmic layer in Table 4.2 and equations (4.1.3), and the buffer layer in equations (4.1.10) to (4.1.17). The problem facing us is the matching of these elements to form a comprehensive velocity variation for the entire flow. Since the discussion of the buffer layer has provided a number of ways of joining the segments of the wall layer, our attention is now concentrated on the matching of wall layer and core flow. One of the matching conditions is the requirement that the velocity vary continuously across the pipe. To determine the matching point itself, a second continuity condition is required; various workers have assumed continuity of: (1) the velocity gradient, (2) the mixing length or (3) the eddy viscosity. It is necessary to have empirical expressions for the variations of the quantity considered both in the wall layer and in the core.

The results obtained from the three matching conditions are not very different. If it is assumed that the eddy viscosity is uniform in the core, as was done in preparing Table 3.3, the third of the conditions is the easiest to apply Using the constant-stress results (4.1.5), we find the matching point to be specified by

$$\epsilon_m = K u_f y = \epsilon_c, \quad \text{the core value.}$$

Then

$$\frac{y_3}{R} = \frac{\epsilon_c}{K u_f R} = \frac{2A}{R_f} = 0{\cdot}192 \tag{4.2.1}$$

where R_f is the parameter introduced in equations (3.3.22), and the values $A = 2{\cdot}6$ and $R_f = 27$ have been used. We are now able to estimate the change in the mean velocity within the core flow more accurately than was done in Table 3.3:

$$\frac{U_c - U_3}{u_f} = \tfrac{1}{4} R_f \left(1 - \frac{y_3}{R}\right)^2 = 4{\cdot}4 \tag{4.2.2}$$

Table 4.3 sets out the limits on the several regions of a pipe flow, and the velocity changes which occur within them, as indicated by equations (3.1.2), (4.2.1, 2) and (4.1.3), the last with $A = 2{\cdot}6$ and $B = 4{\cdot}5$. The results are presented in two ways: in I with the parameter $R^+ = R/y_f$ unspecified; in II with a particular value assigned to R^+. Hence part I of the table indicates how the velocity variation changes as the Reynolds number increases and the viscous layer becomes thinner, while part II indicates the distribution typical of flow at high Reynolds numbers. For the case considered, the velocity change across the core (occupying 80 per cent of the radius) is only 15 per cent of the central velocity.

Table 4.3. Variation in mean velocity within turbulent flow in a smooth-walled pipe

I. General results for $A = 2{\cdot}6$, $B = 4{\cdot}5$, $R_f = 27$

Region	Outer limit y_i/y_f	Outer limit y_i/R	Velocity change $\Delta U/u_f$	Outer velocity U_i/u_f
I. Sublayer	5	$5/R^+$	5	5
II. Buffer	30	$30/R^+$	8·4	13·4
III. Logarithmic	$0{\cdot}192R^+$	0·192	$2{\cdot}6 \ln R^+ - 13{\cdot}3$	$2{\cdot}6 \ln R^+ + 0{\cdot}1$
IV. Core	R^+	1	4·4	$2{\cdot}6 \ln R^+ + 4.5$

II. Results for $R^+ = 8750$, $U_c d/\nu = 5 \times 10^5$, $u_f/U_c = 0{\cdot}035$

Region	Outer limit y_i/y_f	Outer limit y_i/R	$\Delta U/u_f$	U_i/u_f	$\dot{V}/(\pi R^2 u_f)$
I. Sublayer	5	0·0005	5	5	0·002
II. Buffer	30	0·003	8·4	13·4	0·05
III. Logarithmic	1680	0·192	10·3	23·7	7·22
IV. Core	8750	1	4·4	28·1	17·20

The volume flow through each of the four regions is also indicated in part II of Table 4.3. For this high-Reynolds-number flow, the contribution of the viscous layer is virtually negligible.

A second significant feature of the velocity distribution is suggested by the final entry of part I of Table 4.3. The velocity on the axis turns out to be just that which is obtained by evaluating the logarithmic law there. This *exact* agreement is fortuitous, arising from the particular choices made for the empirical constants A, B and R_f, and on the method chosen to match core and wall flows. Other, equally justifiable choices would give predictions differing by $\Delta U/u_f \sim 1$, that is, would change the central velocity by an amount of order u_f. Nevertheless, we may conclude that a simple extrapolation of the logarithmic law gives a fairly realistic prediction of the velocity variation throughout the core.

It is reasonable to extend this simplification to other channel sections, for example, a plane channel, annulus and open channel. This rather cavalier approach is also fairly successful for some boundary layers. However, even for the constant-pressure layer, there is a greater deviation ($\Delta U/u_f \simeq 2$ to 3) between the logarithmic law and the actual velocity in the outer turbulent layer. Moreover, to represent the merging of the boundary layer with the outer flow proper, it is necessary to specify the point at which the logarithmic law gives way to the free-stream velocity.

4.2.2. Friction Laws

Since the velocities of Table 4.3 are specified in terms of the wall stress, through the parameters u_f and y_f, we have the means of relating the flow through a pipe to the friction it generates. We adopt the simplifications suggested by our examination of the tabled results:

(1) neglect of the volume-flow contribution of the viscous layer, and
(2) use of the logarithmic distribution through the entire region of turbulent flow.

The flow through an annular region extending from $y_a = R - r_a$ to $y_b = R - r_b$ with the velocity varying logarithmically, is given by

$$\begin{aligned}\frac{\dot{V}}{\pi R^2 u_f} &= \frac{2}{R^2}\int_{r_a}^{r_b} (A \ln y^+ + B)\, r\, dr \\ &= 2\left[\frac{y}{R}(U^+ - A) - \frac{1}{2}\left(\frac{y}{R}\right)^2 (U^+ - \tfrac{1}{2}A)\right]_{y_a}^{y_b}\end{aligned} \qquad (4.2.3)$$

For $y_a = 0$, the contribution from the lower limit is zero; it remains very small for $y_a < y_2$, the viscous-layer boundary. Accordingly, we adopt a third simplification by neglecting the contribution from the lower limit. Taking the upper limit to be $y_b = R$, we obtain an estimate of the flow through a smooth-walled pipe:

$$\begin{aligned}\frac{\dot{V}}{\pi R^2 u_f} &= U_c^+ - \frac{3A}{2} \\ &= A \ln R^+ + \left(B - \frac{3A}{2}\right)\end{aligned} \qquad (4.2.4)$$

These results can be converted into alternative forms which give information of practical utility:

(1) Introducing the mean velocity $U_a = \dot{V}/(\pi R^2)$, we obtain

$$U_a = U_c - \frac{3A}{2} u_f \qquad (4.2.5)$$

This can be used to estimate the mean velocity U_a and flow rate $\dot{V}$, from a single velocity measurement U_c at the pipe axis. The wall stress or some equivalent information must be available, in order that u_f can be determined.

(2) Since the velocity and length scales are related by $u_f = \nu/y_f = (\tau_w/\rho)^{\frac{1}{2}}$, we have the friction law

$$\frac{U_a}{u_f} = A\ln\left(\frac{Ru_f}{\nu}\right) + \left(B - \frac{3A}{2}\right) \tag{4.2.6}$$

(3) Introducing the Reynolds number $\mathrm{Re}_d = U_a d/\nu$, and the friction parameter $u_f d/\nu$, we obtain an alternative friction law

$$\frac{\mathrm{Re}_d}{u_f d/\nu} = A\ln\left(\frac{u_f d}{\nu}\right) + (B - 2{\cdot}19\,A) \tag{4.2.7}$$

(4) Finally, using the relationships between Reynolds number, friction coefficient and velocity and length scales, namely,

$$\frac{u_f}{U_a} = \left(\frac{c_f}{2}\right)^{\frac{1}{2}} \quad \text{and} \quad \frac{R}{y_f} = \tfrac{1}{2}\mathrm{Re}_d\left(\frac{c_f}{2}\right)^{\frac{1}{2}} \tag{4.2.8}$$

we obtain yet another form of the friction law

$$\frac{1}{\sqrt{c_f}} = \frac{A}{\sqrt{2}}\ln(\mathrm{Re}_d\sqrt{c_f}) + \frac{(B - 2{\cdot}54\,A)}{\sqrt{2}} \tag{4.2.9}$$

This result for turbulent pipe flow may be compared with the much simpler laminar-flow result given in Table 3.2:

$$c_f = \frac{16}{\mathrm{Re}_d} \tag{4.2.10}$$

The results (4.2.6, 7, 9) are essentially equivalent, but differ in the ease with which they provide solutions to the two basic problems:

(1) given the pressure drop or friction (dP_w/dx or τ_w), find the flow rate ($\dot{V}$ or U_a), and

(2) given the flow rate, find the pressure drop or friction.

The formulation (4.2.9) is most widely adopted in practice, despite the fact that each of its parameters, Re_d and c_f, involves the velocity U_a, so that a trial-and-error solution is required for problem (1). This disadvantage is balanced by the fact that the coefficient c_f varies only slowly, so that a fairly accurate first estimate can often be made.

What values should be assigned to the constants A and B in these friction formulae? There are three reasons for expecting that the introduction of the most probable wall-layer values from part IV of Table 4.2 will *not* provide the best possible friction law:

(1) We have seen that there appear to be systematic changes in the constants as the Reynolds number varies.

(2) The results are based on the assumption that the logarithmic velocity variation applies across the entire flow.

(3) The total flow rate and overall pressure drop can be measured more accurately than can the velocity variation near the wall.

Despite these reservations, the friction law obtained using the wall-layer constants ($A = 2{\cdot}5$, $B = 5$) does give good results for very high Reynolds numbers, say, $\mathrm{Re}_d > 10^6$. For the lower Reynolds numbers where the smooth-wall results are more relevant, $3 \times 10^3 < \mathrm{Re}_d < 3 \times 10^6$, the result

$$\frac{1}{\sqrt{c_f}} = 1{\cdot}74 \ln(\mathrm{Re}_d \sqrt{c_f}) - 0{\cdot}40 \tag{4.2.11}$$

gives c_f within about 2 per cent of the mean line of experimental values. This is referred to as the *Prandtl or Kármán–Nikuradse law of friction.* The effective values of the constants

$$A_e = 2{\cdot}46 \quad \text{and} \quad B_e = 5{\cdot}67 \tag{4.2.12}$$

are given in Table 4.2, IV.

4.2.3 Power-Law Approximations

To simplify numerical work and algebraic manipulation, it is sometimes convenient to represent the velocity distribution and friction of pipe flow by power laws:

$$\frac{U}{U_c} = \left(\frac{y}{R}\right)^{1/n} \tag{4.2.13}$$

and

$$c_f = C\,\mathrm{Re}_d^{-1/p} \tag{4.2.14}$$

with *n*, *p* and *C* constants chosen to fit experimental results or, what is very nearly the same thing, the logarithmic formulae (4.2.11, 12). Note that the power-law and logarithmic velocity variations are alike in giving unrealistic predictions very near the wall.

Let us see first how the two velocity distributions are related. For the logarithmic law, the ratio U_c/U_a is given by equations (4.2.5, 8):

$$\frac{U_c}{U_a} = 1 + \frac{3A_e}{2}\frac{u_f}{U_a} = 1 + \frac{3A_e}{2}\left(\frac{c_f}{2}\right)^{\frac{1}{2}} \tag{4.2.15}$$

with A_e the effective constant (4.2.12). For the power-law variation, integration over the cross-section gives

$$\frac{U_c}{U_a} = 1 + \frac{3}{2n} + \frac{1}{2n^2} \tag{4.2.16}$$

Comparing these results, we obtain

$$\frac{1}{n}+\frac{1}{3n^2}=A_e\left(\frac{c_f}{2}\right)^{\frac{1}{2}} \tag{4.2.17}$$

The index n must increase with the Reynolds number, since the friction coefficient decreases. However, since c_f is a rather weak function of Re_d, the index n remains nearly constant over a useful range of Reynolds numbers.

Next consider how the two power laws (4.2.13, 14) are related. Each implies a relationship between U_a/u_f and R/y_f. Using the connections (4.2.8) we can rewrite the friction formula as

$$\left(\frac{U_a}{u_f}\right)^{2p-1}=\left(\frac{2}{C}\right)^p\frac{2R}{y_f} \tag{4.2.18}$$

We argue that the velocity variation in the wall layer (where much of the velocity change occurs) must take the form

$$\frac{U}{u_f}=D\left(\frac{y}{y_f}\right)^{1/n} \quad \text{with } D \text{ constant}$$

in order to be both a power law and a wall layer. Introducing (4.2.16), we obtain

$$\frac{U_a}{u_f}=D\left(\frac{R}{y_f}\right)^{1/n}\bigg/\left(1+\frac{3}{2n}+\frac{1}{2n^2}\right) \tag{4.2.19}$$

If the parameters n, p, C and D change only slowly with the flow conditions (as is suggested by equation (4.2.17) and is found in practice), changes in U_a/u_f arise mostly from variations in R/y_f. Hence the exponents of equations (4.2.18, 19) must be related approximately by

$$n=2p-1 \tag{4.2.20}$$

A relationship between the constants C and D is also implied.

The most used power laws are those for $n=7$ and $p=4$. The *seventh-root velocity profile*

$$\frac{U}{U_c}=\left(\frac{y}{R}\right)^{\frac{1}{7}} \quad \text{or} \quad \frac{U}{u_f}=8{\cdot}74\left(\frac{y}{y_f}\right)^{\frac{1}{7}} \tag{4.2.21}$$

is valid for $Re_d \sim 10^5$. It has also been much used in elementary boundary-layer calculations. The *Blasius friction formula*

$$c_f=0{\cdot}079\,Re_d^{-\frac{1}{4}} \tag{4.2.22}$$

is, for $Re_d<10^5$, as accurate as the logarithmic formula (4.2.11), that is, is within 2 per cent of the experimental trend.

Two other simple friction formulae may be noted:

$$c_f=0{\cdot}046\,Re_d^{-\frac{1}{5}} \tag{4.2.23}$$

used for $10^5<Re_d<10^6$, and

$$c_f=0{\cdot}0014+0{\cdot}125\,Re_d^{-0{\cdot}32} \tag{4.2.24}$$

which is within 3 per cent of the logarithmic result for $\mathrm{Re}_d < 10^6$, and within 7 per cent for $\mathrm{Re}_d < 10^7$.

4.3 THE PROBLEMS OF ROUGHNESS

4.3.1 Classical Analysis

The Velocity Distribution

The experimental trends illustrated in Figure 1.1 show that pipe friction depends strongly on the roughness of the wall; indeed, for high Reynolds numbers the friction is often determined by the roughness alone. We generalize the analysis of Table 4.1 to account for these observations:

$$(1)\quad U \sim \tau_w, \mu, \rho, k, y \tag{4.3.1}$$

for the whole constant-stress layer, with k a length scale for the roughness elements;

$$(2)\quad \frac{dU}{dy} \sim \tau_w, \rho, y \tag{4.3.2}$$

as before, for the fully turbulent part of the constant-stress layer; and

$$(3)\quad \Delta U \sim \tau_w, \rho, R, y \tag{4.3.3}$$

as before, for the turbulent core.

Thus, so far as the fully turbulent part of the flow is concerned, the roughness merely alters the 'slip' near the wall.

The use of a single length to describe the roughness is highly restrictive, but was appropriate for the systematic experiments of Nikuradse, carried out around 1930, and still used as a standard of comparison for other kinds of roughness. He coated the inside of the pipe evenly with carefully graded sand grains, and used the grain size k_s (that is, the mesh size of the grading sieve) as the roughness scale.

A dimensionally homogeneous prescription of the velocity within the wall layer follows from equation (4.3.1):

$$\frac{U}{u_f} = f\left(\frac{y}{y_f}, \frac{k}{y_f}\right) \tag{4.3.4}$$

For the fully turbulent part, equation (4.3.2) gives

$$\frac{y}{u_f}\frac{dU}{dy} = \frac{y}{y_f}\frac{d(U/u_f)}{d(y/y_f)} = A$$

as in equations (4.1.3). It is usual to take the constant A $(= 1/K)$ to have the same value for flow past rough and smooth walls, though the experimental results of Table 4.2 show that this is not strictly so. Integration gives

$$\frac{U}{u_f} = A \ln\left(\frac{y}{y_f}\right) + f\left(\frac{k}{y_f}\right) \tag{4.3.5}$$

the form of the additional function being dictated by equation (4.3.4). This result may equally well be written

$$\frac{U}{u_f} = A \ln\left(\frac{y}{k}\right) + g\left(\frac{k}{y_f}\right) \tag{4.3.5}$$

The two roughness-characterizing functions are related by

$$f\left(\frac{k}{y_f}\right) + A \ln\left(\frac{k}{y_f}\right) = g\left(\frac{k}{y_f}\right) \tag{4.3.6}$$

In these generalizations of the smooth-wall logarithmic formulae (4.1.3), the 'slip velocity' across the viscous region is expressed as a function of the roughness.

We next consider the behaviour of the linked functions $f(k/y_f)$ and $g(k/y_f)$. Note that the parameter k/y_f will increase when the scale of the roughness increases, or when the Reynolds number rises with a corresponding decrease in y_f.

(1) $k/y_f \to 0$. For an effectively smooth wall, the results (4.1.3) are recovered, and

$$f\left(\frac{k}{y_f}\right) \to B \tag{4.3.7}$$

with B the smooth-wall constant.

(2) $k/y_f \to \infty$. For many forms of roughness (including sand grains, as indicated in Figure 1.1), the friction becomes independent of viscosity in these circumstances, the wall being described as *fully rough*. Arguing that the associated velocity variation must exhibit the same behaviour, we have

$$g\left(\frac{k}{y_f}\right) \to B' \tag{4.3.8}$$

a constant specifying the limiting 'slip' for a particular kind of roughness. Values of $B'(= B_s)$ for sand-grain roughness are given in Table 4.2. Others will be given later.

(3) In general, the reduction in the velocity change near the wall may be calculated by subtracting the rough-wall (4.3.5) from the smooth-wall (4.1.3) prediction:

$$\frac{\Delta U}{u_f} = B - f\left(\frac{k}{y_f}\right) = B - g\left(\frac{k}{y_f}\right) + A \ln\left(\frac{k}{y_f}\right) \tag{4.3.9}$$

Figure 4.2 shows this relationship for several kinds of artificial roughness: sand grains, rectangular bars and wire meshes. In each case, the variation $\Delta U/u_f$ for large values of k/y_f has the form suggested by equations (4.3.8, 9), namely

$$\frac{\Delta U}{u_f} = B - B' + A \ln\left(\frac{k}{y_f}\right) \tag{4.3.10}$$

The fact that the limiting forms (4.3.10) do not coincide in Figure 4.2 indicates that the constant B' differs from one kind of roughness to the next. Part of this variation arises from the arbitrary selection of scales for the several shapes of roughness elements. We shall see later how scales can be chosen to bring about the coalescence of all fully-rough (that is, linear) characteristics. However, the curves for smaller values of k/y_f have differing shapes, and cannot be collapsed in this way.

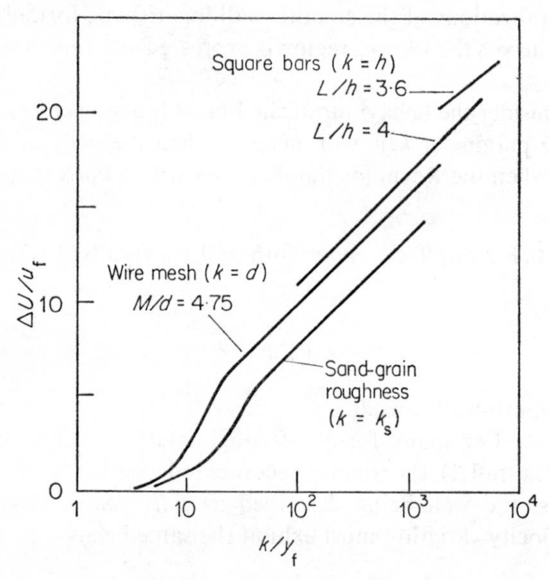

Figure 4.2 Velocity reductions effected by various forms of roughness, as a function of the roughness parameter k/y_f

Concentrating now on the sand-grain roughness which is often used as a standard, we see that it has little effect on the turbulent part of the flow (or on the friction) when

$$\frac{k_s}{y_f} < 4 \tag{4.3.11}$$

Limits of the same order exist for other roughness types. When these conditions are satisfied, the flow and surface are said to be *hydraulically or aerodynamically smooth*. The results of Figure 4.2 suggest further that

$$\frac{k_s}{y_f} > 60 \tag{4.3.12}$$

defines *fully rough conditions*. The limits (4.3.11, 12) are often justified by arguing that the former corresponds to the penetration of the roughness elements into the buffer layer, and the latter to their entry into the fully turbulent layer. These ideas have some utility in a superficial introduction, but are inconsistent in two vital respects:

(1) For sand-grain roughness, the actual roughness height is about $\frac{1}{2}k_s$. Hence the roughness becomes significant for the outer flow when the actual roughness height is less than half the sublayer thickness, $y_1 \simeq 5y_f$.

(2) The flow near the wall is profoundly altered by the penetrating elements. The concept of parallel mean flow near the wall ceases to be relevant, and the distinction between viscous and turbulent layers breaks down.

Looking at the effects of roughness more realistically, we realize that the drag of most roughness elements will have both viscous and inertial contributions. The effects on the flow are mixed: the roughness acts to entrain turbulent fluid and to promote turbulence near the surface; on the other hand, it carries the local effects of viscosity into the flow on the surface of the protrusions. Ultimately, the inertial effects usually become dominant for the drag and for the outer flow.

Friction Laws

To relate the flow through a rough-walled pipe to the friction at the wall, we simply introduce the velocity distribution (4.3.5) into the first of equations (4.2.4), again accepting the simplifying assumptions which lie behind that result. Thus for a rough wall

$$\begin{aligned}\frac{U_a}{u_f} &= U_c^+ - \frac{3A}{2}\\ &= A\ln\left(\frac{R}{k}\right) + g\left(\frac{k}{y_f}\right) - \frac{3A}{2}\end{aligned} \tag{4.3.13}$$

Introducing $(c_f/2)^{\frac{1}{2}} = u_f/U_a$, we obtain the friction law

$$\frac{1}{\sqrt{c_f}} = \frac{A}{\sqrt{2}}\ln\left(\frac{R}{k}\right) + \frac{g(k/y_f) - 3A/2}{\sqrt{2}} \tag{4.3.14}$$

Since $R/y_f = \frac{1}{2}\mathrm{Re}_d(c_f/2)^{\frac{1}{2}}$, this is an implicit expression of the friction law

$$c_f = f\left(\mathrm{Re}_d, \frac{k}{d}\right)$$

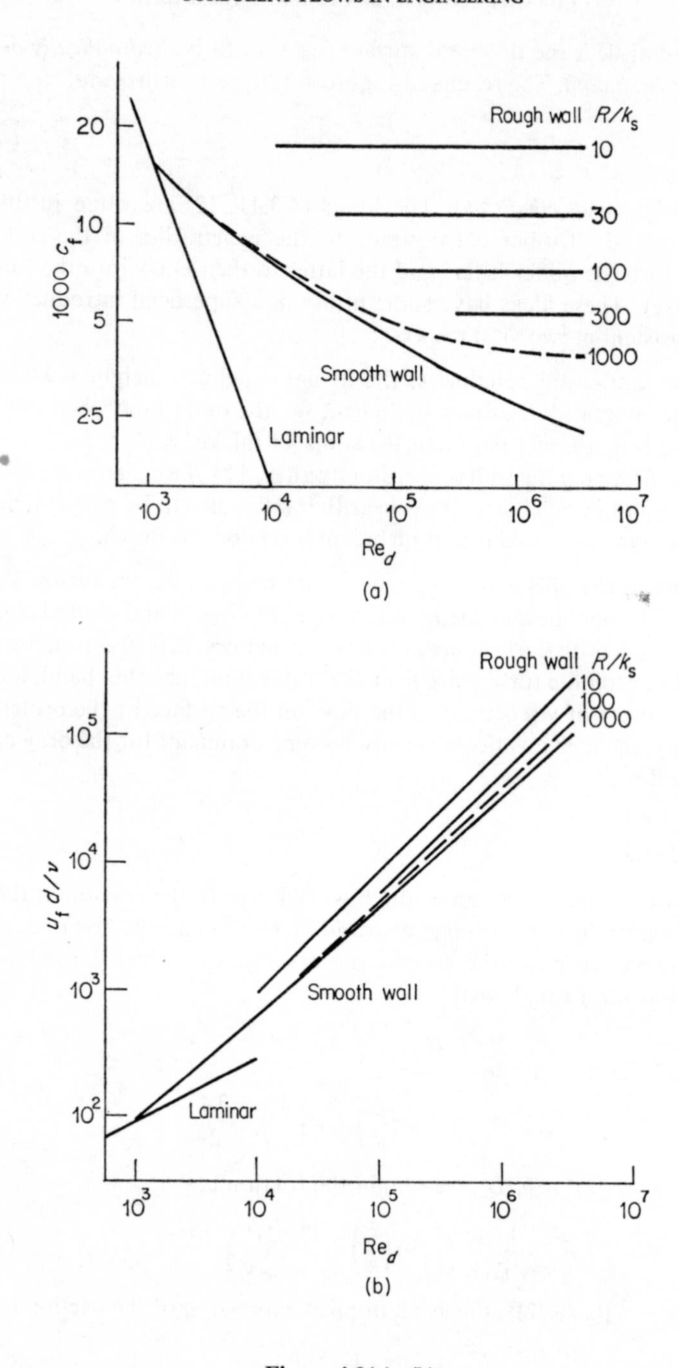
Rough wall R/k_s
10
30
100
300
1000
Smooth wall
Laminar
1000 c_f
20
10
5
2·5
10^3
10^4
10^5
10^6
10^7
Re_d
(a)
Rough wall R/k_s
10
100
1000
Smooth wall
Laminar
$u_f d/\nu$
10^5
10^4
10^3
10^2
Re_d
(b)

Figure 4.3(a), (b)

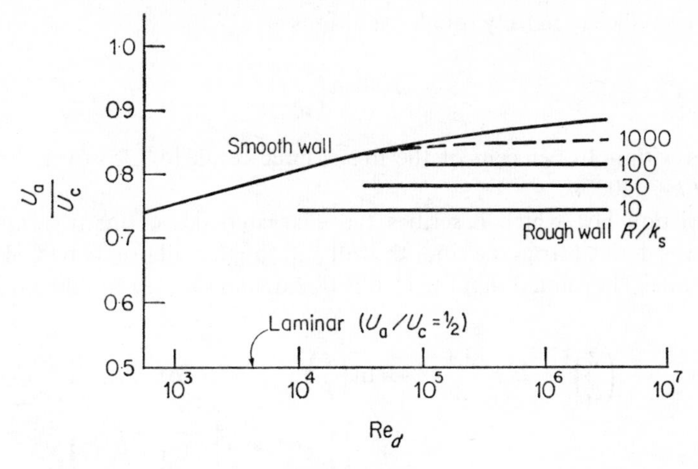

Figure 4.3(c)

Figure 4.3 Characteristics of flow in smooth and rough-walled pipes, as functions of Reynolds number Re_d and relative roughness R/k_s, with k_s the equivalent sand-grain roughness. In each case the dashed curve is obtained from the Colebrook–White formula (4.3.18), with $R/k_s = 1000$ throughout. The laminar, smooth-wall and rough-wall results are derived, respectively, from the friction laws (4.2.10), (4.2.11) and (4.3.15). (a) Variations of friction coefficient $c_f = \tau_w/(\frac{1}{2}\rho U_a^2)$. (b) Variations of the friction parameter $u_f d/\nu = 2R^+$. (c) Variations of velocity profile, measured by U_a/U_c, the ratio of the average velocity to the central velocity, as given by equation (4.2.5)

Given the function $g(k/y_f)$, or one of the functions $f(k/y_f)$ and $\Delta U/u_f$ related to it by equations (4.3.6, 9), we can determine this relationship explicitly.

Comparison of equations (4.2.6) and (4.3.13) provides another interpretation of ΔU, the velocity reduction by roughness:

$$\frac{\Delta U}{u_f} = \Delta\left(\frac{U_a}{u_f}\right) = \Delta\left(\frac{2}{c_f}\right)^{\frac{1}{2}}$$

showing that ΔU is a measure of the associated rise in the friction coefficient.

For sand-grain roughness in fully rough conditions, experiments give

$$\frac{1}{\sqrt{c_f}} = 1{\cdot}74 \ln\left(\frac{R}{k_s}\right) + 3{\cdot}48 \tag{4.3.15}$$

The effective values of the constants of equation (4.3.14) are

$$A_e = 2{\cdot}46 \quad \text{and} \quad B_e' = B_s = 8{\cdot}61 \tag{4.3.16}$$

as indicated in Table 4.2, V. A useful power-law approximation to the pipe-friction coefficient in fully rough conditions is

$$c_f = 0{\cdot}040\left(\frac{k_s}{d}\right)^{0{\cdot}31} \tag{4.3.17}$$

This is within 10 per cent of the logarithmic result (4.3.15) over the range $20 < d/k_s < 2000$.

A friction law which describes the characteristics of many commercial pipes was derived from the smooth-wall and rough-wall results by Colebrook and White. They noted that the results (4.2.6) and (4.3.13) could be rewritten as

$$\left(\frac{2}{c_f}\right)^{\frac{1}{2}} - B' + \frac{3A}{2} = -A\ln\left(\frac{k}{R}\right) \qquad \text{for } y_f/k \ll 1$$

$$= -A\ln\left[\frac{y_f}{R}e^{(B'-B)/A}\right] \qquad \text{for } y_f/k \gg 1$$

on setting $g(k/y_f) = B'$. They arbitrarily combined the two laws to obtain

$$\left(\frac{2}{c_f}\right)^{\frac{1}{2}} = B' - \frac{3A}{2} - A\ln\left(\frac{k}{R} + \frac{C}{\mathrm{Re}_d\sqrt{c_f}}\right) \tag{4.3.18}$$

with $C = \sqrt{8}\exp[(B'-B)/A]$. Introducing the effective values of A, B and $B' = B_s$ given in equations (4.2.12) and (4.3.16), we obtain

$$\frac{1}{\sqrt{c_f}} = 3{\cdot}48 - 1{\cdot}74\ln\left(\frac{k_s}{R} + \frac{9{\cdot}3}{\mathrm{Re}_d\sqrt{c_f}}\right) \tag{4.3.18}$$

for the *Colebrook–White friction law.*

Figure 4.3 displays the results of these friction calculations. Figure 4.3(a) is a theoretical reconstruction (using a few empirical constants) of the experimentally determined Figure 1.1(b). Results are shown for laminar flow, smooth-walled turbulent flow, and for four fully rough walls. A 'transition' curve of the Colebrook–White type is also given. Charts containing a family of such curves—called *Moody diagrams* after their proponent—are widely used in solving practical pipe-flow problems.

Figure 4.3(b) presents the same information in terms of the alternative friction parameter $u_f d/\nu$ introduced in equation (4.2.7). Figure 4.3(c) indicates how the average and central velocities are related for the cases considered.

Equivalent Sand-grain Roughness

The numerical results obtained above relate to sand-grain roughness in the fully rough condition. To apply them to other kinds of roughness, we introduce the equivalent roughness, the sand-grain size which would, in the fully rough condition, give the same friction coefficient. Equation (4.3.14) gives the relationships

$$\left(\frac{2}{c_f}\right)^{\frac{1}{2}} + \frac{3A}{2} = A\ln\left(\frac{R}{k}\right) + g\left(\frac{k}{y_f}\right) = A\ln\left(\frac{R}{k_s}\right) + B_s \qquad (4.3.19)$$

among c_f, the actual friction coefficient, k, a linear scale of the actual roughness, and k_s, the hypothetical sand-grain size or equivalent roughness. If the friction coefficient and actual roughness scale are known, these relationships lead to: the equivalent roughness k_s, the ratio k_s/k, and the value of $g(k/y_f)$ or, in the fully rough condition, the value of the constant B'. Note that

$$\frac{U}{u_f} = A\ln\left(\frac{y}{k_s}\right) + B_s \qquad (4.3.20)$$

for any roughness specified in this way.

Part I of Table 4.4 indicates the ranges of the equivalent roughness commonly adopted in engineering practice. These broad ranges allow for such factors as density of rivets, care of finishing, and aging. The latter term encompasses scale formation, rusting, pitting and vegetation growth, any of which can increase the friction markedly.

Part II of the table gives a variety of dimensionless roughness characteristics, many based on measurements made by Schlichting using artificially roughed surfaces. The two measures of roughness are related by

$$A\ln\left(\frac{k_s}{k}\right) = B_s - B' \qquad (4.3.21)$$

If we set aside the results for two-dimensional roughness elements, which vary in a more complex manner, we may conclude that there is a maximum attainable length scale ratio, $k_s/k \simeq 5$ to 6, corresponding to $B' \simeq 4$ to 5. This phenomenon has a simple explanation: an attempt to roughen the surface further will simply shift its effective position outwards.

The values of Table 4.4, II show that the actual height of roughness elements is a rather poor guide to the effective roughness. Direct identification with the sand-grain equivalent can lead to errors in the mean velocity of order

$$\frac{\Delta U}{u_f} = \Delta B' = \pm 4 \qquad (4.3.22)$$

These correspond to errors in the flow rate of perhaps 20 per cent. Conveniently, the sand-grain values have a central position in the range of roughness characteristics.

Other Measures of Roughness

Equation (4.3.21) demonstrates the equivalence of the two measures of roughness used in Table 4.4, II. Other roughness scales can be obtained by

Table 4.4. Equivalent sand-grain roughness

I. Ranges adopted in engineering practice

Type of surface	k_s in ft/1000	k_s in mm
Drawn glass or metal	up to 0·005	up to 0·0015
Commercial steel or iron	0·15 to 0·85	0·05 to 0·3
Wood staves	0·6 to 3	0·2 to 0·9
Concrete	1 to 10	0·3 to 3
Riveted steel	3 to 30	0·9 to 9

II. Dimensionless measures determined experimentally

Nature of roughness	Configuration	k_s/k	B'
Sand grains ($k = k_s$)	Densely packed	1	8·5
Machined surface	Flow ⊥ grain	0·39	10·9
(k = max to min)	Flow ∥ grain	0·20	12·5
Vegetation (k = typical height)		4	5
Spheres in staggered rows	$L/d = 9{\cdot}75$	0·23	12·2
(separation L, $k = d$)	$= 4{\cdot}88$	0·84	8·9
	$= 2{\cdot}44$	3·07	5·7
	$= 1{\cdot}46$	3·80	5·2
	$= 1{\cdot}00$	0·63	9·7
Hemispheres, as above	$L/d = 5$	0·12	13·8
($k = \frac{1}{2}d$)	$= 3{\cdot}75$	0·19	12·7
	$= 2{\cdot}5$	0·57	9·9
	$= 1{\cdot}0$	1·40	7·7
Interwoven square mesh	$M/d = 4{\cdot}75$	3·5	5·4
(spacing M, $k = d$ of wire)			
Square bars across flow	$L/h = 10$	8·3	3·2
(separation L, $k = w = h$)	$= 4$	3·1	5·7
	$= 3{\cdot}6$	5·4	4·3

assigning different values to the constant B. Thus the velocity profile (4.3.20) is written

$$\frac{U}{u_f} = A \ln\left(\frac{y}{k_i}\right) + B_i$$

with (4.3.23)

$$\frac{k_i}{k_s} = \exp\left(\frac{B_i - B_s}{A}\right)$$

relating the roughness scale k_i to the sand-grain scale. In addition to the choice

$B_i = B_s$ considered above, three ways of selecting the constant B_i have been adopted:

(1) $B_i = B_1 = B$, the smooth-wall constant, giving

$$\frac{U}{u_f} = A \ln\left(\frac{y}{k_1}\right) + B \tag{4.3.24}$$

This choice reduces rough- and smooth-wall results to the same form, with $k_1 \to y_f$ for an effectively smooth wall. We find that

$$k_1 = 0{\cdot}30\, k_s \tag{4.3.24}$$

on taking $A = 2{\cdot}5$, $B_s = 8{\cdot}5$ and $B = 5{\cdot}5$.

(2) $B_i = B_2 = 0$, giving

$$\frac{U}{u_f} = A \ln\left(\frac{y}{k_2}\right) \tag{4.3.25}$$

a form often used to describe the atmospheric boundary layer. The parameter k_2, termed the *roughness length* by meteorologists, is related by

$$k_2 = \frac{k_s}{30} \tag{4.3.25}$$

to the sand-grain scale.

(3) $B_i = B_3 = f(k/y_f)$, as in the first of equations (4.3.5). Here $k_3 \equiv y_f$, and $B_3 \to B$ for effectively smooth-wall flow.

Example 4.15 introduces two further roughness-specifying parameters—an effective and macroviscosity.

4.3.2 Some Unresolved Problems

The methods introduced above are those currently used to account for wall roughness. They leave unresolved a number of practical and conceptual problems; we shall consider five of these, pointing the way to a solution where possible, but more often being able only to define the problem area.

Problem 1. An accurate prediction of friction is possible only if experimental data are available for the particular surface and flow conditions of interest.

Attempts have been made to deduce the friction of rough surfaces by adding the drag coefficients of individual protrusions to the basic smooth-wall friction. This approach is applicable, for example, to a surface with widely spaced rivets. Even there, it is necessary to account for the variation of the drag coefficient as the depth of the viscous layer changes. When the elements are more closely spaced, interference must be taken into account—the influence of an element on the drag of the others, and on the basic wall friction. For these more typical surfaces, realistic predictions require either specific tests or a comprehensive catalogue of results for somewhat similar surfaces.

Problem 2. Some irregular surfaces have friction characteristics unlike those for the sand-roughened and commercially rough surfaces considered earlier.

For such a surface, the friction cannot be specified by a single equivalent roughness, and devices such as the Colebrook–White formula and Moody diagram lose much of their usefulness. Although we cannot at present predict these anomalous characteristics, it is not difficult to explain in general terms how they arise. Figure 4.4 indicates some of the friction characteristics which have been encountered:

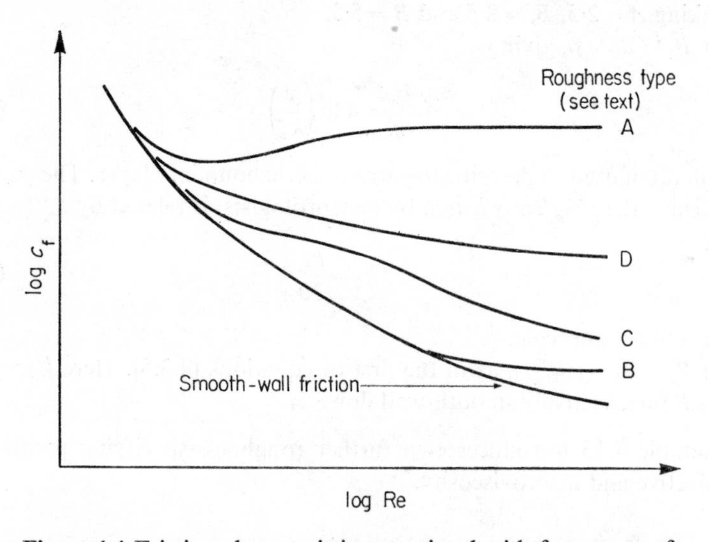

Figure 4.4 Friction characteristics associated with four types of roughness geometry. The relative values of friction are not, in general, those indicated in this figure, where they have been chosen to separate the several curves

Type A. Sand grains and other angular, closely spaced elements. The viscous layer is rapidly eliminated as the Reynolds number rises, and stress is then transmitted by the inertial drag of the elements.

Type B. Protrusions of various sizes, as produced by many manufacturing processes or by aging. The viscous layer is destroyed more gradually, and the elements progressively start to transmit stress by inertial drag.

For the other two surface types, the friction coefficient continues to fall; this may be ascribed to the continued existence of a viscous layer over part of the surface:

Type C. Isolated protrusions on an otherwise fairly smooth wall. The inertial drag coefficient of the large elements takes on a nearly constant value, while the friction coefficient for the surrounding surface continues to fall. As a rough guide, this kind of behaviour may be expected when less than half of the surface area is occupied by the roughness elements.

Type D. Isolated depressions in an otherwise fairly smooth wall. The contribution of the depressions rises slowly as the viscous layer thins.

Some surfaces do not fit into these categories, for example:

(1) the variety of geometries intermediate to them;

(2) the wavy wall, with or without superposed roughness;

(3) surfaces composed of flutings parallel to, or inclined to the mean flow; and

(4) very rough channels with roughness height comparable to the flow width (this situation arises in natural streams with pools and shallows, and in heat-exchange passages).

For such cases, the friction coefficient may be critically dependent on the surface geometry, and may vary in a complicated way with Reynolds number. For very rough channels, transition to turbulence has been reported for $\mathrm{Re}_b = 10$ to 200. The change in the friction characteristic can hardly mark the appearance of true turbulence, with a range of length scales; it is more plausible to ascribe the change to the forming of inertia-dominated separated flows among the roughness elements. (Compare with Figure 1.3.)

Problem 3. Although the mean-velocity distribution is expressed in terms of y, the distance from the wall, the effective wall location is not readily apparent.

This problem can be solved by asking the flow where it thinks the wall is: that virtual origin is selected which reduces an appropriate part of the velocity profile to logarithmic form. Thus

$$U \propto \ln(y + \varepsilon) \tag{4.3.26}$$

where y is measured from some convenient level, and ε specifies the effective wall position relative to that level. For uniform roughness, y may be measured from the tops of the elements; then $\varepsilon > 0$. For isolated elements on a smooth wall, y may be measured outwards from the wall; in some cases it is still found that $\varepsilon > 0$, the effective surface being below the real surface.

Problem 4. The flow very near the wall sometimes dominates heat and mass transfers between wall and fluid, but we have said nothing about the activity among the roughness elements and in the remnants of the viscous layer.

A realistic solution of this problem will require decades of painstaking experimental work. In the meantime, the effect of roughness on heat and mass transfer must be accounted for empirically, with the aid of analogies with momentum transfer. We shall consider the transfer processes themselves in Chapter 5; here we attempt only to establish a general picture of the flows through which the transfer is accomplished. For definiteness, we consider the particular family of surfaces defined in Figure 4.5, surfaces roughened with rectangular bars, any one member being specified by the lengths h, w and L. We assume that the channel is sufficiently wide ($b/h \gg 1$) to allow the development of an essentially parallel mean flow through most of the channel. Among those who have studied this geometry are Perry and his coworkers.

The region near the wall will be termed the *roughness layer*. For the element spacing shown in Figure 4.5, it may be defined tentatively by

$$\frac{y}{h} < 2 \quad \text{and} \quad \frac{y}{y_f} < 30 \tag{4.3.27}$$

The mean flow in the roughness layer is usually three-dimensional, with $U(x, y, z)$. Even for a two-dimensional wall contour $h(x)$, as shown in Figure 4.5, instability of the possible two-dimensional near-wall flow often gives rise to a cellular three-dimensional flow among the roughness elements. The flow around an isolated bump, defined by $h(x, z)$, is often visualized as a 'horseshoe' eddy wrapped around the protrusion.

Within the roughness layer, the time-mean shear stress varies from point to point, although the streamwise average will be independent of distance from the wall in the outer part of the layer (that is, beyond the protrusions, which carry some of the stress themselves). In the nearly parallel flow just beyond the roughness layer, we normally find a reasonable approximation to a *constant-stress layer* with the time-mean stress the same at every point. This is the layer whose existence was postulated in equations (4.3.2, 26).

The *wall layer* near a rough wall has been broken down into only two elements: the parallel-flow constant-stress layer and the roughness layer where the mean flow is normally three-dimensional. However, the structure of the latter is complex:

(1) For large values of y_f, viscosity will influence the mean motion throughout the roughness layer; but if y_f is small enough, the outer portion will be fully turbulent.

(2) Among the roughness elements themselves will be found a variety of separating and reattaching flows, and perhaps concentrated vortices. The character of these flows is strongly dependent on the surface geometry, defined for the case of Figure 4.5 by y_f/h, w/h and L/h.

(3) Right at the surface, a variety of small-scale viscous layers will be found.

Problem 5. For some roughness geometries, no single dimension of the roughness serves to define the friction characteristic or the scale of the logarithmic layer.

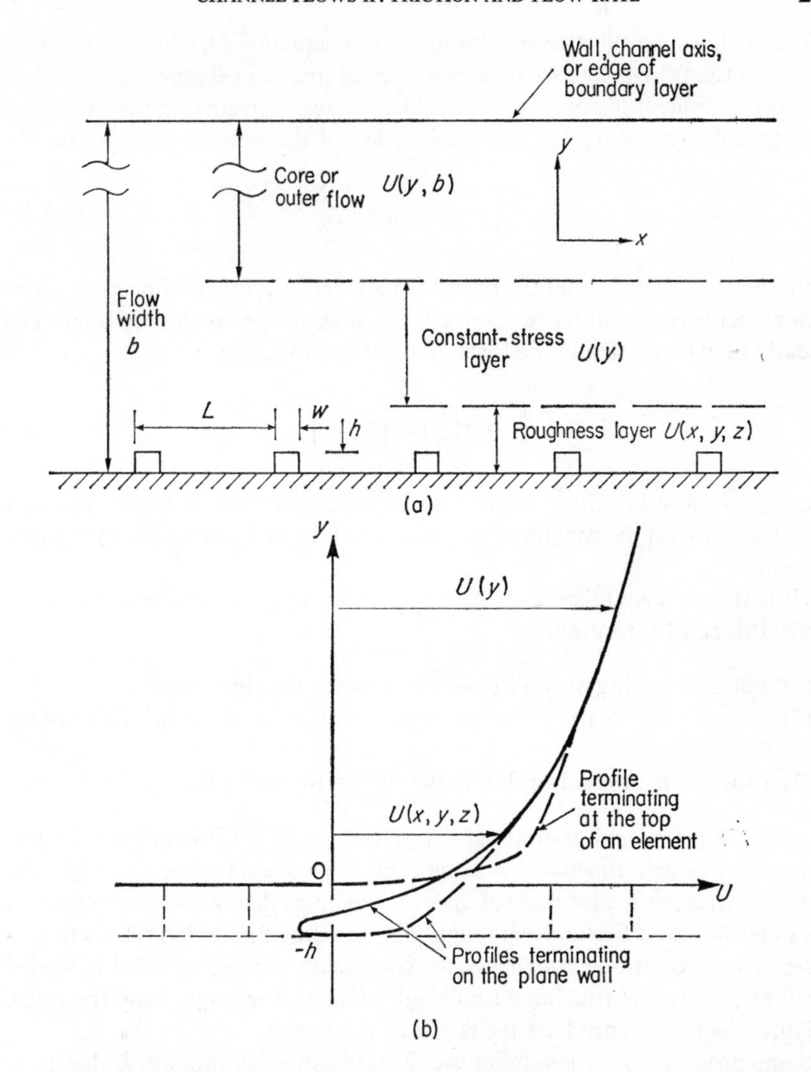

Figure 4.5 Flow near a wall to which rectangular bars are fixed normal to the direction of the mean flow. (a) Geometry of roughness and flow, showing the wall layer composed of roughness layer and constant-stress layer. (b) Time-mean velocity distributions near the wall, for different values of x, the coordinate measured along the wall

In other words, no one of the dimensions l_i will reduce the velocity distributions for various (high) Reynolds numbers to the form

$$\frac{U}{u_{\mathrm{f}}} = A \ln\left(\frac{y + \varepsilon}{l_i}\right) + B'\left(\frac{l_i}{l_1}, \text{etc.}\right) \tag{4.3.28}$$

(Here ε defines the effective wall location, as in equation (4.3.26).) Nor can we represent the fully-rough friction by a straight line as in Figure 4.2.

It is not difficult to see how this problem arises. Considering the roughness of Figure 4.5, we generalize the specification of the logarithmic layer to

$$\frac{\mathrm{d}U}{\mathrm{d}y} \sim \tau_{\mathrm{w}}, \rho, y + \varepsilon, \frac{L}{b} \tag{4.3.29}$$

with $\varepsilon = \varepsilon(h, w, L, y_{\mathrm{f}}, b)$, and the ratio L/b included as a token of possible interactions between the outer motion and the flow in the roughness layer. The integral may be written in the form of equations (4.3.25):

$$\frac{U}{u_{\mathrm{f}}} = A\left(\frac{L}{b}\right) \ln\left(\frac{y+\varepsilon}{k_2}\right) \tag{4.3.30}$$

with $k_2 = k_2(h, w, L, y_{\mathrm{f}}, b)$ a length scale for the constant-stress layer. Note that A may depend on the roughness geometry, and possibly on the flow condition as well.

Hitherto we have taken $k_2 \propto h$, but it is easy to imagine circumstances in which this is inappropriate:

(1) $L/w \simeq 1$, leaving only narrow slots between the elements;
(2) L equal to the scale ($\propto b$) of an important element of the turbulence of the outer flow; and
(3) a surface composed of fluting parallel to the mean flow.

Note too that for sand-grain roughness, $L \simeq 2w \simeq 2h \simeq k_s$; hence it is impossible to say which dimension is significant. We conclude that no single geometric characteristic of varied patterns of roughness determines their effect on the outer flow and friction. Although the preceding discussion relates to periodic or nearly periodic roughness, the conclusion can be extended to varied random patterns of roughness: no single statistical measure (say, the r.m.s. height) adequately correlates the effects of roughness.

Some progress towards a more widely applicable definition of k_2 has been made for the particular roughness geometry of Figure 4.5. When the bars are widely spaced, the roughness height h can be used to standardize the velocity profiles, as in equation (4.3.28), and it is found that $\varepsilon \propto h$. Since $k_2 \propto h$ in these circumstances, and ε and k_2 depend upon the same parameters, it is natural to try $k_2 \propto \varepsilon$ in more varied circumstances. This method of defining the length scale of the logarithmic region has successfully reduced experimental data for closely spaced bars to the pattern of Figure 4.2. Rather surprisingly, the assumption that $k_2 \propto b$, the overall flow width, has also had some success. This suggests that $\varepsilon \propto b$, and that there is sometimes a significant interaction between the flows in the core and roughness layer.

4.4 ASYMMETRIC PARALLEL FLOWS

4.4.1 Experimental Results

We noted earlier that parallel, fully developed flow is possible, not only in the round pipes which have been considered so far, but in a few other channel sections:

(1) the annular passage between concentric cylinders, the inner and outer walls possibly differing in roughness, a situation often encountered in heat exchangers, including nuclear reactors; and

(2) a flat channel between parallel walls, possibly differing in roughness or with a relative motion, as in hydrodynamic bearings, in the clearance passages of turbomachines, and in open channels, where the free surface replaces one of the walls.

In these more varied parallel flows, the stresses at the two walls need not be equal. They will differ when there is a significant relative motion or difference in roughness or, in the case of the annulus flow, when $R_2/R_1 > 1{\cdot}1$, say. Since the flow near a wall is determined by the stress there, we must expect that a difference in the wall stresses will denote asymmetry in nearly every aspect of the flow.

Figure 4.6 illustrates the general character of asymmetric flows. The particular flow, one studied by Hanjalić and Launder, occurs in a flat channel, one of whose walls is effectively smooth, the other being roughened with square bars; very near this wall it is not possible to determine a unique mean velocity $U(y)$. A striking feature of the mean velocity distribution of Figure 4.6(a) is the difference between $y = y_m$, the point of maximum mean velocity, and $y = y_0$, where the shear stress is zero. As was pointed out in Section 3.4.6, this behaviour cannot be represented using the gradient-diffusion model and its mixing-length derivative, which give

$$\frac{\tau}{\rho} = \epsilon_m \frac{dU}{dy} = l_m^2 \left|\frac{dU}{dy}\right| \frac{dU}{dy} \tag{4.4.1}$$

The variations of turbulence intensities shown in Figure 4.6(b) offer an explanation for this failure of the gradient-diffusion model. At $y = y_m$ we may expect $v > 0$ (that is, away from wall 2) to be associated with large values of $-u$, and $v < 0$ to be associated with smaller values; hence $\tau/\rho = -\overline{uv} > 0$ at this point.

The results of Figure 4.6 indicate that there is a region in which $\overline{uv}\,dU/dy > 0$; this corresponds to a negative value for that term of equation (3.3.26) which represents the production of turbulence energy. This is possible because fluid which has been subjected to mean strain of one sign ($dU/dy > 0$, say) is convected into a region where the mean strain has the opposite sign. This process also occurs in symmetrical channel flows, but there the requirements of symmetry ensure that $\overline{uv}\,dU/dy \leqslant 0$ everywhere. In asymmetric flow, there is a

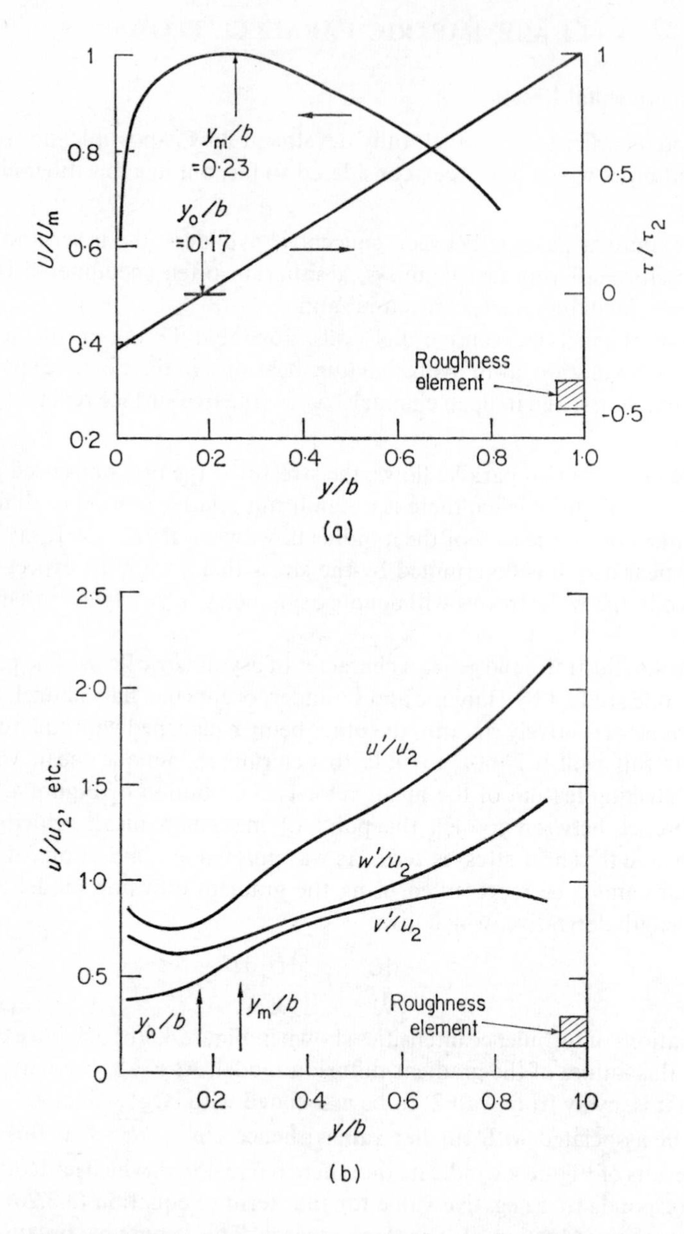

Figure 4.6 Flow in a plane channel with one smooth wall and one rough wall. After Hanjalić and Launder: $Re = 5{\cdot}6 \times 10^4$; $L/h = 10$; $b/h = 17$. (a) Variations of mean velocity and shear stress. (b) Variations of r.m.s. intensities of turbulence components, rendered dimensionless using $u_2 = (\tau_2/\rho)^{\frac{1}{2}}$, the friction velocity for the rough wall

region where the averaged effect of the two types of mean strain is positive. To represent this, we must use a model of turbulence which is considerably more complex than those of Section 3.4. Such models will be discussed in Chapter 9.

4.4.2 Simple Models of the Flow

The absence of symmetry in the mean flow and stress distribution necessitates a reconsideration of the relationship between friction and flow rate. Our earlier calculations tacitly adopted conditions of symmetry about the channel centre-line; we must now introduce an explicit condition to match the flows on the two sides of the channel. Since the precise nature of the velocity variation in the core does not greatly influence the overall flow rate, the choice of matching condition is not too critical for the friction calculation. This argument provides some justification for the simplifying assumption $y_0 = y_m$, that is, coincidence of the positions of zero stress and maximum mean velocity.

For a smooth-walled annulus, the argument has been carried a stage further, and the point $y_0 = y_m$ is commonly assumed to have the same position as in laminar flow through the annulus. In other words, it is assumed that the wall-stress ratio τ_2/τ_1 is a unique function of the annulus shape factor R_2/R_1. This assumption is not unrealistic for high Reynolds numbers ($2U_a\Delta R/\nu > 5000$, say), but is less satisfactory for turbulent flow at lower Reynolds numbers.

For asymmetric flows in plane channels, we shall find that a knowledge of the relative velocity or of the wall roughnesses can be used to predict the wall stresses and the matching point. For an annulus with walls of differing roughness, no way is known to predict the stress ratio and the associated zero-stress point. An estimate might be made by combining the predictions for a smooth-walled annulus and for a plane channel with one or both walls roughened.

The discussion in Section 3.4 of the relationship between shear stress and mean velocity suggests several ways of predicting the velocity variation across an asymmetric flow. The three models to be discussed here illustrate a number of the points made in Section 3.4.6, in connection with the applicability of the eddy-viscosity and mixing-length hypotheses. For definiteness, we consider a planar flow like that of Figure 4.6, but with a few modifications the analysis applies to annulus flow as well.

Model I: a pair of logarithmic layers intersecting at $y = y_i$, so that $y_i = y_m$, the point of maximum velocity. For convenience, we shall later assume that $y_i = y_m = y_0$, the point where $\tau = 0$. However, this additional assumption is not necessary; the model does allow a crude representation of asymmetric flows with $y_m \neq y_0$, although the stress and diffusion in the core are poorly described.

Considering the motion near a smooth, fixed wall, like wall 1 of Figure 4.6, we calculate the maximum velocity as

$$\frac{U_i}{u_1} = A \ln\left(\frac{y_i}{y_1}\right) + B \tag{4.4.2}$$

with $u_1 = \nu/y_1 = (\tau_1/\rho)^{\frac{1}{2}}$ the friction velocity for that wall. The discussion of the other wall and of the total resistance to the flow will be deferred until Section 4.4.3.

Model II: a pair of varying-stress equilibrium layers defined by equation (3.5.10):

$$\frac{dU}{dy} = \frac{A}{y}\left(\frac{\tau}{\rho}\right)^{\frac{1}{2}} \tag{4.4.3}$$

Note that $dU/dy = 0$ for $\tau = 0$, giving $y_0 = y_m = y_i$ for this model. Moreover, $\epsilon_m = 0$ at $y = y_m$, as in the mixing-length results (3.4.60). Hence this model is not suited to the calculation of heat transfer across the flow.

Introducing the linear stress variation

$$\tau = \tau_1 - \alpha y \quad \text{with} \quad \alpha = \frac{\tau_1}{y_0} > 0 \tag{4.4.4}$$

we integrate to find

$$\frac{U}{u_1} = A\left[\ln\left\{\frac{1-(1-y/y_0)^{\frac{1}{2}}}{1+(1-y/y_0)^{\frac{1}{2}}}\right\} + 2\left(1-\frac{y}{y_0}\right)^{\frac{1}{2}}\right] + C \tag{4.4.4}$$

with C a constant of integration. Near the wall, the velocity variation reduces to

$$\frac{U}{u_1} \simeq A \ln\left(\frac{y}{4y_0}\right) + 2A + C$$

provided that $y_0 \gg y_1$. The latter is the requirement for the existence of a logarithmic, fully turbulent region. Comparison with the constant-stress result (4.4.2) indicates that

$$C = A \ln\left(\frac{4y_0}{y_1}\right) + B - 2A \tag{4.4.4}$$

The maximum velocity obtained from equations (4.4.4) is

$$\frac{U_i}{u_1} = C = A \ln\left(\frac{y_i}{y_1}\right) + B - (2 - \ln 4)A \tag{4.4.5}$$

This differs from the constant-stress prediction (4.4.2) by only $\Delta U_i/u_1 = 0{\cdot}613A \simeq 1{\cdot}5$. The use of this more complex model is hardly justified unless the stress at one wall is very small; the flow near that wall will then be more realistically specified by the varying-stress results.

Model III: a pair of logarithmic layers separated by a core where the eddy viscosity is uniform, $\epsilon_m = \epsilon_c$. The relations (4.4.1) indicate that $y_0 = y_m$. This restriction makes possible a finite value of the diffusivity at every point in the core; thus the model is capable of representing heat or mass transfer across the flow.

We take the velocity variation in the core to be

$$\frac{U_m - U}{u_f} = \frac{u_f b}{\epsilon_c}\left(\frac{y}{b} - \frac{y_0}{b}\right)^2 \tag{4.4.6}$$

a generalization of a result (Type A) in Table 3.3, with

$$u_f = \left[\frac{|\tau_1| + |\tau_2|}{2\rho}\right]^{\frac{1}{2}} \tag{4.4.7}$$

a measure of the turbulent activity in the core. For the particular cases $\tau_1 = \pm\tau_2$, this gives the values of u_f used in Table 3.3. The common factor for the flows considered in that table is the velocity range $\Delta U/u_f$. In the absence of more specific results, we determine the more varied asymmetric core flows by taking

$$\frac{\Delta U}{u_f} = 7 \tag{4.4.8}$$

the central value of Table 3.3.

We consider a flow like that of Figure 4.6, with $\tau_1, \tau_2 > 0$, both wall stresses acting to retard the flow. For simplicity, we may now drop the absolute-value bars of equation (4.4.7). We further limit ourselves to cases in which $\tau_2 \geqslant \tau_1$; then

$$\frac{y_0}{b} = \frac{\tau_1}{\tau_t} \leqslant \tfrac{1}{2}$$

with $\tau_t = \tau_1 + \tau_2$ representing the total retardation. The velocity range from the maximum to the further wall is

$$\frac{\Delta U}{u_f} = \frac{u_f b}{\epsilon_c}\left(1 - \frac{y_0}{b}\right)^2 = \frac{u_f b}{\epsilon_c}\left(\frac{\tau_2}{\tau_t}\right)^2 \tag{4.4.9}$$

and the velocity variation (4.4.6) can be written

$$\frac{U_m - U}{u_f} = \left(\frac{\tau_t}{\tau_2}\frac{y}{b} - \frac{\tau_1}{\tau_2}\right)^2 \frac{\Delta U}{u_f} \tag{4.4.10}$$

the final factor being fixed, as indicated in equation (4.4.8).

Again we match wall and core flows by requiring that the eddy viscosity vary continuously across the flow. We equate the wall-layer result (4.1.5) to the core value defined by equation (4.4.9):

$$Ku_1 y_i = u_f b \frac{u_f}{\Delta U}\left(\frac{\tau_2}{\tau_t}\right)^2$$

The matching point is then

$$\frac{y_i}{b} = \frac{A}{\sqrt{2}}\frac{u_f}{\Delta U}\left(\frac{\tau_t}{\tau_1}\right)^{\frac{1}{2}}\left(\frac{\tau_2}{\tau_t}\right)^2 \tag{4.4.11}$$

following the introduction of $u_f/u_1 = (\tau_t/2\tau_1)^{\frac{1}{2}}$. The matching point at the other wall is given by

$$1 - \frac{y_i'}{b} = \frac{A}{\sqrt{2}}\frac{u_f}{\Delta U}\left(\frac{\tau_2}{\tau_t}\right)^{\frac{3}{2}} \tag{4.4.12}$$

For $\tau_2 = \tau_1 = \frac{1}{2}\tau_t$, equations (4.4.11, 12) give the matching point (4.2.1) appropriate to symmetrical pipe or channel flow.

The results (4.4.9) to (4.4.12) apply only when $\tau_2 \geqslant \tau_1 > 0$. Parallel results can be obtained for cases in which τ_2 and τ_1 do not fit into this pattern.

4.4.3 Friction Laws

Having considered a number of ways of describing an asymmetric mean-velocity distribution, we now examine the friction characteristic implied by the simplest of them—Model I, comprising a pair of logarithmic layers. These calculations provide a pattern into which more complicated models may be fitted.

Near the rough wall we have

$$\frac{U}{u_2} = A \ln\left(\frac{b-y}{k_1}\right) + B \tag{4.4.13}$$

on adopting the roughness scale k_1 defined in equations (4.3.24). Here b is the effective width of the channel, increased by an amount ε at the surface to which the roughness elements are fixed, as in equations (4.3.26, 29). For $y = y_i$ this law must also give U_i, the value of equation (4.4.2). Thus the matching condition is

$$u_1\left[A \ln\left(\frac{y_i}{y_1}\right) + B\right] = u_2\left[A \ln\left(\frac{b-y_i}{k_1}\right) + B\right]$$

whence

$$\frac{y_i}{y_1}\left(\frac{k_1}{b-y_i}\right)^{u_2/u_1} = \exp\left[\left(\frac{u_2}{u_1} - 1\right)\frac{B}{A}\right] \tag{4.4.14}$$

The relationships

$$\frac{y_i}{b} = \frac{y_0}{b} = \frac{\tau_1}{\tau_t} \quad \text{and} \quad \frac{u_2}{u_1} = \left(\frac{\tau_2}{\tau_1}\right)^{\frac{1}{2}} \tag{4.4.15}$$

show that condition (4.4.14) is of the form

$$f\left(\frac{\tau_2}{\tau_1}, \frac{k_1}{b}, \frac{y_1}{b}\right) = 0 \tag{4.4.16}$$

This implicitly determines the stress ratio corresponding to a particular roughness and flow condition.

The alternative way of generating asymmetry—through relative motion of the walls—can be fitted into the analysis as follows. Near the wall moving with velocity U_2 the mean velocity is

$$\frac{U}{u_2} = A \ln\left(\frac{b-y}{k_1}\right) + B - \frac{U_2}{u_2} \tag{4.4.17}$$

This may be written

$$\frac{U}{u_2} = A \ln\left(\frac{b-y}{k_1'}\right) + B$$

with

$$k_1' = k_1 \, e^{KU_2/u_2} \tag{4.4.17}$$

an effective roughness which allows for wall motion. The results (4.4.14, 16) can now be rewritten in terms of k_1', thus incorporating the possibility of a moving wall. By setting $k_1 = y_f$, we obtain results appropriate to a moving smooth wall.

For the model adopted here, the flow rate per unit breadth of the channel is

$$\begin{aligned} \dot{v} &= \int_0^b u \, dy \\ &= u_1 \int_0^{y_i} \left[A \ln\left(\frac{y}{y_1}\right) + B\right] dy + u_2 \int_{y_i}^b \left[A \ln\left(\frac{b-y}{k_1}\right) + B\right] dy \\ &= [y(U - Au_1)]_0^{y_i} - [(b-y)(U - Au_2)]_{y_i}^b \end{aligned} \tag{4.4.18}$$

Neglecting the contributions at the limits $y = 0, b$ (as when obtaining the pipe results (4.2.4)), we have

$$U_a = \frac{\dot{v}}{b} = U_i - A\left[u_1 \frac{y_i}{b} + u_2\left(1 - \frac{y_i}{b}\right)\right] \tag{4.4.19}$$

To convert this into a friction law, we introduce a coefficient representing the pressure gradient along the channel:

$$c_t = -\frac{b}{\rho U_a^2} \frac{dP_w}{dx} = \frac{\tau_1 + \tau_2}{\rho U_a^2} = 2\frac{u_f^2}{U_a^2} \tag{4.4.20}$$

for the case $\tau_2, \tau_1 > 0$. The result (4.4.19) can now be written

$$\left(\frac{2}{c_t}\right)^{\frac{1}{2}} = \frac{u_1}{u_f}\left[A \ln\left(\frac{y_i}{b}\frac{b}{y_1}\right) + B - A\frac{y_1}{b}\right] - \frac{u_2}{u_f} A\left(1 - \frac{y_i}{b}\right) \tag{4.4.20}$$

By virtue of the relationships (4.4.15) this has the form

$$f\left(c_t, \frac{\tau_2}{\tau_1}, \frac{y_1}{b}\right) = 0 \tag{4.4.21}$$

Since the conditions at the two walls are specified by the stresses, this result includes both rough-wall and moving-wall cases. Using the results (4.4.14, 17), we can specify the friction characteristics in terms of either roughness or relative wall velocity.

The flow condition is usually specified by the mean-flow Reynolds number $U_a b/\nu$, rather than by y_1/b. The relationship between these parameters can be obtained by rewriting equation (4.4.19) as

$$f\left(\frac{U_a b}{\nu}, \frac{\tau_2}{\tau_1}, \frac{y_1}{b}\right) = 0 \tag{4.4.22}$$

Finally, through the chain of relationships (4.4.16, 17, 21, 22) we can obtain

$$c_t = f\left(\frac{U_a b}{\nu}, \frac{k_1}{b}\right) \tag{4.4.23}$$

for a rough wall, or

$$c_t = f\left(\frac{U_a b}{\nu}, \frac{U_2}{U_a}\right) \qquad (4.4.23)$$

for a moving wall. These are generalizations of the results for a plane channel with smooth, fixed walls.

Although this analysis indicates the general form of the friction laws, we must expect that, as with pipe flow, some adjustment of constants will be required to match experimental results. It will be appreciated that the discussion must be modified in detail to encompass cases in which both walls are rough, and in which $\tau_1/\tau_2 < 0$, as may occur with a moving wall. Somewhat better predictions may be given by the more complex Models II and III which were defined in the previous section; the former is particularly appropriate when τ_1 or $\tau_2 \simeq 0$.

Summing up, we conclude that the velocity distributions developed in Sections 3.4 and 3.5 can be manipulated to give friction predictions for a variety of parallel channel flows. Moreover, models which give the eddy viscosity a constant value through the core are suitable for calculations of heat and mass transfers across these flows.

4.5 MORE VARIED CHANNEL FLOWS

4.5.1 Secondary Flows

The name secondary flow is used for a time-mean motion conceived to be superposed on or embedded in a basic or primary flow. Some judgement is needed in distinguishing the two elements; the secondary flow is usually chosen so that its velocities are considerably smaller than those of the basic motion. Often, when the primary motion is parallel to a wall or to a uniform free stream, the secondary element is taken to be the cross-channel or cross-stream component of the mean flow. One kind of secondary flow has already been noted in Section 3.2.3:

Type 1: the cross-channel component of a fully developed flow in a channel with parallel walls. This develops when the direct mixing stresses in the cross-section are not self-equilibrating, and is of special interest because it derives directly from turbulent activity.

The other types of secondary flow considered below can occur in the absence of turbulence; they may of course interact with turbulence when it is present. Before looking at them we shall consider the effects of the flow defined above.

Figures 4.7(a) and (b), based on results reported by Brundrett and Baines, show the secondary-flow pattern for one half of a rectangular channel, and the

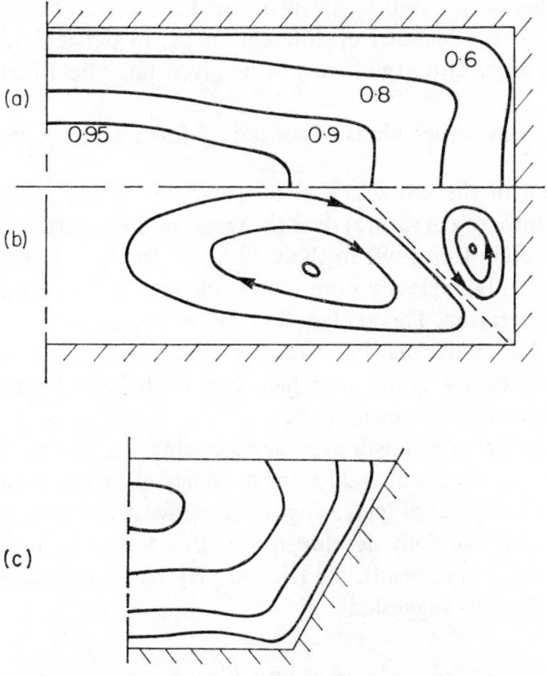

Figure 4.7 Effects of secondary flows (of Type 1) on the velocity distribution in non-circular channels. (a) Contours of constant axial velocity U/U_c in a rectangular channel (3:1), for $U_a d_h/\nu \simeq 6 \times 10^4$, after Brundrett and Baines. (b) Secondary-flow pattern corresponding to (a). (c) Contours of constant axial velocity U/U_c in an open channel of trapezoidal form

corresponding contours of the streamwise component of the mean velocity. Obviously, the convolutions in the velocity contours result from the cross-stream convection of the secondary flows. In Figure 4.7(c) the velocity contours for half of an open channel of trapezoidal form are shown. The secondary flow transfers the maximum in the streamwise velocity to a point below the surface, for typical channel sections, about one-fifth of the way between surface and stream bed.

We have seen that the velocity variation over the core of a turbulent flow is proportionally smaller than that for laminar flow in the same channel. This effect is accentuated by secondary flows, which act to extend the core into the angles of the section. Consequently, for a channel section whose corners are not too sharp, the wall layer and shear stress are fairly uniform around much of the perimeter.

This fortunate circumstance is the basis of a simple way of correlating the friction characteristics for a variety of cross-sections. It is argued that the wall stress in any channel is nearly that which would be generated by the same mean or core velocity in a channel of different shape, in particular, in the much studied round pipe. This argument will be given analytical form in the next section.

Considering now other kinds of secondary flow, we may distinguish:

Type 2: flow in the cross-section of a curved channel, inwards (that is, towards the centre of curvature) near the walls, and outwards in the core. It is not hard to explain this flow: in the wall layer, the radial pressure gradient imposed by the high-velocity core is not adequately balanced by the local centripetal acceleration. This explanation applies in general terms to secondary flow of Type 1 as well: the direct mixing-stress components generate small cross-channel pressure gradients which can be balanced only by a cross-channel component of the mean flow.

Flows of Type 2 are responsible for the deepening of a stream with an erodible bed at the outside of a bend, and for the tendency of such streams to meander. They also have the effect of increasing the streamwise component of friction in a curved channel. For fully developed flow in a long *helical pipe* (to be distinguished from a short bend), the friction may be estimated as follows. For laminar flow, Prandtl suggested

$$\frac{c_f}{c_{f_0}} = 0{\cdot}29\left[\mathrm{Re}_d\left(\frac{d}{D}\right)^{\frac{1}{2}}\right]^{0\cdot36} \tag{4.5.1}$$

when the *Dean number* $\mathrm{De} = \frac{1}{2}\mathrm{Re}_d(d/D)^{\frac{1}{2}}$ lies in the range $40 < \mathrm{De} < 1000$. For turbulent flow, Ito has found

$$\frac{c_f}{c_{f_0}} = \left[\mathrm{Re}_d\left(\frac{d}{D}\right)^2\right]^{0\cdot05} \tag{4.5.2}$$

when the quantity in square brackets is greater than 6. In both cases, d is the tube diameter, D is the helix diameter, and c_{f_0} is the friction coefficient in the absence of curvature. The transition to turbulence is inhibited by curvature, occurring, for example, at $\mathrm{Re}_d \simeq 5000$ when $\mathrm{De} = 40$.

Type 3: a component normal to the outer flow in a boundary layer exposed to a pressure gradient which is not parallel to the free stream. Such skewed layers are found near the tips of wings, on swept wings, on the blades of turbomachinery, and at the bottom of a stirred cup of tea, where they are revealed by the motion of the tea leaves.

Note that the small component of velocity normal to the wall is taken to be an integral part of the primary motion, rather than an aspect of the secondary flow. Since this distinction is arbitrary, the whole flow system may be referred to as a *three-dimensional boundary layer*, with no attempt to distinguish a 'secondary' element.

Type 4: a boundary layer whose fluid is distorted as the primary flow moves around an obstruction. The secondary flow can be discussed in terms of the time-mean vorticity of the boundary layer. The vortex lines are stretched by the distortion, and concentrated vortices trail downstream. Complex separating and reattaching flows will be found in the wake, and also just ahead of the obstruction. Such flows will occur in the roughness layer near a rough wall, the vortex system around an isolated element taking the form of a horseshoe vortex wrapped around the protrusion.

Type 5: the response of a rotating, stratified or electrically conducting fluid when the primary flow is disturbed by an immersed body, by extraction or addition of fluid, or by a change in the rotational speed of a solid boundary. The secondary flow is particularly pronounced in the direction of stratification (or of the vortex or field lines), and a wake-like disturbance may appear ahead of the disturbing element as well as behind.

Type 6: a steady second-order flow associated with a periodic basic flow. Examples are the mean forwards drift associated with waves at the surface of a liquid, and the 'acoustic streaming' generated by a body oscillating in air.

4.5.2 Equivalence of Channel Flows

The prediction of secondary flows of Type 1, those induced by variations of direct mixing stresses in the channel cross-section, and of the associated distributions of axial velocity and wall stress, requires a far more detailed model of turbulence than we have available. Here we seek only to convert the qualitative observations concerning these secondary flows into a working formula relating the well established results for round pipes to the friction in other sections.

Consider a flow which is steady in the mean and uniform along a channel, which may be either closed or open. The net driving force in a section of area A is related to the time-mean shear stresses on the wetted perimeter C_w by

$$-A\frac{d(P_w+\rho gz)}{dx}=\int_{C_w}\tau(s)\,ds=\overline{\tau_w}\,C_w \tag{4.5.3}$$

where z is the surface elevation, s is distance measured around the perimeter, and $\overline{\tau_w}$ is the average stress over the perimeter. Introducing a coefficient to represent the average stress, we have

$$-\frac{d(P_w+\rho gz)}{dx}=c_f\left(\tfrac{1}{2}\rho U_a^2\right)\frac{C_w}{A} \tag{4.5.4}$$

If the coefficient c_f can be specified, we can now calculate the friction for any channel section. We have seen that friction coefficients for pipe flow depend on the scale of the flow, through the Reynolds number $U_a d/\nu$, and on the relative roughness k/d. Our task is to select effective values of these parameters representative of other channel sections.

If the wall stress is indeed fairly uniform around the perimeter, and the velocity is close to U_a through much of the channel, it is reasonable to retain U_a as the velocity scale. It is less obvious how to choose a length scale to replace the diameter d. The problem is usually resolved by basing the scale on A/C_w, which has the dimension of length, and arranging that the friction law incorporating this scale be consistent with known results for pipe and channel flows.

Three ways in which the length scale can be defined are set out in Table 4.5. The three measures of the effective or equivalent or hydraulic size of a channel are related by

$$d_h=2\,R_h'=4\,R_h \tag{4.5.5}$$

The hydraulic diameter d_h is the simplest to introduce into our earlier results; for a round section, it reduces to the actual diameter. Similarly, the radius R_h' reduces to the actual radius when a pipe is considered; this is not the case for the alternative R_h which is more widely used in practice. For the latter, R_h, the term 'hydraulic depth' is more appropriate; unfortunately, this name is also used for the quantity $h_h=A/T$, with T the free-surface width.

Whichever of the scales of Table 4.5 is adopted, the final stage of the argument is that the relationship

$$c_f=f\left(\frac{U_a d_h}{\nu},\frac{k}{d_h}\right) \tag{4.5.6}$$

is independent of the shape of the channel section. The precise length scale adopted is unlikely to be critical, since c_f is a rather weak function of Reynolds number when the flow is turbulent, while the roughness is often so imprecisely known that an erroneous length scale for the flow does not introduce a significant additional error.

The result (4.5.6) requires near uniformity of the shear stress over a large part of the channel perimeter, although the stress can be zero over certain portions which are neglected in calculating the wetted perimeter. The analysis is not applicable to the asymmetric plane flows discussed in Section 4.4, nor to flow in an annulus, save for $R_2/R_1 \simeq 1$, when the annulus is little different from a flat channel with equal wall stresses. Nor will this method of analysis be appropriate to a channel section which has quite different length and velocity scales in its several parts. A classical example is a river which has spilled over onto a shallow flood plain. A rather similar situation is found in heavily

Table 4.5. Effective dimensions of channel sections

Pattern flow	A/C_w	Derived dimensions	Names
Pipe	$d/4$	$d_h = 4A/C_w$	Hydraulic or equivalent diameter
Broad closed channel ($\tau_1 = \tau_2$)	$b/2$	$R_h' = 2A/C_w$	Hydraulic (mean) radius
Broad open channel ($\tau_2 = 0$)	h	$R_h = A/C_w$	Hydraulic (mean) radius or depth

finned heat-exchange passages. Such cases may be treated by summing the contributions of a series of quasi-independent flows, subject to appropriate matching conditions.

4.5.3 Open Channels

Although the preceding discussion encompasses channels with free surfaces at which friction is negligible, some attention must still be given to the special techniques adopted by hydraulic engineers in calculating friction. These differ from the methods developed earlier for two reasons: first, because the empiricism of hydraulics was established before a rational analysis of friction was available; secondly, because the hydraulic engineer must often deal with ill-defined flows, sometimes variable in both space and time, for which a refined analysis would be pointless.

For the calculations of open-channel hydraulics, the result (4.5.4) is recast as

$$S = c_f \frac{U_a^2/2g}{R_h} \tag{4.5.7}$$

where $R_h = A/C_w$ is the hydraulic mean radius, $U_a^2/2g$ is the velocity head and

$$S = -dz/dx$$

is the downwards slope of the free surface; for a fully developed flow, this is also the slope of the bed of the channel. This result may be written as the Chézy formula:

$$U_a = C(R_h S)^{\frac{1}{2}} \tag{4.5.7}$$

with $C = (2g/c_f)^{\frac{1}{2}}$ the *Chézy coefficient*, a measure of the carrying capacity of the channel.

The Chézy coefficient is usually specified by an empirical function which accounts for the nature of the channel and for the scale of the flow. A widely used expression is the *Manning formula*:

$$\begin{aligned} C &= \frac{1}{n} R_h^{\frac{1}{6}} \quad \text{for } R_h \text{ in metres} \\ &= \frac{1\cdot 49}{n} R_h^{\frac{1}{6}} \quad \text{for } R_h \text{ in feet} \end{aligned} \tag{4.5.8}$$

The factor n, called simply Manning's n, accounts for the character of the channel boundaries; later it will be seen to be equivalent to the wall roughness. The ranges of n appropriate to a variety of channels are indicated in Table 4.6.

The measures of friction introduced above are connected by

$$\begin{aligned} \frac{2}{c_f} = \frac{C^2}{g} &= \frac{R_h^{\frac{1}{3}}}{9\cdot 8\, n^2} \quad \text{for } R_h \text{ in metres} \\ &= \frac{R_h^{\frac{1}{3}}}{14\cdot 5\, n^2} \quad \text{for } R_h \text{ in feet} \end{aligned} \tag{4.5.9}$$

Using these, we can express our earlier results in terms of the conventional hydraulic parameters. Alternatively, we can interpret the traditional methods of hydraulics in the context of our analysis of friction. This is done in Table 4.6 for a particular case: a channel with hydraulic radius $R_h = 1$ m and with Reynolds number $\mathrm{Re}_d = 4U_a R_h/\nu = 10^6$. Friction coefficients corresponding to the several ranges of n are given; equation (4.3.15) then leads to relative roughness; finally, the effective roughness size k_s is found. Comparison with part I of Table 4.4 shows that the conventional ranges of n are consistent with the conventional roughness levels.

Finally, we ask why a constant value of the parameter n gives a fairly realistic picture of the dependence of friction on both Reynolds number and roughness. Consider first the fully rough condition normally encountered in hydraulic applications. From equations (4.5.9) we obtain

$$c_f \propto \frac{n^2}{R_h^{\frac{1}{3}}} \tag{4.5.10}$$

But $c_f = f(k_s/R_h)$ under these conditions. Hence

$$c_f \propto \left(\frac{k_s}{R_h}\right)^{\frac{1}{3}} \quad \text{and} \quad n \propto k_s^{\frac{1}{6}} \tag{4.5.11}$$

The first result is close to the power law (4.3.17) which gives the role of roughness for pipe friction. The second result indicates how n will vary with the roughness size; in a hydraulic application, this might be the size of pebbles on the stream bed.

Turning to smooth-wall friction, for which $c_f = f(U_a R_h/\nu)$, we note that the Manning formulae (4.5.7, 8) imply that

$$U_a R_h \propto C R_h^{\frac{3}{2}} S^{\frac{1}{2}} \propto R_h^{\frac{5}{3}} \frac{S^{\frac{1}{2}}}{n}$$

For consistency with the result (4.5.10) we must have

$$c_f \propto \left(\frac{\nu}{U_a R_h}\right)^{\frac{1}{5}} \qquad (4.5.12)$$

with n a constant. This form of power law was noted earlier, equation (4.2.23), as an approximation for smooth-wall flow at high Reynolds numbers.

Table 4.6. Typical values of Manning's roughness parameter n, with corresponding friction coefficient c_f and equivalent sand-grain roughness k_s

		For $R_h = 1$ m, $Re_d = 10^6$		
Type of surface	$1000\,n$	$1000\,c_f$	R_h/k_s	k_s(mm)
Glass, plastic, smooth metal	10	2	20,000	0·05
Timber	11–14	2·4–3·8	8000–800	0·1–1·2
Concrete	12–20	2·8–7·8	3500–45	0·3–20
Earthen canals	20–25	7·8–12·2	45–12	2–80
Natural rivers:				
Straight	25–30	12–18	12–5	80–200
Winding	35–40	24–32		
Very weedy	75–150	110–440	not meaningful	

We may conclude that the traditional empiricism of hydraulics is broadly consistent with the rational analysis of friction presented earlier in this chapter. These simpler methods of friction estimation are adequate for many of the ill-defined situations with which the hydraulic engineer is faced.

4.5.4 Systems of Channels

Hitherto we have considered the dissipation of energy in fully developed flow in a uniform channel. In order to relate the pressures and flow rates in an actual channel or system of channels, we must take into account a number of other factors:

(1) *non-uniform development* of the flow on entering the channel from rest, after passing an obstruction, or on flowing from one channel to another;

(2) *specific losses*, the dissipation at junctions, transitions, fittings and control devices; and

(3) *interaction* between intersecting channels; for closed channels this is expressed by mass-flow balances and by the requirement that the pressure changes through alternative paths be the same.

Since a detailed treatment of these topics would draw us away from the main line of development, we shall confine ourselves to a brief survey incorporating some empirical results of practical utility.

Flow Development

For the purposes of this chapter, the aspect of flow development which is of greatest interest is the departure of the wall friction from its ultimate value for fully developed flow. Fortunately, in turbulent flow in a *uniform closed channel* the friction moves rapidly to its equilibrium value. For example, when fluid flows from a reservoir into a pipe, the fully developed friction is attained when $x/d = 10$ to 20 for high Reynolds numbers. Other aspects of the flow develop more slowly: the mean velocity profile does not attain its final form until $x/d =$ 50 to 80; the fine detail of the turbulence of the core takes much longer to settle down. Hence a definitive study of fully developed channel turbulence requires a very long entrance length indeed. These observations are consistent with the fact that the time scales of the wall turbulence are much smaller than those of the larger motions of the core flow.

The preceding comments apply to high-Reynolds-number flow, which becomes turbulent soon after entering the pipe. At lower Reynolds numbers, the movement towards the final developed state will not begin until transition from laminar to turbulent flow has occurred. For smooth flow into a pipe from rest, the distance required for transition may exceed

$$\frac{x}{d} = \frac{10^5}{\mathrm{Re}_d} \tag{4.5.13}$$

A number of factors shorten the region of laminar flow, for example, unsteadiness before entry, sharp corners at the inlet, and vibration of the system.

For a *uniform open channel*, two forms of development must be considered, possibly with very different time scales. Since the flow depth h can change, the adjustment of the entire flow may extend over very long distances. Whether the flow moves towards uniformity rapidly, slowly, or at all, depends on the downstream conditions, and on whether the flow is *subcritical or supercritical*, that is, whether the local flow velocity is less or greater than the local speed of long waves, $(gh)^{\frac{1}{2}}$. In the terminology of hydraulics, uniform or fully developed flow is referred to as *normal flow*.

The second aspect of development concerns the relationship between the local values of friction and driving force; for a uniform open channel, the latter is dependent on the flow depth, through the hydraulic radius. This relationship, it is usually assumed, is specified by the normal-flow result (4.5.7), except very near a section of discontinuity ($x/h = \pm 10$, say). This procedure neglects accelerations occurring during the prolonged development considered above, and also the departure of the core flow from its equilibrium form. It assumes that the time-scale of the stress-determining wall turbulence is much shorter than the time-scale of the gradually varying flow.

Specific Losses at Discontinuities and Transitions

The rapid dissipation which occurs near a channel irregularity can usually be attributed to *free shear layers* formed when the flow separates from the solid boundaries. The velocity change across such a layer is of the same order as the local average velocity, but the turbulence is no longer restrained by the presence of a solid wall. Hence the rate of energy extraction from the mean flow (which may be estimated as $-\rho\overline{uv}\,\Delta U$ per unit plan area of the layer) is much higher than for the wall layer in the same channel. The total dissipation within the layer is limited by its reattaching to the solid wall; the region of recirculating flow enclosed by the separated layer is typically about five times as long as it is wide. Flows of this kind will be discussed at greater length in Section 7.5.

To calculate specific losses, we need not know exactly how the dissipation is accomplished; we require only its net effect. This can be specified using a *loss coefficient* K or an *equivalent length of pipe* L_K. For constant-density flow, these are related to the change in total pressure by

$$K = 4\,c_f \frac{L_K}{d} = \frac{p_{0_1} - p_{0_2}}{\frac{1}{2}\rho U_0^2} \tag{4.5.14}$$

with U_0 a reference velocity. Table 4.7 gives typical values assigned to these coefficients in engineering calculations relating to pipe flows at high Reynolds numbers. In preparing this table, we assign the value $4c_f = 0{\cdot}025$ in order to relate K and L_K. The values of K in part II of the table are obtained by elementary applications of the continuity, momentum and energy principles for one-dimensional flow.

In the preceding discussion and in the form of Table 4.7, we have tacitly adopted three interrelated assumptions commonly made in engineering practice:

(1) the Reynolds-number dependence of the loss coefficient is negligible;

(2) there is no interference between adjacent transition flows; and

(3) the nature of the inlet flow is irrelevant (this is certainly not true for diffusers, in particular).

These simplifications are justified when the specific losses are a small fraction of the total dissipation in the system. Since this is often the case, they are sometimes referred to as *minor losses*. The conditions under which these are indeed 'minor' are evident from the values of L_K/d presented above. Taking $\sum L/\sum L_K > 4$ to define situations in which great care is not required in estimating the specific losses, we see that there must be, on average, $L/d > 120$ between typical fittings and transitions.

Table 4.7. Specific losses in pipe flows

Nature of distortion	Reference velocity	K	L_K/d
I. Empirical values			
45° elbow	Common U_a	0·4	15
90° standard elbow	Common U_a	0·75	30
90° square elbow	Common U_a	1·5	60
Standard tee:			
Through crosspiece	Common U_a	0·5	20
Through branch	Common U_a	1·5	60
Gate valve:	Common U_a		
Fully open		0·2	8
Half open		5·0	200
Sudden contraction:	Downstream U_2		
$A_1/A_2 = 1{\cdot}5$		0·15	6
$= 2$		0·25	10
$= 5$		0·35	14
Conical diffuser with $m = A_1/A_2$:	Upstream U_1		
$\theta = 5°$		0·14 $(1-m)^2$	6 $(1-m)^2$
$= 10°$		0·18 $(1-m)^2$	7 $(1-m)^2$
$= 20°$		0·43 $(1-m)^2$	17 $(1-m)^2$
$= 40°$		0·86 $(1-m)^2$	34 $(1-m)^2$
II. Theoretical predictions		K	L_K/d
Projecting entry	Downstream U_2	1	40
Flush entry	Downstream U_2	0·5	20
Sudden expansion with $m = A_1/A_2$	Upstream U_1	$(1-m)^2$	$40\,(1-m)^2$

Losses at transition and control structures in *open channels* can be treated in a broadly similar manner, although particular attention must be given to the possibility of changes between supercritical and subcritical flow within the rapidly varying flow. The conversion of a supercritical (or shooting) flow to a subcritical (or tranquil) flow is accomplished through a *hydraulic jump*, a rapid increase in depth analogous to the sudden expansion of a closed channel. See Plate II (facing p. 47). For a hydraulic jump in a broad, flat-bottomed channel, the loss in total head ($H_0 = h + \frac{1}{2}U_a^2/g$) is given by a one-dimensional analysis as

$$\frac{H_{0_1} - H_{0_2}}{h_1} = \left(\frac{h_2}{h_1} - 1\right)^3 \bigg/ \left(\frac{4h_2}{h_1}\right) \tag{4.5.15}$$

with h_1 and h_2 the upstream and downstream depths. This kind of coefficient is convenient for the calculations of open-channel hydraulics. Alternatively, the loss can be related to the velocity head of the stream ($U_1^2/2g$ upstream or $U_2^2/2g$ downstream), using the relationship

$$F_1^2 = \frac{U_1^2}{gh_1} = \frac{h_2}{2h_1}\left(1 + \frac{h_2}{h_1}\right) > 1 \tag{4.5.16}$$

which connects the *Froude number* F_1 of the upstream flow to the change in depth across the jump.

Analysis of Networks

Consider a number of closed channels connected to form a network. Using the methods described above (or in less clear-cut cases, experimental characteristics) we can relate the dissipation within each limb of the network to the flow through the limb. These relationships can be approximated with fair accuracy by

$$\Delta(P + \rho g z) \propto \dot{V}^n \tag{4.5.17}$$

where z is the vertical coordinate, and the index $n = 1$ (or a little more) for laminar flow, $=2$ (or a little less) for turbulent flow.

For constant-density flow through a network of *closed channels*, each limb has some influence in every other limb.* However, the influence of more distant parts of the network is slight, and the highly developed numerical technique of relaxation can be used to determine the pattern of flows and pressures which satisfy the laws (4.5.17) and the conditions

$$\sum \dot{V}_i = 0 \quad \text{at each junction} \tag{4.5.18}$$

$$\sum (\Delta P)_i = 0 \quad \text{around each closed circuit} \tag{4.5.19}$$

with $\dot{V}_i$ and $(\Delta P)_i$ the several volume flows at a junction and pressure changes around a circuit.

* This is also true for a varying-density flow which is everywhere subsonic, but not for one which is supersonic in some places.

For a system of *open channels*, or with some open and some closed channels, the analysis is potentially more difficult. We have noted already that the flow may move only slowly towards its normal depth and velocity profile. A further complication arises if parts of the flow are supercritical, for the regions upstream will not be directly influenced by downstream conditions. Hence the well proved relaxation methods are not appropriate at every point in the network, and one of the problems of analysis is to discover where they can be used, that is, where the flow is subcritical and where supercritical. For open-channel flow, the conditions (4.5.19) are replaced by

$$\sum (\Delta h)_i = 0 \quad \text{around each closed circuit} \qquad (4.5.20)$$

with $(\Delta h)_i$ the several changes in surface elevation. Where there are abrupt transitions between supercritical and subcritical flow (hydraulic jumps), or in the opposite direction, the requirements (4.5.20) must take account of them.

FURTHER READING

Chow, V. T. *Open-channel Hydraulics*, McGraw-Hill, New York (1959): Chapters 1, 2, 5, 6 and 8
Henderson, F. M. *Open Channel Flow*, Macmillan, New York (1966): Chapters 4 and 5
Knudsen, J. S. and D. L. Katz. *Fluid Dynamics and Heat Transfer*, McGraw-Hill, New York (1958): Chapters 5 to 9
Morris, H. M. *Applied Hydraulics in Engineering*, Ronald Press, New York (1963): Chapters 3, 4 and 5
Prandtl, L. Reference 8: Chapter III, Sections 8, 11 and 12
Schlichting, H. Reference 11: Chapter XX
Sutton, O. G. *Atmospheric Turbulence*, Methuen, London (1955): Chapters II to IV
Sutton, O. G. *Micrometeorology*, McGraw-Hill, New York (1953): Chapter 3
Townsend, A. A. Reference 13: Chapter 9

SPECIFIC REFERENCES

Brundrett, E. and W. D. Baines. 'The production and diffusion of vorticity in duct flow', *J. Fluid Mech.*, **19**, pp. 375–394 (1964)
Hanjalić, K. and B. E. Launder. 'Fully developed asymmetric flow in a plane channel', *J. Fluid Mech.*, **51**, pp. 301–335 (1972)
Hinze, J. O. 'Turbulent pipe-flow', in Reference 14
Patel, V. C. 'Calibration of a Preston Tube and limitations on its use in pressure gradients', *J. Fluid Mech.*, **23**, pp. 185–208 (1965)
Perry, A. E., W. H. Schofield and P. N. Joubert. 'Rough wall turbulent boundary layers', *J. Fluid Mech.*, **37**, pp. 383–413 (1969)
Reynolds, A. J. 'Analysis of turbulent bearing films', *J. Mech. Eng. Sci.*, **5**, pp. 258–272 (1963)

EXAMPLES

4.1 An analysis parallel to that of Table 4.1 can be used to investigate the zero-stress layer, the limiting form of the equilibrium wall layer which has the stress variation $\tau = \alpha y$ near the smooth wall $y = 0$.

(a) Show that $U = \frac{1}{2}(\alpha/\mu)y^2$ in the viscous sublayer and that $(\rho y/\alpha)^{\frac{1}{2}} \mathrm{d}U/\mathrm{d}y = 1/K_0$, a constant, in the fully turbulent part of the flow. How does the mixing length vary in the latter region?

(b) By matching the two components (a) at the point where $\epsilon_m = \nu$, show that

$$U = \frac{2}{K_0}\left(\frac{\alpha y}{\rho}\right)^{\frac{1}{2}} + C_0\left(\frac{\nu\alpha}{\rho}\right)^{\frac{1}{3}}$$

with C_0 a constant representing the velocity change near the wall, where the effective viscosity is small.

(c) Compare the result (b) with that of Example 3.28(c) to see how the constants A, $3B_1/2A$ and K_0 are related, and estimate the value of K_0, given that $3B_1/2A \simeq -0{\cdot}2$.

(d) Would you expect these results to apply near the free surface of a broad open channel?

4.2 (a) What variations of the eddy viscosity are implied by the models (4.1.10, 11)?

(b) How do the eddy-viscosity formulae (4.1.12) to (4.1.17) vary near the wall? Which are most realistic in this respect?

4.3 Although it does not have the proper form (4.1.9) very near the wall, Rannie's assumption (4.1.13) for the eddy viscosity has the advantage of providing explicit formulae for other features of the flow.

(a) Show that it implies that

$$U^+ = y_s^+ \tanh\left(\frac{y^+}{y_s^+}\right)$$

and that the viscous-dissipation term of equation (3.3.26) has the form

$$\left(\frac{\mathrm{d}U^+}{\mathrm{d}y^+}\right)^2 = \mathrm{sech}^4\left(\frac{y^+}{y_s^+}\right)$$

for the coordinates of Figure 3.8(b).

(b) Show that the direct viscous dissipation ε_v (rendered dimensionless as in (a)) gives $9{\cdot}7\,y_f$ when integrated over the region of validity of Rannie's formula, and $0{\cdot}2\,y_f$ when integrated over the logarithmic part of the wall layer. Note: $\int \mathrm{sech}^4 X\,\mathrm{d}X = \frac{1}{3}\tanh X(2 + \mathrm{sech}^2 X)$

(c) Finally, show that the ratio of the direct viscous dissipation to the total dissipation in a channel is

$$\frac{\int \varepsilon_v\,\mathrm{d}y}{U_a\,\tau_w/\rho} = \frac{9{\cdot}9\,u_f}{U_a}$$

Does it seem that the viscous element of dissipation is insignificant in high-Reynolds-number flows?

4.4 The dissipation in pipe flow is described by the variations of four distinct energy-conversion rates:

(1) the total extraction of energy from the mean flow, $\tau\,\mathrm{d}U/\mathrm{d}y$,
(2) direct viscous dissipation, $\rho\varepsilon_v = \mu(\mathrm{d}U/\mathrm{d}y)^2$,
(3) production of turbulence energy, $-\overline{uv}\,\mathrm{d}U/\mathrm{d}y$, and
(4) dissipation within the turbulence, $\varepsilon_t = \varepsilon - \varepsilon_v$.

These contributions may be identified in equations (3.3.25, 26).

(a) Show that (1) is given by τ_w^2/μ in the sublayer and by $(\tau_w^2/\mu)/Ky^+$ in the logarithmic layer.

(b) Use these results, together with those of the preceding example and the data of Figure 3.8, to sketch the variations of the four conversion rates over the pipe radius.

(c) Where is the energy extracted from the mean flow? Where is it dissipated? Where does the production of turbulence take place? What proportion of the energy extracted goes to the turbulence?

4.5 What locations for the matching point between the wall layer and core of a pipe flow are determined by requiring continuity of: (1) the velocity gradient at the junction between logarithmic layer and a region of constant eddy viscosity, and (2) the mixing-length variation comprising a linear element near the wall and a constant value in the core? Are the differences from the result (4.2.1) likely to be significant?

4.6 (a) For what Reynolds number would the viscous layer in a pipe occupy the entire wall layer?

(b) At what point in a pipe cross-section can the mean velocity U_a be measured?

(c) Derive power laws relating the parameters R^+, Re_d and u_f/U_a, any one of which defines the flow condition in a pipe.

(d) Does the velocity distribution in a pipe become more uniform or less uniform when the wall is roughened, the mean velocity being unchanged?

4.7 The methods used for pipe flow may be adapted to build up a velocity distribution for Couette or pure shearing flow, and to predict its friction characteristic.

(a) Show that continuity of any one of eddy viscosity, mixing length and velocity gradient implies continuity of all.

(b) Show that the friction law is $U_2/u_f = 5\ln(b/y_f) + 6{\cdot}5$.

4.8 An analysis of experimental data undertaken by Nunner revealed the following relationship between the index n of equation (4.2.13) and the pipe friction coefficient:

Table 4.8.

n	10	7·5	5	3·3	2·5
c_f	0·0025	0·004	0·010	0·030	0·060

The first pair of results apply to smooth walls, the last pair to rough walls. Compare these results with equation (4.2.17).

4.9 For the purposes of structural design, the variation of wind speed with height is usually represented by $U/U_1 = (y/\delta)^{1/n}$. Values of n and δ representative of neutrally stable conditions are given in Table 4.9 (these are conditions in which neither thermal instability nor convective damping is important, and are normally encountered when the wind speed exceeds, say, 6 m/sec). The values of δ, the thickness of the surface layer, are only a rough guide, since wind-profile data extending over a great range of heights are rare.

Estimate the corresponding friction coefficients and roughness scales.

Table 4.9.

Terrain	n	δ (m)
Coastal waters	10	150
Open country, grassland, tundra	7	270
Agricultural land with hedges and walls	5	340
Wooded country, towns, suburbs, rough coastal belts	3·5	400
Centres of large cities	2	550

4.10 The near-universality of the form of the velocity distribution near a wall provides a means of determining the wall stress.

(a) Show that this can be done, for either a rough or a smooth wall, by plotting the measured distribution of mean velocity in a semi-logarithmic form.

(b) For a smooth wall, show that

$$\frac{U}{U_1} = (\tfrac{1}{2}c_f)^{\frac{1}{2}} \left[A \ln \left(\frac{U_1 y}{\nu} \right) + A \ln (\tfrac{1}{2}c_f)^{\frac{1}{2}} + B \right]$$

where U_1 is the reference velocity used in defining c_f. Explain how to prepare a universal chart on which a measured distribution can be plotted to determine c_f.

(c) Can the technique proposed in (b) be extended to rough walls?

(d) Measurements in a smooth-walled pipe of radius 10 cm give the results of Table 4.10; what is the friction coefficient based on the average velocity?

Table 4.10.

y (mm)	0·5	1	1·5	2	2·5	5	10
U (m/sec)	1·50	2·47	3·06	3·44	3·64	4·16	4·55
y	15	20	25	30	35	100	
U	4·81	5·06	5·26	5·40	5·60	6·50	

4.11 Using the power laws (4.2.22) and (4.3.17), show that the Colebrook-White formula (4.3.18) implies that the change from effectively smooth to fully rough conditions takes place over the range of Reynolds numbers given by

$$\mathrm{Re}_d \left(\frac{k_s}{d} \right)^{1 \cdot 15} = 4 \text{ to } 120$$

4.12 What error might be introduced into an estimate of the flow rate by the variability noted in equation (4.3.22)?

4.13 Show that, when power laws of the form $U \propto y^{1/n}$ and $c_f \propto (k/d)^{1/p}$ are used for a rough-walled pipe, the result (4.2.17) is still applicable, but the indices are related by $n = 2p$. Is the approximate result (4.3.17) broadly consistent with these predictions?

4.14 The values of the roughness length (of equations (4.3.25)) given in Table 4.11

have been found to represent the velocity distribution above various crops. Are they consistent with the value $k_s/k \simeq 4$ for vegetation which is given in Table 4.4, II?

Table 4.11.

Type of vegetation	Roughness length k_2 (cm)
Fully grown root crops	10
Wheat: light winds	9
strong winds	4
Downland: in summer	3
in winter	1·5
Short grass	0·3

4.15 We have seen that the roughness of a surface can be represented by a variety of length scales; here we investigate two ways of describing it using an effective viscosity.

(a) Show that equation (4.3.23) can be written

$$\frac{U}{u_f} = A \ln\left(\frac{yu_f}{\nu_e}\right) + B$$

with ν_e an effective viscosity for the roughness layer, related to other measures of roughness by

$$\frac{\nu_e}{\nu} = \frac{k_1}{y_f} = \exp\left(\frac{K\,\Delta U}{u_f}\right)$$

(b) Meteorologists introduce an allied quantity, which they term the macroviscosity N, by writing equation (4.3.23)

$$\frac{U}{u_f} = A \ln\left(\frac{yu_f}{N}\right)$$

How is N related to the effective viscosity of (a)?

(c) Sutton proposed the formula

$$\frac{U}{u_f} = A \ln\left(\frac{yu_f}{N + \nu/9}\right)$$

as applicable to both smooth and rough surfaces. Why? Show that the Colebrook–White friction formula (4.3.18) is implied by this postulate.

(d) Yet another form of velocity distribution used by meteorologists was proposed by Rossby:

$$\frac{U}{u_f} = A \ln\left(\frac{u_f\, y + N}{N + \nu/9}\right)$$

Show that the corresponding mixing-length variation is $l_m = K(y + k_2)$. Is this sensible?

4.16 List the various ways of specifying the roughness of a surface and show how they are related.

4.17 Following a more thorough examination of the consequences of the overlapping of the wall layer and the core region, Tennekes and Lumley proposed that the constant A in the logarithmic law of the wall will change with the flow condition:

$$A = 3 - 5\left(\frac{1}{R^+}\right)^{\frac{1}{3}}$$

Is this result consistent with the values for smooth and rough walls given in Table 4.2? The comparison for rough walls can be made by rewriting the result given above in terms of the friction coefficient, and in terms of the effective viscosity of Example 4.15.

4.18 How should the criteria (4.3.27), which tentatively define the extent of the roughness layer of Figure 4.5, be changed as the ratios b/h, w/h and L/h vary?

4.19 (a) Show that the friction coefficient of a surface to which are fixed a number of identical roughness elements may be estimated as

$$c_\mathrm{f} = c_{\mathrm{f}_0} + \frac{nS_\mathrm{e}}{S}\left(\frac{U}{U_\mathrm{a}}\right)^2 C_\mathrm{d}$$

with c_{f_0} the basic friction coefficient, nS_e/S the frontal area of the elements per unit area of surface, U the effective velocity of the flow around the elements, and C_d the drag coefficient of an element.

(b) A steel pipe of inner diameter 30 cm has external strengthening rings fitted every 50 cm of its length, each being fastened by ten rivets. The hemispherical rivet heads within the pipe have diameter 2 cm. Estimate the variation of c_f with Re_d for this pipe, taking $C_\mathrm{d} = 0{\cdot}4$, and assuming the pipe surface to have a typical roughness.

4.20 The varied friction characteristics of Figure 4.4 are categorized in Section 4.3.2 in terms of four types (A to D) of roughness geometry. Morris approached the problem of classification in another way, by referring to the nature of the flow near the wall. He introduced the terms: *subnormal turbulence*, for flow near a smooth wall; *hyperturbulent or wake-interference flow*, for roughness Types A and B; *semi-smooth turbulent flow*, for roughness Type C; and *skimming or quasi-smooth flow*, for roughness Type D.

(a) Explain why he adopted this terminology.

(b) Morris has only one classification for the wide range of surfaces within Types A and B. Show that, for moderate Reynolds numbers, there is a difference in the flow near the two kinds of surface. How can this be described in Morris's terminology? Why does the friction characteristic for roughness Type A at first fall below its high-Reynolds-number level?

4.21 (a) How do the mixing length and eddy viscosity vary across the flow in the three asymmetric-flow models considered in Section 4.4.2?

(b) Comment on the usefulness of these models in calculating heat transfer: (1) into the flow, the walls being kept at the same temperature, and (2) across the flow, the walls having widely differing temperatures.

4.22 Calculate the average velocity for asymmetric duct flow, using equations (4.4.6) to (4.4.12). Is the result significantly different from (4.4.19), obtained using only logarithmic layers?

4.23 Write a computer program, based on the results of Section 4.4.3, to predict the friction in a flat channel with one rough wall.

4.24 Estimate the friction coefficient for flow in a helical pipe with $d/D = 0{\cdot}1$, for $\mathrm{Re}_d = 1000$ and 10,000. What can you conclude about the sensitivity of laminar and turbulent friction to channel curvature? Why is this?

4.25 Only very small secondary-flow velocities are needed to produce significant changes in the stress distribution within a non-circular channel section. This may be shown using the Reynolds-flux model of Example 3.22(b), which implies that

$$V_m = V_0 + \tfrac{1}{2}V$$

where $V_m = G_m/\rho = \tau/\rho U_a$ is the local value of the net transfer velocity, with V_0 the effective velocity associated with the turbulent mixing, and V the actual secondary-flow velocity (measured positive in the direction of V_0).

(a) Show that a doubling of the effective shear stress will be achieved if $V/U_a = \tfrac{1}{2}c_f$.

(b) Will secondary convection influence V_0, the direct contribution of the turbulence? By what mechanisms does secondary-flow convection influence the wall stresses?

4.26 Roughness of a given size has a much greater retarding effect on a pipe flow than on an open channel whose flow depth equals the pipe radius. Why is this? Does the use of a hydraulic diameter or radius take this effect into account?

4.27 The distribution of velocity in parallel *laminar* flow through a channel of elliptical section is given by

$$U = F\left(1 - \frac{y^2}{a^2} - \frac{z^2}{b^2}\right)$$

and the flow rate by

$$\dot{V} = \tfrac{1}{2}FA$$

where $A = \pi ab$ and $F = -\tfrac{1}{2}\mu(\mathrm{d}P/\mathrm{d}x)a^2b^2/(a^2 + b^2)$, a and b being the semi-axes.

(a) How is the discharge related to that through a circular pipe with the same area and pressure gradient?

(b) What is the ratio of stresses at the ends of the major and minor axes? For turbulent flow in a channel with $a/b = 2$, the ratio of stresses is found to be $1{\cdot}12$; how does this compare with the laminar result?

(c) How successful is the introduction of the hydraulic diameter in relating the laminar elliptical-section friction to that in a round tube?

4.28 (a) Considering a concentric annulus, express the hydraulic diameter in terms of the two diameters, and show that the average friction coefficient of equation (4.5.4) is related to the coefficients for the two walls by

$$c_f = \frac{d_2\, c_{f_1} + d_2\, c_{f_2}}{d_1 + d_2}$$

(b) A correlation of experimental results led Davis to propose

$$c_f = 0{\cdot}055\left(1 - \frac{d_1}{d_2}\right)^{0{\cdot}1} \mathrm{Re}^{-0{\cdot}2}$$

for the average friction coefficient, with $\mathrm{Re} = (d_2 - d_1)\, U_a/\nu$. Why does the use of the hydraulic diameter not, in this case, eliminate the effect of channel shape? Does the result behave as it should for $d_1/d_2 \to 0$ and 1?

(c) The friction characteristic can be estimated by introducing the laminar-flow ratio

$$\frac{\tau_1}{\tau_2} = \frac{d_2(d_0^2 - d_1^2)}{d_1(d_2^2 - d_0^2)}$$

into the result (a), with $d_0^2 = (d_2^2 - d_1^2)/\ln(d_2/d_1)^2$ giving the point at which $\tau = 0$. Show that, on this basis

$$\frac{c_f}{c_{f_2}} = \frac{1 - d_1/d_2}{1 - d_0^2/d_2^2}$$

(d) Using the pipe-flow result (4.2.23) to give c_{f_2}, compare the prediction (c) with the empirical result (b).

4.29 In an air-conditioning system, air at 80 °F is to be conveyed a distance of 200 ft at the rate of 1500 cfm (ft^3/min). The duct must run through a passage of rectangular section 1 ft × 2·5 ft; the passage contains three sharp right-angle bends.

Estimate the pressure drop which would occur in three proposed ducts:

(a) a sheet-metal fabrication just filling the passage; it is estimated that $k_s = 0{\cdot}03$ in;

(b) two smooth-walled circular tubes, each of diameter one foot; and

(c) a smooth-walled elliptical tube just filling the passage.

4.30 A shell-and-tube heat exchanger consists of twenty unfinned tubes (outer diameter 1 cm) evenly spaced through a cylindrical shell (diameter 8 cm; length 1 m). Inlet and outlet tubes (diameter 4 cm) are normal to the axis of the shell.

(a) Neglecting scale formation, estimate the flow of cool water which will be achieved between shell and tubes for an overall pressure drop of 0·2 bar.

(b) By how much might the flow be reduced when scale forms on the tubes?

(c) To ensure that the performance of (a) is maintained after scale formation, a larger shell is to be provided, the tubes being evenly spaced through it. What diameter is required?

4.31 A number of friction formulae of the general form

$$U_a = C R_h^p S^q$$

are used in engineering practice. Some of these are specified in Table 4.12.

(a) What are the advantages and disadvantages of these alternatives, compared with the Colebrook–White formula? What does each imply about the variation of friction with surface roughness? Why do the indices change from formula to formula?

(b) A steel pipe of diameter 4 ft and length one mile connects two reservoirs whose elevation differs by 50 ft. Find the discharge when it flows full and half-full, using the three empirical formulae (with central values for the coefficient C), and using equation (4.3.15) with an appropriate value of k_s.

Table 4.12.

Originators	Usual application	C (English units)	p	q
Manning	Open and closed channels	$1{\cdot}49/n$	$\frac{2}{3}$	$\frac{1}{2}$
Hazen–Williams	Water supply and sanitary engineering	100–170	0·63	0·54
Scobey	Irrigation: concrete	95–130	$\frac{5}{8}$	$\frac{1}{2}$
	steel	120–155	0·58	0·526
	wood	113	0·65	0·555

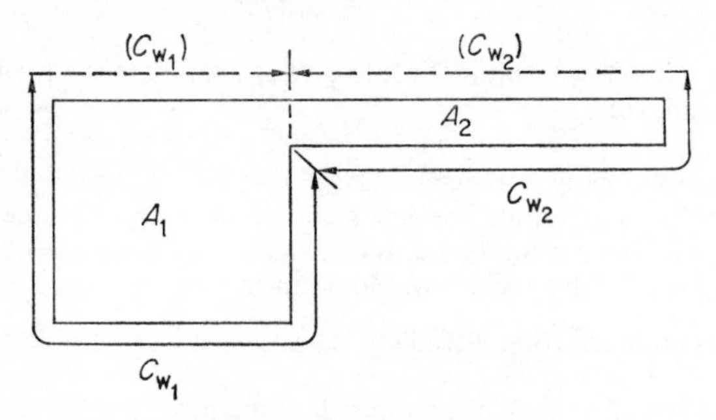

Figure 4.8 See Example 4.32

4.32 Figure 4.8 shows schematically a channel section (which may be either open or closed) with two very different length scales.

(a) Show that

$$\rho g S = \frac{\tau_1 C_{w_1} + F_L}{A_1} = \frac{\tau_2 C_{w_2} - F_L}{A_2}$$

hold for the two parts of the flow, with F_L the interaction force between the two sections, and $S = -\mathrm{d}(z + P_w/\rho g)/\mathrm{d}x$ the hydraulic gradient in the channel.

(b) Assuming the interaction force to be negligible, show that the flow through the channel is given by

$$\dot{V} = S^{\frac{1}{2}} \sum C_i A_i R_{h_i}^{\frac{1}{2}} = S^{\frac{1}{2}} \sum K_i$$

where C_i and K_i are, respectively, the Chézy coefficient and the conveyance for the ith section of the channel.

(c) What is the ratio of the conveyances for $A_1 = 2A_2$ and $C_{w_1} = C_{w_2}$, with k_s the same for the two sections? Find the effect on the conveyances, and on the total flow, of an interaction force $F_L = 0{\cdot}1A_1(\rho g S)$.

4.33 The most economical design for an excavated canal is often that whose cross-sectional area is the smallest able to convey the required flow. The bed slope is commonly fixed by the topography, and cannot be varied to increase the velocity.

(a) Show that, for a given bank and bed material, it is appropriate to select the channel section which, for a given flow area, has the largest possible hydraulic radius, and the smallest possible wetted perimeter.

(b) Show that the most economical rectangular section is one whose width is twice the water depth. What is the most economical section of all?

(c) Suppose that the cost of lining the channel is about equal to the cost of excavation. What is then the most economical rectangular section?

4.34 A concrete culvert, 50 m long, 2 m wide, and 1·5 m deep, has a uniform slope of 1 per cent.

(a) For a water flow of 4 m³/sec, what is the normal depth? Does this correspond to supercritical or subcritical flow?

(b) If the culvert runs full, what is the change in pressure from end to end when: (1) the water filling it is still, (2) the flow is 4 m³/sec, and (3) the flow is 12 m³/sec?

4.35 The results obtained here are the basis for the varying-flow calculations required to predict surface profiles in steadily flowing rivers and canals.

(a) Assuming one-dimensional flow (that is, U = constant over the entire flow area), show that the rate of change of the total head H_0 along an open channel is given by

$$\dot{m}\frac{\mathrm{d}(gH_0)}{\mathrm{d}x} = -U\,\tau_w\,C_w$$

whence

$$\frac{\mathrm{d}H_0}{\mathrm{d}x} = -\frac{\tau_w}{\rho g R_h} = -S_f$$

with S_f the friction slope, representing the local dissipation.

(b) Expressing the total head as $H_0 = z + h + U^2/2g$ with h the flow depth measured from the bed level $z(x)$, show that

$$U\frac{\mathrm{d}U}{\mathrm{d}x} = -\frac{\tau_w}{\rho R_h} - g\frac{\mathrm{d}(h+z)}{\mathrm{d}x}$$

Interpret this as a momentum equation relating the convective acceleration to the friction and pressure gradient.

(c) Show that the continuity requirement $UA = \dot{V}$ implies, for a uniform channel, that

$$\frac{\mathrm{d}U}{\mathrm{d}x} = -\frac{U}{A}\frac{\mathrm{d}A}{\mathrm{d}x} = -\frac{U}{h_h}\frac{\mathrm{d}h}{\mathrm{d}x}$$

where $h_h = A/T$ (T being the surface width) is the hydraulic depth. Use this result to show that

$$(1 - F^2)\frac{\mathrm{d}h}{\mathrm{d}x} = S_0 - S_f$$

where $S_0 = -\mathrm{d}z/\mathrm{d}x$ is the bed slope and $F^2 = U^2/gh_h$ is a generalized Froude number. (The definition of equations (4.5.16) is appropriate only for a channel with a flat bottom and vertical walls.) Interpret the last result physically.

(d) Considering a broad channel with flat bottom and vertical walls, and assuming that the local dissipation is that given by the Manning formula for normal flow with the same depth and velocity, show that the result (c) can be written

$$\frac{\mathrm{d}h}{\mathrm{d}x} = S_0\frac{1 - (h_0/h)^{\frac{10}{3}}}{1 - (h_c/h)^3}$$

where $h_c = (\dot{v}^2/g)^{\frac{1}{3}}$ is the critical depth for the flow, and h_0 is the normal depth for the flow $\dot{v}$ on the slope S_0.

(e) Under what circumstances will the flow develop towards the normal depth? Is the precise form of the friction law important for calculations using this formula?

4.36 The one-dimensional continuity and momentum results $A_1 U_1 = A_2 U_2$ and $(P_2 - P_1) A_2 = -\rho U_1 A_1 (U_2 - U_1)$ lead to the loss coefficient for a sudden expansion which is given in Table 4.7, II.

(a) Justify the use of these relationships.

(b) Use them to calculate the change in total pressure and the loss coefficient $(1 - A_1/A_2)^2$.

(c) Show that the corresponding results for flow in a broad, flat-bottomed open channel can be obtained by replacing A by h, the water depth, and P by $\frac{1}{2}\rho g h$, the average gauge pressure.

(d) Manipulate the open-channel conservation laws to obtain the result (4.5.16) and then (4.5.15).

4.37 (a) Two pipes in series carry the same flow. Show that the ratio of the pressure drops is

$$\frac{L_2}{L_1}\left(\frac{d_1}{d_2}\right)^5 \frac{c_{f_2}}{c_{f_1}}$$

Find also the ratio of the velocities, Reynolds numbers and wall shear stresses.

(b) What special forms can be developed for the pressure-drop ratio subject to the friction laws $c_f \propto (1/Ud)^{1/p}$ and $(k/d)^{1/p}$?

(c) Repeat the analysis for two pipes in parallel, finding the ratio of flow rates corresponding to equal pressure drops.

(d) Use these results to find the ratios appropriate to laminar flows in the two pipes.

5

Channel Flows III: Heat and Mass Transfer

The aims of this chapter parallel those of the preceding chapter, but its structure is more complex, since the transport of any entity within a turbulent flow depends on the velocity field, mean and fluctuating. Thus the analysis of a transfer process requires some understanding of momentum transfer and the associated velocity variation. In compensation, we are able to establish analogies linking rates of heat and mass transfer to friction generation.

In some circumstances, there is a reciprocal dependence of the velocity field on the heat or mass transfer. The temperature variation associated with rapid heat transfer or with very high velocities gives rise to changes in the molecular transport properties of the fluid; these modify the structure of the part of the wall layer where molecular diffusion is important. Concentration variations associated with mass transfer can have a similar effect. Moreover, large mass-transfer rates induce a mean velocity normal to the wall; this modifies the momentum balance and diffusive properties throughout the flow.

These relationships will be examined in the following order:

(1) consideration of the analogies between transfers of momentum, heat and mass, and of the turbulent Prandtl and Schmidt numbers which link these processes;

(2) application of dimensional analysis to transfers in flow near a wall, including high-speed flow, where significant temperature changes occur in the absence of heat transfer at the wall;

(3) development of transfer laws for a smooth-walled pipe for the cases of moderate, high and low Prandtl and Schmidt numbers, using a combination of analytical and empirical results;

(4) generalization to include the effects of large temperature differentials, non-uniform wall flux, wall roughness, and more varied channel shapes; and

(5) modification to incorporate a component of velocity normal to the wall, which may arise spontaneously during rapid mass transfer, or may be imposed for cooling (blowing through a porous wall) or for boundary-layer control (suction through the wall).

This discussion omits some important classes of transfer processes, notably those occurring in: flows in which chemical reactions take place, flows in which buoyancy is important, two-phase flows, and flows containing solid particles or, what is nearly the same, small droplets. While it is true that these processes are less well understood than those in single-phase, non-reactive fluids, the compelling reason for their omission is the necessity of restricting the length of this introductory treatment.

It is easy to produce an extensive list of interesting and commercially important processes involving more complex forms of turbulence, for example: boiling heat transfer; atmospheric dispersion; combustion of gaseous, solid and liquid fuels; pneumatic and hydraulic transport of solid particles; and erosion by wind, waves or running water. The most urgent problems which turbulent activity currently poses for the engineer are those arising in these 'complex fluids'. Although certain fundamental aspects of turbulence in a homogeneous fluid have not yet been fully elucidated, this chapter will show that our understanding of them is adequate for many of the transfer calculations which arise in engineering design.

5.1 TRANSFER ANALOGIES

5.1.1 Reynolds Analogy

While discussing the Reynolds fluxes representing transfers of momentum, heat and mass, in Section 3.4.4, we noted that the transfer coefficients—the friction coefficient and Stanton numbers—were related by

$$\tfrac{1}{2}c_f = \mathrm{St} = \mathrm{St}_c = \frac{G_m}{\rho U_0} \tag{3.4.30}$$

under certain special circumstances, with G_m the Reynolds flux for momentum. In dimensional parameters, this analogy pointed out by Reynolds is

$$\frac{\tau}{U_0} = \frac{\dot{q}_h}{\Delta H} = \frac{\rho N}{\Delta C} = G_m \tag{3.4.30}$$

The conditions under which these results apply are either:

(1) the Reynolds flux G_s is the same for each transferred entity S, or

(2) the profiles of the time-mean properties U, H and C are similar and, if the flow is fully turbulent, the eddy diffusivities ϵ_s are equal; if part of the flow is influenced by molecular diffusion, both eddy and molecular diffusivities must be equal.

Symbolically, these requirements may be expressed:

(1) For friction and heat transfer

$$G_h = G_m \quad \text{or} \quad \frac{1}{\Delta H}\frac{\partial H}{\partial y} = \frac{1}{U_0}\frac{\partial U}{\partial y} \quad \text{and} \quad \mathrm{Pr_t}\,(=\mathrm{Pr}) = 1 \tag{5.1.1}$$

(2) For friction and mass transfer

$$G_c = G_m \quad \text{or} \quad \frac{1}{\Delta C}\frac{\partial C}{\partial y} = \frac{1}{U_0}\frac{\partial U}{\partial y} \quad \text{and} \quad \mathrm{Sc_t}\,(=\mathrm{Sc}) = 1 \tag{5.1.2}$$

(3) For heat and mass transfer

$$G_h = G_c \quad \text{or} \quad \frac{1}{\Delta C}\frac{\partial C}{\partial y} = \frac{1}{\Delta H}\frac{\partial H}{\partial y} \quad \text{and} \quad \mathrm{Le_t}\,(=\mathrm{Le}) = 1 \tag{5.1.3}$$

The molecular and turbulent Prandtl and Schmidt numbers were introduced in Section 1.6. Here we use also the molecular *Lewis number*

$$\mathrm{Le} = \frac{\mathrm{Pr}}{\mathrm{Sc}} = \frac{D}{\kappa} \tag{5.1.4}$$

and its turbulent* counterpart

$$\mathrm{Le_t} = \frac{\mathrm{Pr_t}}{\mathrm{Sc_t}} = \frac{\epsilon_D}{\epsilon_h} \tag{5.1.5}$$

which may be called the *turbulent Lewis number*.

For compactness, much of the following discussion relates to the analogy between friction and heat transfer. The relationships (5.1.1, 2. 3) indicate how the results can be applied to other pairs of transfers.

Another result which bears upon this analogy can be extracted from Example 3.18, which deals with heat transfer in *laminar flow* within a round tube. For a uniform rate of heat input (implying a uniform temperature differential between wall and axis) and constant fluid properties, it was found that

$$\mathrm{Nu} = \frac{\dot{q}_w d}{k(T_w - T_f)} = \tfrac{8}{3}, 6, \tfrac{48}{11} \tag{5.1.6}$$

the three values of the Nusselt number holding for

(1) $T_f = T_c$, the temperature on the tube axis,

(2) $$T_f = T_a = \frac{2\pi}{A}\int_0^R T r\,\mathrm{d}r \tag{5.1.7}$$

the average temperature over the section, and

(3) $$T_f = T_b = \frac{2\pi}{\dot{V}}\int_0^R uTr\,\mathrm{d}r \tag{5.1.8}$$

the bulk or mixing-cup temperature.

* Note that ϵ_D is used for the turbulent mass diffusivity, to avoid confusion with ϵ_c, which denotes the constant diffusivity sometimes adopted.

In terms of the quantities in which Reynolds analogy is expressed in equations (3.4.30), the results (5.1.6) are

$$\frac{\mathrm{St}}{\frac{1}{2}c_f} = \frac{\mathrm{Nu}}{8\,\mathrm{Pr}} = \frac{1}{3\,\mathrm{Pr}} = \frac{3}{4\,\mathrm{Pr}} = \frac{6}{11\,\mathrm{Pr}} \tag{5.1.9}$$

where $\mathrm{St} = \dot{q}_w/[\rho c_p U_a(T_w - T_f)]$ and $c_f = \tau_w/(\frac{1}{2}\rho U_a^2)$, and $c_f = 16/\mathrm{Re}_d$ for this flow, as indicated in Table 3.2.

We see that Reynolds analogy gives a fairly accurate relationship between friction and heat transfer for laminar pipe flow, provided that

(1) an appropriate choice is made for the temperature and velocity scales of the transfer coefficients, and

(2) the Prandtl (or Schmidt or Lewis) number is not too far from unity, the analogy being most accurate (for $T_f = T_a$ or T_b) when $\mathrm{Pr} \simeq 0{\cdot}7$.

By coincidence, this value represents many common gases (see Table A.1 of the Appendix) over a considerable range of temperature, say, 200 to 600 °K; a theoretical prediction by Eucken gives

$$\mathrm{Pr} = \frac{4\gamma}{9\gamma - 5}$$

with $\gamma = c_p/c_v$ the specific-heat ratio.

The first of the conditions noted above indicates that the simple analogy applies only to fluxes with similar boundary conditions, that is, fluxes from and to the same places, with the terminal values of the mean properties defining the scales U_0, ΔH and ΔC. When these conditions are not satisfied, we may still be able to use

$$\mathrm{St} \propto c_f \tag{5.1.10}$$

for a particular fluid and flow type, and

$$\mathrm{St} \propto \frac{c_f}{\mathrm{Pr}} \tag{5.1.10}$$

for a particular type of laminar flow and set of thermal boundary conditions. These results are convenient for presenting and extrapolating experimental results, even when the constant of proportionality is not known initially.

5.1.2 Extensions of Reynolds Analogy

Returning to the conditions (5.1.1, 2, 3) for the applicability of Reynolds analogy in turbulent flow, we ask whether the analogy can be extended by relaxing some of these requirements. For compactness, we consider the relationship between momentum and enthalpy transfers; the extension to mass transfer is obvious.

For a parallel mean flow $U(y)$, the two fluxes are given by

$$\rho(\epsilon_{\mathrm{m}} + \nu)\frac{\mathrm{d}U}{\mathrm{d}y} = \tau = G_{\mathrm{m}} U_0 \tag{5.1.11}$$

and

$$-\rho(\epsilon_{\mathrm{h}} + \kappa)\frac{\partial H}{\partial y} = \dot{q}_{\mathrm{h}} = G_{\mathrm{h}} \Delta H \tag{5.1.12}$$

whence

$$-\frac{\epsilon_{\mathrm{m}} + \nu}{\epsilon_{\mathrm{h}} + \kappa}\frac{\Delta H\, \mathrm{d}U/\mathrm{d}y}{U_0\, \partial H/\partial y} = \frac{\tau \Delta H}{\dot{q}_{\mathrm{h}} U_0} = \frac{G_{\mathrm{m}}}{G_{\mathrm{h}}} \tag{5.1.13}$$

These are particular forms of equations (3.4.8), with the diffusivities shown explicitly as sums of molecular and turbulent-mixing components. In equations (5.1.11) the conventional signs have been introduced, as in equations (3.2.10) and (3.4.25).

If we assume only that the mean-property profiles are similar, we find, as in equations (3.4.10)

$$\frac{\tau \Delta H}{\dot{q}_{\mathrm{h}} U_0} = \frac{G_{\mathrm{m}}}{G_{\mathrm{h}}} = \frac{\epsilon_{\mathrm{m}} + \nu}{\epsilon_{\mathrm{h}} + \kappa} = \mathrm{Pr_e} \tag{5.1.14}$$

with $\mathrm{Pr_e}$ the *effective Prandtl number*. The simple Reynolds analogy (3.4.30) can now be generalized to

$$\frac{\tau}{U_0} = \mathrm{Pr_e}\frac{\dot{q}_{\mathrm{h}}}{\Delta H} = G_{\mathrm{m}} = \mathrm{Pr_e}\, G_{\mathrm{h}} \tag{5.1.15}$$

or

$$\tfrac{1}{2}c_{\mathrm{f}} = \mathrm{Pr_e}\, \mathrm{St} = \frac{G_{\mathrm{m}}}{\rho U_0} = \mathrm{Pr_e}\frac{G_{\mathrm{h}}}{\rho U_0} \tag{5.1.15}$$

These and similar relationships involving mass-transfer parameters are referred to as *modern or strong forms of Reynolds analogy*. The requirement that

$$\mathrm{Pr_e} = f(\mathrm{Pr}) \tag{5.1.16}$$

be constant across the flow (and like restrictions on $\mathrm{Sc_e}$ and $\mathrm{Le_e}$) is often appended. Equations (5.1.15) then imply that

$$\frac{G_{\mathrm{m}}}{G_{\mathrm{h}}} = \text{constant} \quad \text{and} \quad \frac{\tau}{\dot{q}_{\mathrm{h}}} = \text{constant} \tag{5.1.16}$$

across the flow (with similar relationships involving G_{c} and N, the mass flux), that is, that the several fluxes vary in the same way across the flow. These additional restrictions prove very convenient in the theoretical analysis of transfer processes. While they are not strictly valid, as will be seen shortly, such simplifications are often admissible in view of the lack of precision in empirical data concerning transfer processes.

The extended Reynolds analogy (5.1.15) has the form of the laminar-flow result (5.1.9). Paralleling equations (5.1.10), we have

$$\mathrm{St} \propto \frac{c_\mathrm{f}}{\mathrm{Pr_e}} \tag{5.1.17}$$

for a particular flow type, with the constant of proportionality dependent on the selections made for the scales U_0 and ΔH.

Another way of generalizing Reynolds analogy is to make an assumption concerning $\mathrm{Pr_t}$ and $G_\mathrm{m}/G_\mathrm{h}$; this leads to a specification of the velocity and enthalpy profiles, no longer assumed similar. For fully turbulent flow, equations (5.1.13) give

$$-\frac{\mathrm{d}(H/\Delta H)}{\mathrm{d}(U/U_0)} = \mathrm{Pr_t}\frac{G_\mathrm{h}}{G_\mathrm{m}} \tag{5.1.18}$$

If the right-hand combination is assumed constant across the flow, we find

$$\frac{H_1 - H}{\Delta H} = \mathrm{Pr_t}\frac{G_\mathrm{h}}{G_\mathrm{m}}\frac{U - U_1}{U_0} \tag{5.1.19}$$

with H_1 and U_1 reference values. The two profiles have the same basic shape, but do not overlap when scaled using simply U_0 and ΔH.

5.1.3 Validity of Assumptions

In the preceding pages we have obtained simple relationships among the fluxes of momentum, heat and mass, from postulates whose justification is the simplicity of the resulting formulae, rather than a foundation in experiment or rational analysis. We shall now subject these assumptions to a variety of empirical and theoretical tests.

The simple Reynolds analogy (3.4.30) depends only on the equality of Reynolds fluxes: $G_\mathrm{m} = G_\mathrm{h} = G_\mathrm{c}$. In reality, the results

$$\mathrm{St} = \tfrac{1}{2}c_\mathrm{f} \quad \text{and} \quad \mathrm{St_c} = \tfrac{1}{2}c_\mathrm{f} \tag{5.1.20}$$

prove reasonably accurate (within, say, 20 per cent) for low-velocity wall flows with Pr, $\mathrm{Sc} \simeq 1$, in particular, for heat transfer in gases, for which $\mathrm{Pr} \simeq 0{\cdot}7$. The analogies (5.1.20) cannot be expected to apply for ranges of temperature or concentration large enough to produce significant property variations, nor for markedly inhomogeneous fluids, such as arise in boiling heat transfer, nor for the transport of finite particles which do not immediately adopt the local fluid velocity. Even when these extreme cases are set aside, the simple relationships (5.1.20) can be grossly misleading for Prandtl and Schmidt numbers much greater or much less than unity. We conclude that the Reynolds fluxes associated with the several transfers are not equal in general, although they can become so in special circumstances, notably, when Pr or Sc $\simeq 1$ and the transferring medium is fairly homogeneous.

The analogy between heat and mass transfers

$$\mathrm{St_c} = \mathrm{St} \tag{5.1.21}$$

is more precisely accurate for Le = 1 than are the results (5.1.20) for Pr, Sc = 1, provided that we again confine attention to nearly homogeneous fluids. This can be justified as follows. It is plausible to suppose that the transport processes —molecular and turbulent—are essentially similar for heat and for a passively convected substance.* We may expect the transfer laws

$$\mathrm{St} = f(c_\mathrm{f}, \mathrm{Pr}) \quad \text{and} \quad \mathrm{St_c} = f(c_\mathrm{f}, \mathrm{Sc})$$

to have the same functional form:

$$\mathrm{St} \text{ or } \mathrm{St_c} = f_1(c_\mathrm{f}, \mathrm{Pr} \text{ or } \mathrm{Sc}) = f_2(\mathrm{Re}, \mathrm{Pr} \text{ or } \mathrm{Sc}) \tag{5.1.21}$$

This result gives St = $\mathrm{St_c}$ for Le = Pr/Sc = 1. It also has important implications for experimental technique, suggesting that data for heat and mass transfer can be combined into a single law, applicable to the transport of any passive entity in a nearly homogeneous fluid.

Deissler found heat and mass-transfer measurements for pipe flow to coalesce for $0{\cdot}5 < \mathrm{Pr}, \mathrm{Sc} < 60$ and for $10{,}000 < \mathrm{Re}_d < 50{,}000$. Although such systematic comparisons have not been made for other flow species, much indirect evidence supports the results (5.1.21). This broad range of agreement suggests that, when Le = 1 and the fluid is homogeneous, the other requirements of equations (5.1.3) are satisfied: the turbulent Lewis number $\mathrm{Le_t} = 1$; the Reynolds fluxes vary in the same way; and the temperature and concentration profiles are similar.

Results paralleling equations (5.1.21) can be set down for other pairs of transfers—for example, those of heat and momentum—but they provide only a rough collapse of data relating to the two transfer processes.

Turning now to the extended Reynolds analogy (5.1.15), we note that the key assumption is similarity in the profiles of mean properties. We shall see later, in Section 5.2.3, that similarity of velocity and temperature profiles can be achieved only when Pr, $\mathrm{Pr_t} \simeq 1$. Hence the generality of the strong form of Reynolds analogy is somewhat illusory, although the results (5.1.15) prove useful in indicating trends and suggesting orders of magnitude.

Flux Variations

Even though we know the fundamental postulate of profile similarity to be incorrect, we must consider the other restrictions noted in equations (5.1.16),

* This argument is convincing for the internal-energy component of the enthalpy flux, $\overline{ve}$, but not for the work flux $\overline{pv}$; see equations (3.2.17). However, the latter contributes only a fraction of order $(c_p - c_v)/c_p = 1 - 1/\gamma$ to the enthalpy transfer—very small for most liquids, and around one-third for gases.

since they are often used to simplify the analysis of transfer processes. The conservation laws for parallel, linearly developing flow show how the fluxes vary across the flow. From equations (3.3.10, 12, 20, 35) we have:

(1) the momentum balance

$$\frac{\mathrm{dP_w}}{\mathrm{d}x} = \rho \frac{\mathrm{d}(\nu\, \mathrm{d}U/\mathrm{d}y - \overline{uv})}{\mathrm{d}y} = \frac{\mathrm{d}\tau}{\mathrm{d}y} \tag{5.1.22}$$

(2) the energy balance

$$\rho U \frac{\partial H}{\partial x} = \rho \frac{\partial(\kappa \partial H/\partial y - \overline{vh})}{\partial y} + \frac{\mathrm{d}(\tau U - J_q)}{\mathrm{d}y}$$

$$= -\frac{\mathrm{d}\dot{q}_\mathrm{h}}{\mathrm{d}y} + \frac{\mathrm{d}(\tau U - J_q)}{\mathrm{d}y} \tag{5.1.23}$$

(3) the mass balance

$$U \frac{\partial C}{\partial x} = \frac{\partial(D\, \partial C/\partial y - \overline{vc})}{\partial y} = -\frac{\mathrm{d}N}{\mathrm{d}y} \tag{5.1.24}$$

The possibility of a source term in the mass balance has been rejected in the mass balance (5.1.24); the final term of the energy equation, that representing dissipation and diffusion, is also negligible in most low-speed flows. Even when these sources are absent, there is a fundamental difference between the momentum balance and the other two conservation laws, for they contain the mean velocity $U(y)$ in their convective derivatives. The parallel structure of the last two balances is consistent with the far-ranging detailed analogy noted in equations (5.1.21).

Integrating these equations from the wall outwards, with the source terms omitted, we obtain

$$\tau = G_\mathrm{m} U_0 = \tau_\mathrm{w} + \rho \frac{\mathrm{d}P_\mathrm{w}}{\mathrm{d}x} y \tag{5.1.25}$$

$$\dot{q}_\mathrm{h} = G_\mathrm{h} \Delta H = \dot{q}_\mathrm{w} - \rho \frac{\partial H}{\partial x} \int_0^y U\, \mathrm{d}y \tag{5.1.26}$$

$$N = \frac{G_\mathrm{c} \Delta C}{\rho} = N_\mathrm{w} - \frac{\partial C}{\partial x} \int_0^y U\, \mathrm{d}y \tag{5.1.27}$$

The nature of the variations in a symmetrical plane channel, with equal fluxes into the flow at the two walls, is indicated in Figure 5.1. Note the departure of the enthalpy and mass flux distributions from linearity near the wall, where the velocity changes rapidly. In a pipe this is even more marked, since the heat (or mass) flows through a reducing area as it moves away from the wall. In the core of the flow, where the velocity is fairly uniform, the fluxes $\dot{q}_\mathrm{h}$ and N vary nearly linearly. The distribution will become more nearly linear as the Reynolds number increases and the velocity profile flattens. It appears that the assumption that

$$\frac{\dot{q}_h}{\tau} \text{ or } \frac{N}{\tau} = \text{constant across the flow}$$

is not seriously in error at high Reynolds numbers.

Turbulent Prandtl and Schmidt Numbers

We turn to another aspect of the simplifying assumptions (5.1.16), the supposed constancy of the effective Prandtl and Schmidt numbers, still remembering that the fundamental assumption of profile similarity is seldom valid.

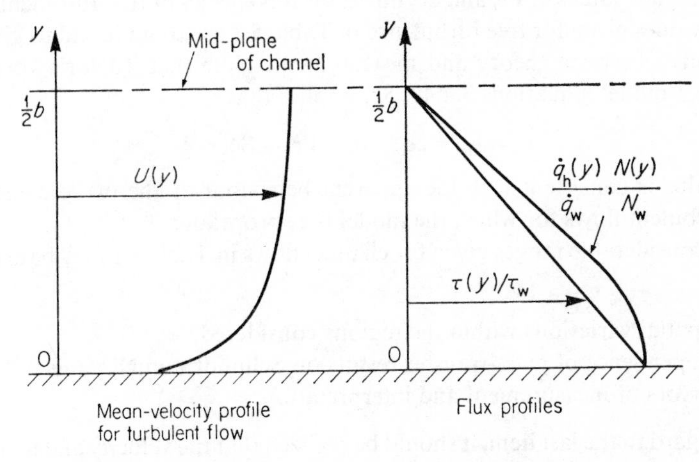

Figure 5.1 Variations of cross-stream fluxes (net shear stress τ, enthalpy flux $\dot{q}_h$, and mass flux N) in symmetrical plane flow in a broad flat channel

Note first that the arguments leading to equations (5.1.21) imply also that

$$\text{Pr}_t = f(\text{Re}, \text{Pr}) \quad \text{and} \quad \text{Sc}_t = f(\text{Re}, \text{Sc})$$

depend in the same way on the respective ratios of molecular diffusivities. Hence

$$\text{Pr}_t = \text{Sc}_t \quad \text{for} \quad \text{Le} = \frac{\text{Pr}}{\text{Sc}} = \frac{D}{\kappa} = 1$$

Further, since Pr_t and Sc_t are rather weak functions of Pr and Sc, and since the measurements through which they are determined are rather imprecise, we may suppose that Pr_t and Sc_t are indistinguishable for a wider range of transfers with $\text{Le} \sim O(1)$, that is, for molecular diffusivities of the same order. This condition is satisfied in the mingling of gases with molecular weights that are not too different, a fact often put to use in studying the mechanisms of heat transfer in gases by means of the analogous mass transfer.

Table 5.1 presents values of the turbulent Prandtl and Schmidt numbers measured, or inferred, for flows of air and mercury, typical fluids of moderate and very small Prandtl number. Throughout, $0{\cdot}5 < Pr_t, Sc_t < 2$; hence the simple Reynolds analogy (3.4.30) is unlikely to be *grossly* misleading when applied to the *turbulent* diffusion within any fluid.

The diffusivity ratios are found to depend on the Reynolds number specifying the level of turbulent activity; the approximate dependence is indicated for the core regions of the channel flows. Round jets commonly have higher Reynolds numbers than do the other free turbulent flows considered in Part III of the table, for reasons of experimental convenience, and this may account in part for the higher values of Pr_t and Sc_t noted for this species of free-turbulent flow.

The values given for free turbulence in Table 5.1 are those found to give the best match between theory and measurement. Note that Taylor's vorticity-transport model (equations 3.4.49, 52) implies that

$$\epsilon_h = \epsilon_\omega = 2\epsilon_m \quad \text{or} \quad Pr_t = Sc_t = \tfrac{1}{2} \tag{5.1.28}$$

This value is consistent with the apparent behaviour of the two-dimensional free-turbulent flows for which the model is appropriate.

The considerable ranges given for channel flows in Table 5.1 may be attributed to:

(1) spatial variations within the regions considered,
(2) dependence of mass-transfer results on Schmidt number, and
(3) errors of measurement and interpretation.

With regard to the last item, it should be realized that the velocity and temperature (or concentration) profiles are usually measured separately, and that each is then differentiated before the results are combined. Near the wall, the results are very sensitive to errors in y and, in the core, to errors in ΔT or ΔC. Hence precise values are hard to obtain in both regions.

At present, our understanding of the spatial variation of the turbulent Prandtl number near a wall is quite unsatisfactory. Distributions which either decrease or increase as the wall is approached have been used; for example:

$$Pr_t = \frac{1}{1 + 400^{-y/R}} \tag{5.1.29}$$

for air flowing in a pipe, and

$$Pr_t = 1{\cdot}75 - 1{\cdot}25\frac{y}{\delta} \tag{5.1.30}$$

for a boundary layer in air. It is possible that these forms were successful because they cancel other imperfections in the theoretical models adopted. Useful results can often be obtained with a turbulent Prandtl or Schmidt number which is uniform across the flow. It is necessary, however, to adopt a value consistent with the molecular transport properties and with the local level of turbulent activity. In the following pages, we shall investigate the dependence of the turbulent Prandtl number on its molecular counterpart and on Reynolds number, setting aside the dependence on flow species and on position within the flow.

Table 5.1. Turbulent Prandtl and Schmidt numbers

I. Air ($\mathrm{Pr} = 0{\cdot}7$) *flowing in pipes and channels* ($\mathrm{Re}_d = 30{,}000$)	$\mathrm{Pr_t}$ (or $\mathrm{Sc_t}$)
Wall layer ($30 < y^+ < 300$)	$0{\cdot}9 \pm 0{\cdot}2$
Matching region ($y/R \sim 0{\cdot}2$)	$0{\cdot}8 \pm 0{\cdot}2$
Core ($\mathrm{Pr_t} \propto \mathrm{Re}^{0{\cdot}15}$)	$0{\cdot}7 \pm 0{\cdot}2$
II. Mercury ($\mathrm{Pr} = 0{\cdot}024$) *flowing in pipes* ($\mathrm{Re}_d = 300{,}000$)	$\mathrm{Pr_t}$
Wall layer	$1{\cdot}7 \pm 0{\cdot}3$
Core ($\mathrm{Pr_t} \propto \mathrm{Re}^{-0{\cdot}46}$)	$1{\cdot}4 \pm 0{\cdot}2$
III. Air and other gases ($\mathrm{Pr} = 0{\cdot}7$) *in free-turbulent flow: values giving the best overall fit*	$\mathrm{Pr_t}$ (or $\mathrm{Sc_t}$)
Mixing layer	0·5
Plane wake	0·55
Plane jet	0·5 to 0·55
Edge of boundary layer	0·5
Round jet	0·7 to 0·75

A Simple Transfer Model

The analysis is based on the gradient-diffusion model of the eddy diffusivity, equation (3.4.17). When the correlation coefficient is assumed to have the form $R_{22}(\tau) = \exp(-\tau/T_\mathrm{L})$, with T_L the Lagrangian integral time scale for the lateral velocity fluctuations, the calculations of Example 3.21 indicate that

$$\mathrm{Pr_t} = \frac{\epsilon_\mathrm{m}}{\epsilon_\mathrm{h}} = \frac{1 + T_\mathrm{L}\,\delta_\mathrm{h}}{1 + T_\mathrm{L}\,\delta_\mathrm{m}} \tag{5.1.31}$$

with δ_h and δ_m specifying the transfers between a moving element of fluid and its surroundings, as in equation (3.4.11). Note that $1/\delta_h$ and $1/\delta_m$ are time scales for the small-scale processes through which the element's energy and momentum are altered. To establish plausible expressions for these time scales, in terms of the diffusivities of the fluid and the scales of the mean motion, we return to the relationship

$$\frac{dP}{d_t} = -\delta(P - \bar{P}) \tag{3.4.11}$$

through which δ was introduced. Here P is any property of the fluid. We shall use a simple model of the transfer process to find values for each quantity except δ, and thus see how it must depend on the parameters defining the model.

Figure 5.2(a) shows the model: a circulating flow, with scales L and U, which conveys fluid across the differential ΔP. Multiplying equation (3.4.11) by the volume of the circulating fluid, we estimate the right-hand side $\sim \delta L^3 \Delta P$. In Example 1.25 it is shown that the advance of a diffusive front normal to the streamlines is given by $y \sim (Kt)^{\frac{1}{2}}$, with K the diffusivity and t the time allowed for diffusion. Here $t \sim L/U$, and the amount transferred may be estimated as $\sim (KL/U)^{\frac{1}{2}} L^2 \Delta P$. The convective derivative on the left-hand side of equation (3.4.11) is now seen to have the form $U dP/dx \sim (U/L)(KL/U)^{\frac{1}{2}} L^2 \Delta P$. The estimates of the two sides of the equation will be consistent if

$$\delta \sim \left(\frac{KU}{L^3}\right)^{\frac{1}{2}}$$

Taking the Lagrangian time scale $T_L \sim L/U$, we have

$$T_L \delta \sim \left(\frac{K}{UL}\right)^{\frac{1}{2}} \tag{5.1.32}$$

Finally, we argue that the transfer is dominated by the largest scales of the turbulence, those defined by the mean motion, and obtain

$$T_L \delta \sim \left(\mathrm{Re}\frac{\nu}{K}\right)^{-\frac{1}{2}} \tag{5.1.32}$$

with Re a mean-flow Reynolds number.

Applying this result to transfers of momentum and heat, we are able to express the turbulent Prandtl number (5.1.31) as

$$\mathrm{Pr}_t = \frac{1 + C_1 \mathrm{Pe}^{-\frac{1}{2}}}{1 + C_2 \mathrm{Re}^{-\frac{1}{2}}} \tag{5.1.33}$$

with $\mathrm{Pe} = \mathrm{Pr}\,\mathrm{Re}$ the mean-flow Péclet number. The constants C_1 and C_2 will be chosen below to match the empirical values of Table 5.1. However, we note immediately that

$$\mathrm{Pr}_t \to 1 \quad \text{as} \quad \mathrm{Re} \to \infty$$

in accord with the trends indicated in that table.

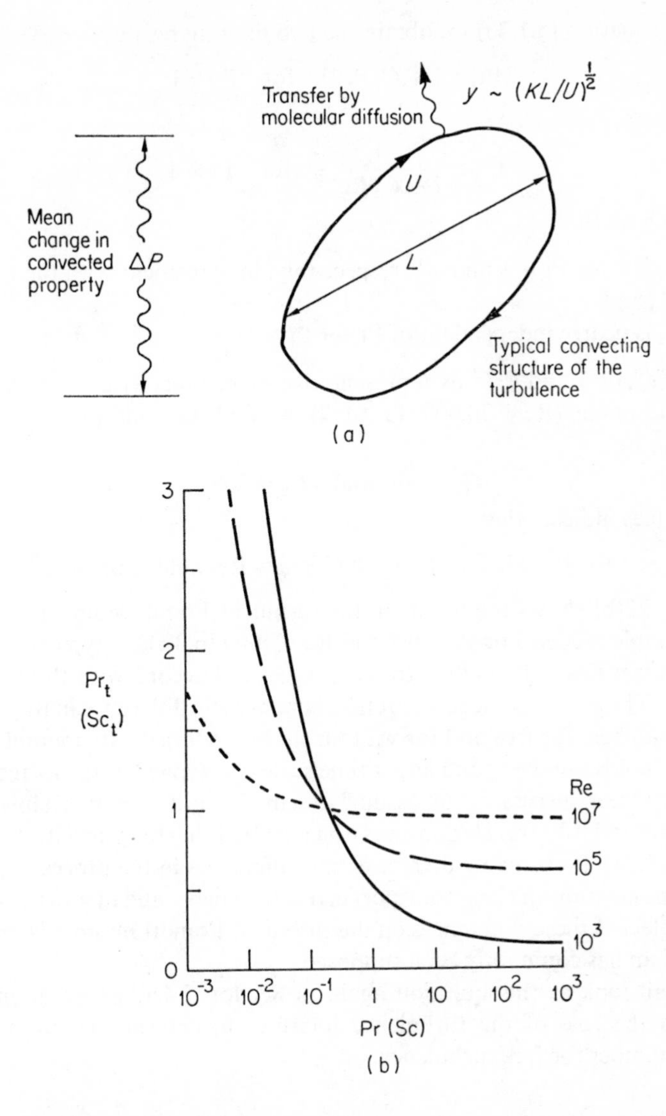

Figure 5.2 Turbulent transport processes and the turbulent Prandtl number. (a) A simple model illustrating the transport of a general property P by a combination of turbulent mixing (with scales L and U) and molecular diffusion (with diffusivity K). (b) Dependence of turbulent Prandtl number Pr_t on Reynolds number Re and molecular Prandtl number Pr, suggested by the model (a) and empirical data. This is equivalent to $\mathrm{Sc}_t = f(\mathrm{Re}, \mathrm{Sc})$, with Sc the molecular Schmidt number. The variations from flow to flow and from point to point within a flow are not considered

From equation (5.1.33) we obtain the two limiting results

$$\mathrm{Pr_t} \simeq 1 + C_1 \mathrm{Pe}^{-\frac{1}{2}} \quad \text{for} \quad \mathrm{Pr} \ll 1 \tag{5.1.34}$$

and

$$\mathrm{Pr_t} \simeq \frac{1}{1 + C_2 \mathrm{Re}^{-\frac{1}{2}}} \quad \text{for} \quad \mathrm{Pr} \gg 1 \tag{5.1.34}$$

These suggest that

(1) $\mathrm{Pr_t} \gtrless 1$ for $\mathrm{Pr} \ll 1$ and $\gg 1$, respectively, in agreement with the trends of Table 5.1, and
(2) $\mathrm{Pr_t}$ is nearly independent of Pr for $\mathrm{Pr} \gg 1$.

Taking $\mathrm{Pr_t} = 1{\cdot}5$ and $0{\cdot}75$ as representative of mercury ($\mathrm{Re} = 300{,}000$, $\mathrm{Pr} = 0{\cdot}024$) and of air ($\mathrm{Re} = 30{,}000$, $\mathrm{Pr} = 0{\cdot}7$), we find the constants of equation (5.1.33) to be

$$C_1 = 86 \quad \text{and} \quad C_2 = 200 \tag{5.1.33}$$

These values indicate that

(3) $\mathrm{Pr_t} < 1$ for $\mathrm{Pr} \simeq 1$, for all but the highest Reynolds numbers.

Figure 5.2(b) shows the trends in the turbulent Prandtl number suggested by this simple model. For $\mathrm{Pr} = 0{\cdot}7$ and $\mathrm{Re} = 1000$ to 3000, a typical range for free-turbulent flows, $\mathrm{Pr_t} = 0{\cdot}58$ to $0{\cdot}62$, in general accord with the results of Table 5.1. These results suggest that the characteristic difference in the effective Prandtl numbers for free and for wall turbulence is largely attributable to the lower Reynolds numbers, and larger time scales T_L, typical of the former class. It has not been necessary to assume essentially different mechanisms for momentum and heat transfers, as was done in obtaining the vorticity-transport results (5.1.28). Presumably there are real differences in the processes responsible for momentum and heat transport in free turbulence and in wall turbulence, but the effect of these differences on the turbulent Prandtl number is probably smaller than has commonly been supposed.

We shall look at this question again in Section 7.4.1, where attention is drawn to the role of the turbulence interface in determining the effective Prandtl number for free turbulence.

5.2 DIMENSIONAL ANALYSIS AND SIMILARITY

5.2.1 Heat-transfer Laws

The general nature of the relationships among the parameters which determine the transfer of heat or mass between channel walls and the fluid within can be deduced from dimensional arguments like those applied in Chapter 4 to friction and the velocity distribution. The analysis is somewhat more complicated, since parts of the treatment of friction are embedded in it. Friction laws of quite wide applicability can be obtained even with the neglect of the influence of heat transfer on the velocity field, that is, by assuming uniform, temperature-independent fluid properties. The converse is most certainly not true. The dominant influence of the velocity field is usually through the eddy diffusivity, but it also affects the transfer rate by modifying the flux variation, as indicated in equations (5.1.26, 27).

Most of the following discussion relates directly to heat transfer; analogous results for mass transfer are pointed out in Section 5.2.4.

Consider first the heat-transfer law for fully developed pipe flow, that matching the friction law

$$c_f = \frac{\tau_w}{\frac{1}{2}\rho U_a^2} = f\left(\frac{\rho U_a d}{\mu}, \frac{k_s}{d}\right) \tag{5.2.1}$$

which relates the wall stress τ_w to

(1) the geometric parameters U_a, d and k_s, the average velocity, pipe diameter and sand-grain roughness; and
(2) the fluid properties μ and ρ, the viscosity and density.

The heat transfer at the wall, $\dot{q}_w$ in the unit time through unit area, may be expected to depend upon

(1) the temperatures T_w and T_f of the wall and fluid;
(2) the geometric parameters U_a, d and k_s; and (5.2.2)
(3) the fluid properties k, c_p, ν and κ—the thermal conductivity, the specific heat and the diffusivities of momentum and heat.

The following paragraphs discuss the implications of the choices (5.2.2).

Although the form of the transfer law is broadly independent of the temperatures chosen to represent wall and fluid, the numerical constants are sensitive to them. In equations (5.1.6, 9) we have an extreme example of the influence of the fluid temperature. More fundamentally, we have failed to specify the way in which the temperature field develops along the pipe; hence our analysis encompasses a variety of cross-stream temperature profiles as well. The boundary conditions most often considered are:

(1) uniform heat flux, for which $T_w - T_f \simeq$ constant, and the temperature field is linearly developing, and
(2) uniform wall temperature T_w.

We shall consider the former case for the most part, later (Section 5.3.3) noting the adjustment required when the wall temperature is maintained constant.

We have seen that the effect of roughness on friction cannot always be characterized by a single length scale, and the same must be expected for transfer processes. Hence the parameter k_s represents an appropriate number of roughness-specifying dimensions.

Finally, we consider the selection of fluid properties in the list (5.2.2). Other, equivalent choices could be made (for example, k, c_p, μ and ρ), without altering the essential nature of the results. More significant are the parameters which have been omitted, notably, the mean-free-path and the speed of sound. The inclusion of the former introduces the *Knudsen number* Kn into the heat-transfer law, as in equation (2.4.1), applicable to a fine wire. The addition of the sound speed introduces the *Mach number* $\mathrm{Ma} = U/a$, or perhaps the specific-heat ratio $\gamma = c_p/c_v$. The parameters Kn and Ma are not usually important for channel flows, though they can be significant for boundary layers on vehicles travelling at high speeds through the atmosphere, particularly at high altitudes.

Yet another fluid property which has been omitted from the list (5.2.2) is the coefficient of expansion or isobaric compressibility β. The present analysis relates to *forced* convection, that is, to velocity fields whose existence does not depend on heat transfer. In *free or natural convection* the motion is induced by the changes in fluid weight which arise from the temperature variations associated with heat transfer. There is a spectrum of *combined or mixed convection*, the flow modified by buoyancy but not entirely dependent on it. For channel flows, buoyancy usually manifests itself in combined convection; this will be considered briefly in Section 5.3.3. In a less restricted fluid, pure natural convection is often found; in Chapter 7 free-turbulent flows of this kind will be considered, and in Chapter 8, boundary layers. Equations governing vertical buoyant flows are developed in Section 6.2.6.

The dependence of the wall flux on the parameters (5.2.2) is commonly expressed as

$$\mathrm{Nu} = \frac{\dot{q}_\mathrm{w} d}{k(T_\mathrm{w} - T_\mathrm{f})} = f\left(\mathrm{Re}_d, \frac{k_s}{d}, \mathrm{Pr}, \frac{T_\mathrm{w}}{T_\mathrm{f}}, \mathrm{Ec}\right) \tag{5.2.3}$$

The presence of the first three groups on the right requires no comment, in view of the preceding discussion of Reynolds analogy. The last group

$$\mathrm{Ec} = \frac{U_\mathrm{a}^2}{c_p(T_\mathrm{w} - T_\mathrm{f})} \tag{5.2.4}$$

is the *Eckert number*; its form suggests that it represents kinetic heating, that is, the rise in temperature as the kinetic energy of the fluid is converted into thermal energy upon deceleration. This interpretation is not apposite for flow in a uniform channel, with U_a constant; in such cases the combinations

$$\begin{aligned} \tfrac{1}{2}c_\mathrm{f}\,\mathrm{Ec} &= \frac{u_\mathrm{f}^2}{c_p\,\Delta T} \\ \tfrac{1}{2}c_\mathrm{f}\frac{\mathrm{Ec}}{\mathrm{St}} &= \frac{\tau_\mathrm{w} U_\mathrm{a}}{\dot{q}_\mathrm{w}} \end{aligned} \tag{5.2.4}$$

and

$$(\tfrac{1}{2}c_f)^{\frac{3}{2}}\frac{\text{Ec}}{\text{St}} = \frac{\rho u_f^3}{\dot{q}_w} \tag{5.2.4}$$

are more significant. They relate the heat transfer to the dissipation within the channel, which acts somewhat in the manner of a thermal source.

The remaining dimensionless parameter of equation (5.2.3), the temperature ratio T_w/T_f, accounts in a formal way for the temperature dependence of the fluid properties. A comprehensive representation would require that several coefficients of the equations of state be represented in dimensionless form. When $\Delta T/T_w$ is small, these effects can be neglected; for somewhat larger values of $\Delta T/T_w$, they can be accounted for approximately by evaluating the fluid properties at the *mean film temperature* $T_m = \frac{1}{2}(T_w + T_f)$, or at some other representative fluid temperature. This matter is considered further in Section 5.3.3.

When the temperature differential, $T_w - T_f$, is neither so large that property changes are significant, nor so small that 'heating' from dissipation is important, we can simplify the law (5.2.3) to

$$\text{Nu} = f\left(\text{Re}_d, \frac{k_s}{d}, \text{Pr}\right) \tag{5.2.5}$$

Here a 'pure' heat-transfer coefficient (Nu) is related to parameters specifying the flow condition (Re_d and k_s/d) and the fluid species (Pr).

Convective heat transfer can sometimes be specified more simply by a law of the form

$$\text{St} = \frac{\text{Nu}}{\text{Pr}\,\text{Re}_d} = f\left(\text{Re}_d, \frac{k_s}{d}, \text{Pr}\right) \tag{5.2.6}$$

Using the friction law (5.2.1) we can rewrite these results as

$$\text{Nu} = f\left(c_f, \frac{k_s}{d}, \text{Pr}\right) \quad \text{and} \quad \text{St} = f\left(c_f, \frac{k_s}{d}, \text{Pr}\right) \tag{5.2.7}$$

We may indulge in wishful thinking by wondering whether the introduction of the friction coefficient (itself dependent on k_s/d) takes adequate account of wall roughness. This hypothesis will be examined in Section 5.4.1, and will be found to be unrealistic.

The Case of Very Small Prandtl Number

There is one situation in which the heat-transfer laws can be simplified further, that in which the thermal conductivity is very large in comparison with the viscosity, as is typical of liquid metals. Here the molecular thermal diffusivity may remain comparable with the turbulent diffusivity through much of the channel or, at least, through much of the wall layer, where the gradients of velocity and temperature are largest. Thus the molecular viscosity ν and roughness k_s lose their most important roles as definers of the effective diffusivity; they still have an effect through the velocity distribution, but this is accounted for in part by the retention of the scales U_a and d. On dropping the viscosity and roughness from the list (5.2.2), we find that the results (5.2.5, 6) simplify to

$$\mathrm{Nu} = \mathrm{Pe}\,\mathrm{St} = f(\mathrm{Pe}) \tag{5.2.8}$$

where $\mathrm{Pe} = \mathrm{Pr}\,\mathrm{Re}_d = U_a d/\kappa$ is the *Péclet number*, the ratio of mean-flow convection to molecular thermal diffusion. We shall see later that the effects of Reynolds number cannot be wholly absorbed in this way, but the result does indicate the dominant relationship.

Practical Heat-transfer Coefficients

A variety of dimensioned measures of heat-transfer capability are encountered in engineering practice; here we shall see how they are related to the dimensionless groups introduced above.

The performance of heat exchangers is often quoted in terms of an *overall heat-transfer factor or coefficient*

$$\frac{\dot{Q}}{A(T_1 - T_2)} = U \quad \text{(a commonly used symbol)} \tag{5.2.9}$$

with A the total transfer area and $\dot{Q}$ the rate of heat transfer. A *film coefficient* of similar form

$$\frac{\dot{Q}}{A(T_w - T_f)} = h \quad \text{(a commonly used symbol)} \tag{5.2.9}$$

is used to specify the transfer between wall and fluid. The conventional units for these coefficients are W/m^2 °K or Btu/ft^2 hr °R.

The characteristics of solid components of complex form can be described using a *shape factor*

$$S = \frac{\dot{Q}}{k(T_1 - T_2)} \tag{5.2.9}$$

with k the thermal conductivity of the material. This factor has the dimensions of length; for example

$$S = \frac{2\pi L}{\ln(R_2/R_1)}$$

for a circular cylinder of length L and inner and outer radii R_1 and R_2.

Alternatively, a *thermal resistance* may replace any of these coefficients; it is related to them by

$$R = \frac{A\,\Delta T}{\dot{Q}} = \frac{1}{U}, \frac{1}{h}, \frac{A}{kS} \tag{5.2.9}$$

These 'practical' coefficients are related to the dimensionless heat-transfer parameters by

$$\begin{aligned} \mathrm{Nu} &= \frac{\dot{Q}d'}{Ak(T_1 - T_2)} \\ &= \frac{Ud'}{k}, \frac{hd'}{k}, \frac{Sd'}{A}, \frac{d'}{kR} \end{aligned} \tag{5.2.10}$$

where d' is a characteristic dimension of the system.

5.2.2 Recovery Effects

Until now, equation (5.2.3) has been thought of as a heat-transfer law. It can be given another interpretation, of particular relevance for the boundary layers formed on vehicles travelling through the air at high speeds. For an insulated wall, $\mathrm{Nu} = \dot{q}_w = 0$, and

$$\begin{aligned} r &= \frac{T_r - T_f}{U_f^2/2c_p} = \frac{2}{\mathrm{Ec}_r} \\ &= f\left(\mathrm{Re}_d, \frac{k_s}{d}, \mathrm{Pr}, \frac{T_r}{T_f}\right) \end{aligned} \tag{5.2.11}$$

Here T_r is the *recovery or adiabatic wall temperature*;

U_f is a representative fluid velocity, for a channel, the average velocity; for a boundary layer, the free-stream velocity;

$r = (T_r - T_f)/(T_0 - T_f)$ is the *recovery factor*, with $T_0 = T_f + U_f^2/2c_p$ the total temperature of the stream; and

Ec_r is the Eckert number for recovery conditions, that is, for an adiabatic wall.

We see that the recovery factor specifies the amount by which the temperature at the wall rises above the static temperature of the fluid. The rise is proportional to the kinetic energy $\frac{1}{2}U_f^2$, with U_f in general the change in velocity between wall and outer fluid. Once 'kinetic heating' does become significant it is strongly dependent on this velocity change. The *enthalpy recovery factor*

$$r = \frac{H_r - H_f}{\frac{1}{2}U_f^2} = \frac{H_r - H_f}{H_0 - H_f} \tag{5.2.12}$$

with $H_0 = H_f + \frac{1}{2}U_f^2$ the total enthalpy, is more appropriate for large temperature changes, when c_p varies significantly.

The temperature and enthalpy recovery factors relate the actual difference in thermal energy between wall and outer flow to that which would be achieved by adiabatic deceleration. The recovery factor represents the ability of a fluid to retain dissipated energy near the wall, rather than transferring it into the body of the flow. Thus the factor is strongly dependent on the Prandtl number $\mathrm{Pr} = \nu/\kappa$. The approximate formulae for constant-pressure boundary layers on flat plates give an idea of this dependence:

(1) for laminar flow

$$r \simeq \mathrm{Pr}^{\frac{1}{2}} \quad \text{for} \quad \tfrac{1}{2} < \mathrm{Pr} < 5 \tag{5.2.13}$$

(2) for turbulent flow

$$r \simeq \mathrm{Pr}^{\frac{1}{3}} \quad \text{for} \quad \tfrac{1}{2} < \mathrm{Pr} < 2 \tag{5.2.14}$$

The recovery factor depends also on the way in which the flow near the wall is developing. For parallel laminar flow, a study of simple examples shows that

$$r \propto \mathrm{Pr} \tag{5.2.15}$$

which may be compared with equation (5.2.13). For a turbulent wall layer, the effect of flow development is not as marked, since the wall-layer flow is nearly parallel, except in extreme cases. However, a Reynolds-number effect must be expected, since the thickness of the viscous layer depends on this factor. The general nature of the dependence is suggested by a theoretical prediction of Shirokow:

$$r = 1 - 4{\cdot}55\,(1 - \mathrm{Pr})\,\mathrm{Re}_x^{-0{\cdot}2} \tag{5.2.16}$$

(In this result for a boundary layer, $\mathrm{Re}_x = U_1 x/\nu$.) This is the simplest of various proposals which have been made to improve upon the estimate (5.2.14).

For all of the situations considered above, the recovery factors for gases lie in the range

$$r = 0{\cdot}7 \text{ to } 1{\cdot}0 \tag{5.2.17}$$

Hence, for the cases of greatest practical importance, we may estimate the recovery temperature as

$$T_r \simeq T_f + 0{\cdot}85\frac{U_f^2}{2c_p} = T_0 - 0{\cdot}15\frac{U_f^2}{2c_p} \tag{5.2.17}$$

This shows that the 'kinetic heating' will raise the wall temperature nearly to the total temperature of the stream, unless the surface is cooled. The calculation of the recovery temperature is sometimes referred to as the *thermometer problem*. When a thermometer is placed in a flowing fluid, a pattern of recovery temperatures is established over its surface, the temperature registered being some integrated value of these, usually not far below the total temperature.

Figure 5.3 indicates the way in which the temperature varies within a developing laminar boundary layer; we shall treat the turbulent wall layer later (Section 5.3.1 and Figure 5.4). Two situations are considered: a variety of transfer rates for a particular fluid, illustrating the effect of dissipation; and an adiabatic wall past which flow fluids of various Prandtl numbers. In each case, the velocity profile is slightly modified by the viscosity variations within the fluid, but these changes are negligible when $(T_w - T_f)/T_f$ and $(T_r - T_f)/T_f$ are small. In Figure 5.3(a) we note that the distortion of the temperature profile by the recovery phenomena will become negligible (save for its effects on the velocity profile) when $|T_w - T_f| \gg T_r - T_f$. In Figure 5.3(b) we see that the

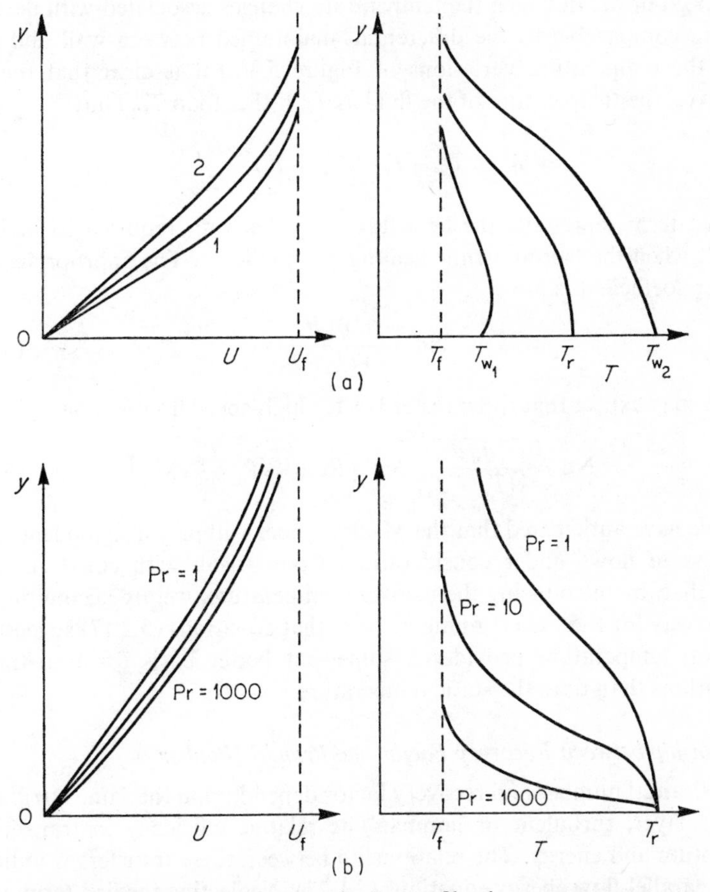

Figure 5.3 Temperature and velocity profiles in constant-pressure laminar boundary layers, after van Driest. (a) Effects of heat transfer and dissipation for $\mathrm{Pr} \simeq 1$: profiles for an adiabatic wall (recovery conditions, r), for a cooled wall ($T_{w_1} < T_r$) and for a heated wall ($T_{w_2} > T_r$). (b) Role of Prandtl number for an adiabatic wall

'momentum boundary layer' is thicker than the 'thermal boundary layer' for $\mathrm{Pr} > 1$, that is, when the diffusion of momentum is more rapid than that of heat. Together, the figures suggest that a close approximation to similarity of the velocity and temperature profiles will be achieved only when all of the following conditions are satisfied:

(1) $|T_\mathrm{w} - T_\mathrm{f}| \gg T_\mathrm{r} - T_\mathrm{f}$
(2) $\mathrm{Pr} \simeq 1$
(3) $(T_\mathrm{w} - T_\mathrm{f})/T_\mathrm{w} \ll 1$

These considerations suggest a way of estimating the heat-transfer rate for high-speed flows, in which the temperature changes associated with deceleration are comparable to the differential maintained between wall and fluid. From the temperature variations of Figure 5.3(a) it is clear that the wall 'perceives' the temperature of the fluid as T_r, rather than T_f. Thus

$$\dot{q}_\mathrm{w} \propto T_\mathrm{w} - T_\mathrm{r} = (T_\mathrm{w} - T_\mathrm{f}) - \frac{U_\mathrm{f}^2}{2c_p}$$

The last term represents the heat flux (into the wall) required to maintain $T_\mathrm{w} = T_\mathrm{f}$ when the 'aerodynamic heating' is significant. The appropriate heat-transfer coefficient is now

$$\mathrm{Nu} = \frac{\dot{q}_\mathrm{w}\, d}{k(T_\mathrm{w} - T_\mathrm{r})} \tag{5.2.18}$$

and we may expect that the transfer law for high-speed flow will have the form

$$\mathrm{Nu} = \frac{\dot{q}_\mathrm{w}\, d}{k(T_\mathrm{w} - T_\mathrm{r})} = f\left(\mathrm{Re}_d, \frac{k_\mathrm{s}}{d}, \mathrm{Pr}, \frac{T_\mathrm{w}}{T_\mathrm{f}}, \mathrm{Ma}\right) \tag{5.2.18}$$

Here we have anticipated that the Mach number will play a significant role in the class of flows under consideration. Comparison with equation (5.2.3) shows that, by introducing the recovery temperature, we are accounting in a specific way for the Eckert number. Note that equations (5.2.17) suggest that the total temperature provides a somewhat better basis for heat-transfer calculations than does the static temperature.

Relationship between Recovery Factor and Prandtl Number

The Prandtl number and recovery factor depend upon the same attribute of a shear layer, turbulent or laminar—its relative efficiency in transmitting momentum and energy. The relationship between these transfers is indicated by the parallel-flow energy equations (5.1.23). Neglecting the first term, which represents mean-flow convection, and the contribution of J_q, the lateral flux of kinetic energy, we have

$$\frac{\mathrm{d}\dot{q}_\mathrm{h}}{\mathrm{d}y} = \rho \frac{\mathrm{d}}{\mathrm{d}y}\left(-\kappa \frac{\mathrm{d}H}{\mathrm{d}y} + \overline{vh}\right) = \frac{\mathrm{d}(\tau U)}{\mathrm{d}y}$$

Integration gives

$$-\kappa\frac{dH}{dy}+\overline{vh}-\frac{\tau U}{\rho}=\frac{(\dot{q}_h-\tau U)_1}{\rho}=C_1$$

the constant C_1 being evaluated at a reference position 1.

Introducing the diffusivities of momentum and enthalpy to represent the turbulent transfers, we have

$$(\kappa+\epsilon_h)\frac{dH}{dy}+(\nu+\epsilon_m)\frac{d(\frac{1}{2}U^2)}{dy}=-C_1$$

and a second integration gives

$$H+\mathrm{Pr}_e\tfrac{1}{2}U^2=-\int\frac{C_1}{\kappa+\epsilon_h}\,dy \tag{5.2.19}$$

subject to the assumption that the effective Prandtl number Pr_e is constant. This result relates the temperature and velocity changes in a variety of flows, laminar and turbulent. The integral on the right gives the temperature differential required to sustain the energy flux $(\dot{q}_h-\tau U)_1$; the contributions of dissipation $(\tau\, dU/dy)$ and the allied diffusion process $(U\, d\tau/dy)$ are given in the explicit integral on the left.

We focus our attention on the particular case of $C_1=0$, which will be achieved:

(1) at an insulated wall, where $\dot{q}_h=U=0$, so that

$$H+\mathrm{Pr}_e\tfrac{1}{2}U^2=H_r \tag{5.2.20}$$

(2) on the plane of symmetry of a core flow, where $\dot{q}_h=0$ and $U=U_c$, so that

$$H+\mathrm{Pr}_e\tfrac{1}{2}U^2=H_c+\mathrm{Pr}_e\tfrac{1}{2}U_c^2 \tag{5.2.21}$$

Comparing equations (5.2.20, 12), we see that

$$r=\mathrm{Pr}_e \tag{5.2.22}$$

subject to the assumption that Pr_e is constant in the region over which the major velocity and temperature changes occur. For parallel laminar flows in which this assumption is valid, $r\propto \mathrm{Pr}$, as noted in equation (5.2.15). For the turbulent flow of a gas near a wall, the assumption is fairly realistic, since $\mathrm{Pr}=0{\cdot}7$, while Table 5.1 suggests that $\mathrm{Pr}_t=0{\cdot}7$ to $0{\cdot}9$ in the wall layer. Experimental values of r for gases lie in the range 0·85 to 0·90, consistent with the result (5.2.22).

The result (5.2.21) cannot be expected to apply to free-turbulent flows or in the outer part of a developing boundary layer, even though the boundary conditions appear to be satisfied in such cases. Although the diffusivity ratio may be reasonably constant, the neglect of the terms representing mean-flow convection cannot be justified.

5.2.3 The Temperature Distribution

We return to the situation which concerned us in Section 5.2.1, taking the heat transfer at the wall to be large compared with the internal dissipation, so that we may neglect the temperature-recovery effects associated with the Eckert number. Setting aside also the effects of varying properties (represented by the ratio T_w/T_f in equations (5.2.3)), we may suppose that the temperature distribution in fully developed pipe flow is given by

$$T_w - T \sim y, R, k_s, \dot{q}_w, u_f, \rho, c_p, \nu, \kappa \tag{5.2.23}$$

Once again, the choice of parameters (for example, u_f and ρ instead of τ_w and μ) is made for convenience. A dimensionless representation is

$$\frac{T_w - T}{\theta_f} = f\left(\frac{y}{y_f}, \frac{R}{y_f}, \frac{k_s}{y_f}, \mathrm{Pr}, \frac{u_f^3}{\dot{q}_w}\right) \tag{5.2.24}$$

The temperature scale

$$\theta_f = \frac{\dot{q}_w}{\rho c_p u_f} \tag{5.2.25}$$

is sometimes termed the *friction temperature*, by analogy with the friction velocity, though the label 'friction' is not as appropriate here.

The first three of the groups appearing on the right of equation (5.2.24) define the velocity and diffusivity distributions, and require no further comment; nor does the Prandtl number. The final group is one of the equivalents of the Eckert number identified in equations (5.2.4). It represents the effect of dissipation and diffusion on the temperature profile, and will be omitted henceforth.

Subject to this restriction to relatively low-speed motions, arguments like those summarized in Table 4.1 give results applicable to limited parts of the flow:

(1) for the wall layer as a whole

$$\frac{T_w - T}{\theta_f} = f\left(\frac{y}{y_f}, \frac{k_s}{y_f}, \mathrm{Pr}\right) \tag{5.2.26}$$

(2) for the part of the wall layer (if any) which is fully turbulent, in the sense that both $\nu \ll \epsilon_m$ and $\kappa \ll \epsilon_h$ apply

$$y\frac{d(T/\theta_f)}{dy} = f(\mathrm{Pr}) \tag{5.2.27}$$

(3) for the core flow

$$\frac{T_w - T_c}{\theta_f} = f\left(\frac{y}{R}, \mathrm{Pr}\right) \tag{5.2.28}$$

More explicit results can be obtained for the sublayer, the fully turbulent layer, and the core. For the *linear sublayer* on a smooth wall

$$\dot{q}_w = \frac{k(T_w - T)}{y}$$

whence

$$\frac{T_w - T}{\theta_f} = \Pr y^+ \tag{5.2.29}$$

with $y^+ = y/y_f$. However, in this linear layer $y^+ = U^+ = U/u_f$, and

$$\frac{T_w - T}{\theta_f} = \Pr \frac{U}{u_f} \tag{5.2.30}$$

For the *fully turbulent layer*

$$y^+ \frac{d(T/\theta_f)}{dy^+} = -A_T \tag{5.2.31}$$

paralleling the first of equations (4.1.3). This result assumes that $\dot{q}_w$ is constant, an assumption which is seen in Figure 5.1 to be realistic for the wall region when $\mathrm{Ec} \simeq 0$. The relationship of the constant A_T to that appearing in the corresponding velocity distribution can be found as follows. Note that

$$\epsilon_h = -\frac{\dot{q}_w}{\rho c_p \, dT/dy} = \frac{u_f y}{A_T} \tag{5.2.32}$$

Combined with $\epsilon_m = u_f y/A$, valid in the logarithmic layer, this gives

$$\Pr_t = \frac{\epsilon_m}{\epsilon_h} = \frac{A_T}{A} = \text{constant} \tag{5.2.33}$$

for the region in which the two fully turbulent layers overlap. We have seen that the turbulent Prandtl number is a weak function of the molecular Prandtl number, and it follows that the constant A_T will be also. This is consistent with equation (5.2.27).

Integration of equation (5.2.31) gives

$$\frac{T_w - T}{\theta_f} = A_T \ln y^+ + f_T(k_s^+, \Pr) \tag{5.2.34}$$

The form of the additional function has been selected in conformity with equation (5.2.26). Introducing the logarithmic velocity distribution (4.3.5), we obtain

$$\begin{aligned} \frac{T_w - T}{\theta_f} &= \frac{A_T}{A}\frac{U}{u_f} + f_T(k_s^+, \Pr) - \frac{A_T}{A} f(k_s^+) \\ &= \Pr_t \frac{U}{u_f} + f_c(k_s^+, \Pr) \end{aligned} \tag{5.2.35}$$

with the use of the relationship (5.2.33).

Finally, we turn to the *core region*, taking the eddy diffusivities (ϵ_h and $\epsilon_m = \epsilon_c$) to be constant there:

$$\rho c_p \epsilon_h \frac{dT}{dy} = -\dot{q}_h = -\dot{q}_c \left(1 - \frac{y}{R}\right) \tag{5.2.36}$$

with $\dot{q}_c$ the effective value of the heat flux for the core. From Figure 5.1 we find that $\dot{q}_c > \dot{q}_w$ by virtue of the slow fall in $\dot{q}_h$ in the wall layer, and see that this quantity may be assumed constant in the turbulent core without serious error. Integration gives

$$\frac{T - T_c}{\theta_f} = \tfrac{1}{4}\mathrm{Pr}_c R_f \left(\frac{r}{R}\right)^2 \frac{\dot{q}_c}{\dot{q}_w} \tag{5.2.37}$$

with $R_f = u_f d/\epsilon_c$ as in equations (3.3.22), and $\mathrm{Pr}_c = \epsilon_c/\epsilon_h$ the value appropriate to the core. Introducing the velocity deficit from Table 3.3, we have

$$\frac{T - T_c}{\theta_f} = \mathrm{Pr}_c \frac{\dot{q}_c}{\dot{q}_w} \frac{U_c - U}{u_f} \tag{5.2.38}$$

for the core region.

Each of the results (5.2.30, 35, 38) displays a species of similarity for parts of the velocity and temperature profiles. Complete similarity can be achieved (for a smooth wall) only when the following conditions are satisfied:

$$\begin{gathered} \mathrm{Pr} = \mathrm{Pr}_t \\ f_c = f_T(\mathrm{Pr}) - \mathrm{Pr}_t B = 0 \\ \mathrm{Pr}_c \frac{\dot{q}_c}{\dot{q}_w} = 1 \end{gathered} \tag{5.2.39}$$

The two profiles are then related everywhere by

$$\frac{T_w - T}{\theta_f} = \mathrm{Pr} \frac{U}{u_f} \tag{5.2.40}$$

There follows the transfer law

$$\frac{\rho c_p U(T_w - T)}{\dot{q}_w} = \mathrm{Pr} \frac{U^2}{u_f^2}$$

or

$$\frac{\mathrm{St}}{\tfrac{1}{2}c_f} = \frac{1}{\mathrm{Pr}} = \frac{1}{\mathrm{Pr}_t} \tag{5.2.41}$$

with St and c_f defined using the corresponding values of U and T for any value of y. This generalization of Reynolds analogy was obtained earlier in equations (5.1.15); the present derivation gives a clearer idea of the requirements for similarity.

The second of the conditions (5.2.39) is satisfied by $\mathrm{Pr} = \mathrm{Pr}_t = 1$. The two regions where molecular activity is significant then coincide, and $f_T(1) = B$

relates the 'slip constants' of the velocity and temperature distributions. More generally, if we take $f_T \simeq B\text{Pr}$, as seems reasonable since a considerable part of the 'slip' occurs in the linear layer (5.2.30), we have for the second of the conditions (5.2.39):

$$f_c \simeq B(\text{Pr} - \text{Pr}_t)$$

which is satisfied when $\text{Pr} = \text{Pr}_t$. Hence we may expect a close approximation to wall-layer similarity (as in equation (5.2.40)) for gas flows, in which Pr ($\simeq 0{\cdot}7$) and Pr_t ($0{\cdot}9$ to $0{\cdot}7$) are not far from equality, nor from unity.

For large and moderate Prandtl numbers, the contribution of the core to the temperature change will not be very significant, and similarity in the wall layer alone will justify the result (5.3.41). However, for gases, with $\text{Pr}_c \sim 1$ and $\dot{q}_c/\dot{q}_w > 1$, the third of the conditions (5.2.39) should not in fact be violated too rudely.

These considerations suggest that the result (5.2.41) be replaced by

$$\frac{\text{St}}{\frac{1}{2}c_f} = \frac{1}{\text{Pr}_a} \tag{5.2.42}$$

with Pr_a an average value estimated as

$$\text{Pr}_a = \frac{0{\cdot}9\,n + \text{Pr}}{n+1} \tag{5.2.42}$$

The weighting factor n ($= 2$ to 4) accounts for the division of mean-property changes between the viscous and turbulent regions. For gases, with $\text{Pr} = 0{\cdot}7$, we obtain $\text{Pr}_a = 0{\cdot}80$ and $0{\cdot}85$ for $n = 2$ and 4. The formulae (5.2.42) give reasonably accurate results (within 10 per cent, say) for $0{\cdot}5 < \text{Pr} < 5$, and are therefore appropriate for all gases and for some liquids as well; see Table A.2 of the Appendix.

5.2.4 Mass Transfer

Many of the results of the preceding pages can be converted for application to mass transfer using the parallels drawn in Table 3.1:

$$\begin{aligned} &\dot{q}_h \to N, \quad \text{the mass flux} \\ &\rho H = \rho c_p T \to C, \quad \text{the mass concentration} \\ &\kappa \to D, \quad \text{the mass diffusivity} \\ &\text{Pr} \to \text{Sc}, \quad \text{the Schmidt number} \end{aligned} \tag{5.2.43}$$

Thus equations (5.2.5, 6, 8) give:

(1) the *Sherwood number*, the equivalent of the Nusselt number, as

$$\text{Sh} = \frac{N_w d}{D\Delta C} = f\left(\text{Re}_d, \frac{k_s}{d}, \text{Sc}\right) \tag{5.2.44}$$

(2) the *mass-transfer Stanton number* as

$$\mathrm{St_c} = \frac{N_w}{U_a \Delta C} = f\left(\mathrm{Re}_d, \frac{k_s}{d}, \mathrm{Sc}\right) \tag{5.2.44}$$

(3) and for large Schmidt numbers

$$\mathrm{Sh} = \mathrm{Pe_c}\,\mathrm{St_c} = f(\mathrm{Pe_c}) \tag{5.2.45}$$

with $\mathrm{Pe_c} = U_a d/D$, the mass-transfer analogue of the Péclet number.

Not only are analogous sets of parameters connected by these equations, but they are connected by the same functional relations which apply to heat transfer, provided, of course, that the transferred substance follows the fluid as does heat, and that neither 'contaminant' modifies the fluid motion significantly.

Turning to the more general result (5.2.3), we find that the parallels are not so neat. Using the transformations (5.2.43) we can convert the parameters (5.2.2) to mass-transfer analogues, save that $c_p = (\partial H/\partial T)_p$ has no obvious counterpart in mass transfer. The dimensionless result paralleling (5.2.3) is then

$$\mathrm{Sh} = \frac{N_w d}{D \Delta C} = f\left(\mathrm{Re}_d, \frac{k_s}{d}, \mathrm{Sc}, \frac{C_w}{C_f}\right) \tag{5.2.46}$$

There is no counterpart of the Eckert number, provided that we rule out sources of the transferred substance within the flow. It follows that there is no phenomenon directly analogous to temperature recovery, and that the analogy between heat and mass transfer does not extend to high-speed flows in which this phenomenon is of significance.

Another difficulty is suggested by the presence in equation (5.2.46) of the parameter C_w/C_f, the ratio of the concentration at the wall to that in the body of the fluid. It is hardly to be expected that the dependence of the fluid properties on the concentration $C + c$ will be similar to that for temperature $T + \theta$. We cannot expect that a 'temperature loading' factor in a heat-transfer law will have an exact counterpart in the analogous mass-transfer result. We conclude then that the analogy between heat and mass transfer cannot be carried beyond the level of the results (5.2.44, 45).

Turning to the variation of concentration in a pipe flow, we obtain from equation (5.2.24):

$$\frac{C_w - C}{N_w/u_f} = f\left(\frac{y}{y_f}, \frac{R}{y_f}, \frac{k_s}{y_f}, \mathrm{Sc}\right) \tag{5.2.47}$$

the dissipation parameter being omitted once again. The functional forms of the temperature distribution apply once more; for example, the functions $A_T(\mathrm{Sc})$ and $f_T(k_s^+, \mathrm{Sc})$ of equation (5.2.34) will appear in the concentration distribution.

The similarity requirements analogous to (5.2.39) are less often satisfied than are the heat-transfer conditions (5.2.39) themselves. As a consequence, the similarity law

$$\frac{C_w - C}{N_w/u_f} = \mathrm{Sc}\frac{U}{u_f} \tag{5.2.48}$$

and modified Reynolds analogy

$$\frac{\mathrm{St}_c}{\frac{1}{2}c_f} = \frac{1}{\mathrm{Sc}} = \frac{1}{\mathrm{Sc}_t} \tag{5.2.49}$$

are less likely to be useful approximations. We noted earlier that $\mathrm{Sc}_t \simeq \mathrm{Pr}_t$ unless Sc and Pr are very different. The real difficulty is with the molecular Schmidt number, which depends on the ratio of molecular weights of the mingling substances. As an example, consider air and hydrogen: although $\mathrm{Pr} \simeq 0{\cdot}7$ for both at room temperature, $\mathrm{Sc} \simeq 0{\cdot}25$ when hydrogen diffuses through air.

There are, however, some important cases of diffusion in air for which Sc has a value conveniently near the range $\mathrm{Sc}_t = 0{\cdot}7$ to $0{\cdot}9$ for a gaseous wall layer: water vapour ($\mathrm{Sc} \simeq 0{\cdot}6$), carbon dioxide (1·0), oxygen (0·8), and methane (0·8). For such cases, the result (5.2.49) will be useful, with an average Sc_a calculated as in equations (5.2.42).

5.3 TRANSFER RATES IN PIPE FLOW

5.3.1 Survey of the Problem

Analogies and similarity arguments have suggested relationships between the coefficients defining the transfer rates at a smooth wall (St, St_c, Nu and Sh) and the friction coefficient c_f, for cases in which the profiles of velocity and of the transferred entity are not too different in shape. We have seen that such situations arise when Pr, $\mathrm{Sc} \sim 1$. Still confining our attention primarily to fully developed flow in a round pipe, we broaden the discussion, first, by including dissimilar profiles corresponding to Pr, $\mathrm{Sc} \ll 1$ and $\gg 1$, and then by considering the effects of a large or non-uniform temperature differential between wall and fluid. The role of wall roughness will be considered in Section 5.4.1.

The Role of the Prandtl or Schmidt Number

Figure 5.4 reveals the range of problems to be treated. In Figure 5.4(a) are given variations of the effective diffusivities of momentum $(\epsilon_m + \nu)$ and enthalpy $(\epsilon_h + \kappa)$, based on the eddy-viscosity variation of Figure 3.3(b). Molecular Prandtl numbers of $\mathrm{Pr} = 0{\cdot}1$, 1 and 10 are considered; the turbulent

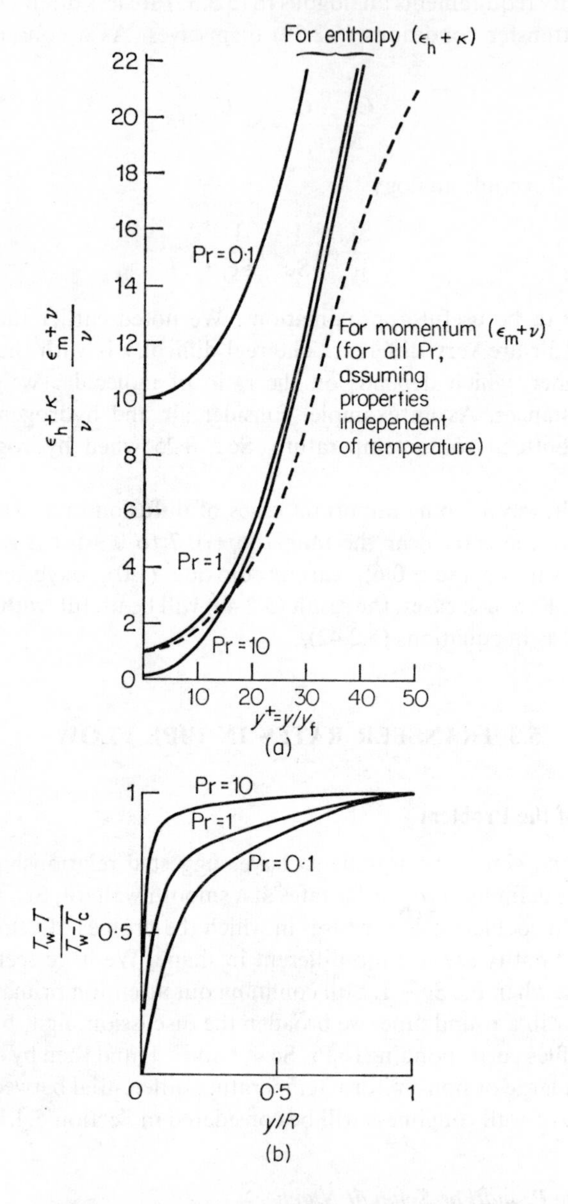

Figure 5.4 Role of the Prandtl number in turbulent flow near a wall. (a) Variations of diffusivities of momentum and enthalpy in the viscous layer, for $Pr_t = 0{\cdot}7$. (b) Temperature profiles across the radius of a pipe corresponding to the diffusivities of (a), for $Re_d \simeq 10^4$

Prandtl number has been assigned the value $Pr_t = 0{\cdot}7$ throughout. The profiles of the two diffusivities are not essentially different when $Pr = 1$. This is also so for $Pr \gg 1$, except very near the wall, where the effective Prandtl number $Pr_e \simeq Pr \gg 1$. When $Pr \ll 1$, the two profiles differ still more profoundly, with $Pr_e \ll 1$ throughout the viscous layer.

The consequences of these diffusivity variations appear in the temperature distributions of Figure 5.4(b). For $Pr = 1$, the mean temperature varies much as does the mean velocity, following the pattern of Table 4.3. For $Pr \gg 1$, the temperature rise occurs almost entirely in the viscous layer; hence the distribution of ϵ_h in the fully turbulent region is of little importance, the heat transfer being dominated by the eddy-diffusivity variation near the wall. Thus predictions for $Pr \gg 1$ are strongly influenced by the selection made from the various eddy-viscosity formulae (4.1.10) to (4.1.17).

For $Pr \ll 1$, the temperature rise is distributed over the cross-section. Thus for the smallest Prandtl numbers, the turbulent diffusivity will be unimportant, except for its effects on the velocity distribution, and hence on the variation of $\dot{q}_h$ across the flow. For somewhat larger Prandtl numbers, the growing importance of the eddy diffusivity causes a deviation from the model of pure molecular conduction.

The dominance of the molecular diffusivity when $Pr \ll 1$ complicates the prediction of transfer rates and the interpretation of experiments, particularly for laminar flows. In analysing parallel flows, we can usually adopt conservation laws of the form (3.3.5, 6), obtained by dropping the molecular-conduction term from the streamwise flux J_x given in the first of equations (3.3.2). But when the molecular diffusivity is large, this term can be significant compared to the mean-flow convection.

Table 5.2. Ranges of molecular Prandtl and Schmidt numbers

Class of fluid	Pr	Sc
Gases	0·65 to 1·1	(0·004) 0·1 to 4
'Common' liquids (water, liquid hydrogen, refrigerants, alcohols)	0·9 to 50	200 to 4000
'Viscous' liquids (oils, kerosene, liquid oxygen, glycerine, solutions, hydraulic fluids)	20 to 10^5	—
Liquid metals	0·003 to 0·03	—

Table 5.2 indicates the ranges of Prandtl and Schmidt numbers associated with several classes of fluids; the bracketed quantity is an atypical extreme value. We see that the gases and 'common' liquids have moderate Prandtl numbers; for heat-transfer calculations, they fall within the compass of a

developed Reynolds analogy; for mass-transfer calculations, only the gases do so. The 'viscous' liquids, typically oils, require results for $Pr \gg 1$, and the 'common' liquids require analogous results for $Sc \gg 1$. Only the liquid metals require the results for $Pr \ll 1$, although some mass transfers in gases fall into the domain $Sc \ll 1$. In summary, there are applications for transfer laws over the entire range $10^{-3} < Pr, Sc < 10^5$, with the possible exception of a small gap around $Pr, Sc = 0{\cdot}1$.

Analytical Models

So many attempts have been made to predict heat transfer in pipe flow, and so complicated is the algebra of the more advanced models, that it is not expedient to work through the whole history of the problem. We shall survey the analytical models which have been used, and then examine a few of the simpler and more fruitful examples.

(a) Prandtl and, independently, Taylor improved upon Reynolds' analysis by distinguishing a region of molecular diffusion near the wall (of the same thickness for velocity and temperature) and a fully turbulent outer flow where, in effect, Reynolds analogy held, with $Pr_t = 1$. Thus Prandtl obtained

$$St = \frac{\frac{1}{2}c_f}{1 + 5(Pr - 1)(\frac{1}{2}c_f)^{\frac{1}{2}}} \tag{5.3.1}$$

Although the assumptions are clearly unrealistic, this model illustrates an approach which is capable of considerable development, and provides a framework to which empirical results can be fitted. It will be considered in detail below.

(b) von Kármán advanced the discussion by distinguishing a buffer layer, specified by equation (4.1.11), in which both molecular and turbulent diffusion were active. He maintained the same boundaries for the several layers of the velocity and temperature variations, and took $Pr_t = 1$ in the outer turbulent flow, obtaining

$$St = \frac{\frac{1}{2}c_f}{1 + 5(\frac{1}{2}c_f)^{\frac{1}{2}}[Pr - 1 + \ln\{1 + 5(Pr - 1)/6\}]} \tag{5.3.2}$$

(c) Martinelli retained von Kármán's distribution of eddy viscosity in the viscous layer, but took more realistic account of the turbulent region, allowing $Pr_t \neq 1$, and taking the flux $\dot{q}_h$ to vary linearly, that is, in the same way as the shear stress.

(d) Lyon improved upon Martinelli's model by taking account of the non-linearity of the $\dot{q}_h$ distribution (as shown in Figure 5.1), but still retained von Kármán's eddy diffusivities. We shall consider Lyon's approach in more detail below, and see what it implies for the limiting case of $Pr \ll 1$.

(e) Rannie used his continuous diffusivity variation (4.1.13) and thus allowed the regions where molecular diffusion is important to be different for momentum and heat. However, he did not deal so carefully with the outer flow, taking $\mathrm{Pr}_t = 1$ and $\dot{q}_h = \text{constant}$. This analyis, like that of Martinelli, leads to formulae generally similar to, but more complicated than, von Kármán's (5.3.2).

(f) Deissler gathered together some of the loose ends of the preceding models, introducing a more realistic diffusivity variation (4.1.15) near the wall, and providing a consistent variation of $\dot{q}_h$ using equation (5.1.26). However, he still took $\epsilon_m = Ku_f y$ and $\mathrm{Pr}_t = 1$ through the entire fully turbulent region, and this reduces the accuracy of his calculations for $\mathrm{Pr} \ll 1$. We shall later consider the limit of $\mathrm{Pr} \gg 1$, for which an explicit formula can be obtained, in contrast to the general case, for which numerical integration is necessary.

(g) Rohsenow and Cohen calculated $\mathrm{Pr}_t = f(\mathrm{Pr})$ and introduced this into Martinelli's formulae to obtain more realistic results for $\mathrm{Pr} \ll 1$.

These are only a few of the many investigations of heat transfer in wall turbulence, but they represent the main line of development. Taken together, they give a fairly realistic picture of heat and mass transfer over the entire range defined by Table 5.2, except for the region around Pr, Sc = 0·1, for which there are few practical applications.

5.3.2 Transfer Formulae for a Smooth Wall

We now derive some simple results for the ranges of moderate, small and large Prandtl numbers. These introduce methods which have been developed into more complete analyses, and indicate the patterns to which empirical results can best be fitted in the three régimes. Rough walls will be considered in Section 5.4.1.

Moderate Prandtl Numbers: the Prandtl–Taylor Analysis

The fundamental idea, applied in more detailed models by von Kármán, Martinelli, and many others, is to add the thermal resistances of the several regions through which the heat passes. In general, the thermal resistance is $R_{12} = (T_1 - T_2)/\dot{q}_h$. For the two-layer model of Prandtl and Taylor, the total resistance is

$$R = R_m + R_t \tag{5.3.3}$$

the sum of the resistances of the regions of molecular and turbulent transport. Assuming that $\dot{q}_h = \dot{q}_w$ for both, which implies that the wall layer is a small fraction of the pipe radius or boundary-layer thickness, and that the analysis is more appropriate for high Reynolds numbers, we rewrite the resistance sum as

$$\frac{T_w - T_f}{\dot{q}_w} = \frac{T_w - T_s}{\dot{q}_w} + \frac{T_s - T_f}{\dot{q}_w} \tag{5.3.3}$$

with $T_s = T(y_h)$ the temperature at the edge of the 'sublayer' where molecular conduction of heat is concentrated.

We shall generalize the original model by allowing:

(1) $y_h \neq y_m$ for the boundaries of the 'sublayers' for heat and momentum, and

(2) $\mathrm{Pr}_t \neq 1$ in the turbulent region.

Hence the results will give a general idea of the effects of these factors on the transfer law. For the molecular layers, we have

$$\dot{q}_w = \frac{k(T_w - T_s)}{y_h} \quad \text{and} \quad \tau_w = \frac{\mu U_s}{y_m} \tag{5.3.4}$$

with $U_s = U(y_m)$. Then

$$R_m = \frac{T_w - T_s}{\dot{q}_w} = \mathrm{Pr}\frac{y_h}{y_m}\frac{U_s}{c_p \tau_w} \tag{5.3.4}$$

For the turbulent region, we apply the generalized Reynolds analogy (5.1.15), neglecting the fact that U_s and T_s are not evaluated at the same point:

$$\frac{\tau_w c_p(T_s - T_f)}{\dot{q}_w(U_f - U_s)} = \mathrm{Pr}_t \tag{5.3.5}$$

whence

$$R_t = \frac{T_s - T_f}{\dot{q}_w} = \mathrm{Pr}_t\frac{U_f - U_s}{c_p \tau_w} \tag{5.3.5}$$

Adding the resistances as in equations (5.3.3), we obtain

$$\frac{\tau_w c_p(T_w - T_f)}{\dot{q}_w U_f} = \frac{\frac{1}{2}c_f}{\mathrm{St}} = \mathrm{Pr}_t + \frac{U_s}{U_f}\left[\mathrm{Pr}\frac{y_h}{y_m} - \mathrm{Pr}_t\right] \tag{5.3.6}$$

The whole of the right-hand side may be thought of as an average Prandtl number, like that of equations (5.2.42).

Turning to the simple Prandtl–Taylor model, we take $\mathrm{Pr}_t = 1$ and $y_h = y_m$, and introduce $u_f/U_f = (\frac{1}{2}c_f)^{\frac{1}{2}}$, to obtain

$$\mathrm{St} = \frac{\frac{1}{2}c_f}{1 + (U_s/u_f)(\mathrm{Pr} - 1)(\frac{1}{2}c_f)^{\frac{1}{2}}} \tag{5.3.7}$$

The only parameter which can be adjusted to match the behaviour of real wall layers is $U_s/u_f = y_s/y_f = y_s^+$. The most plausible selection might seem to be $y_s^+ \simeq 11$, the matching point of equations (4.1.10). However, the range of application usually sought for laws of this form is $0{\cdot}7 < \mathrm{Pr} < 20$, accommodating many common liquids as well as gases. To weight the law towards $\mathrm{Pr} > 1$, for which $\kappa < \nu$ and $y_h < y_m$, a smaller value of y_s^+ is appropriate. Prandtl suggested

$$y_s^+ = 5{\cdot}6 \tag{5.3.8}$$

and this was later generalized by Hoffman to

$$y_s^+ = 7{\cdot}5\,\mathrm{Pr}^{-\frac{1}{6}} \tag{5.3.9}$$

The value $y_s^+ = 5$ is often adopted, as in equations (5.3.1, 2), with the unconvincing justification that this is the nominal sublayer boundary of equations (3.1.2) and (4.1.11).

We have already noted some unrealistic features of the Prandtl–Taylor model. For channel flows, there is yet another reason for the departure of its predictions from the pattern of experimental results. The latter are usually expressed in terms of the average velocity U_a and bulk temperature T_b, while the theoretical model relates the velocity and temperature at the same point. It is apparent from equations (5.1.7, 8) that $T_b > T_a$, so that

$$\mathrm{St}(T_b) < \mathrm{St}(T_a) \quad \text{and} \quad \mathrm{Nu}(T_b) < \mathrm{Nu}(T_a)$$

as in the laminar results (5.1.6). This difficulty does not arise for boundary layers, for which the free-stream values provide suitable references, nor for $\mathrm{Pr} \gg 1$, for which $T_a \simeq T_b$, as is evident from Figure 5.4.

The remarks of the preceding paragraph are consistent with the empirical heat-transfer law suggested for pipe flow by Petukhov and Popov:

$$\mathrm{St} = \frac{\dot{q}_w}{\rho U_a c_p (T_w - T_b)} = \frac{\frac{1}{2}c_f}{1{\cdot}07 + 12{\cdot}7(\mathrm{Pr}^{\frac{2}{3}} - 1)(\frac{1}{2}c_f)^{\frac{1}{2}}} \tag{5.3.10}$$

with $\frac{1}{2}c_f = \tau_w/\rho U_a^2$. This is accurate within a few per cent for $0{\cdot}7 < \mathrm{Pr} < 50$. Thus the Prandtl–Taylor model, though inaccurate in detail, proves helpful by providing a pattern to which empirical results can be fitted.

General Results for Pipe Flow

It is easy to see why the bulk temperature is commonly used to characterize the fluid passing along a channel. The heat passing into a basically steady, radially symmetric pipe flow between two stations x_1 and x_2 is related to the temperature distributions at these stations by

$$\begin{aligned} \dot{Q} &= 2\pi R \int_{x_1}^{x_2} \dot{q}_w(x)\,\mathrm{d}x \\ &= 2\pi \int_0^R \rho c_p U(T_2 - T_1)\,r\,\mathrm{d}r \\ &= \dot{m}\,c_p[T_b(x_2) - T_b(x_1)] \end{aligned} \tag{5.3.11}$$

Constant fluid properties have been assumed; $\dot{m} = \rho\dot{V} = \rho U_a A$ is the mass flow along the pipe, and

$$T_b(x) = \frac{2\pi}{\dot{V}} \int_0^R U(r)\,T(r,x)\,r\,\mathrm{d}r$$

is the bulk temperature. This is the temperature which is achieved if the flow through a section is thoroughly mixed; hence it is sometimes termed the *mixing-cup temperature*. The simple result

$$\Delta T_b = \frac{\dot{Q}}{c_p \dot{m}} \tag{5.3.11}$$

relates the mass flow and total heat-transfer rate.

These results apply for any flux distribution $\dot{q}_w(x)$. For the particular case of constant heat input and a linearly developing temperature field:

$$\frac{\partial T}{\partial x} = \frac{T_b(x_2) - T_b(x_1)}{x_2 - x_1} = \text{constant} \tag{5.3.12}$$

throughout the flow, and

$$\begin{aligned} \dot{q}_w &= \rho c_p U_a \frac{A}{C_w} \frac{T_{b_2} - T_{b_1}}{x_2 - x_1} \\ &= \rho c_p U_a (\tfrac{1}{4} d) \frac{\partial T}{\partial x} \end{aligned} \tag{5.3.12}$$

with C_w the wetted perimeter of the pipe. For this special case, the heat flow can be measured using the temperature change between any two points with the same radial position, but this is not true for other patterns of heat input, for example, that for T_w constant.

We now consider the related fields of velocity, diffusivity, temperature and heat flux which must be specified or calculated in an accurate analysis of heat transfer. It is on these results that the predictions of Reichardt, Lyon and Deissler, for example, are based. The calculation centres around the temperature difference:

$$T_w - T_b = \frac{\int_0^R rU(T_w - T)\,dr}{\int_0^R rU\,dr} \tag{5.3.13}$$

when use is made of the definition of the bulk temperature. Evidently, we must find the temperature and velocity variations. The former is obtained from

$$\dot{q}_h = -\rho(\kappa + \epsilon_h)\frac{\partial H}{\partial r}$$

as

$$c_p(T_w - T) = H_w - H = \int_r^R \frac{\dot{q}_h\,dr'}{\rho(\kappa + \epsilon_h)} \tag{5.3.14}$$

This integral involves the diffusivity variation, which is assumed to be known, and the flux variation, which can be calculated from

$$-\rho U \frac{\partial H}{\partial x} = \frac{1}{r}\frac{\partial(r\dot{q}_h)}{\partial r} \tag{5.3.15}$$

on neglecting streamwise molecular diffusion and the dissipation-diffusion contribution of equations (5.1.23). Thus

$$\frac{\dot{q}_h}{\rho} = \frac{1}{r}\int_0^r r' U \frac{\partial H}{\partial x} \mathrm{d}r' \tag{5.3.15}$$

Here the velocity distribution appears again, as in equation (5.3.13).

For the particular case of constant heat input, for which $\partial H/\partial x$ is uniform, the problem is now fully defined. Given the radial variations of ϵ_m and Pr_t (which define U and ϵ_h), we can, by a series of routine integrations, find:

(1) the radial variations of $\dot{q}_h$ and $T_w - T$, and

(2) an expression for $T_w - T_b$ in terms of the parameters defining the flow, that is, the heat-transfer law.

Small Prandtl Numbers: the Slug-flow Approximation

To illustrate the method set out above, we consider the simple case of constant heat input into 'slug flow' with constant diffusivity:

$$\frac{\partial H}{\partial x} = \text{constant}$$
$$U = U_c = \text{constant}$$
$$\kappa + \epsilon_h \simeq \kappa = \text{constant}$$

The results provide a fair approximation for turbulent flow (where U does not vary too much) of a fluid with $\mathrm{Pr} \ll 1$. From equations (5.3.15, 14, 13) we obtain

$$\begin{aligned} \dot{q}_h &= \tfrac{1}{2}\rho U_c r \frac{\partial H}{\partial x} \\ T_w - T &= \frac{1}{4}\frac{\rho U_c}{k}(R^2 - r^2)\frac{\partial H}{\partial x} \\ T_w - T_b &= \frac{1}{8}\frac{\rho U_c}{k} R^2 \frac{\partial H}{\partial x} \end{aligned} \tag{5.3.16}$$

Introducing the wall-flux equations (5.3.12), and setting $U_c = U_a$, we have

$$\mathrm{Nu} = \frac{\dot{q}_w d}{k(T_w - T_b)} = 8 \tag{5.3.16}$$

What deviations from this simple result will arise if more realistic velocity and diffusivity distributions are specified? Since the velocity profile depends on the Reynolds number, a more exact treatment would give a transfer law in which this parameter appeared. However, the results (5.1.6) show that $\mathrm{Nu} = 48/11 \simeq 4{\cdot}4$ for the extreme case of laminar flow with a parabolic velocity distribution. Hence the introduction of the turbulent velocity profile should reduce the Nusselt number only a little below 8; moreover, the effect of changes in Reynolds number should be small.

A more important role of the Reynolds number was noted in deriving equation (5.2.8), its effect on the ratio ϵ_h/κ. For the core, it is reasonable to expect that $\epsilon_h \propto u_f d$, as in Table 3.3. Then

$$\frac{\epsilon_h}{\kappa} \propto \frac{u_f}{U_a} \mathrm{Pe} \propto \mathrm{Re}^{-1/m} \mathrm{Pe} \tag{5.3.17}$$

with $m = 8$ to 11. This weak Reynolds-number dependence has often been neglected by taking $\mathrm{Nu} = f(\mathrm{Pe})$ simply, as in equation (5.2.8). With more accurate data available, it is now possible to improve on this. Whatever the exact form of the departures from the limiting result for $\epsilon_h/\kappa \ll 1$, their effect will be to increase the heat-transfer rate.

These expectations are borne out by the semi-empirical formulae which have been proposed:

(1) Rohsenow and Cohen have given

$$\mathrm{Nu} = 6{\cdot}7 + 0{\cdot}0041 \exp(41{\cdot}8\,\mathrm{Pr})\,\mathrm{Pe}^{0{\cdot}793} \tag{5.3.18}$$

(2) Dwyer has suggested

$$\mathrm{Nu} = 7{\cdot}0 + 0{\cdot}025\left[\mathrm{Pe} - 1{\cdot}82\,\mathrm{Re}_d\left(\frac{\epsilon_m}{\nu}\right)_{\max}^{-1{\cdot}4}\right]^{0{\cdot}8} \tag{5.3.19}$$

for $\mathrm{Pe} > 400$. The omission of the second term in the square brackets gives a result suggested earlier by Lyon, and widely recommended.

The considerable variability in heat-transfer results for $\mathrm{Pr} \ll 1$, and consequently in the formulae used for this régime, has a number of possible causes:

(1) oxide formation on the pipe wall,
(2) failure of some liquid metals to wet the wall, particularly in earlier experiments where this problem was not appreciated,
(3) failure to account for streamwise molecular conduction in the fluid, and
(4) departures from uniformity of heat flux along the wall.

The final point will be considered further in Section 5.3.3.

Large Prandtl Numbers: Deissler's Limiting Result

Deissler pointed out that the heat-transfer law for a fluid with a large Prandtl number can be found by using the fact that the temperature rise occurs very close to the wall, where $\dot{q}_h$ is essentially constant and, even more important, ϵ_h can be expressed simply. Using Deissler's specification of the wall layer (4.1.15), or Hama's (4.1.16) or van Driest's (4.1.17), we obtain

$$\frac{\epsilon_m}{\nu} \simeq n^4 y^{+4} \quad \text{for small } y^+ \tag{5.3.20}$$

For this calculation, only equation (5.3.14) is required; the constancy of $\dot{q}_h$ in the region of interest makes the results (5.3.15) unnecessary; the fact that $T \to T_b$ within a small distance of the wall makes the calculation (5.3.13) irrelevant. Hence

$$c_p(T_w - T_b) = \frac{\dot{q}_w y_f}{\rho} \int_0^{y_b^+} \frac{dy^+}{\kappa + (\nu/\mathrm{Pr}_t) n^4 y^{+4}}$$

with y_b^+ a point in the region where $T \simeq T_b$. Since the 'tail' of the integral, where $y^+ > y_b^+$, contributes little, we adopt the asymptotic value

$$c_p(T_w - T_b) \sim \frac{\dot{q}_w \mathrm{Pr}_t}{\rho u_f n^4} \int_0^{\infty} \frac{dy^+}{a^4 + y^{+4}} \tag{5.3.20}$$

The definite integral is of standard form, and has the value $\pi/(\sqrt{2}\, a)^3$. Hence

$$\mathrm{St} = \frac{\dot{q}_w}{\rho c_p U_a (T_w - T_b)} = \frac{2n}{\pi}\left(\frac{2u_f^2}{U_a^2}\right)^{\frac{1}{2}} (\mathrm{Pr}^3 \mathrm{Pr}_t)^{-\frac{1}{4}}$$

$$= \frac{2n}{\pi \mathrm{Pr}_t^{\frac{1}{4}}} c_f^{\frac{1}{2}} \mathrm{Pr}^{-\frac{3}{4}} \tag{5.3.21}$$

This result depends only weakly on Pr_t. Taking $\mathrm{Pr}_t = 0{\cdot}9$ and setting $n = 0{\cdot}124$, the value suggested by Deissler, we obtain

$$\mathrm{St} = 0{\cdot}077\, c_f^{\frac{1}{2}} \mathrm{Pr}^{-\frac{3}{4}} \tag{5.3.21}$$

Introducing the Blasius friction formula (4.2.22), for example, we have

$$\mathrm{St} = 0{\cdot}022\, \mathrm{Re}^{-\frac{1}{8}}\, \mathrm{Pr}^{-\frac{3}{4}} \quad \text{or} \quad \mathrm{Nu} = 0{\cdot}022\, \mathrm{Re}^{\frac{7}{8}}\, \mathrm{Pr}^{\frac{1}{4}} \tag{5.3.22}$$

These results are in good agreement with experiment for $\mathrm{Pr} > 200$. Since the index of the Prandtl number depends on the form of the variation $\epsilon_m \propto y^m$ adopted near the wall, this provides some support for the variation adopted, namely, $m = 4$.

Transfer laws of this general form are applied over a broad range of Prandtl numbers and for many kinds of boundary geometries. A representative result for turbulent flow in a smooth-walled pipe is

$$\mathrm{Nu} = 0{\cdot}023\, \mathrm{Re}_d^{0{\cdot}8}\, \mathrm{Pr}^m \tag{5.3.23}$$

with fluid properties evaluated at the bulk temperature, and

$m = 0{\cdot}4$ when the fluid is heated ($T_w > T_b$), and
$ = 0{\cdot}3$ when the fluid is cooled ($T_w < T_b$).

This result may be expected to be accurate within ten per cent for $0{\cdot}5 < \mathrm{Pr} < 120$ and $2000 < \mathrm{Re}_d < 10^7$. Some variations which have been suggested are:

(1) Dittus and Boelter took different constants for heating and cooling (0·0243 and 0·0265).

(2) Colburn took $m=\frac{1}{3}$, and evaluated the transport properties at the mean film temperature.

(3) Sieder and Tate took $m=\frac{1}{3}$, and evaluated the properties at the bulk temperature, save for the factor

$$\left(\frac{\mu_b}{\mu_w}\right)^{0.14} \tag{5.3.24}$$

introduced on the right to encompass fluids, such as oils, whose viscosity changes rapidly with temperature.

(4) Hoffman suggested slightly different constants: 0·024, and $m=0.37$ for heating.

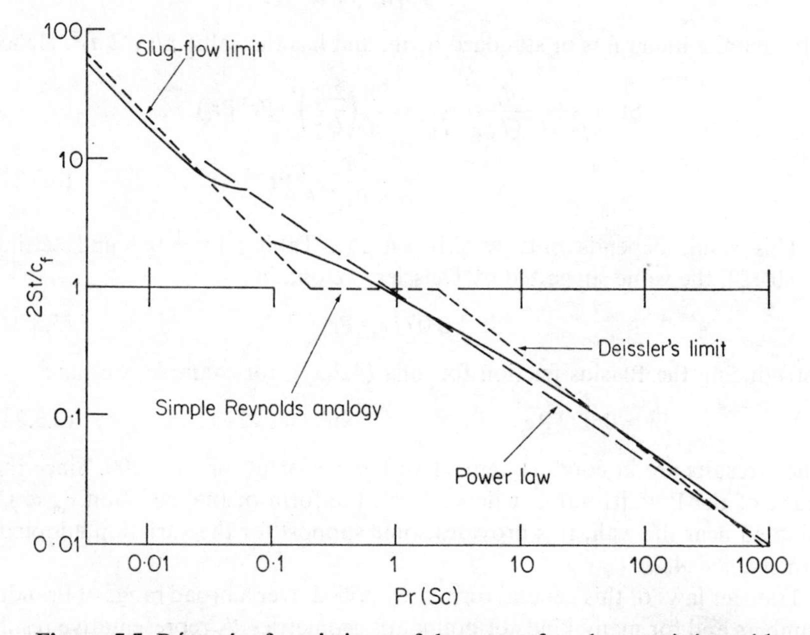

Figure 5.5 Résumé of variations of heat-transfer characteristics with Prandtl number, for pipe flow with $\mathrm{Re}_d=10^4$. The results also apply for mass transfer in the form $2\mathrm{St}_c/c_f=f(\mathrm{Sc})$, with Sc the molecular Schmidt number. Short-dashed lines: limiting results. Long-dashed line: simple power law. Solid lines: semi-empirical results of limited applicability

Figure 5.5 puts the varied heat-transfer formulae into perspective. It presents the parameter $2\mathrm{St}/c_f$ (whose reciprocal is sometimes termed the *Reynolds analogy factor*) as a function of the Prandtl or Schmidt number for a fixed value of the Reynolds number, $\mathrm{Re}_d=10^4$. The variation is largely defined by three limiting results shown by dashed lines: Reynolds simple analogy (3.4.30); the slug-flow result (5.3.16) for Pr, $\mathrm{Sc} \ll 1$; and Deissler's result (5.3.22) for Pr, Sc $\gg 1$. Two semi-empirical results—equation (5.3.18) for Pr, Sc $\ll 1$ and the more generally valid equation (5.3.10)—indicate how the measured values depart from the three limits. As was anticipated, there is some uncertainty around Pr, Sc $= 0{\cdot}1$, fortunately a region in which there is little practical interest. But even here it is possible to make a reasonably accurate estimate of the parameter $2\,\mathrm{St}/c_f$, say, within 20 per cent. It should be remembered that these results apply only to flow in an effectively smooth-walled pipe.

The simple power law (5.2.23) is also shown in Figure 5.5, with $m = 0{\cdot}35$. We see clearly how such a law is able to bridge the various régimes with a fair degree of accuracy. It is also clear why there is some variation in the values assigned to the constants, for an improvement in one régime inevitably results in less satisfactory predictions elsewhere. Nevertheless, a single power law does indicate the trends in the transfer characteristics (within, say 50 per cent) over the entire range $0{\cdot}001 < \mathrm{Pr}, \mathrm{Sc} < 1000$, for this Reynolds number at least.

It is instructive to consider the effect which an increase in Reynolds number has on this diagram. The right-hand portion is little altered; since

$$\frac{\mathrm{St}}{c_f} \propto \mathrm{Re}^{\frac{1}{8}} \quad \text{for Deissler's result (5.3.22)}$$

$$\propto \mathrm{Re}^{\frac{1}{20}} \quad \text{for the power law (5.3.23)}$$

these curves shift only a little to the right. However, the region where Pr $\ll 1$ changes markedly, since

$$\frac{\mathrm{St}}{c_f} \propto \mathrm{Re}^{-\frac{3}{4}} \quad \text{for the slug-flow result} \qquad (5.3.16)$$

Hence the gap around Pr $= 0{\cdot}1$ becomes more pronounced, while the laws for Pr $\ll 1$ deviate more widely from the slug-flow pattern, and become less certain.

5.3.3 Effects of the Temperature Differential

The primary role of the temperature differential between wall and fluid is taken into account in a heat-transfer law in which the heat flux is scaled with the local value of the differential. Here we consider three respects in which this procedure is inadequate. We shall discuss in turn:

(a) methods of accounting for the temperature-dependence of the fluid properties,

(b) the role of fluid buoyancy, and

(c) the effect of variations of heat flux and temperature differential along the pipe, in particular, the case of a constant wall temperature.

The first two become important only when the temperature differential is sufficiently large, but the third is broadly independent of the magnitude of the local differential. Each problem has an analogue in mass transfer, but only for the last can the results be transformed directly.

Temperature-dependence of the Fluid Properties

This problem can be approached in several ways:

(1) A 'loading factor' introduced into the transfer law, for example, the forms $(T_b/T_w)^m$ and $(\mu_b/\mu_w)^m$ used in equations (2.4.2) and (5.3.24).

(2) The evaluation of certain of the fluid properties at a *reference temperature* T_R. The most widely used value is the mean film temperature, the arithmetic average $T_m = \frac{1}{2}(T_b + T_w)$.

(3) A formal transfer calculation based on equations (5.3.13, 14, 15), with appropriate temperature-dependence of the fluid properties introduced into the integrals. The results of such calculations can be interpreted in terms of either loading factors or reference temperatures.

It is not to be expected that the corrections will have the same form for gases and liquids, nor for fluids of the same phase but widely differing Prandtl numbers. For gases, the temperature dependence may be approximated by

$$\frac{\mu}{\mu_w} = \frac{k}{k_w} = \left(\frac{T}{T_w}\right)^{0 \cdot 7} \quad \text{and} \quad \frac{\rho}{\rho_w} = \frac{T}{T_w} \tag{5.3.25}$$

the pressure being assumed constant across the wall layer. For liquids

$$\frac{\mu}{\mu_w} = \left(\frac{T}{T_w}\right)^{-n} \quad \text{with } n = 1 \text{ to } 4 \tag{5.3.26}$$

and

$$\rho = k = \text{constant}$$

are more realistic. Finally, the property whose variation is significant will depend upon the Prandtl number, and on the entity transferred, momentum or heat.

These points are illustrated in Figure 5.6, which gives some reference temperatures for liquids, predicted by Deissler, using the pattern (5.3.26). Note that the differences between heating and cooling are rather small, and also that the reference temperatures for friction and heat-transfer calculations are quite different.

Deissler also found that the reference temperature

$$T_R = 0 \cdot 4\, T_w + 0 \cdot 6\, T_b \tag{5.3.27}$$

was appropriate to gas flows, for both heat transfer and friction. However, in high-speed gas flows, the temperature distribution near the wall is affected by dissipation and diffusion, as was explained in Section 5.2.2. For turbulent boundary layers, Eckert has suggested

$$T_R = \tfrac{1}{2}(T_w + T_f) + 0{\cdot}22\, r \frac{U_f^2}{2c_p} \tag{5.3.28}$$

with T_f and U_f free-stream values, T_w the wall temperature, and r the recovery

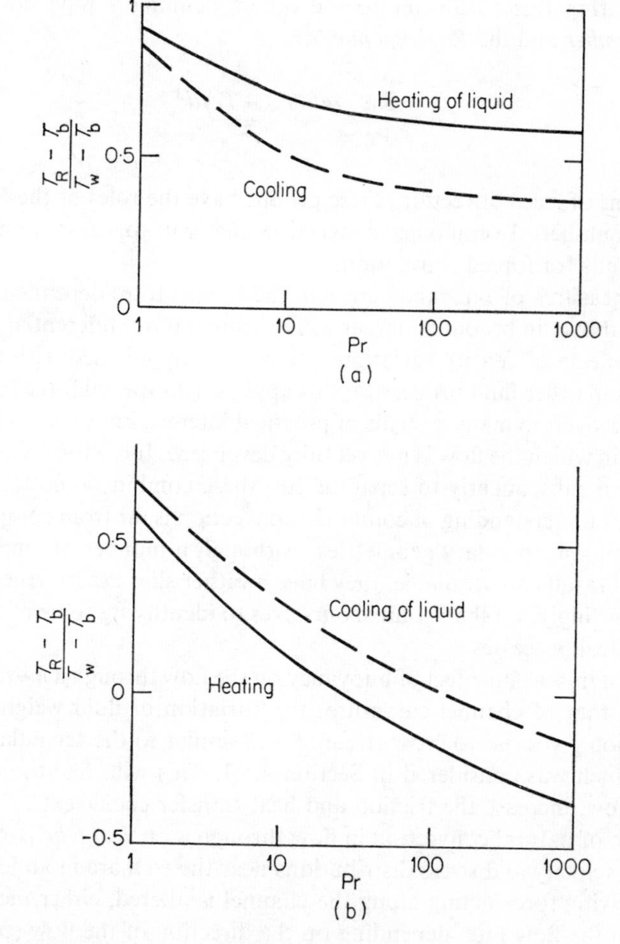

Figure 5.6 Reference temperatures for the evaluation of fluid properties in turbulent wall layers in flows of liquids, as calculated by Deissler. (a) Values for use in heat-transfer laws. (b) Values for use in friction laws

factor. The heat-transfer coefficients are defined in terms of $T_w - T_r$, as in equations (5.2.18).

Combined or Mixed Convection

The introduction of an additional parameter into the list (5.2.2) to represent fluid buoyancy will add another dimensionless group to the law (5.2.3). Buoyancy is usually represented by $g\beta$, the product of the acceleration of gravity and the coefficient of cubical expansion, $-(1/\rho)(\partial\rho/\partial T)_p$. Dimensionless groups relating this parameter to the others defining a pipe flow are the *Grashof number* and the *Rayleigh number*:

$$\mathrm{Gr} = \frac{\mathrm{Ra}}{\mathrm{Pr}} = \frac{g\beta(T_w - T_b)\,d^3}{\nu^2} \tag{5.3.29}$$

In problems of free convection, these groups have the roles of the Péclet and Reynolds numbers. In combined convection, they will appear in terms modifying the results for forced convection.

These measures of buoyancy are like the temperature-dependence factors discussed above, in becoming larger as the temperature differential increases. Thus the effects of density variations are commonly mingled with the effects of changes in other fluid properties; this applies with special force for laminar flows. Moreover, in many systems of practical interest, buoyancy is important in regions in which the flow is not yet fully developed. It is a formidable task to predict, and subsequently to separate out, these combined effects. In consequence, our understanding of combined convection is far from complete, even for the simplest boundary geometries. Although a number of empirical and theoretical results are available, they have a rather slim experimental foundation. Accordingly, we shall confine ourselves to identifying some of the underlying physical processes.

The most important effect of buoyancy on the flow through a *horizontal tube* is akin to that of channel curvature; the variation of fluid weight over the cross-section gives rise to cross-stream flows similar to the secondary flow of Type 2 which was considered in Section 4.5.1. This will, like the curvature-induced flow, increase the friction and heat-transfer coefficients.

The role of natural convection in flow through a *vertical tube* is more complex. The velocity and stress distributions near the wall are modified, and the overall driving force acting along the channel is altered, either increasing or decreasing the flow rate, depending on the direction of the flow (upwards or downwards) and of the heat transfer (to or from the fluid). When the forced convection is very weak, free convection patterns may develop, with motion both up and down the channel at different points in the cross-section.

In the empirical laws governing these situations, the magnitude of the buoyancy correction is usually measured by the quantity

$$\frac{\text{Gr}}{\text{Re}^m} \quad \text{with } m = 1 \text{ to } 2$$

Typically, the correction is negligible when this parameter is less than about 0·05.

Non-uniform Heat Flux

Although the case of uniform heat flux is that most easily dealt with theoretically, the assumption of constant wall temperature provides a better representation for many experiments and practical situations. In general, a heat-transfer coefficient based on the local temperature differential will be lower for a constant wall temperature than for constant flux, since the large differentials upstream leave the temperature near the wall higher than it would be for a constant differential. However, when the temperature change is concentrated near the wall ($\text{Pr} \gg 1$), the gradient at the wall reacts quickly to the changing conditions, and there is little difference between the results for T_{w} constant and $\dot{q}_{\text{w}}$ constant. When $\epsilon_{\text{h}}/\kappa \ll 1$ through much of the flow, that is, when Re, Pr and Pe are small, the lag of the temperature gradient will be greater, and the difference between the two transfer coefficients is more significant.

Table 5.3. Ratio of heat-transfer coefficients for constant wall temperature and constant wall flux, for turbulent pipe flow

Pe	10^6	10^5	10^4	10^3	10^2	10
Nu_T/Nu_q	0·96	0·95	0·90	0·83	0·75	0·73

Table 5.3 indicates the relationship between the transfer coefficients for uniform wall temperature and uniform heat flux, Nu_T and Nu_q, for a range of Péclet numbers. These results are based on predictions of Seban and Shimazaki over the ranges $\text{Re}_d = 10^4$ to 10^6 and $\text{Pr} = 0{\cdot}001$ to 1. Note that they fall plausibly between $\text{Nu}_T/\text{Nu}_q = 1$ and $5{\cdot}75/8 = 0{\cdot}72$, the latter applying to slug flow with constant thermal conductivity. These results suggest another source of error in interpreting heat-transfer measurements for $\text{Pr} \ll 1$.

For turbulent flow Table 5.3 indicates that, except for small Prandtl numbers, the difference in the transfer coefficients is virtually negligible, compared with other sources of uncertainty. This conclusion does not extend to parallel laminar flows, for which the ratio Nu_T/Nu_q is independent of the Reynolds number, since the velocity profile and diffusivity are fixed. Even for laminar flows, however, $\text{Nu}_T/\text{Nu}_q > 0{\cdot}8$ is usual.

5.4 MORE VARIED BOUNDARIES

5.4.1 Transfer at a Rough Surface

Although roughness has the same qualitative effect on heat and mass transfer as it has on momentum transfer—namely, an increase—our attitude to the change is quite different. While it is usual to seek the lowest possible friction, an increase in heat- and mass-transfer rates is often economically desirable. In consequence, we have now to consider not only *natural roughness* (for example, that arising spontaneously from vegetation, in an eroding stream, or during a manufacturing process), but *artificial roughness*, designed to maximize the transfer rate in some specific piece of equipment. Moreover, in fundamental studies of roughness it may be expedient to use a well-defined artificial surface, even when natural roughness is of primary interest.

It is possible to imagine situations in which both natural and artificial roughness occur, but these are uncommon in practice. In heat-transfer equipment the flow passages are usually small, in order that a large transfer area be available between wall and fluid; hence the Reynolds numbers are normally low enough for commercial finishes to be effectively smooth. However, aging may bring the surface to the effectively rough condition.

The protruding elements used to promote heat transfer fall into several classes:

(1) Fins lying parallel to the mean flow to provide an *extended surface*. This gives 'more of the same', offering a larger surface area (with a fairly uniform temperature, since the fin conductivity is much greater than that of the fluid), without altering the fundamental nature of the flow near the surface.

(2) *Separation inducers* lying normal to the mean flow. Although these do increase the contact area, their vital function is the stirring of the fluid near the wall; especially high transfer rates are achieved in the highly non-steady zones of reattachment downstream of the protrusions.

(3) *'Turbulence' inducers*, isolated elements establishing vortices and secondary flows which stir the fluid near the wall. These are used to inhibit boundary-layer separation on aircraft wings; small blades are set at an angle to the flow, so that strong tip vortices form. They are less useful in heat-transfer applications, since 'hot spots' are produced.

(4) Elements intermediate to the preceding classes, for example, spiralling fins in an annular flow passage.

Extended surfaces (1) can be treated, at least approximately, using the methods developed for channel flows and boundary layers. We shall direct our attention to the stirring devices (2, 3), which have something in common with natural roughness.

Although separation and turbulence inducers act to increase both friction and the transfer rate, different aspects of the motion in the roughness layer are responsible for the transfers of momentum and of a passive entity. The zone of reattachment may not contribute much to the drag, but it is of great importance for heat and mass transfer. The front and rear faces of protruding elements may contribute little to the transfer rate, while dominating friction generation. It must be recognized also that the degree of contact between wall and roughness elements can significantly influence the overall thermal resistance, while having no effect on the friction. We conclude that the laws governing the several transfers will often be significantly different.

In practice, the parallel between artificial stirring elements and 'natural' roughness is not as close as might be expected. For optimum performance, artificial elements must often be rather widely spaced, as we shall see later, and the flow in the roughness layer is quite different from that near closely packed sand grains. Moreover, artificial protrusions are usually identical and periodically spaced, for ease of design and manufacture.

Reference has just been made to the 'optimum' performance of a roughened surface. The significance of this term varies from one application to the next. Sometimes the pressure drop across a heat exchanger is not very important, but a compact installation is vital. Here the designer is free to adopt narrow, highly convoluted flow passages. More often, a balance must be sought between the power required to pump the fluid through the unit, the size of the plant, and the first cost. Then the ratio of the heat transferred to the pumping power

$$F = \frac{\dot{q}_w}{\tau_w U_a}$$

is an important figure of merit, and the performance of a roughened surface may be measured by the *effectiveness*

$$\eta = \frac{F_r}{F_s} = \frac{(\mathrm{St}/c_f)_r}{(\mathrm{St}/c_f)_s} = \frac{(\mathrm{Nu}/c_f)_r}{(\mathrm{Nu}/c_f)_s} \tag{5.4.1}$$

where the subscripts r and s denote rough and smooth-wall coefficients for the same Reynolds and Prandtl numbers.

Artificial roughness is often used in nuclear reactors, and here the economic balance is an especially complex one. The neutron-absorbing capacity of the material of the roughness elements plays a part in the neutron economy of the reactor core, and thus influences the degree of enrichment required for the nuclear fuel. Hence the assessment of roughness must extend beyond such mundane matters as the pressure drop across the heat-transfer channels.

Concentrated Roughness

We consider first roughness of the Types A and B defined in Section 4.3.2; this comprises most natural roughness and also closely spaced artificial elements, including the classical sand-grain pattern. The limiting cases of very large and very small Prandtl or Schmidt number will be examined first.

For Pr, Sc $\ll 1$, the transfer rate is nearly independent of Reynolds number and roughness; for example, Nu $= 8$ for slug flow, as for a smooth wall. Hence

$$\mathrm{Nu_r} \simeq \mathrm{Nu_s} \quad \text{and} \quad \eta = \frac{(c_\mathrm{f})_\mathrm{s}}{(c_\mathrm{f})_\mathrm{r}} < 1 \quad \text{for Pr, Sc} \ll 1 \tag{5.4.2}$$

Figure 4.3(a) indicates that values as low as $\eta = 0{\cdot}1$ are possible, for high Reynolds numbers and very rough surfaces.

For Pr, Sc $\gg 1$, the thermal resistance is concentrated in the roughness layer, and must be expected to depend on the roughness type. Clearly, we may normally expect that

$$\mathrm{Nu_r} > \mathrm{Nu_s} \quad \text{for Pr, Sc} \gg 1 \tag{5.4.3}$$

since the surface area is increased, and the fluid near the wall is stirred by the protrusions. It is not at all obvious that this improvement will overbalance the rise in friction; hence

$$\eta > 1 \quad \text{or} \quad \eta \simeq 1 \quad \text{for Pr, Sc} \gg 1 \tag{5.4.3}$$

is all that can be said without a detailed investigation.

We can discuss transfers for moderate Prandtl and Schmidt numbers by modifying the Prandtl–Taylor model. Retaining the results (5.3.5) for the resistance of the fully turbulent region, we generalize equation (5.3.6) to

$$\tfrac{1}{2}\frac{c_\mathrm{f}}{\mathrm{St}} = \mathrm{Pr_t}\left(1 - \frac{U_\mathrm{s}}{U_\mathrm{f}}\right) + \frac{\tau_\mathrm{w} c_p}{U_\mathrm{f}} R_\mathrm{m} \tag{5.4.4}$$

Here $R_\mathrm{m} = (T_\mathrm{w} - T_\mathrm{s})/\dot{q}_\mathrm{w}$ is the thermal resistance of the region where molecular conduction is important, and U_s and T_s are values at its outer boundary. We shall concentrate on the fully rough condition, for which the resistance R_m arises entirely from the roughness layer, there being no viscous region between it and the outer, fully turbulent flow. For the fully rough condition, $U_\mathrm{s}/U_\mathrm{f} \ll 1$; the term containing this ratio will be omitted from equation (5.4.4) henceforth.

Our task is to assign to the resistance R_m a value representing the energy-transferring activity within the roughness layer. This must depend, in detail, on the nature of the roughness, and it is unrealistic to seek a single result describing all roughness layers. However, the general nature of the effect of roughness can be discerned by considering a few simple models of the transfer. The simplest hypothesis is that the fluid is stagnant over a distance comparable to k_s, the sand-grain scale, so that the transfer depends only on the molecular conductivity of the fluid, k. Thus $R_\mathrm{m} = C_1 k_\mathrm{s}/k$, with C_1 a constant, and equation (5.4.4) gives

$$\tfrac{1}{2}\frac{c_\mathrm{f}}{\mathrm{St}} - \mathrm{Pr_t} \simeq C_1\left(\frac{\tau_\mathrm{w} c_p k_\mathrm{s}}{U_\mathrm{f} k}\right) = C_1(\tfrac{1}{2}c_\mathrm{f})^{\frac{1}{2}}\,\mathrm{Pr}\frac{k_\mathrm{s}}{y_\mathrm{f}} \tag{5.4.5}$$

Note that the factor k_s/y_f is an index of the Reynolds number, through

$$\frac{k_s}{y_f} = \frac{k_s\,U_f}{\nu}\frac{u_f}{U_f} = \frac{k_s}{d}\frac{U_f\,d}{\nu}(\tfrac{1}{2}c_f)^{\frac{1}{2}}$$

A somewhat more realistic model is sketched in Figure 5.7; the transfer is accomplished by a recirculating flow within the roughness layer, its velocity being of order u_f. Heat can enter the recirculating flow, to be carried to the outer flow, only by diffusion at the solid boundary. The distance the heat diffuses in the circulation time, $t \sim k_s/u_f$, is given by

$$y \sim (\kappa t)^{\frac{1}{2}} \sim \left(\frac{\kappa k_s}{u_f}\right)^{\frac{1}{2}}$$

(See Example 1.25.) The energy flux from unit breadth (normal to the plane

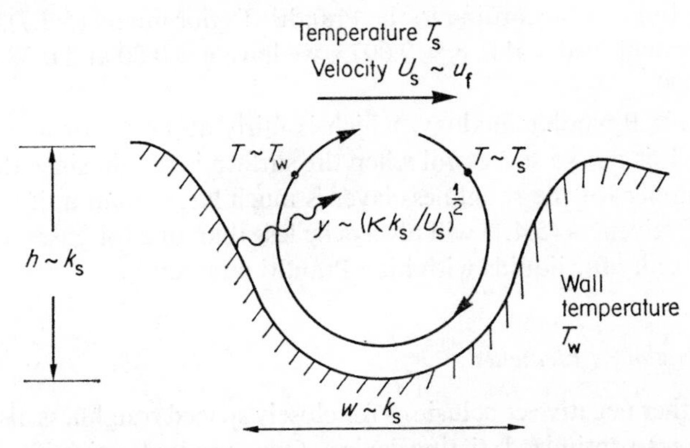

Figure 5.7 A simplified model of the heat-transfer processes in the roughness layer, in which energy diffused from the wall is carried to the outer flow by recirculating flows among the roughness elements

shown) of this cavity is of order $\rho c_p(T_w - T_s)\,y u_f$. Assuming that this is the major component of the heat flux, we have

$$\dot{q}_w \sim \rho c_p(T_w - T_s)\,y\frac{u_f}{k_s}$$

whence

$$R_m \sim \frac{k_s}{\rho c_p y u_f} = \frac{(\mathrm{Pr}\,k_s/y_f)^{\frac{1}{2}}}{\rho c_p u_f} \tag{5.4.6}$$

on introducing the diffusion length found above. The general result (5.4.4) now gives

$$\frac{1}{2}\frac{c_f}{\mathrm{St}} - \mathrm{Pr}_t \simeq C_2(\tfrac{1}{2}c_f)^{\frac{1}{2}}\,\mathrm{Pr}^{\frac{1}{2}}\left(\frac{k_s}{y_f}\right)^{\frac{1}{2}} \tag{5.4.7}$$

which may be compared with equation (5.4.5).

Other simple models lead to different powers of Pr and k_s/y_f, but the basic trends are those indicated by the two models considered. We may expect that the transfer in any one roughness layer is accomplished by a combination of such simple processes. Owen and Thomson analysed heat and mass-transfer data relating to two-dimensional roughness, pyramids, and sand indentations (not sand roughness itself, in view of the wall-resistance problem), and provided the estimate

$$\frac{1}{2}\frac{c_f}{\mathrm{St}} = 1 + 0{\cdot}52\,(\tfrac{1}{2}c_f)^{\frac{1}{2}}\,\mathrm{Pr}^{0\cdot 8}\left(\frac{k_s}{y_f}\right)^{0\cdot 45} \tag{5.4.8}$$

on the assumption that $\mathrm{Pr}_t = 1$. For the data considered, Pr, Sc $= 0{\cdot}7$ to 7 and $k_s/y_f > 100$.

Let us see what this result implies for the effectiveness (5.4.1) when Pr $= 1$, so that $(\mathrm{St}/\frac{1}{2}c_f)_s = 1$, according to the Prandtl–Taylor model (5.3.7). Assigning the typical rough-wall value $c_f = 0{\cdot}0075$, we have $\eta = 0{\cdot}80$ and $0{\cdot}58$ for $k_s/y_f =$ 100 and 1000.

The simple Reynolds analogy, which is fairly accurate for a smooth wall when Pr $= 1$, is not so successful when the surface is rough, since the effective Prandtl number for the roughness layer is much larger than unity. It appears that the effectiveness (5.4.1) will always be less than one for gases, although it can exceed unity for liquids with high Prandtl numbers.

Isolated Roughness Elements

These rather negative conclusions for closely spaced roughness also apply in general terms to optimized stirring devices. One cannot often justify roughening on the grounds that the effectiveness is improved, but only because the unit can be more compact, and thus cheaper.

Figure 5.8 gives an example of the information required to optimize the transfer characteristics for a particular kind of roughening element. These data were reported by Wilkie and relate to the utility of transverse ribs on the steel fuel pins of a nuclear reactor. In this application, friction is of less importance than an increase in heat-transfer rate, which must be achieved with the minimum addition of material to the pins. For this reason, only relatively small ribs have been considered.

It is apparent that the transfer coefficient can be increased by a factor of three or more by a suitable choice of transverse ribs. However, in these gas flows, the increase in friction is roughly twice as great, a situation much like that for 'natural' roughness. For this application, it is appropriate to select relatively widely spaced ribs, with $h/L = w/L \simeq 10$, say.

The differences in the augmentation factors of Figures 5.8(a) and (b) demonstrate that different mechanisms are indeed responsible for the changes in friction and in heat transfer. Figure 5.8(c) shows the patterns of local flux

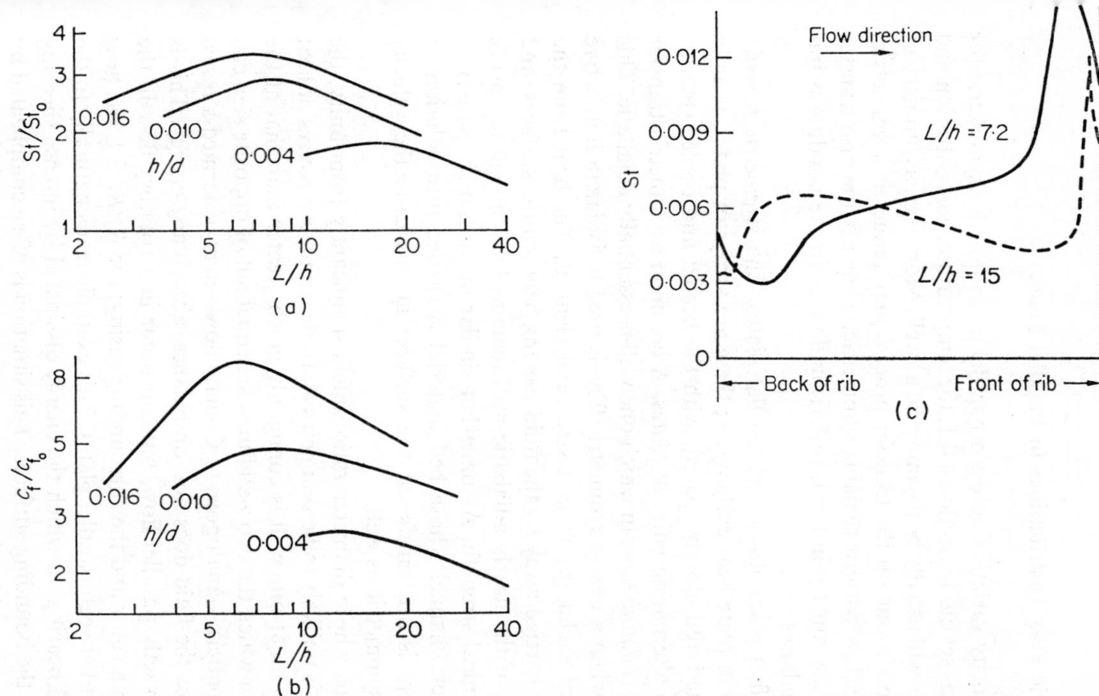

Figure 5.8 Transfer characteristics of surfaces artificially roughened by transverse ribs, as measured by Wilkie in annulus flows of air. Symbols as in Figure 4.5: rib pitch L, height h, width w (maintained constant in these tests at 0·025 in). The Reynolds number is 10^5, calculated using a hydraulic diameter d based on the area between the rough inner cylinder and the cylindrical surface on which the mean velocity is highest. (a) Ratio of Stanton number to that for a smooth surface, $St_0 = 0{\cdot}0029$. (b) Ratio of friction coefficient to that for a smooth surface, $c_{f_0} = 0{\cdot}0054$. (c) Variation of local Stanton number between ribs for two roughness geometries, inferred from study of the transfer of naphthalene from the surface. $Re_d \simeq 2 \times 10^5$; $St_0 = 0{\cdot}0025$. Mean values: $St = 0{\cdot}0068$ and $0{\cdot}0057$ for $L/h = 7{\cdot}2$ and 15, respectively

between the ribs for two spacing ratios; an analogous mass transfer was examined to facilitate the experiments. Comparison of the markedly different variations shows how important are the details of the flow around the elements in determining the overall transfer rate.

5.4.2 Asymmetric Flux Distributions in Parallel Flows

When the velocity variation across a parallel mean flow is asymmetric for any of the reasons set out in Section 4.4.1, the temperature, concentration and flux distributions will usually be asymmetric as well. Moreover, asymmetry of the boundary conditions on the transfer process may render it asymmetric, even though the flow pattern remains symmetric. Thus asymmetric transfer can result from asymmetry in the transferring flow or in the conditions imposed at its boundaries.

We consider first plane flows between flat plates. The *symmetric transfer* pattern is found in plate heat exchangers; these are arrays of plates, usually heavily corrugated and closely spaced, with the heated and cooled streams passing between alternating pairs of plates. A common asymmetric transfer is that within an *isolating flow*, in which one wall is essentially adiabatic. This situation arises when a nearly symmetric flow is used to isolate a hot, active element of a heat exchanger from a structural element. The heat from the active element is carried away by the fluid passing between the surfaces, and the temperature of the nearly adiabatic wall remains low enough to ensure adequate mechanical strength. A somewhat similar mass-transfer pattern is found in an open channel whose bed material is carried into solution or suspension. There is no transfer at the surface; in this case, the velocity distribution is asymmetric as well.

For an annulus whose diameter ratio differs significantly from unity, the flux distribution is unlikely to be symmetrical. In the annular passages of heat exchangers, the flux at one wall is commonly much greater than that at the other. The case in which the two wall fluxes are equal but of opposite sign can also arise. This pattern—analogous to Couette flow—may be termed a *transmitting flow*, since the fluid does not, on balance, absorb any energy. This is unlikely in an exactly parallel flow, but can occur in a rotatory flow in the annulus between a rotor and the surrounding casing; for $R_2/R_1 \simeq 1$, the flow is essentially that between parallel plates. This example leads naturally to the flow in a journal bearing, in which the energy dissipated by intense shearing is transmitted to the bounding surfaces. This situation is often complicated by wall deformations caused by the large temperature gradients.

From these examples, it will be clear that there are many technically important situations in which asymmetric transfer patterns occur. A comprehensive investigation of the varied boundary conditions on the flow and on the transfer processes would include the effects of: roughness; relative velocity of the walls; the diameter ratio, for an annulus; and differing wall temperatures and fluxes. Although accurate predictions may require specific experiments, the semi-empirical models that have been developed for pipe flows can often be adapted to parallel flows subject to other boundary conditions. To see how this is done, we shall discuss one simple class of transfer problems.

We consider symmetrical plane flow between fixed walls, allowing the ratio of the wall fluxes, $\dot{q}_1/\dot{q}_2$ (both positive when into the flow), to take on values

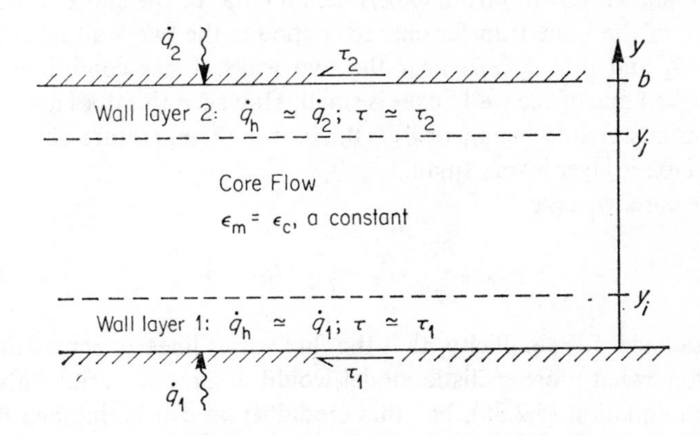

Figure 5.9 Model of turbulent flow in a broad flat channel

other than unity, so that the analysis includes the *isolation problem* ($\dot{q}_1$ or $\dot{q}_2 = 0$) and the *transmission problem* ($\dot{q}_1 = -\dot{q}_2$), as well as *symmetric transfer* ($\dot{q}_1 = \dot{q}_2$). As is indicated in Figure 5.9, we divide the flow into three regions: two wall layers and a core flow, meeting at y_i and $y_j (= b - y_i$, for this symmetrical flow). In the core we take the eddy viscosity to be constant, $\epsilon_m = \epsilon_c$, as in Model III of Section 4.4.2. This provides a sensible variation of diffusivity across the channel, and allows heat transfer from wall to wall to be included in the discussion.

It will be assumed that the major part of the temperature variation occurs in the wall layers; this rules out application of the results for $\mathrm{Pr} \ll 1$. We shall not specify a particular model for the wall layer, but define its role through

$$\frac{\tau_w c_p (T_1 - T_i)}{\dot{q}_1 U_i} = \frac{\tau_w c_p (T_2 - T_j)}{\dot{q}_2 U_i} = \left(\frac{1}{2}\frac{c_f}{\mathrm{St}}\right)_w = f\left(c_f, \mathrm{Pr}, \frac{k_s}{b}\right) \tag{5.4.9}$$

Here

$$(c_f)_w = \frac{\tau_w}{\frac{1}{2}\rho U_i^2}$$

and (5.4.10)

$$\mathrm{St}_w = \frac{\dot{q}_1}{\rho U_i c_p (T_1 - T_i)} = \frac{\dot{q}_2}{\rho U_i c_p (T_2 - T_j)}$$

characterize the wall layers, $U_i = U_j$ being the velocity at the junctions between wall layers and core. The form of the wall-region transfer laws (5.4.9) can be predicted using one of the analyses discussed in Section 5.3.2, or can be deduced (as will be shown shortly) from experimental data for the entire flow.

The use of the same transfer characteristic for the two wall layers implies that $\dot{q}_h \simeq \dot{q}_1$ and that $\dot{q}_h \simeq \dot{q}_2$ near the two walls. These conditions are not satisfied when one of the wall fluxes is small. However, this does not appear to lead to serious error when $\dot{q}_1$ or $\dot{q}_2 \simeq 0$, since the temperature change in the nearly adiabatic layer is very small.

For the core, we take

$$-\rho c_p \epsilon_h \frac{\partial T}{\partial y} = \dot{q}_h = \dot{q}_1 - (\dot{q}_1 + \dot{q}_2)\frac{y}{b} \tag{5.4.11}$$

having assumed, for simplicity, that the flux varies linearly across the entire flow. A somewhat more realistic model would assume linearity only in the core, as in equation (5.2.36), but this modification can be lumped with the turbulent Prandtl number, as is evident in equation (5.2.38). The flux variation (5.4.11) does not assume $\dot{q}_h/\tau =$ constant across the flow: $\tau = 0$ for $y/b = \frac{1}{2}$, while the minimum (or maximum) temperature T_m is achieved for

$$\frac{y_m}{b} = \frac{\dot{q}_1}{\dot{q}_1 + \dot{q}_2} \tag{5.4.12}$$

Integrating equation (5.4.11), with ϵ_h assumed constant, we obtain

$$T - T_m = \frac{\dot{q}_1 + \dot{q}_2}{2b\rho c_p \epsilon_h}(y - y_m)^2 \tag{5.4.13}$$

for the temperature distribution in the core, provided that $\dot{q}_1 + \dot{q}_2 \neq 0$.

Let us apply these results first to *symmetrical transfer*, for which $\dot{q}_1 = \dot{q}_2 = \dot{q}_w$ and $y_m = \frac{1}{2}b$. The temperature rise from wall to centre-line is found from equations (5.4.9, 13):

$$T_w - T_m = \frac{\dot{q}_w U_i}{\tau_w c_p}\left(\frac{1}{2}\frac{c_f}{\mathrm{St}}\right)_w + \frac{\dot{q}_w}{\rho b c_p \epsilon_h}(y_i - \tfrac{1}{2}b)^2 \tag{5.4.14}$$

On introducing $\mathrm{Pr}_c = \epsilon_c/\epsilon_h$ and $R_f = u_f b/\epsilon_c$, as in equations (3.3.22), we obtain

$$\frac{\tau_w c_p (T_w - T_m)}{\dot{q}_w U_i} = \left(\frac{1}{2}\frac{c_f}{\mathrm{St}}\right)_w + \mathrm{Pr}_c R_f \frac{u_f}{U_i}\left(\frac{y_i}{b} - \frac{1}{2}\right)^2 \tag{5.4.15}$$

This can be used to separate out the wall-layer contribution $(\frac{1}{2}c_f/\mathrm{St})_w$ from an empirical overall law $\frac{1}{2}c_f/\mathrm{St}$. For high Reynolds numbers and a smooth wall, the two will not be very different.

Consider next the case of an *isolating flow*, with $\dot{q}_2=0$, $\dot{q}_1=\dot{q}_w$ and $y_m=b$. We assume that the core extends right to the wall $y=b$, neglecting the 'zero-flux' layer near it, and the recovery effects discussed in Section 5.2.2. The temperature difference between the walls is

$$T_1-T_2=T_w-T_m=\frac{\dot{q}_w U_i}{\tau_w c_p}\left(\frac{1}{2}\frac{c_f}{\mathrm{St}}\right)_w+\frac{\dot{q}_w}{2b\rho c_p \epsilon_h}(y_i-b)^2$$

whence

$$\frac{\tau_w c_p(T_w-T_m)}{\dot{q}_w U_i}=\left(\frac{1}{2}\frac{c_f}{\mathrm{St}}\right)_w+\frac{1}{2}\mathrm{Pr}_c R_f\frac{u_f}{U_i}\left(\frac{y_i}{b}-1\right)^2 \tag{5.4.16}$$

Two ways of using this result are:

(1) to assess the effectiveness of an isolating flow, by predicting the temperature drop for a given heat flux and flow rate, or by predicting the coolant flow required for a specified temperature drop; and

(2) to predict heat-transfer for symmetric heating ($\dot{q}_1=\dot{q}_2$) from measurements obtained when only one wall is heated; the latter situation is easier to achieve experimentally.

The accuracy of the applications of the results (5.4.15, 16) depends on the magnitude of the wall-layer contribution $(\frac{1}{2}c_f/\mathrm{St})_w$ relative to the core element. If fluids with $\mathrm{Pr} \ll 1$ are excluded, $\mathrm{Pr}_c=1$ or less. From Table 3.3, we obtain the estimate $R_f=25$, whence $y_i/b=A/R_f=0{\cdot}1$; compare with equations (4.2.1). For high Reynolds numbers, $u_f/U_i \simeq (\frac{1}{2}c_f)^{\frac{1}{2}}=0{\cdot}1$ to $0{\cdot}03$. Hence the core contributions can be estimated as:

$$\begin{aligned}&0{\cdot}4 \text{ to } 0{\cdot}12 \text{ for equation (5.4.15), with } \dot{q}_1=\dot{q}_2\\&1{\cdot}0 \text{ to } 0{\cdot}3 \text{ for equation (5.4.16), with } \dot{q}_2=0\end{aligned} \tag{5.4.17}$$

The higher values correspond to very rough walls and to low Reynolds numbers.

Noting in Figure 5.5 that $\frac{1}{2}c_f/\mathrm{St} \simeq 1$ for $\mathrm{Pr} \simeq 1$, we conclude that the difference in the transfer coefficients (5.4.15, 16) is of order 20 per cent, for gases flowing between smooth walls. As the Prandtl number increases, the difference becomes smaller, since $\frac{1}{2}c_f/\mathrm{St}>1$. As the friction coefficient increases, the difference becomes somewhat larger. The relationship between the two heat-transfer characteristics also depends on the temperature chosen to represent the fluid, and will be modified if the bulk temperature is used in one or both coefficients.

Finally, we consider heat transfer across the flow, a generalization of the *transmission problem*. For arbitrary fluxes at the two walls, we find

$$T_1-T_2=\frac{U_i}{\tau_w c_p}\left(\frac{1}{2}\frac{c_f}{\mathrm{St}}\right)_w(\dot{q}_1-\dot{q}_2)+\frac{\dot{q}_2-\dot{q}_1}{2\rho c_p \epsilon_h}(y_i-y_j)$$

This leads to the cross-flow transfer law

$$\frac{\tau_w c_p(T_1 - T_2)}{U_i(\dot{q}_1 - \dot{q}_2)} = \left(\frac{1}{2}\frac{c_f}{\mathrm{St}}\right)_w + \mathrm{Pr}_c R_f \frac{u_f}{U_i}\frac{y_j - y_i}{2b} \tag{5.4.18}$$

This is a generalization of equation (5.4.16), and the core term takes on the second set of values given in (5.4.17). The results (5.4.16, 18) are slightly different because the former extends the core distribution to the wall in the limiting case of $\dot{q}_2 = 0$.

5.4.3 Non-circular and Irregular Channels

The problems of predicting transfer rates in non-circular and irregular channels are generally similar to those which arise in calculating friction and dissipation. Although not as much empirical information is available, this deficiency is partially offset by the analogies between transfer processes and friction.

Consider first a transfer between the wall of a channel with a *non-circular section* and the fluid passing through it. It was found that the introduction of the hydraulic diameter (or the equivalent radius; see Table 4.5) into round-pipe formulae gives reasonable predictions of friction, so long as the stress is fairly uniform over the wetted perimeter. The use of this procedure in transfer calculations adds another restriction: the major changes of mean temperature (or concentration) must occur, like those of mean velocity, in a wall layer which is uniform over much of the wetted perimeter. This is not the case for fluids with Pr, Sc $\ll 1$, where the transfer is more like that in laminar flow, with an additional 'shape effect' not accounted for by the equivalent diameter.

In a *curved channel* the transfer rates are augmented by secondary flows, as is friction; see equations (4.5.1, 2). For a fully developed turbulent flow of air in a helical pipe, the formula

$$\frac{\dot{q}_w}{(\dot{q}_w)_0} = 1 + 3{\cdot}5\frac{d}{D} \tag{5.4.19}$$

with d and D the pipe and helix diameters, was obtained by Jeschke, but the dependence on Reynolds and Prandtl numbers has not been investigated systematically. As was pointed out in Section 5.3.3, a secondary convection pattern is established in a horizontal channel by the density differences associated with heat transfer. This too acts to increase the heat transfer, and presumably the friction, but full experimental information is lacking.

The variation of heat transfer in the *developing flow* near the entrance to a channel is often estimated, following Nusselt, by

$$\frac{\dot{q}_w}{(\dot{q}_w)_0} = \left(\frac{d}{x}\right)^{0{\cdot}054} \quad \text{for} \quad 10 < \frac{x}{d} < 400 \tag{5.4.20}$$

However, the rate of approach to the equilibrium value $(\dot{q}_w)_0$ (or N_0 for mass transfer) is strongly dependent on the Prandtl (or Schmidt) number. Since the part of the flow nearest the wall has the smallest time scales, it moves quickly to its ultimate form. We conclude that the equilibrium value is adopted most rapidly for Pr, Sc $\gg 1$. As a rough guide, we may take

$$\dot{q}_w \simeq (\dot{q}_w)_0 \quad \text{for} \quad \frac{x}{d} > 20 \quad \text{when Pr} > 1 \text{ or } \simeq 1 \tag{5.4.21}$$

For liquid metals, we may expect that

$$\dot{q}_w \simeq (\dot{q}_w)_0 \quad \text{for} \quad \frac{x}{d} > 60 \quad \text{when Pr} \ll 1 \tag{5.4.21}$$

When considering the extra dissipation near *abrupt discontinuities* in a channel, we were able to give systematic results (Table 4.7) for the losses associated with common irregularities. Less attention has been given to the transfer rates at abrupt discontinuities. This is an anomalous situation, since the need for the information is more stringent where transfer processes are concerned. To calculate the effective friction in a duct—the net effect of wall friction and specific losses—we require only the total dissipation in the regions of non-uniform flow. While analogous results are useful in predicting overall heat- and mass-transfer rates, local variations are also important, for reasons that are set out below.

An area where the heat transfer is low will become a 'hot spot'; in extreme cases, local boiling or even melting or burning may occur. These processes can lead to further reduction in the local heat-transfer rate. Spatially rapid variations in heat flux can produce damaging thermal stresses in the wall material, while rapid fluctuations in time can lead to fatigue of the material, through varying thermal stress. Figure 5.8(c) illustrates the variability of the heat flux near a boundary discontinuity, and its sensitivity to the detailed geometry.

Local variations in mass-transfer rates correspond to non-uniform erosion and deposition, most pronounced when the boundary is composed of discrete particles which can be moved by the fluid. Thus, near an obstruction in a stream with an erodible bed—a bridge pier, for example—the bed may be greatly distorted. Similar phenomena can be observed in drifting snow. When the surface is significantly deformed by erosion or deposition, the flows which provide the transfers are modified in turn. Thus in a 'uniform' channel, it is possible for the stream and the erodible bed to settle into a pattern of continuous development, with regular waves of the bed material progressing slowly along the stream.

At present our ability to predict the complex flow patterns and transfer processes described in the last few paragraphs is extremely limited. In critical cases, it is necessary to undertake specific tests, or to adopt pessimistic assumptions as a basis for design.

5.5 FLOW NEAR A PERMEABLE BOUNDARY

5.5.1 Applications

Hitherto we have discussed transfers between impermeable walls and the fluid flowing along them. In most cases of turbulent flow, the process is dominated by the wall layer, where the mean flow is parallel to the wall—exactly so in fully developed channel flow, nearly so in a developing boundary layer or wall jet. Such unity as exists among the various wall flows depends on this common feature, as do the analogies linking the transfers of momentum, mass and heat. We now seek to generalize the analysis of flow near a wall, still requiring that the tangential component of the mean velocity drop to zero on the surface, but allowing the wall to be permeable, so that there is a finite normal component of velocity in the wall layer. Some situations in which flow towards or away from the surface can arise are noted below.

Large Mass-transfer Rates

The close parallels which we have supposed to exist between heat and mass transfers require that neither process alters the mean flow and turbulence of the convecting fluid; in many circumstances this condition is satisfied to an adequate approximation. We noted earlier that large ranges of temperature or concentration may lead to significant changes in fluid properties, and hence in the transferring flows. A more profound change in these flows occurs when there is a net flux of material at the flow boundary. This can occur when the fluid has a single component, for example, when steam condenses onto a wall, or is formed by evaporation there. It can also occur in binary fluids, as when water condenses from air flowing over a surface, or evaporates into an air stream. Such processes often involve large concentration gradients; hence changes in fluid properties may be important as well.

Care is required in defining the mass-transfer rate and coefficient for large transfer rates, since the flux at the wall may be quite different from the normal flux relative to the moving fluid. This effect is particularly important in laminar flows and in the parts of turbulent flows where molecular conduction is important, since even a small lateral convection is comparable to the diffusive flux from molecular activity. Considerable progress has been made in analysing these laminar flows: results are available—including the 'film' and 'penetration' theories—to correct mass-transfer coefficients for the convection induced by the transfer itself; and predictions have been made of the effects of blowing and suction on recovery factors, friction coefficients and transfer rates.

In the situations described above, changes do occur in the stress distribution within the flow and in its heat-transfer characteristics. But in most cases these are incidental to the mass transfer. For the following applications, our point of view is reversed: mass transfer is imposed in order to modify the stress or heat-transmitting capacity of the flow near the wall.

Mass-transfer Cooling

In situations where the wall temperature would otherwise rise to an unacceptable level—combustion chambers, exhaust nozzles and turbine blades—the 'heat blocking' effect of a mass transfer away from the surface ($V_w > 0$) may be put to use.

The term *transpiration cooling* (or sometimes sweat cooling) is used when the outwards flow is uniformly distributed over the surface; this can be achieved by injection through a porous material, such as a sintered metal. When a liquid is transpired into a gas, the term evaporative transpiration cooling is used. The injected fluid then alters the wall temperature in three ways: by absorbing heat while passing through the wall, by accepting thermal energy during evaporation, and by establishing an outwards convection within the gas.

Film cooling is similar in essence, but injection takes place at discrete orifices or slots. This technique is particularly appropriate for local 'hot spots'. When a liquid is used, the fluid may spread over the wall and evaporate as would a transpired liquid. The action of an energetic gas flow may produce waves on the liquid film, or may cause it to disintegrate prior to evaporation.

For the most extreme conditions encountered in rocket nozzles and on re-entry vehicles, *ablation cooling* is utilized. The surface is made of an ablative material (for example, quartz, glass or Teflon) which is removed by melting, evaporation or sublimation. This obviates the necessity of supplying a cooling fluid.

Boundary-layer Control

Suction ($V_w < 0$) through slots or distributed pores can be used to prevent separation in diffusers or on wings, by removing the least energetic part of the boundary layer. For diffusers, this has the effect of reducing losses and unsteadiness of the flow; for lifting surfaces, it leads to increased lift and reduced pressure drag. By delaying transition to turbulence, suction may also reduce the friction drag: although the boundary layer is thinned, the increase in laminar friction is less than that which would follow transition. It should be noted, in connection with mass-transfer cooling, that an increase in the injection rate which caused transition could have the effect of increasing the heat transfer to the surface.

Fluid injection is also used for boundary-layer control. Blowing along the surface re-energizes the boundary layer by introducing a wall jet in the least active region. The wall jet acts on the outer fluid somewhat in the manner of suction, since fluid is entrained into it. This technique has been used most often in the field of aeronautics, but can also be applied to the rudder of a ship. This mode of injection differs fundamentally from those considered above: here it is the momentum of the injected fluid which modifies the outer flow; there it was the volumetric injection rate which was significant.

Looking back over these applications, we note that flows towards and away from the surface are both of practical importance, though the former are somewhat more varied. We note also that the definition of the 'surface' at which transfer occurs is somewhat arbitrary. Should just-condensed water be treated as part of the flow, or as part of the wall? How do the pore flows merge with the main body of the flow? We conclude that, while the outer wall layers for blowing and suction form a continuous spectrum encompassing the several classes of applications, the innermost regions will vary from one case to the next. In the following discussion, we shall adopt the simplest possible description of the flow at the wall.

5.5.2 Transfers in the Generalized Wall Layer

Mass transfer, transpiration cooling and boundary-layer control all provide examples of wall layers that are essentially uniform in the streamwise direction, the velocity normal to the wall being specified as $V_w(y)$. This generalization of the parallel-flow wall layer may be termed a *streamwise-uniform layer*. In order to investigate easily the major effects of the flow through the wall, we further restrict the discussion by taking the fluid properties to be uniform through the layer. This assumption is fairly realistic for a boundary layer subjected to suction, but is less satisfactory for large mass transfers and for transpiration cooling.

A general result governing the transfer of any quantity across a streamwise-uniform wall layer can be obtained from the conservation statement (3.3.1), by retaining only the gradient normal to the wall:

$$\frac{dJ_y}{dy} = \dot{S}_s \tag{5.5.1}$$

The rejected term expresses the change in the flux in the direction of the wall; its form is indicated by equations (3.3.2):

$$\frac{\partial J_x}{\partial x} = -\rho K \frac{\partial^2 S}{\partial x^2} + \rho \frac{\partial(\overline{su})}{\partial x} + \rho U \frac{\partial S}{\partial x} \tag{5.5.2}$$

The diffusion terms, the first two on the right, are usually negligible near a wall. The convection term, the last on the right, increases as U rises away from the wall (unless $\partial S/\partial x \equiv 0$), while the gradient retained in equation (5.5.1) may decrease markedly. Hence for a developing distribution of the property S (that is, for $\partial S/\partial x \neq 0$), the simple result (5.5.1) holds only in a region sufficiently close to the wall. This is in fact the wall layer, for the preceding discussion sets out the arguments for the existence of a layer broadly independent of conditions in the outer flow.

As was pointed out in Section 3.3.1, the source term $\dot{S}_s$ may represent: the pressure gradient, in the momentum equation; dissipation, in the mechanical or thermal energy equations; or a chemical reaction, in a mass balance. For simplicity, we shall neglect this term in the following analysis, remembering that the contributions it represents are undoubtedly important in some situations; namely, in a boundary layer near separation, in high-speed flow near an insulated wall, and in a flow where combustion takes place.

Taking the term $\dot{S}_s$ to be negligible in the generalized wall layer, we have simply

$$J_y(y) = J_y(0) \tag{5.5.3}$$

or, with the use of equations (3.3.2)

$$-\rho K \frac{\partial S}{\partial y} + \rho S V + \rho\, \overline{sv} = -\rho K \left(\frac{\partial S}{\partial y}\right)_{\mathrm{w}} + \rho S_{\mathrm{w}} V_{\mathrm{w}} \tag{5.5.3}$$

The subscript w denotes values at the wall $y = 0$, or, more realistically, in a region of molecular conduction a little removed from the kaleidoscopic activity at the wall.

For the parallel flows considered in Chapter 3, the cross-stream flux was simply the sum of two diffusion terms. To account for mean convection in the y-direction, we now use the symbol J_T to represent the total flux of S in the y-direction, and the symbol J to represent the sum of molecular and turbulent diffusion. Thus

$$J_T = J + \rho S V \quad \text{with} \quad J = -\rho K \frac{\partial S}{\partial y} + \rho\, \overline{sv} \tag{5.5.4}$$

and equations (5.5.3) can be rewritten

$$J_T(y) = J_T(0) \tag{5.5.5}$$

and

$$J + \rho S V = J_{\mathrm{w}} + \rho S_{\mathrm{w}} V_{\mathrm{w}} \tag{5.5.5}$$

We apply these results first to the transfer of the basic fluid itself, taking $J = J_{\mathrm{w}} = 0$, since there is no diffusion in this case. Hence

$$V = V_{\mathrm{w}} \tag{5.5.6}$$

showing that the normal component of velocity is uniform throughout the wall layer. This result can also be obtained from the continuity equation (3.3.7). The variation of the diffusion flux is now seen to be governed by

$$J = J_{\mathrm{w}} + \rho V_{\mathrm{w}}(S_{\mathrm{w}} - S) \tag{5.5.6}$$

in a streamwise-uniform wall layer where the various source terms are negligible.

Results for particular transfers can be obtained using the substitutions of Table 3.1:

(1) For momentum, $S \to U$ and $J \to -\tau$ give

$$\tau = \tau_w + \rho V_w U = \mu \frac{dU}{dy} - \rho \overline{uv} \tag{5.5.7}$$

When the pressure gradient is taken into account

$$\tau = \tau_w + \rho V_w U + \frac{dP_w}{dx} y \tag{5.5.8}$$

This indicates, in a formal manner, how suction ($V_w < 0$) acts to compensate for a separation-inducing pressure gradient ($dP_w/dx > 0$).

(2) For enthalpy, $S \to H$ and $J \to \dot{q}_h$ give

$$\dot{q}_h = \dot{q}_w + \rho V_w (H_w - H) = -\rho\kappa \frac{\partial H}{\partial y} + \overline{vh} \tag{5.5.9}$$

This indicates that outwards convection ($V_w > 0$) into a hot fluid ($H > H_w$) reduces the inwards diffusion near the wall ($-\dot{q}_w$) below the value ($-\dot{q}_h$) for the outer flow. This is the essence of the 'heat blocking' achieved by transpiration, although modification of the diffusivity variation also plays a part.

(3) For mass transfer, $S \to C/\rho$ and $J \to N$ give

$$N = N_w + V_w (C_w - C) = -D \frac{\partial C}{\partial y} + \overline{vc} \tag{5.5.10}$$

Although these have the form of the results for heat transfer, the normal convection velocity cannot, in some circumstances, be specified arbitrarily, since it depends on the transfer rate.

5.5.3 The Velocity Distribution in the Locally Determined Layer

In Section 3.5 we derived an equation (3.5.8) for the velocity variation in the fully turbulent part of a wall layer with varying shear stress, by scaling the terms of the turbulence-energy equation (3.4.2) with the local stress and the distance from the wall. Thus, for a locally determined or equilibrium layer

$$\frac{dU}{dy} = \frac{A}{y} \left(\frac{\tau}{\rho} \right)^{\frac{1}{2}} \left(1 + \frac{3B_1}{2A} \frac{y}{\tau} \frac{d\tau}{dy} \right) \tag{5.5.11}$$

with A and B_1 constants introduced in equations (3.5.7). Setting aside boundary layers near separation and other cases in which $|(y/\tau) d\tau/dy|$ is large, we found two simplifying features:

(1) the final term of equation (5.5.11), which represents energy diffusion, is insignificant; and

(2) the stress variation associated with the pressure gradient has a small effect in the locally determined layer close to a wall.

We can extend the hypothesis of locally determined flow to wall layers with lateral convection, by introducing into equation (5.5.11) the expression (5.5.8), which prescribes the stress variations arising from convection and from a pressure gradient. However, the simplifications possible in the absence of convection suggest that the forms

$$\frac{dU}{dy} = \frac{A}{y}\left(\frac{\tau}{\rho}\right)^{\frac{1}{2}} \quad \text{and} \quad \tau = \tau_w + \rho V_w U$$

will be adequate for a wide range of channel and boundary-layer flows. Thus the velocity variation is specified by

$$\frac{dU}{dy} = \frac{A}{y}(u_f^2 + V_w U)^{\frac{1}{2}} \tag{5.5.12}$$

with $u_f = (\tau_w/\rho)^{\frac{1}{2}}$ still based on the stress at the wall. An exact integral can be found for this equation, but not when the full stress variation (5.5.8) is retained.

The last result is commonly referred to as the *mixing-length model*, since it was first obtained by introducing $l_m = Ky$ (as for an impermeable boundary) into the mixing-length results (3.4.55). The corresponding eddy viscosity is found from equations (3.4.56):

$$\epsilon_m = l_m\left(\frac{\tau}{\rho}\right)^{\frac{1}{2}} = Ky\left(\frac{\tau_w}{\rho} + V_w U\right)^{\frac{1}{2}} \tag{5.5.13}$$

One may question the use of local scales, in view of the cross-stream convection V_w. In partial justification, we note that the velocities of the lateral flow are usually very small; typically, $V_w < u_f$ or even $\ll u_f$. Although they can profoundly alter the flow very near the wall, their direct effect should be slight in the turbulent part of the flow.

Before integrating equation (5.5.12), let us consider the wall layer from a more general point of view. We may expect that

$$U \sim y, \tau_w, \rho, \mu, V_w, k$$

whence

$$\frac{U}{u_f} = f\left(\frac{y}{y_f}, \frac{k}{y_f}, \frac{V_w}{u_f}\right) \tag{5.5.14}$$

Here k is a length scale acknowledging the roles of wall roughness and of the scale and structure of the pores in the wall; in practice, a number of parameters may be required. Most models of the flow near a permeable boundary neglect entirely the complex structure of the flow very near the wall; they assume (for the case of blowing) that the pore flows mix instantaneously to give spatially uniform lateral convection. When the pores are large, or when the flows from several combine, blowing will have an effect very like that of solid roughness elements. Considering the solid roughness itself, we may expect its role to be enhanced by suction ($V_w < 0$) and to be reduced by blowing ($V_w > 0$). The pore structure will, of course, contribute to the surface roughness.

These remarks suggest that there will often be formed a *roughness-mixing layer*, where the effects of roughness and of the pore flows are inextricably combined. Beyond this region of three-dimensional mean flow, it will sometimes be possible to distinguish a streamwise-uniform viscous layer. This is the situation usually assumed to exist. However, in the fully rough condition, the roughness-mixing layer will extend to the fully turbulent region.

Integrating equation (5.5.12), we find

$$\frac{2}{V_w}(u_f^2 + V_w U)^{\frac{1}{2}} = A \ln\left(\frac{y}{y_f}\right) + f\left(\frac{k}{y_f}, \frac{V_w}{u_f}\right) \tag{5.5.15}$$

for the fully turbulent part of the streamwise-uniform region. The form of the additive function is dictated by the general result (5.5.14). Alternatively

$$\frac{2}{V_w^+}[(1 + V_w^+ U^+)^{\frac{1}{2}} - 1] = A \ln y^+ + B(k^+, V_w^+) \tag{5.5.16}$$

for the scaled coordinates (4.1.4), and with the additive function written so that the familiar logarithmic laws are incorporated: as $V_w \to 0$, equation (4.3.5) emerges; as V_w, $k \to 0$, equation (4.1.3).

The function $B(k^+, V_w^+)$—a constant for any one flow—specifies the changes in stress and velocity across the roughness-mixing and viscous layers. Several ways of determining this constant have been suggested:

(1) Stevenson brought forward evidence justifying the use of the value for a smooth impermeable wall for a considerable range of blowing and suction.

(2) An analysis of further experimental data led Simpson to suggest that

$$U^+ = K_0 \quad \text{for} \quad y^+ = K_0 \tag{5.5.17}$$

is a point on every velocity profile, with $y^+ = K_0 \simeq 11$, the intersection of linear and logarithmic laws determined by equations (4.1.10). Although its physical basis is not obvious, the assumption works well.

(3) A number of ways of matching the fully turbulent and fully viscous layers have been proposed. One of these will be discussed in the next section.

None of these suggestions takes account of the roughness-mixing layer; all assume that $B = B(V_w^+)$ only.

Whatever method is used to specify the 'slip' constant B, equation (5.5.16) can have only a restricted range of application. This can be seen by considering two limiting cases:

(1) the *asymptotic suction layer*, for which $V_w < 0$ and $\partial U/\partial x = 0$ throughout the flow; and

(2) the *blow-off limit*, for which $V_w > 0$ and $\tau_w = 0$.

An asymptotic suction layer is a boundary layer which has ceased to expand because a balance is established between its outwards spread and the inwards convection. The continuity equation (3.3.7) indicates that $V = V_w$ throughout the flow. At the outer edge of the boundary layer

$$\tau = 0 \quad \text{giving} \quad \tau_w = -\rho V_w U_1 \tag{5.5.18}$$

with U_1 the free-stream velocity. For larger values of $|V_w|$ equation (5.5.16) fails to give meaningful results for the outer part of the flow, since one element becomes imaginary; the region of unrealistic behaviour moves rapidly towards the wall as $-V_w^+$ is increased.

The blow-off limit, beyond which the diffusive activity of the outer flow can no longer reach the wall, is more difficult to analyse, but Coles has found equation (5.5.16) to be unsatisfactory for this case as well. We conclude that this equation is meaningful for moderate transfer rates only, that is, for lateral flows well inside the range defined by asymptotic suction and blow-off, roughly $V_w/U_1 = -0{\cdot}008$ to $0{\cdot}03$. Coles has suggested that

$$-0{\cdot}004 < \frac{V_w}{U_1} < 0{\cdot}01 \tag{5.5.19}$$

defines conditions in which the 'mixing-length' results are realistic.

5.5.4 The Use of Eddy Diffusivities

The results obtained above provide predictions of the friction at a permeable wall, since the only attribute of the viscous layer relevant to friction is the change in velocity. Other transfer calculations require, as for the case of a solid wall, a more detailed model of the flow very near the surface. It is convenient to define this region by a formula giving the effective diffusivity.

Returning to the general results (5.5.4, 6), we introduce molecular and eddy diffusivities K and ϵ_s:

$$J = -\rho\left[K + \frac{\epsilon_s}{\epsilon_m}\epsilon_m\right]\frac{dS}{dy} = J_w + \rho V_w(S_w - S) \tag{5.5.21}$$

In this differential equation the diffusive effect of the turbulence is expressed in terms of the eddy viscosity ϵ_m and the diffusivity ratio ϵ_m/ϵ_s. According to the application, the latter is unity or a turbulent Prandtl or Schmidt number not too different from unity. The diffusivity ratio can be adequately represented by a simple function of distance from the wall, usually a constant or a linear variation.

For the case of an impermeable wall and constant stress ($V_w = 0$ and $\tau = \tau_w$), the eddy-viscosity variations (4.1.10) to (4.1.17) make possible the solution of equation (5.5.21), as indicated in equations (4.1.18, 19). Even for this simple case, a numerical solution will be required when a realistic diffusivity variation is specified. To encompass situations in which the stress varies significantly, as a result of lateral convection or a strong pressure gradient, we must modify the prescription of the eddy viscosity. A popular way of generalizing the wall layer, used in particular by Spalding and his coworkers, is through modification of van Driest's formula (4.1.17). In his model

$$\epsilon_m = l_m^2 \frac{dU}{dy} = l_m \left(\frac{\tau}{\rho}\right)^{\frac{1}{2}} \tag{5.5.22}$$

with the mixing length 'damped' near the wall:

$$l_m = Ky\left[1 - \exp\left(-\frac{y^+}{y_s^+}\right)\right] \tag{5.5.23}$$

Modifications have been achieved by introducing factors into the argument of the exponential term. The simplest are $(\tau/\tau_w)^{\frac{1}{2}}$ and τ/τ_w, proposed by Patankar and Spalding and by Launder and Jones, respectively. With the stress given by equation (5.5.7) or equation (5.5.8), these factors introduce the required dependence on the stress-determining factors.

The introduction of the factor $(\tau/\tau_w)^{\frac{1}{2}}$ replaces the length scale $y_f = \nu(\tau_w/\rho)^{-\frac{1}{2}}$ by the local value $\nu(\tau/\rho)^{-\frac{1}{2}}$; this is consistent with the hypothesis that diffusion is locally determined. Thus for $V_w > 0$ or $dP_w/dx > 0$, both giving $\tau > \tau_w$, the fully turbulent result $l_m = Ky$ appears to be approached *more quickly* than when the stress is constant. In other words, the viscous layer is made thinner (as a multiple of y_f) by outwards convection! It must be remembered, however, that the wall stress is reduced by blowing, so that $y_f = \nu/u_f$ is itself increased.

A variety of other multiplying factors have been inserted into the damping exponent of equation (5.5.23), their complexity steadily increasing to represent a wider range of experimental results. For heat- and mass-transfer predictions, the ratio ϵ_m/ϵ_s can also be tailored to suit measurements. Through these artifices, it has proved possible to represent a progressively wider range of phenomena; unfortunately, without adding much to our understanding of them, or to our confidence in predicting beyond the range of existing experimental results.

All of the proposed modifications of equation (5.5.23) have one feature in common:

$$l_m = Ky \quad \text{for} \quad y^+ \gg 1$$

Thus in the fully turbulent region we always recover the 'mixing-length' result (5.5.13), or its generalization which includes the pressure gradient. Hence the limitations noted earlier (5.5.19) apply to this entire class of wall-layer models.

The various developments of the van Driest model, and of other viscous-layer descriptions, give

$$\frac{\epsilon_m}{\nu} = f\left(y, U, \frac{dU}{dy}; V_w, \frac{dP_w}{dx}\right) \tag{5.5.24}$$

The last two parameters are constants for any one layer. The velocity* can be calculated from a particular case of equation (5.5.21):

$$\rho(\nu+\epsilon_m)\frac{dU}{dy}=\tau_w+\rho V_w U+\frac{dP_w}{dx}y \tag{5.5.25}$$

For completeness, the pressure-gradient term has been included in this generalization of equation (4.1.18). Once the velocity variation is known, an explicit eddy-viscosity profile is defined; this can be introduced into other specific cases of equation (5.5.21) to predict other properties and transfer rates. Thus

$$-\rho\left(\kappa+\frac{\epsilon_m}{Pr_t}\right)\frac{dH}{dy}=\dot{q}_w+\rho V_w(H_w-H) \tag{5.5.26}$$

defines the enthalpy and temperature profiles (neglecting recovery phenomena) and ultimately the heat flux.

The results (5.5.25, 26) provide generalizations of the analogies developed earlier for impermeable walls. Neglecting the pressure gradient, we can combine them to obtain

$$-Pr_e\frac{dU}{dH}=\frac{\tau_w+\rho V_w U}{\dot{q}_w+\rho V_w(H_w-H)}$$

For a region in which the effective Prandtl number Pr_e is constant, integration gives

$$\frac{\dot{q}_w+\rho V_w(H_w-H)}{\dot{q}_w+\rho V_w(H_w-H_1)}=\left[\frac{\tau_w+\rho V_w U}{\tau_w+\rho V_w U_1}\right]^{Pr_e} \tag{5.5.27}$$

with the subscript 1 identifying a point in the region. The two profiles are similar when $Pr_e=1$ or when $V_w\to 0$. Departures from similarity will not be large over a range of adjacent conditions.

FURTHER READING

Bird, R. B., W. E. Stewart and E. N. Lightfoot. Reference 1: Chapters 12, 13, 20 and 21

Davies, J. T. *Turbulence Phenomena*, Academic Press (1972)

Deissler, R. G. and R. H. Sabersky. 'Convective heat transfer and friction in flow of liquids', Section E of Reference 7

Eckert, E. R. G. and R. M. Drake. Reference 3: Chapters 8, 11 and 12.

Henderson, F. M. *Open Channel Flow*, Macmillan, New York (1966): Chapter 10

Hinze, J. O. Reference 5: Chapter 5 and Sections 7–11 and 7–12

Knudsen, J. G. and D. L. Katz. *Fluid Dynamics and Heat Transfer*, McGraw-Hill, New York (1958): Part III

Patankar, S. V. and D. B. Spalding. *Heat and Mass Transfer in Boundary Layers*, Morgan-Grampian, London (1967)

* The 'slip' constant $B(V_w^+)$ can be determined from the velocity profile; thus it is implicit in the equation defining the eddy viscosity.

Prandtl, L. Reference 8: Chapter V, Sections A and D
Rohsenow, W. M. (editor). *Developments in Heat Transfer*, Edward Arnold, London (1964)
Rohsenow, W. M. and H. Y. Choi. Reference 9: Chapters 8, 11 and 16
Scheidegger, A. E. *Theoretical Geomorphology*, Springer, Berlin (1961): Parts IV, VI and VIII
Schubauer, G. B. and C. M. Tchen. 'Turbulent flow', Section B of Reference 7
van Driest, E. R. 'Convective heat transfer in gases', Section F of Reference 7
Welty, J. R., R. E. Wilson and C. E. Wicks. *Fundamentals of Momentum, Heat and Mass Transfer*, Wiley, New York (1969): Chapters 19 to 22 and 28 to 31
Yuan, S. W. 'Cooling by protective fluid films', Section G of Reference 7

SPECIFIC REFERENCES

Coles, D. 'A survey of data for turbulent boundary layers with mass transfer', in Reference 17
Launder, B. E. and W. P. Jones. 'A note on Bradshaw's hypothesis for laminarization', *Amer. Soc. Mech. Eng.*, Paper No. 69-HT-12 (1969)
Owen, P. R. and W. R. Thomson. 'Heat transfer across rough surfaces', *J. Fluid Mech.*, **15**, pp. 321–334 (1963)
Rohsenow, W. M. 'Heat transfer to liquid metals', in *Developments in Heat Transfer*, Edward Arnold, London (1964)
Wilkie, D. 'Forced convection heat transfer from surfaces roughened by transverse ribs', in *Proc. Third International Heat Transfer Conference*, Chicago (1966)

EXAMPLES

5.1 (a) Show that $(d\dot{q}_h/dy)_0 = 0$ at the wall of a very broad, flat channel.
(b) For a pipe, $(d\dot{q}_h/dy)_0 > 0$ is possible. Under what circumstances?
(c) How is the result (a) modified when dissipation and diffusion (that is, recovery effects) are taken into account?

5.2 Consider a class of geometrically similar heat exchangers, with tubes of diameter d and length L, through which a gas is pumped, its thermal resistance controlling the heat transfer.

(a) Taking the temperature differential and the mass flow to be fixed, show that the heat transfer and pressure drop vary approximately as follows:

$$\dot{Q} \propto Ld^{-0.8} \quad \text{and} \quad \Delta p \propto Ld^{-4.8}$$

(b) It may be assumed that the first cost of the unit $\propto Ld$, and that the present value of the operating costs $\propto \Delta p$. Show that, for a particular heat load, the total cost of a unit of the class considered varies as $C_1 d^{1.8} + C_2 d^{-4}$, with C_1 and C_2 constants. On what factors do these constants depend? Are the assumptions regarding the variation of the cost reasonable?

(c) Under optimum conditions, what proportion of the total cost is first cost? What is the effect of using a diameter twenty per cent above or below the optimum? Is it better to go a little above or a little below the optimum? Are these considerations much influenced by the form selected for the friction law?

5.3 The motion of a small spherical particle (of volume V and density ρ_p) within a moving fluid is related to the local fluid velocity v by

$$\rho_p V \frac{dv_p}{dt} = 3\pi\mu d(v - v_p) + \rho V \frac{dv}{dt} + \tfrac{1}{2}\rho V \left(\frac{dv}{dt} - \frac{dv_p}{dt} \right)$$

The drag (first term on the right) has been estimated using Stokes' law; the flow pattern has been assumed to be nearly that of steady motion; and non-linear convective accelerations have been neglected. The second and third terms on the right represent 'buoyancy' arising from the pressure gradient in the fluid, and the 'apparent mass' of the fluid around the sphere. Since the equation is linear, the motion induced by the particle's buoyant weight does not influence the response to the motion of the surrounding fluid.

(a) Show that this result can be written

$$\frac{dv_p}{dt} + av_p = av + b\frac{dv}{dt}$$

with $a = 36\nu/[(2r+1)d^2]$ and $b = 3/(2r+1)$, where $r = \rho_p/\rho$. In what way is the analysis simpler for a sand grain moving through the air, rather than through water?

(b) The general nature of the response to turbulence can be inferred by taking $v = Ae^{i\omega t}$. Show that the response is of the form $v_p = Be^{i\omega t} + Ce^{-at}$, and that the time taken for the particle oscillation to settle into a steady pattern is of order $1/a$. Estimate this time for a sand grain carried in air and in water. What can you conclude?

(c) Show that the harmonic part of v_p is specified by

$$\frac{B}{A} = \frac{1 + ibf}{1 + if} \quad \text{with} \quad f = \frac{\omega}{a}$$

Examine the behaviour for $r = 0, 1, \infty$ and for $f = 0, \infty$. What do these limits represent? Is the response what you would expect? What do these results imply about the diffusion of the particles compared with that of 'elements' of the fluid itself? (Remember that the dispersion after a long time is determined primarily by the largest 'eddies'.)

5.4 Equation (5.2.19) was derived from equations (3.3.33) and (5.1.23), which govern the transfer of enthalpy across a turbulent layer. Using equations (3.3.37) for the transfer of total enthalpy, develop the more complete result

$$(\kappa + \epsilon_h)\frac{dH}{dy} + (\nu + \epsilon_m)\frac{d(\frac{1}{2}U^2)}{dy} + (\nu + \epsilon_k)\frac{d(\frac{1}{2}\overline{q^2})}{dy} = \text{constant}$$

What generalization of the result (5.2.20) is implied?

5.5 (a) Use the results (5.2.20, 21) to calculate the temperature differential between pipe wall and centre-line for a high-velocity adiabatic air flow.

(b) Why do the temperature *vs* velocity relationships (5.2.21, 38) have quite different forms? How can they be combined?

5.6 Two series of experiments give the following values for the constants specifying the logarithmic portions of the profiles of mean velocity and temperature in turbulent flow of air near a smooth wall:

Table 5.4

A	B	A_T	f_T
2·46	5·5	2·22	3·3
2·65	6·5	2·30	1·1

What values of the turbulent Prandtl number are implied? How well are the conditions for profile similarity (5.2.39) satisfied?

5.7 Equations (5.2.41, 42) indicate how heat transfer and friction are related when the velocity and temperature profiles have nearly the same shape.

(a) Using the result (4.2.7) show that the heat-transfer law implied for pipe flow is

$$\mathrm{Nu} = \frac{F}{A \ln F + B - 2{\cdot}19\,A} \quad \text{with} \quad F = \mathrm{Re}_d(\tfrac{1}{2}c_f)^{\frac{1}{2}}$$

when the profiles are exactly similar. (Here T_f and U_f relate to the same point in the flow.)

(b) Show that an additional factor $\mathrm{Pr}/\mathrm{Pr_a}$ appears on the right when there is a departure from exact similarity of the profiles. How do these results compare with those of the Prandtl–Taylor analysis?

5.8 How does the factor n of equations (5.2.42) change as the Prandtl and Reynolds numbers vary?

5.9 (a) Deduce from equations (5.1.32, 33)

$$\mathrm{Le_t} = \frac{1 + C_1\,\mathrm{Pe}^{-\frac{1}{2}}}{1 + C_1\,\mathrm{Pe_c}^{-\frac{1}{2}}}$$

How can the result be used to estimate turbulent Prandtl numbers from measured values of the turbulent Schmidt number?

(b) Under what circumstances do the limiting results $\mathrm{Le_t} = 1$ and $\mathrm{Le_t} = \mathrm{Le}^{-\frac{1}{2}}$ apply? What are the analogous results for $\mathrm{Pr_t}$ and $\mathrm{Sc_t}$? Why do these differ from the corresponding results for $\mathrm{Le_t}$?

(c) For gases (see Table 5.2 for typical characteristics) show that

$$\begin{aligned} \mathrm{Le_t} &= 0{\cdot}9 \pm 0{\cdot}2 \quad \text{for Re} = 10^5 \\ &= 1 \quad \pm 0{\cdot}4 \quad \text{for Re} = 10^4 \end{aligned}$$

5.10 Even when the conditions (5.1.21) are satisfied, the conjugate relationships

$$\mathrm{St_c} = f(\mathrm{St}, \mathrm{Le}) \quad \text{and} \quad \tfrac{1}{2}c_f = f(\mathrm{St}, \mathrm{Pr})$$

are not precisely equivalent, either for laminar or for turbulent flow.

(a) For laminar flow in a round tube, show that

$$\mathrm{St_c} = \mathrm{Le\,St}$$

Under what circumstances does the friction coefficient follow this pattern? Why are the results generally different?

(b) Compare the result (5.1.33) with that of Example 5.9(a). Why do not

$Le_t = f(Le)$ and $Pr_t = f(Pr)$ have the same form? Why is Pr_t usually less than Le_t? Why are equations (5.1.32) less convincing for momentum transfer than for heat or mass transfer?

5.11 What ranges of Nusselt number and film coefficient (see equations (5.2.9)) may normally be expected for flows in smooth-walled pipes, for the four classes of fluids considered in Table 5.2? Why are liquid metals particularly suitable for the extremely compact heat exchangers of certain nuclear reactors?

5.12 Compare the results (5.3.6, 10) to see what the latter implies about $(T_w - T_f)/(T_w - T_b)$, y_h/y_m, y_s^+ and Pr_t. Are these suggestions sensible?

5.13 Show that the maximum eddy viscosity in pipe flow may be estimated as $\epsilon_m/\nu = 0{\cdot}006\, Re_d^{0.9}$ and that the effect of turbulent mixing on the heat transfer in a pipe may be expected to be negligible (save for its effect on the velocity profile) when

$$\frac{Pe}{1 + 86\, Pe^{-\frac{1}{4}}} < 60 \quad \text{or} \quad Pe < 340$$

5.14 For a certain configuration of tubes lying across a turbulent air stream, it is known that the overall heat-transfer factor of equations (5.2.9) is $U = 8$ Btu/ft² hr °R. However, part of the thermal resistance, estimated as $R = 0{\cdot}03$ ft² hr °R/Btu, arises within the tube and across its wall.

(a) Find the film coefficient for the external flow, and estimate the mass transfer factor for an alcohol with $Sc = 1{\cdot}3$ which wets the outside of the tube, assuming flow conditions identical to those of the heat-transfer measurements.

(b) If the velocity is increased by a factor of 2 and the scale is reduced by a factor of 3, what is the effect on the mass flux?

5.15 (a) Assuming that $\dot{q}_h$ and τ are constant, show that the heat-transfer characteristic of a wall layer can be calculated as

$$\frac{1}{St} = \frac{\rho c_p U_1 (T_w - T_1)}{\dot{q}_w}$$

$$= \left(\frac{2}{c_f}\right)^{\frac{1}{2}} \int_0^{y_1^+} \left[\frac{\nu}{\kappa + \epsilon_h}\right] dy^+$$

$$= \frac{2}{c_f} + \left(\frac{2}{c_f}\right)^{\frac{1}{2}} \int_0^{y_1^+} (Pr_e - 1)\, dU^+$$

(b) Find the heat-transfer law for $Pr > 1$ which is implied by Rannie's diffusivity (4.1.13), taking $Pr_t = 1$ and neglecting molecular conduction for $y^+ > y_s^+$.

(c) Taking $\epsilon_m/\nu = n^4 y^{+4} = n^4 U^{+4}$ as in equations (5.3.20), show that

$$\frac{1}{2}\frac{c_f}{St} = 1 + \frac{\pi}{4n} c_f^{\frac{1}{2}} \left(\frac{Pr}{Pr_t}\right)^{-\frac{1}{4}} (Pr - Pr_t) \quad \text{for } Pr \gg 1$$

Why does this result differ from equations (5.3.21)? Is the difference significant? Which behaves more reasonably for moderate values of Pr?

5.16 (a) What is the form of the plot shown in Figure 5.5 when the flow is laminar? Draw this curve, and sketch the variations you would expect during transition to turbulence and during development of the turbulence.

(b) How is Figure 5.5 modified if we consider heat transmission across plane parallel turbulent flow instead of heat transfer into pipe flow?

5.17 For mass transfer in gases, the driving potential is usually expressed in terms of the partial pressure. For the component labelled A, this is $p_A = (m_A/V)R_A T$.

(a) Show that the mass-transfer Stanton number may now be expressed as

$$\mathrm{St_c} = \frac{N_A R_A T}{U_0 \Delta p_A} = \frac{\mathbf{N}_A \mathbf{R} T}{U_0 \Delta p_A}$$

with bold symbols denoting molar quantities.

(b) For the many problems involving the transfer of water vapour in air, it is convenient to introduce instead the specific humidity $s = \rho_v/\rho$, where $\rho = p/RT$ is the total density, usually very nearly that of the gaseous component. Show that

$$\mathrm{St_c} = \frac{N_v}{\rho U_0 \Delta s}$$

(c) The readings of a psychrometer, a combination of dry-bulb and wet-bulb thermometers, are usually interpreted on the assumption that (1) St = $\mathrm{St_c}$. Show that this leads to

$$\frac{\Delta H}{\Delta s} = H_{fg}$$

on taking (2) $\dot{q}_w = N_v H_{fg}$, with H_{fg} the enthalpy change (per unit mass of water) during evaporation. Justify the assumptions (1) and (2), and the introduction of

$$\Delta = (\text{wet-bulb or saturation value}) - (\text{dry-bulb or free-air value})$$

(d) How can the results of (c) be modified to measure the concentration of the vapour of a liquid fuel for which Sc = 2·6? Is this kind of modification significant for water vapour?

(e) Apply the result of (c) to a particular case, using steam tables to determine the required properties of water vapour.

5.18 Considering combined forced and free convection in a vertical pipe, show that the ratio of buoyancy forces to the friction force at the wall of the pipe is of the form $g(\rho_b - \rho_w)d/\tau_w$. Hence show that the condition for the effects of buoyancy to be negligible is

$$\frac{\mathrm{Gr}}{\mathrm{Re}} \ll 1 \quad \text{for laminar flow}$$

$$\frac{\mathrm{Gr}}{\mathrm{Re}^{1\cdot 8}} \ll 1 \quad \text{for turbulent flow}$$

5.19 Why does equation (5.4.8) not give Reynolds analogy when Pr → 1? In what fundamental way are the simple models leading to equations (5.4.5, 7) faulty? How is this set right, in a crude way, in equation (5.4.8)?

5.20 (a) Use equations (5.3.21) and (5.4.8) to show that the effectiveness (5.4.1) of roughness in promoting heat transfer may be estimated from

$$\eta = 17\cdot 5\,\mathrm{Pr}^{-0\cdot 05}(k_s^+)^{-0\cdot 45}$$

when the Prandtl number is large.

(b) Does the indicated dependence on Prandtl number seem reasonable, in view of the argument leading to the results (5.4.3)?

(c) What is the largest value which the effectiveness can be expected to take on?

5.21 Rewrite equation (5.4.15) in terms of the standard coefficients c_f and St, those based on U_a and T_b. Make a realistic estimate of the contribution of the wall layer $(\frac{1}{2}c_f/\mathrm{St})_w$ for air flowing in a smooth channel with $U_a b/\nu = 10^5$.

5.22 (a) Show that Nu = 4, 2 and 1, respectively, for slug flow of a fluid with a very low Prandtl number, for the cases of (1) symmetric transfer, (2) isolating flow, and (3) transmitting flow.

(b) Why is the heat-transfer law so much more strongly dependent on the boundary conditions than are equations (5.4.15, 16, 18)? How can those equations be used to find the slug-flow results quoted above?

5.23 (a) Derive a working formula for the heat transfer at the solid boundaries of an open channel with flat bottom and vertical sides. Indicate the limits on its applicability.

(b) How would you go about predicting heat transfer at the free surface?

5.24 In an air-conditioning unit, dry air (30 °C, 1 atm) flows between moistened plates; the average velocity in the 3 cm gap between the plates is 20 m/sec.

(a) Sketch the variation of mass flux along one of the plates.

(b) Calculate a typical transfer rate using the result (5.3.23).

(c) How long can the plates usefully be made?

5.25 The heat transfer through the inner wall of an annulus is often estimated using $\mathrm{Nu}/\mathrm{Nu}_1 = (d_2/d_1)^m$ with the index $m = 0{\cdot}5 \pm 0{\cdot}05$, and Nu_1 the value for $d_2/d_1 = 1$. Compare this correction with that for friction given in Example 4.28. Are the two results consistent with the analogy between friction and heat transfer? (Remember that the stresses at the two walls differ.)

5.26 Show that the reduction in the heat transfer *to* a wall as a result of transpiration is given by

$$\rho V_w(H_1 - H_w) - (\dot{q}_{h_0} - \dot{q}_{h_1})$$

with $\dot{q}_{h_0}$ the outwards flux for $V_w = 0$, and with the subscript 1 denoting a point just beyond the wall layer. What is the significance of these two terms? What signs may they be expected to have?

5.27 State whether the terms of equations (5.5.10) are positive, zero, or negative when steam condenses on a wall from

(1) a flow of pure steam, and

(2) a flow of steam and air.

For the second case, consider separately the fluxes of the two components, and sketch the concentration gradients and profiles.

5.28 Many of the phenomena discussed in this chapter can be illustrated by the simple case of uniform heat transmission and mass transfer across a laminar Couette flow whose properties are assumed uniform. The results are of wider practical significance in defining the mean properties in the sublayer (no longer linear, but still viscous) at a smooth permeable wall.

(a) Show that

$$\dot{q}_h - \tau U = \dot{q}_w + \rho V_w(H_w - H)$$

with $\tau = \tau_w + \rho V_w U$, when the contribution of dissipation and diffusion is included.

(b) Derive the velocity distribution

$$\frac{U}{U_2} = \frac{\exp(\mathrm{Re}_b\, y/b) - 1}{\exp(\mathrm{Re}_b) - 1}$$

with $\mathrm{Re}_b = V_w b/\nu$ a Reynolds number characterizing the mass transfer. Sketch the profiles for several values of $\mathrm{Re}_b \gtrless 0$, noting the development of boundary layers at higher Reynolds numbers.

(c) Neglecting dissipation, and assuming that $\partial T/\partial x = 0$, show that the heat transfer at the bottom plate is given by

$$\mathrm{Nu} = \frac{\mathrm{Pe}_b}{\exp(\mathrm{Pe}_b) - 1} \quad \text{with} \quad \mathrm{Pe}_b = \frac{V_w b}{\kappa}$$

and that it is reduced by blowing. Under what circumstances are the velocity and temperature profiles similar?

(d) For $V_w = 0$ and $\partial T/\partial x = 0$, but with dissipation included, show that

$$\mathrm{Nu} = 1 - \tfrac{1}{2}\mathrm{Pr}\ \mathrm{Ec} \quad \text{with} \quad \mathrm{Ec} = \frac{U_2^2}{c_p(T_1 - T_2)}$$

when Nu is based on $T_1 - T_2$. (The combination Pr Ec—an index of the importance of 'viscous heating'—is sometimes referred to as the Brinkman number.) Show also that the recovery factor $r = \mathrm{Pr}$, and finally that $\mathrm{Nu} = 1$ when based on $T_1 - T_r$.

(e) Derive the general result for this flow:

$$r = \mathrm{Pr}\left[1 + \mathrm{Re}_b \int_0^1 \frac{2c_p(T_1 - T)}{U_2^2}\, \mathrm{d}\left(\frac{y}{b}\right)\right]$$

Show that the temperature distribution can be found by an iterative procedure, and find the first two approximations. How is the recovery factor changed by blowing and by suction? Reconcile these predictions with the results of (c) above. For a permeable boundary, how meaningful are the concepts of an adiabatic wall and the recovery factor?

5.29 (a) For a generalized wall layer on a permeable wall, show that the 'slip' constant is

$$B = B_0 - K_0 + \frac{2}{V_w^+}[(1 + V_w^+ U^+)^{\frac{1}{2}} - 1] - A \ln\left(\frac{y^+}{K_0}\right)$$

with $B_0 = K_0 - A \ln K_0$ the value for an impermeable wall.

(b) Show that, according to Simpson's assumption that $U^+ = y^+ = K_0$ is a point on every profile, B varies between a minimum of $B_0 - K_0$ (at the blow-off limit) and a maximum of $B_0 + K_0$ (at the limit of asymptotic suction, and with $\frac{1}{2}c_f = V_w^{+2} = 1/K_0^2$).

(c) An alternative assumption is that ϵ_m/ν has the same value ($K_0/A \simeq 4{\cdot}4$), for all values of V_w^+, where the fully turbulent region meets the viscous profile

$$U^+ = \frac{\exp(V_w^+ y^+) - 1}{V_w^+}$$

Show that the matching point is given by

$$y_i^+ \exp(\tfrac{1}{2} V_w^+ y_i^+) = K_0$$

and that

$$B = B_0 - K_0 + \frac{2}{V_w^+}\left(\frac{K_0}{y_i^+} - 1\right) + \tfrac{1}{2} A V_w^+ y_i^+$$

How does B behave as the two limits are approached?

(d) Plot the three suggested variations of B as functions of V_w^+—Stevenson's $B = B_0$, and the two considered above. Which appears to be the most realistic choice?

5.30 Equation (5.5.16) implies a friction law of the form $c_f = f(\text{Re}, V_w^+)$, which can be obtained in the manner of equation (4.2.7).

(a) Taking Re $= 10^4$, plot this law over the range (5.5.19), determining the constant B according to the assumptions of Simpson and Stevenson. How important is the choice between these two hypotheses?

(b) In Example 3.22 a simple argument suggested that the Reynolds flux for momentum was modified by lateral convection to $G_m = G_0 + \frac{1}{2}G_L$. Show that this postulate is equivalent to $c_f = c_{f_0} + \text{constant} \times V_w/u_f$. What values of the constant are suggested by the results of (a)?

5.31 In principle, equation (5.5.25) can be solved for any eddy-viscosity variation of the form (5.5.24). However, the techniques adopted depend on the nature of the function specifying the diffusivity.

(a) For Deissler's distributions (4.1.14, 15), show that

$$\frac{dU^+}{dy^+} = \frac{\tau/\tau_w}{1 + \epsilon_m/\nu}$$

can be integrated directly.

(b) For van Driest's model, and developments of it which leave $l_m^+ = f(y^+, U^+)$, show that

$$\left(1 + l_m^{+2}\frac{dU^+}{dy^+}\right)\frac{dU^+}{dy^+} = \frac{\tau}{\tau_w}$$

is more easily integrated when written as

$$\frac{dU^+}{dy^+} = \frac{2\tau/\tau_w}{1 + [1 + 4 l_m^{+2}(\tau/\tau_w)]^{\frac{1}{2}}} = f\left(y^+, U^+; V_w^+, \frac{d(P_w/\tau_w)}{dx^+}\right)$$

(c) Does the method of calculation of the enthalpy variation from equation (5.5.26) depend on the way in which the eddy viscosity is related to the other characteristics of the flow? What is the form of the equation to be solved?

5.32 (a) Write a computer program to predict the heat-transfer to fully developed turbulent pipe flow in a smooth-walled pipe. Aim for an accuracy of ten per cent over the range of feasible Reynolds and Prandtl numbers.

(b) Extend this program to cover rough walls and flow development, including an initial laminar section. What accuracy can be achieved in these calculations?

6

Developing Flows I: Fundamentals

The preceding chapters concern fully developed channel flows, whose mean motion is exactly parallel to the channel wall. Now we extend the discussion to nearly parallel motions, both those near walls—boundary layers and wall jets—and those embedded in a body of fluid—the free-turbulent flows: wakes, jets and mixing layers. Much of our knowledge of wall layers is relevant to developing wall flows: very near the wall the mean motion is almost parallel, and the transfer processes are like those near the wall in fully developed flow. However, the outer part of a wall-bounded flow is akin to a free-turbulent motion, and is dominated by the spreading of the mean motion and turbulence.

The following chapters deal for the most part with 'thin' flows, greatly elongated in the direction of the mean motion, by virtue of the relative slowness of lateral diffusion. Chapters 7 and 8 look at two classes of thin flows—free-turbulent flows and boundary layers—while Chapter 9 takes tentative steps towards the prediction of more complex motions. The present chapter provides the basis for these specific studies by:

(1) surveying the general features of the outer turbulent region, of the outer flow proper, and of the turbulence interface which lies between;

(2) developing the 'boundary-layer' or 'thin-flow' equations—momentum and energy equations, both differential and integral, simplified in accordance with the flow geometry; and

(3) surveying the methods of representing and analysing developing flows which will be used in the succeeding chapters.

Most of the discussion relates to the two simplest mean-flow geometries: *plane flows* with U, $V=f(x,y)$; and axisymmetric, non-swirling flows with U, $V=f(x,r)$. For compactness, the latter class will be referred to as *round flows*. The foundation for the study of these flow geometries is provided in Chapter 3. We shall adopt the conservation laws established in Section 3.3, and also make use of the gradient-diffusion transfer models of Section 3.4—those involving the eddy viscosity and mixing length. However, the most powerful tools of analysis—particularly for the simple patterns of Chapters

7 and 8—are similarity arguments. These often reveal the pattern of development without requiring a detailed understanding of the transfer processes.

6.1 THIN SPREADING FLOWS

6.1.1 Examples

We shall be concerned for the most part with flows in which one mean-velocity component is much greater than the others, with the region of large velocity gradients and significant turbulent activity elongated in the direction of the large component. This geometrical restriction has vital mathematical consequences: each cross-stream derivative will greatly exceed the corresponding derivative in the direction of the mean flow. We shall now consider some of the ways in which this situation can arise. It will be possible to discuss in detail only a few of these flows, but this survey will testify to the wide field of application of the 'boundary-layer' or 'thin-flow' approximations.

Plane Flows

A numerous class of thin flows are those concentrated near a plane, usually taken to be $y=0$; here $U(x,y) \gg V(x,y)$ and $\partial/\partial x \ll \partial/\partial y$. Note that the x, y-plane to which the mean flow is confined is normal to this plane of concentration.

Free-turbulent flows of this class include: the wake behind a cylinder lying across a uniform stream where $U = U_1$; the jet emerging from a slot into still fluid; and the mixing layer between two streams initially having uniform velocities U_1 and U_2. These examples can be generalized by allowing the convecting streams to be non-uniform. Thus we have the wake in shearing flow, with $U_1(y)$ outside the wake, and the jet in streaming flow, possibly with varying $U_1(x)$. When the plane of concentration is a fixed wall, we have two species of plane flow, the boundary layer and the wall jet, the latter with or without an external streaming flow. In either case, the free-stream velocity may vary along the flow.

It is possible to have plane flows concentrated near curved surfaces; for example, a boundary layer or wall jet on a curved wall, and a curved jet, across which the pressure and velocity change. The thin-flow conditions still apply, provided that x is measured along the surface and y normal to it.

For boundary layers and wall jets exposed to an adverse (that is, decelerating) pressure gradient, whether on a curved or plane wall, we must anticipate the possibility of separation. Near the point of separation (or near reattachment, if this occurs subsequently) the conditions $U \gg V$ and $\partial/\partial x \ll \partial/\partial y$ are not obeyed. However, these boundary-layer requirements may be satisfied again following separation or reattachment.

Axisymmetric Flows without Swirl

Round free-turbulent flows concentrated near an axis include the jet from a round orifice, and the wake of a body of revolution. As jets and wakes from non-circular orifices and bodies spread, they will move towards round forms, but a long time may elapse before the effects of asymmetric generating conditions are erased.

The class of axisymmetric thin flows also includes flows concentrated near a plane, for example, the radially directed 'fan' jet from a circular slot, the radial wall jet, and the boundary layer formed at a stagnation point. We may also consider flows concentrated near a cylinder, for example, the cylindrical mixing layer at the beginning of a round jet, and the boundary layers and wall jets outside a circular cylinder and inside a pipe. Finally, we have the important cases of boundary layers on a body of revolution and inside axisymmetric nozzles and diffusers; in these cases, $U_1 = U_1(x)$ is usual.

Three-dimensional Flows

Axisymmetric swirling flows form a numerous group, many of which are thin in the radial direction, with $\partial/\partial x \ll \partial/\partial r$. This can be achieved in two ways: in strongly swirling flows, with $W(x,r) \gg V(x,r)$; and in weakly swirling flows, with $U(x,r) \gg V(x,r)$. The first type is found in cyclone separators and wing-tip vortices, and the second in the jet leaving a fan or other turbomachine. Two wall-bounded flows of this general class are the boundary layers on a rotating plate in an otherwise still fluid, and on a fixed plate bounding a rotating fluid. In the latter case, there may develop elongated axisymmetric cells of flow towards and away from the plate.

Some more varied three-dimensional but thin flows are:

(1) the boundary layers on blades and wings, with $V \ll U, W$ and $\partial/\partial y \gg \partial/\partial x, \partial/\partial z$;

(2) the wake behind a body yawed with respect to the mean flow, again with $V \ll U, W$; and

(3) the non-circular wake or jet, with $V, W \ll U$ and $\partial/\partial y, \partial/\partial z \gg \partial/\partial x$.

Flows with Varying Density

Each of the motions considered above can be generalized by taking into account density variations. These may arise in many ways: from mass or heat transfer, from fluid compressibility in high-velocity flows, from differing fluid species, or from temperature or concentration stratification, as in the atmosphere or in lakes and oceans. We shall consider only one example of the effects of density variation—free or natural convection, in which the motion is driven by the buoyancy of the fluid. Thin flows of this class include buoyant jets, plumes and wakes, and the buoyancy-generated boundary layer (or is it a wall jet?) on a vertical plate or cylinder.

6.1.2 Spreading Turbulence

Conditions at the Free Boundary

The streamwise development of the flows mentioned above is a consequence of the existence of a free boundary between the actively turbulent region and the outer fluid. We shall concentrate on the more typical case in which the outer fluid is nearly still or is in nearly uniform motion. Thus we set aside a variety of situations in which the outer flow displays turbulent activity independent of the spreading shear flow; for example, small-scale turbulence in a wind tunnel, very large eddies in the atmosphere, or intermediate scales in the core of a diffuser or nozzle.

Subject to this limitation, we may expect most measures of the activity within the shear layer to become small as the outer flow is approached. For any thin flow:

$$\frac{\partial U}{\partial y}, \frac{\partial T}{\partial y} \to 0$$

and

$$\tau, \dot{q}_{\mathrm{h}} \to 0 \tag{6.1.1}$$

For a turbulent flow, we have in addition:

$$\overline{u^2}, \overline{q^2} \to 0$$

and

$$\rho\overline{uv} = -\tau_{\mathrm{t}} \to 0, \qquad \rho\overline{uh} = (\dot{q}_{\mathrm{h}})_{\mathrm{t}} \to 0 \tag{6.1.2}$$

and so forth. One aspect of the spreading flow extends into the outer flow; in general, we must allow

$$V \neq 0 \tag{6.1.3}$$

at the free boundary.

The Turbulence Interface

The time-averaged boundary conditions (6.1.1, 2, 3) apply to both laminar and turbulent shear flows. But the approach to the free-stream values is accomplished in quite different ways in the two kinds of flow. A laminar shear layer merges smoothly with the outer flow, as indeed do the mean values of a turbulent layer. However, the instantaneous picture is quite different. Figure 1.4 and Plate I (facing p. 46) reveal a clearly defined interface between the nearly irrotational outer fluid and the highly rotational fluid of the actively turbulent region.

The smoothness of the time-averaged merging of shear layer and outer flow is a consequence of the unsteadiness and highly convoluted shape of the interface, characteristics resulting from the activity of the larger elements of the turbulence. The fact that the larger convolutions often display a measure of periodicity suggests that it may be possible to identify a large-scale structure typical of each species of turbulent flow. Plate I reveals that the smaller scales within the fully turbulent region are fairly homogeneous. This is reasonable, since unrestrained large 'eddies' can carry the smaller elements across a free shear layer or the outer part of a wall-bounded flow.

So far we have described the abrupt transition between turbulent shear flow and the outer fluid merely as an 'interface'. What is its structure, and how does it advance into the outer fluid to accomplish the spreading of the flow? The answer to the first question is fairly simple. The turbulence boundary is a highly sheared layer with thickness comparable to the scale of the smallest eddies of the turbulence: steeper gradients cannot be maintained in the flow, while efficient cross-stream mixing brings some of the smallest elements of the dissipating structure to the boundary. The interface is often referred to as a *viscous superlayer*, by analogy with the thin sublayer on a wall, but the two viscous regions are so unlike that this usage seems inappropriate.

We turn now to the question of the spreading of turbulence. Although the lateral expansion of a laminar flow is controlled by viscosity, this is not the case for the mean turbulence boundary and the associated profiles of mean properties, such as velocity and temperature. Free-turbulent flows commonly exhibit viscosity-independence or Reynolds-number similarity: the angle of spread of a turbulent jet, for example, is nearly independent of Reynolds numbers used to define the flow. Seemingly, the rate at which the outer fluid becomes turbulent is determined by the viscosity-independent parts of the flow, that is, by the mean motion and largest elements of the turbulence. In this respect the spreading of turbulence is like turbulent dissipation, whose rate is nearly independent of the ultimate viscous processes.

Although a number of processes which contribute to the spreading of turbulence have been identified, agreement has not yet been reached on their relative importance. Perhaps this is because the 'entrainment' process has no counterpart in analytically tractable isotropic and homogeneous turbulence. There follow brief descriptions of two ways in which fluid beyond the turbulence interface might be incorporated with fluid which became turbulent earlier.

(1) The outer fluid is engulfed by deforming and folding of the boundary, through the combined action of the mean flow and largest elements of the turbulence. The entrained fluid is ultimately absorbed (that is, rendered vortical, so that it can take part in the characteristic turbulent interactions) by subsequent, internal advance of the interface.

(2) The outer fluid is rendered slightly vortical by viscous advance of the interface, the rate being enhanced by the small convolutions. The larger scales then play their part by stretching the interface, thus amplifying the vorticity there by the vortex-stretching process outlined in Section 1.4.

For both proposals, the rate of conversion to turbulence is strongly dependent on the mean motion, consistent with the observed Reynolds-number similarity of spreading.

Although the turbulence interface has a fundamental role in many aspects of the dynamics of developing flows, fairly realistic predictions of mean properties can often be obtained without taking explicit account of its existence. This is possible only because its major effects are represented by empirical coefficients.

Example of the Plane Wake

We shall now see how these ideas apply to a particular flow, the plane wake behind a circular cylinder in a uniform stream, a flow which has been carefully studied by Townsend and others. Here the turbulence can be measured relatively easily, since (except very near the cylinder) the mean velocity is nearly uniform and is large compared to the turbulent fluctuations, which are linked to the steadily diminishing velocity defect. These conditions are ideal for hot-wire measurements, in contrast to conditions at the edge of a jet in still air, or at the still-air edge of a mixing layer.

Figures 6.1(a) and (b) display profiles of mean velocity and of the intensity of the streamwise fluctuation for four stations along the wake. The mean velocity settles quickly into a nearly self-preserving, bell-shaped profile, approximated by

$$\frac{U_1 - U(x,y)}{U_0(x)} = \exp\left[-k\left(\frac{y}{l_0}\right)^2\right] \tag{6.1.4}$$

with $U_0 = U_1 - U(x,0)$ a scale for the velocity variation, and $l_0(x)$ a length scale for the flow breadth. For this flow, similarity arguments to be presented later suggest that $U_0 \propto x^{-\frac{1}{2}}$ and $l_0 \propto x^{\frac{1}{2}}$, when x is measured from an appropriate virtual origin. This scaling is used in Figure 6.1 and is very successful, for the mean velocity at least. The profile (6.1.4) is also found to apply to the viscous wake behind a flat plate or other body which does not shed vortices. This reinforces the impression that the mean-velocity profile is only weakly dependent on the details of the transfer mechanism.

In Figure 6.1(b) we see that, although the fluctuations ultimately adopt a self-preserving pattern with the same scales as the mean velocity, they do so only far downstream. In this respect the wake is somewhat unusual, for the dissipation of the extra energy introduced during wake formation is slow compared to the convection imposed by the outer flow. Nevertheless, these observations provide widely applicable warnings: we cannot assume that every aspect of a turbulent flow has attained self-preserving form on finding that one profile has done so; in particular, the mean-velocity profile is an insensitive indicator of turbulent activity.

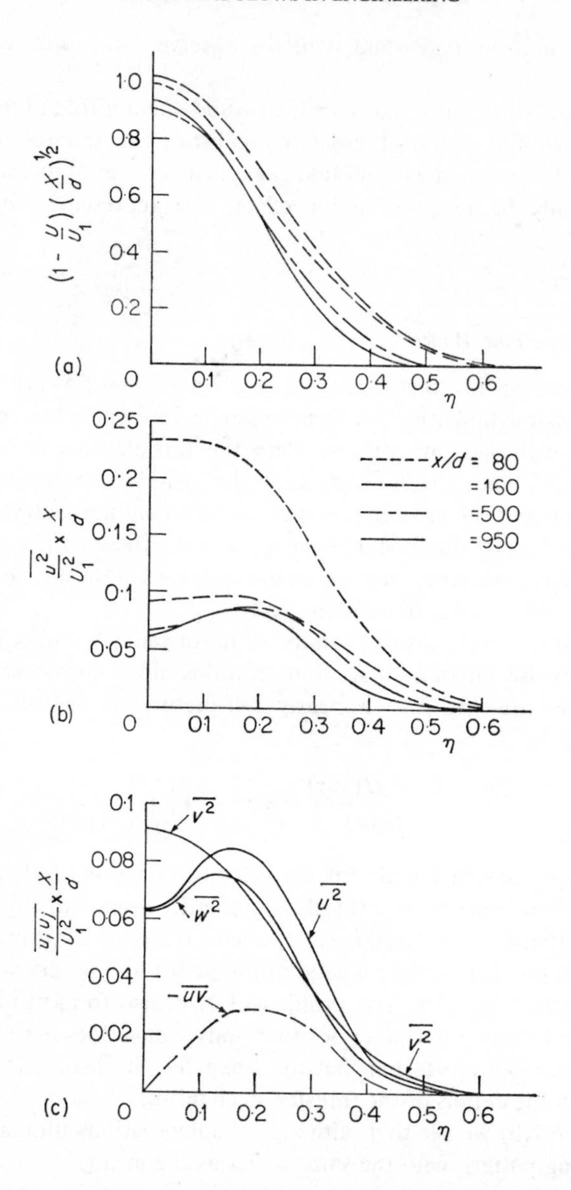

Figure 6.1 Turbulent flow in a plane wake, according to the measurements of Townsend. Results for $Re_d = 8400$ ($x/d = 80$ and 160) and for $Re_d = 1360$ ($x/d = 500$ and 950). In each case the streamwise coordinate x is measured from a virtual origin a little ahead of the cylinder. The scaled lateral coordinate is $\eta = y/(xd)^{\frac{1}{2}}$. (a) Development of the velocity defect. The results for the two Reynolds numbers are not exactly comparable, but do illustrate the general pattern of change. (b) Development of the streamwise component of the velocity fluctuation. (c) Final forms ($x/d = 950$) of component intensities and mixing stress

Figure 6.1(c) shows the ultimate self-preserving forms of component intensities and mixing stress. Comparison with Figure 3.2 shows that the intensities are more nearly equal than in wall turbulence. However, the components in the plane of the mean motion are somewhat larger than the third component, suggesting that the turbulence contains significant two-dimensional elements. Note that the mixing stress and two of the components attain their maximum values in the region where the mean rate of strain ($\partial U/\partial y$) and turbulence production ($-\overline{uv}\,\partial U/\partial y$) are largest. However, the cross-stream component is largest near the central plane, indicating that there is efficient transfer across this plane from one side of the wake to the other.

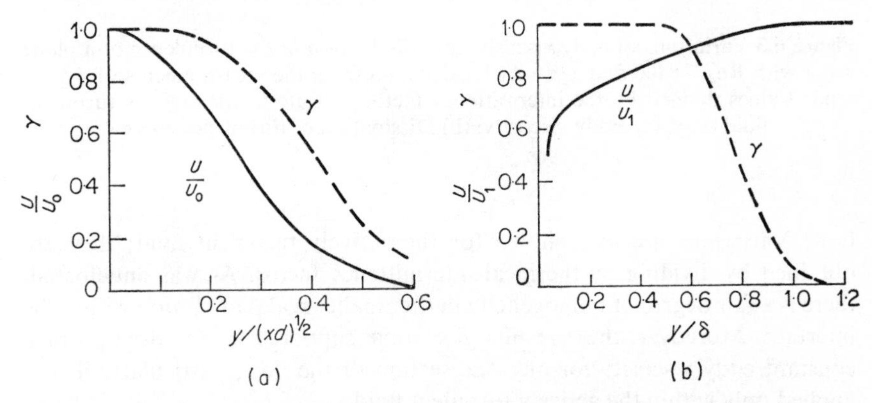

Figure 6.2 Variations of intermittency factor γ and mean velocity. (a) Plane wake. (b) Constant-pressure boundary layer

Figures 6.2(a) and (b) show the variations of mean velocity and intermittency factor γ across a plane wake and a constant-pressure boundary layer. There is hardly any part of the wake which is not visited occasionally by effectively non-turbulent fluid. The zone of marked intermittency is proportionally narrower in the boundary layer; the no-penetration condition at the wall makes itself felt throughout the layer. In both Figures 6.2(a) and (b) the regions of intermittent turbulence extend well beyond the mean-velocity variations, suggesting that the furthest projections of the interface move very nearly at the convection velocity U_1. The intensity profiles of Figure 6.1 extend even further, since fluctuations are induced in the outer fluid by activity at and within the turbulence interface. The differing shapes and relationships of velocity and intermittency profiles for the wake and boundary layer indicate the difficulty of establishing a unified theory encompassing the whole range of free-turbulent and outer-turbulent flows.

Figure 6.3 concerns two aspects of the small-scale activity in wake turbulence. It shows how the eddy viscosity ϵ_m and turbulent dissipation ε_t vary across the

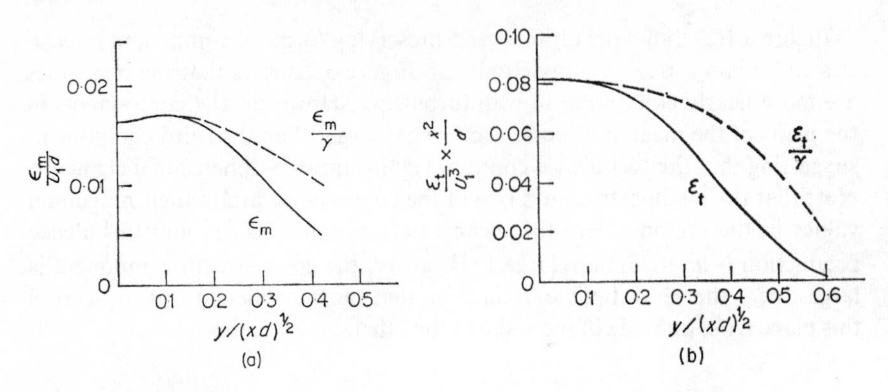

Figure 6.3 Variations of eddy viscosity and dissipation in the turbulence of a plane wake with $\mathrm{Re}_d = 8400$ and $x/d = 160$, calculated from the measurements of Townsend. Values divided by the intermittency factor γ relate to the actively turbulent fluid only. (a) Eddy viscosity. (b) Dissipation of turbulence energy

flow. Variations are also shown for the actively turbulent fluid; they are obtained by dividing by the local intermittency factor. As was anticipated, there is a fair degree of homogeneity in the smaller scales of motion within the interface. Moreover, these results give some support for the adoption of a constant eddy viscosity for any one section of the flow, particularly if it is applied only within the actively turbulent fluid.

6.1.3 Measures of Flow Width

The width of a flow is used in many ways, for example, in calculating a local Reynolds number to specify the dynamic condition of the flow and hence its friction-generating potential; or, more directly, in defining the region occupied by the flow and the changes in mean properties. For a channel flow, the channel breadth itself is an obvious measure of width. For a flow with a free boundary, the situation is not so clear-cut. The lateral extent of the mean profiles differs from one property to the next, and each profile merges asymptotically with the outer flow. Hence a variety of measures of width are used, although those most often encountered are based on the mean-velocity profile.

For *boundary layers* the most obvious measure of thickness is the distance over which the mean velocity achieves some specified large fraction of the free-stream velocity. The thicknesses δ_{99} and δ_{995}, corresponding to $U = 0{\cdot}99U_1$ and $0{\cdot}995U_1$, are often used. Of course, for Prandtl numbers (Pr or $\mathrm{Pr_t}$) significantly different from unity, these values may not be an accurate index of the extent of the temperature variation.

Other measures of boundary-layer width in common use are based on weighted averages of the velocity variation:

(1) the displacement thickness

$$\delta_1 = \int_0^{y_1} \left(1 - \frac{\rho U}{\rho_1 U_1}\right) dy \tag{6.1.5}$$

(2) the momentum thickness

$$\delta_2 = \int_0^{y_1} \frac{U}{U_1}\left(1 - \frac{\rho U}{\rho_1 U_1}\right) dy \tag{6.1.6}$$

(3) the energy thickness

$$\delta_3 = \int_0^{y_1} \frac{\rho U}{\rho_1 U_1}\left(1 - \frac{U^2}{U_1^2}\right) dy \tag{6.1.7}$$

Here y_1 is taken sufficiently far from the wall to encompass the whole of the velocity and density variations; this limit is sometimes written as ∞ for compactness. In the applications we shall consider, the density will be sensibly constant, and these formulae can be simplified by striking out the densities.

The 'thicknesses' (6.1.5, 6, 7) are not, strictly speaking, measures of the width of the flow, but are expressions of certain of its dynamic attributes. The displacement thickness gives the lateral shifting of the outer flow by the shear layer; the momentum thickness measures the reduction in momentum flux; and the energy thickness gives the reduction in the kinetic energy of the fluid passing through the boundary layer.

If the velocity and density profiles are similar along the flow, with

$$\frac{U}{U_1} = f\left(\frac{y}{\delta_0}\right) \quad \text{and} \quad \frac{\rho}{\rho_1} = g\left(\frac{y}{\delta_0}\right) \tag{6.1.8}$$

where δ_0 is any measure of width, we find that

$$\frac{\delta_{99}}{\delta_1}, \frac{\delta_1}{\delta_2}, \frac{\delta_1}{\delta_3}, \text{etc. are constants} \tag{6.1.9}$$

In these circumstances, any one of the thicknesses can serve as an indicator of changes in width. The constants of (6.1.9) differ from one set of similar profiles to the next, and may be thought of as indices of the profile shape. The *shape factor*

$$H = \frac{\delta_1}{\delta_2} \tag{6.1.10}$$

is often used to specify the shape of a boundary-layer profile. It is only when several such ratios are specified that the profile is fully defined.

The width of a *free-turbulent flow* or turbulent wall jet (or of an analogous laminar motion) can also be measured using weighted averages like equations

(6.1.5, 6, 7). Changes in sign will be necessary when $U < U_1$, as in a wake, and another scale velocity must replace U_1 when there is no external velocity. A more common technique is to define the width in terms of a particular point on the velocity profile. The point at which one-half of the velocity change has occurred is often adopted. For the profile given by equation (6.1.4), this interpretation implies that the constant $k = \ln 2 = 0{\cdot}693$. Other values which are sometimes used are $k = 1$ and $\frac{1}{2}$, for which l_0 corresponds to $(U_1 - U)/U_0 = e^{-1} = 0{\cdot}368$ and $e^{-\frac{1}{2}} = 0{\cdot}606$. Obviously, it is important to know which definition is used in any particular case.

The velocity range across the flow provides a convenient scale for the velocity variation: U_1 for a boundary layer, $U_1 - U(x,0)$ for a wake, and $U(x,0)$ for a jet. A more stringent test of similarity is provided by selecting length and velocity scales external to the profile. This is done in Figure 6.1, where d, x and U_1 provide scales for a plane wake. For a round jet, scaling might be effected using x and $U_0 = (F/\rho x^2)^{\frac{1}{2}}$, where F is the jet momentum. If the development is actually of self-preserving form, the external scales will change in proportion to those implicit in the profile. The relationship between these two kinds of scale will be considered again at the end of Section 7.3.3.

6.1.4 The Outer Flow

Our attention has so far been given to processes within the turbulence interface. Now we consider three aspects of the region beyond:

(1) the mean flows induced by the development of the shear layer;
(2) the non-steady motion impressed by the activity of the interface; and
(3) the dynamic boundary condition at the edge of the thin flow.

Displacement Flows and Entrainment

We have noted already, in introducing the displacement thickness, that boundary-layer growth is associated with an outwards mean velocity at the edge of the layer; that is, $V = V_1 > 0$ there. On the other hand, the deceleration of a jet, with axial momentum remaining constant, requires that the mass flux along the jet increase steadily. (This will be seen more clearly in Section 7.1.) Hence at the edge of the jet $V_1 < 0$, and the actual rate of advance of the turbulence interface is greater than would be inferred from an observation of the mean jet boundary.

The process of absorption of fluid into a turbulent flow is termed entrainment, but there is no general agreement on the precise meaning of the word. It might be defined in terms of:

(1) the mean fluid velocity V_1 at the edge of the turbulent region; $V_1 > 0$

implies an increase in the mass flux of the flow, and this could happen even in laminar motion;

(2) the mean velocity of the turbulent interface relative to the moving fluid, V_i, positive outwards; $V_i > 0$ implies an increase in the flux of turbulent fluid; and

(3) the absolute mean velocity of the interface, $V_1 + V_i$; $V_1 + V_i > 0$ implies that the boundary appears to move in the direction of V positive.

However it is defined, the rate of lateral spreading is small in a thin flow:

$$U \gg V_1,\ V_i,\ V_1 + V_i \tag{6.1.11}$$

Since the term entrainment is so loosely defined, we shall use it only in a general way to allude to the spreading of the flow. In reading, it is necessary to check carefully the way (or ways) in which a writer uses the term.

Fluctuations in the Outer Flow

Although random velocity fluctuations do occur beyond the turbulence interface, they are not turbulence in the strict sense, since the fluid is nearly irrotational and no production through vortex stretching can occur. The irrotational motion can be described using a *velocity potential* $\phi(x,y,z)$, related to the velocity vector and to the individual components by

$$\mathbf{u} = -\nabla\phi \quad \text{and} \quad u_i = -\frac{\partial \phi}{\partial x_i} \tag{6.1.12}$$

with ∇ the gradient operator. Phillips studied the statistical properties of the irrotational motion resulting from stationary random fluctuations in a fixed plane ($y = 0$, say), and the results have been found to be valid for the motion beyond actual turbulence interfaces. Among the simple relationships which Phillips derived are

$$\overline{v^2} = \overline{u^2} + \overline{w^2}, \quad \overline{uv} = 0, \quad \text{etc.} \quad \text{and} \quad \overline{q^2} \propto y^{-4} \tag{6.1.13}$$

The last result suggests that the forced motion dies away very rapidly within the non-turbulent fluid.

Dynamics of the Outer Flow

In a non-steady potential flow of constant-density fluid, the pressure and velocity are related at any instant by

$$\frac{\partial \phi}{\partial t} + \frac{P_1 + p}{\rho} + \tfrac{1}{2}(U_1 + u)^2 + \tfrac{1}{2}(V_1 + v)^2 + \tfrac{1}{2}w^2 = \text{constant}$$

In this generalization of *Bernoulli's theorem*, the varying velocity components are given in a form appropriate to the outer boundary of a turbulent motion with plane mean flow. Averaging over time for a statistically steady flow, we have

$$\frac{P_1}{\rho} + \tfrac{1}{2}U_1^2 + \tfrac{1}{2}V_1^2 + \tfrac{1}{2}\overline{q^2} = \text{constant} \qquad (6.1.14)$$

The last of the results (6.1.13) shows that, by going a little distance beyond the interface, we can render the final term negligible. Also, in a thin flow we may expect that $V_1^2 \ll U_1^2$. Hence the outer dynamic boundary condition is simply

$$\frac{P_1}{\rho} + \tfrac{1}{2}U_1^2 = \text{constant} \qquad (6.1.15)$$

and the pressure gradient acting on the boundary layer or other developing flow is related to the external velocity variation by

$$\frac{1}{\rho}\frac{\mathrm{d}P_1}{\mathrm{d}x} = -U_1\frac{\mathrm{d}U_1}{\mathrm{d}x} \qquad (6.1.15)$$

We assume that the length scales of the mean outer flow are large compared with those of the shear layer, so that no difficulty arises in defining the outer pressure as a function of the distance x measured along the flow.

To solve the thin-flow equations which will be developed shortly, we require a specification of one of the gradients (6.1.15). In an experiment on the development of a shear layer, the pressure or free-stream velocity can be measured. However, in analysing the flow around a structure which does not yet exist—say, the wing of an aeroplane at the design stage—one must predict these distributions before the production of the structure can proceed. Powerful methods of calculation are available to find the potential flow around a wing of specified shape. But we have seen that the growth of a boundary layer will shift the outer flow away from the surface, and this displacement may be comparable to the wing thickness. Hence a more appropriate free-stream velocity variation is that in the potential flow around a hypothetical body, the actual wing augmented by the displacement thickness of the boundary layer. Since the thickness itself depends upon the pressure distribution, an iterative solution may be required.

6.2 EQUATIONS FOR BOUNDARY LAYERS AND OTHER THIN FLOWS

6.2.1 Order-of-magnitude Analysis

We consider in the first instance an elongated flow of constant density whose mean motion is confined to the x, y-plane. Variations in velocity and in a general property S which is convected and diffused through the flow are governed by the continuity equation (3.3.7)

$$\frac{\partial U}{\partial x} + \frac{\partial V}{\partial y} = 0 \qquad (6.2.1)$$

and by a general conservation law obtained from equations (3.3.1, 2, 9)

$$\frac{\mathrm{D}S}{\mathrm{D}t} = U\frac{\partial S}{\partial x} + V\frac{\partial S}{\partial y} = \frac{\partial(SU)}{\partial x} + \frac{\partial(SV)}{\partial y}$$
$$= \frac{\dot{S}_s}{\rho} + \frac{\partial}{\partial x}\left(K\frac{\partial S}{\partial x} - \overline{su}\right) + \frac{\partial}{\partial y}\left(K\frac{\partial S}{\partial y} - \overline{sv}\right) \tag{6.2.2}$$

Here K is the molecular diffusivity of S, and $\dot{S}_s$ is its source strength, for which a number of interpretations were noted following equations (3.3.2).

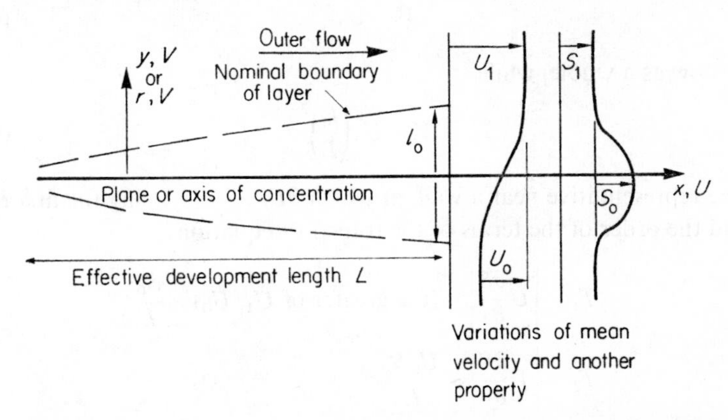

Figure 6.4 Basic geometry of a developing shear layer, showing length, velocity and property scales

Figure 6.4 depicts the basic geometry of the flow; it could be a boundary layer, wake, mixing layer, or jet in still or moving air, depending on the interpretation given to the *velocity scales*: U_1 for the outer flow, and U_0 for the variation and fluctuations in velocity. The linking of the mean velocity variation and turbulence assumes Reynolds-number similarity of the relevant scales.

The *length scales* are the distance L from the effective start of the developing layer to the station considered, and the distance l_0 over which the velocity changes by U_0. The criterion

$$\frac{l_0}{L} < 0{\cdot}1 \tag{6.2.3}$$

encompasses the whole of a developing flow such as a jet. The more stringent

$$\frac{l_0}{L} < 0{\cdot}01 \tag{6.2.3}$$

is representative of the wall layer, if any, l_0 then being measured from the wall to the edge of the wall layer.

The *property scales* are S_0, the change in the mean value over a distance comparable to l_0, and s_0, the typical magnitude of turbulent fluctuations s. The relationship between these scales depends on the property considered, as we shall see later. Note that in using l_0 for both velocity and property variations, we eschew cases such as $\mathrm{Pr} \ll 1$.

We now use these scales to express the orders of magnitude of the terms in the continuity and transport equations (6.2.1, 2). The former gives us the order of the lateral velocity. Its two terms must be comparable; hence for a flow near an impermeable wall or plane of symmetry

$$V \sim U_0 \frac{l_0}{L} \tag{6.2.4}$$

in the flow as a whole, while

$$V \sim U_0 \left(\frac{l_0}{L}\right)^2 \tag{6.2.4}$$

may be representative near a wall, at which $\partial U/\partial x = 0$. Using the first result, we find the order of the terms of the transport equation:

$$\begin{aligned}
T_1 &= \left| U\frac{\partial S}{\partial x} \right| \sim \text{(the greater of } U_1,\, U_0) \times \frac{S_0}{L} \\
T_2 &= \left| V\frac{\partial S}{\partial y} \right| \sim \frac{U_0 S_0}{L} \\
T_3 &= \left| \frac{\partial}{\partial x}\left(K\frac{\partial S}{\partial x}\right) \right| \sim \frac{KS_0}{L^2} \\
T_4 &= \left| \frac{\partial}{\partial y}\left(K\frac{\partial S}{\partial y}\right) \right| \sim \frac{KS_0}{l_0^2} \\
T_5 &= \left| \frac{\partial(\overline{su})}{\partial x} \right| \sim \frac{U_0 s_0}{L} \\
T_6 &= \left| \frac{\partial(\overline{sv})}{\partial y} \right| \sim \frac{U_0 s_0}{l_0}
\end{aligned} \tag{6.2.5}$$

Looking at these in pairs, we see that:

$$\text{(a)} \quad \frac{T_2}{T_1} \sim 1 \quad \text{or} \quad \frac{U_0}{U_1} \; (< 1, \text{ by hypothesis}) \tag{6.2.6}$$

The latter result shows that T_2 is small for a small-deficit wake and small-increment convected jet.

$$\text{(b)} \quad \frac{T_3}{T_4} \sim \left(\frac{l_0}{L}\right)^2 \tag{6.2.7}$$

Hence T_3 is nearly always negligible, although we must bear in mind extreme cases such as $\Pr \ll 1$, for which streamwise molecular conduction may be significant and l_0 is not a suitable scale for S.

(c) $$\frac{T_5}{T_6} \sim \frac{l_0}{L} \tag{6.2.8}$$

This is not a very conclusive result, since the correlation coefficient R_{su} may be considerably larger than R_{sv} (for example, when $s = u$). We shall try to keep T_5 if possible, but it will be dropped when it proves too inconvenient, and when the accuracy required does not justify a refined calculation.

Some further conclusions from the list (6.2.5) are:

(d) T_1, T_4 and T_6 are normally the more important terms. They must balance one another or the source term $\dot{S}_s$, whose likely magnitude depends on the property considered.

(e) For free turbulence or an outer turbulent region

$$T_4 \ll T_1, T_6 \text{ (and perhaps } \dot{S}_s) \tag{6.2.9}$$

(f) For a wall layer, the second equations of the pairs (6.2.3, 4) imply that

$$T_1, T_2 \ll T_4, T_6 \text{ (and perhaps } \dot{S}_s) \tag{6.2.10}$$

Here mean-flow convection is normally insignificant, although T_2 can be important near a permeable wall, a situation considered in Section 5.5, but not in the present analysis.

The simplified transport equations suggested by this analysis are given below. For *thin flows in general*:

$$\begin{aligned} U\frac{\partial S}{\partial x} + V\frac{\partial S}{\partial y} &= \frac{\partial (SU)}{\partial x} + \frac{\partial (SV)}{\partial y} \\ &= \frac{\dot{S}_s}{\rho} - \frac{\partial(\overline{sv})}{\partial y} + \frac{\partial}{\partial y}\left(K\frac{\partial S}{\partial y}\right) - \frac{\partial(\overline{su})}{\partial x} \end{aligned} \tag{6.2.11}$$

It may seem that not much has been achieved, but we know that the last term is probably small in many circumstances, that the penultimate term may be omitted outside the wall layer, and that the second mean-convection term is small in wake-like flows.

For *wall layers*, we have a more conclusive result:

$$\frac{\partial}{\partial y}\left(\overline{sv} - K\frac{\partial S}{\partial y}\right) = \frac{\dot{S}_s}{\rho} \tag{6.2.12}$$

This is the law on which our earlier analysis of transport in wall layers was based: the inner part of a wall flow is still 'locally determined'. Complete wall flows can be described by splicing an appropriate outer turbulent flow to the wall layer. Where necessary, the latter may include the effects of: a pressure gradient (Sections 3.5 and 5.5); roughness (Sections 4.3 and 5.4); or flow through the wall (Section 5.5). The outer flow does, of course, influence the wall layer by determining the wall stress, the stress gradient and the effective roughness; it is only the *form* of the wall laws which is independent of the mode of development.

Since the viscous part of a turbulent wall flow is described by these simpler laws, it should seldom be necessary to use equation (6.2.11) with the molecular-conduction term retained. In other words, molecular conduction and mean-flow convection are unlikely to be important together, in the same part of a turbulent wall flow (unless the wall is porous).

Laminar Flow

The simpler case of purely molecular diffusion is of interest for its own sake, and as a standard of comparison for turbulent processes. For thin laminar flows

$$U\frac{\partial S}{\partial x}+V\frac{\partial S}{\partial y}=\frac{\dot{S}_s}{\rho}+\frac{\partial}{\partial y}\left(K\frac{\partial S}{\partial y}\right) \tag{6.2.13}$$

If the source term is negligible, the equation will balance only if $T_1 \sim T_4$. Hence

$$l_0 \sim \left[\frac{U_1 \text{ or } U_0}{KL}\right]^{-\frac{1}{2}} \tag{6.2.14}$$

and the 'thinness' of the flow is specified by

$$\frac{l_0}{L} \sim \text{Pe}_L^{-\frac{1}{2}} = \left[\frac{\nu}{K}\text{Re}_L\right]^{-\frac{1}{2}} \qquad (6.2.14)$$

where Pe_L and Re_L are Péclet and Reynolds numbers based on L, the flow length. The final result suggests, for example, that

$$\frac{l_U}{l_T} \simeq \left(\frac{\nu}{K}\right)^{\frac{1}{2}} = \text{Pr}^{\frac{1}{2}} \tag{6.2.15}$$

relates the widths of the temperature and velocity profiles. While this result is exactly correct for Pr = 1, it indicates only the general nature of the relationship for Pr $\gtrless$ 1, since the role of convection will in general vary with Pr. Results to be obtained in Section 7.4.3 imply that (6.2.15) does in fact apply for a laminar wake; a more precise result for a laminar boundary layer will be developed in Section 8.3.

It is apparent from equations (6.2.14) that the thinness of a laminar flow increases, and the accuracy of the boundary-layer approximation improves, as the Reynolds number increases. This conclusion does not extend to all turbulent flows. It is true that $l_0/L \propto \text{Re}_L^{-\frac{1}{2}}$, approximately, for a constant-pressure boundary layer. But for a jet in still air, the angle of spread is Reynolds-number independent; at least an increase in Reynolds number does not worsen the thin-flow approximation.

Returning now to the laminar transport equation (6.2.13), we note that profiles of properties S_1 and S_2 (for example, U and H) will be similar when $K_1 = K_2$ (for example, when Pr = 1), provided that

(1) the boundary conditions are similar for the two transfers, and
(2) the source terms $\dot{S}_s$ have the same profile, or, more realistically, are negligible.

This conclusion cannot be carried over to turbulent flows; it is clear from Figure 5.2(b) that $\overline{s_1 v} \neq \overline{s_2 v}$ is possible, even though $K_1 = K_2$. Nor is it to be expected that the influence of the turbulence interface will be the same for transfers of momentum and of a passive entity such as heat.

6.2.2 Momentum Equations

Our first specific applications of the general result (6.2.11) are to the transport of cross-stream and streamwise momentum. For the former

$$S, S_0 \to V \sim U_0 l_0 / L \quad \text{and} \quad s, s_0 \to v \sim U_0$$

When these scales are introduced into the magnitudes (6.2.5), it is apparent that equation (6.2.11) may be reduced dramatically:

$$\frac{\partial(\overline{v^2})}{\partial y} = \frac{\dot{S}_s}{\rho} = -\frac{1}{\rho}\frac{\partial P}{\partial y} \tag{6.2.16}$$

on introducing the source term appropriate to momentum transfer in the y-direction. Integration gives

$$P + \rho\overline{v^2} = P_1(x) \tag{6.2.17}$$

with the arbitrary function fixed using conditions at the outer edge of the shear layer.

Formally, this result is equivalent to that (3.2.8) for parallel flow. But here it is not exact, since some normally small terms have been omitted. The error in the pressure change across the flow has elements

$$\frac{\Delta P}{\rho} \sim U_0^2\left(\frac{l_0}{L}\right), \quad U_0^2\left(\frac{\nu}{U_0 L}\right) \quad \text{and} \quad (\text{max. of } U_1, U_0)^2 \frac{l_0}{R} \tag{6.2.18}$$

The first two are neglected in equation (6.2.11); the last represents the centripetal pressure gradient in a flow concentrated near a surface with radius of curvature R, measured in the x,y-plane.

The interpretation of equation (6.2.17) also differs from that for parallel flow because the turbulence varies along the stream. Here

$$\frac{\partial P}{\partial x}=\frac{\mathrm{d}P_1}{\mathrm{d}x}-\rho\frac{\partial(\overline{v^2})}{\partial x} \tag{6.2.19}$$

showing that the pressure gradients outside the turbulent region and within are no longer precisely equal.

Turning now to the streamwise momentum, with $S \to U$ and $s \to u$, we note that $S_0 = U_0$ and $s_0 \sim U_0$. Hence no further terms of equation (6.2.11) become negligible. The momentum source term can be related to free-stream conditions using equations (6.2.19) and (6.1.15):

$$\begin{aligned}\dot{S}_s &= -\frac{\partial P}{\partial x}=-\frac{\mathrm{d}P_1}{\mathrm{d}x}+\rho\frac{\partial(\overline{v^2})}{\partial x}\\ &=\rho U_1\frac{\mathrm{d}U_1}{\mathrm{d}x}+\rho\frac{\partial(\overline{v^2})}{\partial x}\end{aligned} \tag{6.2.20}$$

Hence we obtain the *Reynolds momentum equation* for a boundary layer or other thin flow:

$$\begin{aligned}U\frac{\partial U}{\partial x}+V\frac{\partial U}{\partial y}+\frac{\partial}{\partial x}(\overline{u^2}-\overline{v^2}) &= U_1\frac{\mathrm{d}U_1}{\mathrm{d}x}+\frac{\partial}{\partial y}\left(\nu\frac{\partial U}{\partial y}-\overline{uv}\right)\\ &=-\frac{1}{\rho}\frac{\mathrm{d}P_1}{\mathrm{d}x}+\frac{1}{\rho}\frac{\partial\tau}{\partial y}\end{aligned} \tag{6.2.21}$$

The corresponding result (3.3.20) for exactly parallel flow retains only the last pair of terms. This approximation is still appropriate in a wall layer, as indicated in equation (6.2.12).

The convection terms on the left of equation (6.2.21) prevent explicit integration to find the shear-stress profile. In a developing flow, the shear-stress variation need not be linear (the fact that it is nearly so in a constant-pressure boundary layer is 'accidental'); its form is one of the flow characteristics which is to be determined. Note also that the dynamic balance is influenced by two direct components of the stress tensor. However, these normal-stress terms tend to cancel one another in equation (6.2.21), and may sometimes be neglected.

Laminar Flow

Restricting equation (6.2.21) to laminar flow, we obtain the *Prandtl momentum equation*:

$$\begin{aligned}\frac{\mathrm{D}U}{\mathrm{D}t} &= \frac{\partial U}{\partial t}+U\frac{\partial U}{\partial x}+V\frac{\partial U}{\partial y}\\ &=-\frac{1}{\rho}\frac{\mathrm{d}P_1}{\mathrm{d}x}+\frac{\partial}{\partial y}\left(\nu\frac{\partial U}{\partial y}\right)\end{aligned} \tag{6.2.22}$$

where the local acceleration $\partial U/\partial t$ has been added for generality. This, with the continuity equation

$$\frac{\partial U}{\partial x}+\frac{\partial V}{\partial y}=0 \qquad (6.2.22)$$

defines a closed mathematical problem, once $P_1(x)$, $\nu(x,y)$ and appropriate boundary conditions have been specified.

These laminar-flow results make clear the profound change in the problem of flow prediction which has been effected by the 'thin-flow' approximation. Now the pressure gradient is imposed on the flow, rather than arising spontaneously within it. Moreover, the rejection of one of the higher-order stress terms—$\partial(\nu\,\partial U/\partial x)/\partial x$—reduces the number of boundary conditions which can (or must) be satisfied. These changes imply that the thin layer is independent of conditions downstream; hence its behaviour can be predicted by a once-for-all streamwise integration, in which account is taken of the 'history' of the flow, but not of its 'future'. (This behaviour is consistent with the engineer's dictum: 'You can't push on a rope'.) Near a point of separation or stagnation, where U, $V=0$, the flow is no longer nearly parallel, and the thin-flow approximations break down. The pressure and velocity fields are then connected by their usual, more complex, relationship.

These comments apply, in a general way, to thin turbulent flows, although the nature of the mathematical problem is less obvious and does, indeed, depend on the model chosen to represent turbulent transfers.

6.2.3 Energy Equations

Consider first the *turbulence kinetic energy*. Since the source term $\dot{S}_s$ has a rather complicated form, as is evident from the parallel-flow results (3.3.24, 25, 26), we shall not work directly from the basic result (6.2.11). Instead, we add the mean-flow convection to the net cross-stream flux for parallel flow:

$$\frac{\mathrm{D}(\tfrac{1}{2}\overline{q^2})}{\mathrm{D}t}=\tau\frac{\partial U}{\partial y}-\frac{\partial}{\partial y}(\dot{W}_p+J_q)-\rho\varepsilon \qquad (6.2.23)$$

Note that a term of form $\partial(\overline{su})/\partial x$ is neglected by this procedure.

For the fully turbulent region, we obtain

$$\frac{\mathrm{D}(\tfrac{1}{2}\overline{q^2})}{\mathrm{D}t}+\overline{uv}\frac{\partial U}{\partial y}+\frac{\partial}{\partial y}\left[\overline{v\left(\frac{p}{\rho}+\tfrac{1}{2}q^2\right)}\right]+\varepsilon_t=0 \qquad (6.2.24)$$

The last three terms were discussed following equation (3.3.26). The first represents *mean-flow convection or advection.* Figure 6.5 shows how these contributions vary across a turbulent wake, and in the outer part of a constant-pressure boundary layer. These values were obtained, like those of Figure 3.8, by the somewhat untrustworthy procedure of estimating the dissipation from limited measurements and then finding the 'pressure diffusion' as the balancing item. However, they demonstrate the general nature of the energy balance. The corresponding variations in mean velocity and shear stress are given for reference.

Some expected features common to Figures 6.5(a) and (c) are:

(1) The dissipation is positive everywhere.

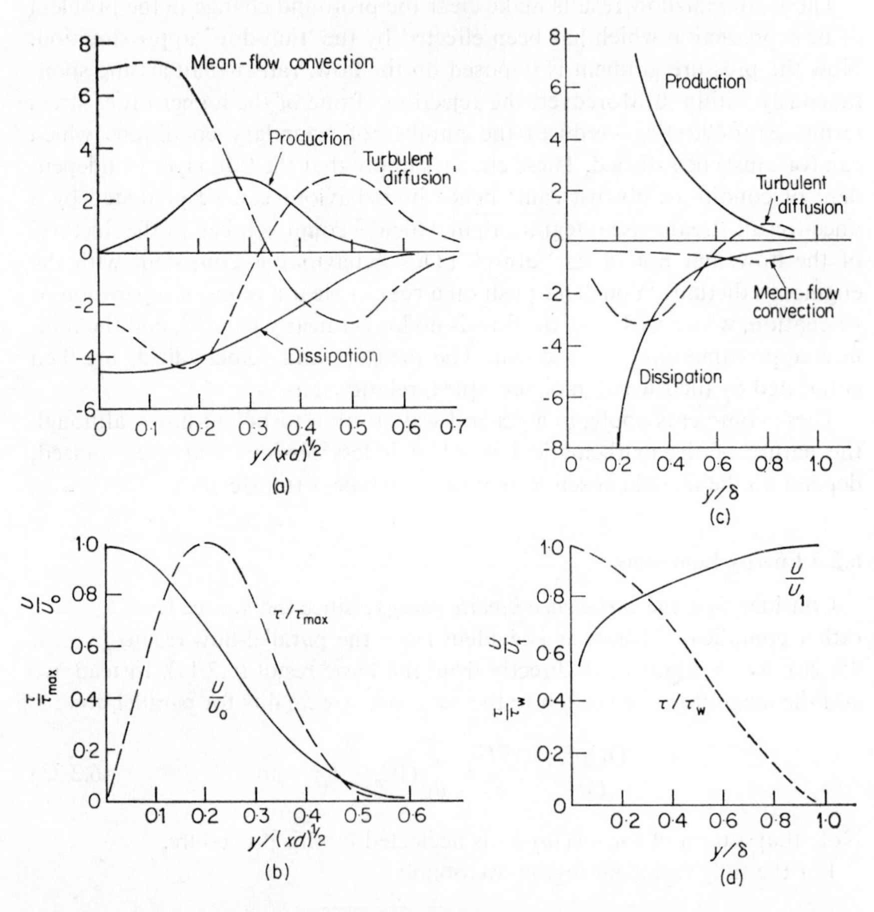

Figure 6.5 The turbulence energy equation and associated variations of mean velocity and shear stress. (a) Energy balance for a plane wake, according to Townsend, for $\mathrm{Re}_d = 8400$, $x/d = 160$, with virtual origin $25d$ ahead of the generating cylinder. The terms of equation (6.2.24) are made dimensionless by division by the scale U_1^3/d. (b) Velocity and shear-stress variations for a plane wake, corresponding to (a). (c) Energy balance for a constant-pressure boundary layer, according to Townsend, for $U_1\delta_1/\nu \simeq 4000$ and $u_\tau/U_1 \simeq 0{\cdot}043$. The terms of equation (6.2.24) are made dimensionless by division by the scale u_τ^3/δ. (d) Velocity and shear-stress variations for a constant-pressure boundary layer, corresponding to (c)

(2) Production has the opposite sign, and is a maximum where the velocity gradients are largest.

(3) Diffusion is, generally speaking, from a region of high intensity to one of low intensity; in each case, it is towards the spreading boundary, in the outer part of the shear layer.

(4) The relationship between the several terms is much the same near each free boundary.

Some significant differences in the two energy balances are:

(5) In the wake, there is no region where dissipation and production are dominant, and hence nearly in balance.

(6) Mean-flow convection is significant only in the outer half of the boundary layer, but is important throughout the wake.

As has been anticipated in a number of places, we may conclude that the inner, say, twenty per cent of the boundary layer is a 'locally determined' or 'equilibrium' layer in the sense of Section 3.5. This is consistent with Figure 6.5(d) which shows that the shear stress changes by a relatively small fraction in this layer, while the greater part of the velocity change occurs there.

For a developing flow the balance of *mean-flow kinetic energy* is also of interest. One way of looking at this (there are several) is to multiply the momentum equation (6.2.21) by the mean velocity:

$$\frac{\mathrm{D}(\frac{1}{2}U^2)}{\mathrm{D}t}+U\frac{\partial}{\partial x}(\overline{u^2}-\overline{v^2})=-\frac{U}{\rho}\frac{\mathrm{d}P_1}{\mathrm{d}x}+\frac{U}{\rho}\frac{\partial \tau}{\partial y} \tag{6.2.25}$$

This relates the change in kinetic energy to work done by the normal stresses, by the pressure gradient and by the shear stress.

Turning now to *thermal energy equations*, we generalize the convective derivatives of equations (3.3.33, 35, 37) to obtain results for thin developing flows.

(1) The internal-energy equation for constant density is

$$\rho\frac{\mathrm{D}E}{\mathrm{D}t}+\frac{\partial \dot{q}_\mathrm{e}}{\partial y}=\rho\varepsilon \tag{6.2.26}$$

(2) The enthalpy equation for low-speed flow of any fluid is

$$\rho\frac{\mathrm{D}H}{\mathrm{D}t}+\frac{\partial \dot{q}_\mathrm{h}}{\partial y}=\frac{\partial}{\partial y}(\tau U-J_q) \tag{6.2.27}$$

(3) The total-enthalpy equation is

$$\rho\frac{\mathrm{D}H_0}{\mathrm{D}t}+\frac{\partial \dot{q}_{\mathrm{h}_0}}{\partial y}=\rho\frac{\partial}{\partial y}\left[\tfrac{1}{2}\nu\left(1-\frac{1}{\mathrm{Pr}}\right)\frac{\partial(U^2+\overline{q^2})}{\partial y}\right] \tag{6.2.28}$$

requiring only that density changes *across* the flow be small.

In all of these equations, streamwise diffusion is neglected.

In each case, the source term on the right is often negligible; thus, for example

$$U\frac{\partial H}{\partial x}+V\frac{\partial H}{\partial y}=-\frac{1}{\rho}\frac{\partial \dot{q}_h}{\partial y}$$
$$=\frac{\partial}{\partial y}\left(\kappa\frac{\partial H}{\partial y}-\overline{vh}\right) \tag{6.2.29}$$

As anticipated in Section 6.2.1, this has the form of the momentum equation (6.2.21), provided that the pressure-gradient and normal-stress terms are also negligible. If the shear stress and enthalpy flux vary in the same way (that is, for $Pr=1$ and/or $Pr_t=1$, depending on the application), the profiles $U(y)$ and $H(y)$ can be similar. Then the solution of the momentum equation gives the enthalpy distribution immediately.

More generally, the momentum equation must be solved before the heat-transfer equation, in order that the convection terms can be specified for the latter calculation. Of course, even for a wall layer with negligible mean convection, the velocity variation must be known, in order that the turbulent diffusion ($\overline{vh}$ or ϵ_h) can be calculated.

6.2.4 Integral Results

Many interesting and useful relationships among the gross properties of thin flows can be obtained by integrating the differential transport equations *across* the flow. The resulting equations give information about streamwise development, sometimes in very simple form owing to the vanishing of certain contributions during integration.

In the following derivations, we shall consider the plane $y=0$ to represent either a fixed, impermeable wall or a plane of symmetry: in either case, $V(0)=0$. We integrate the general result (6.2.11) from $y=0$ to $y=y_1$, a point within the free stream, taking y_1 constant along the flow. On this latter plane, $S=S_1(x)$ and the diffusive flux $J(y_1)=J_1=0$ (a particular instance is $\tau_1=0$). Thus we obtain the general integral

$$\frac{d}{dx}\int_0^{y_1}(SU+\overline{su})\,dy+S_1V_1=\int_0^{y_1}\frac{\dot{S}_s}{\rho}\,dy+\frac{J_0}{\rho} \tag{6.2.30}$$

The flux at $y=0$ is

$$J_0=-\rho K\left(\frac{\partial S}{\partial y}\right)_w \quad \text{for a wall}$$
$$=0 \quad \text{for a plane of symmetry in free turbulence.}$$

To reduce the integrated result further, we note that the continuity equation (6.2.1) implies that

$$V_1=-\int_0^{y_1}\frac{\partial U}{\partial x}\,dy=-\frac{d}{dx}\int_0^{y_1}U\,dy \tag{6.2.31}$$

Using the manipulation

$$S_1 V_1 = -\frac{\mathrm{d}}{\mathrm{d}x}\int_0^{y_1} S_1 U\,\mathrm{d}y + \frac{\mathrm{d}S_1}{\mathrm{d}x}\int_0^{y_1} U\,\mathrm{d}y$$

we obtain

$$\frac{\mathrm{d}}{\mathrm{d}x}\int_0^{y_1} U(S - S_1)\,\mathrm{d}y = \int_0^{y_1}\left[\frac{\dot{S}_s}{\rho} - \frac{\partial(\overline{su})}{\partial x} - U\frac{\mathrm{d}S_1}{\mathrm{d}x}\right]\mathrm{d}y + \frac{J_0}{\rho} \tag{6.2.32}$$

This relates the flux of the deficit in S to the wall flux and a complicated 'source' term.

Momentum Integrals

For the particular case of the streamwise momentum equation:

$$S \to U,\ s \to u,\ J_0 \to -\tau_w \quad \text{and} \quad \dot{S}_s/\rho = U_1\,\mathrm{d}U_1/\mathrm{d}x + \partial\overline{v^2}/\partial x$$

The last result follows from equations (6.2.20). When these are introduced, the integral equation becomes

$$\frac{\mathrm{d}}{\mathrm{d}x}\int_0^{y_1} U(U_1 - U)\,\mathrm{d}y + \frac{\mathrm{d}U_1}{\mathrm{d}x}\int_0^{y_1}(U_1 - U)\,\mathrm{d}y + \frac{\mathrm{d}}{\mathrm{d}x}\int_0^{y_1}(\overline{v^2} - \overline{u^2})\,\mathrm{d}y = \frac{\tau_w}{\rho} \tag{6.2.33}$$

In terms of the displacement and momentum thicknesses (6.1.5, 6):

$$\frac{\mathrm{d}(\rho U_1^2\delta_2)}{\mathrm{d}x} = \tau_w + \frac{\mathrm{d}P_1}{\mathrm{d}x}\delta_1 + \frac{\mathrm{d}}{\mathrm{d}x}\int_0^{y_1}(\overline{u^2} - \overline{v^2})\,\mathrm{d}y \tag{6.2.33}$$

These results connect the change in momentum flux along the flow to the wall stress and pressure force, with a normal-stress correction. This correction is most likely to be significant when the other terms are small, for example, near separation.

The momentum integral is often written, with the normal-stress term omitted, as

$$\frac{\mathrm{d}\delta_2}{\mathrm{d}x} + \frac{H+2}{U_1}\frac{\mathrm{d}U_1}{\mathrm{d}x}\delta_2 = \tfrac{1}{2}c_f \tag{6.2.34}$$

where $H = \delta_1/\delta_2$ is the shape factor of equation (6.1.10), and c_f is the wall friction coefficient. This result is referred to as the *von Kármán integral momentum equation.* It shows clearly the significance of the integral formulation: a partial differential equation has been reduced to an ordinary differential equation prescribing the overall pattern of development. To illustrate the power of this result, we consider some simple applications immediately.

(1) For a *free-turbulent flow*, equation (6.2.34) gives

$$\frac{d\delta_2}{\delta_2} = -(H+2)\,\frac{dU_1}{U_1}$$

If the flow is self-preserving, with H constant as in equations (6.1.8, 9), integration gives

$$\delta_2\, U_1^{H+2} = \text{constant} \tag{6.2.34}$$

This relates the flow width to changes in the free-stream velocity along the flow.

(2) For a *constant-pressure boundary layer*, laminar or turbulent, integration along the stream gives

$$\rho U_1^2\,\delta_2(L) = \int_0^L \tau_w(x)\,dx = D'(L)$$

or (6.2.35)

$$2\frac{\delta_2(L)}{L} = \frac{1}{L}\int_0^L c_f\,dx = C_f$$

where D' is the total drag applied on the surface, and C_f is the *average friction coefficient* over the length L. The first result shows how δ_2 represents the momentum deficit of the layer.

Table 6.1. Approximate laws of growth and friction for boundary layers with uniform external velocity U_1 and uniform wall roughness height k

Flow type	Proposed variation of c_f	Predicted variations	
		δ	δ/x and c_f
Laminar	Re_δ^{-1}	$(\nu/U_1)^{\frac{1}{2}}x^{\frac{1}{2}}$	$Re_x^{-\frac{1}{2}}$
Turbulent, smooth wall	$Re_\delta^{-\frac{1}{4}}$	$(\nu/U_1)^{\frac{1}{5}}x^{\frac{4}{5}}$	$Re_x^{-\frac{1}{5}}$
Turbulent, fully rough	$(\delta/k)^{-\frac{1}{3}}$	$k^{\frac{1}{4}}x^{\frac{3}{4}}$	$(x/k)^{-\frac{1}{4}}$

(3) Another way of using the momentum integral to investigate a constant-pressure boundary layer is to note that

$$\frac{d\delta_2}{dx} = \tfrac{1}{2}c_f \tag{6.2.35}$$

specifies either the friction variation or the thickness variation, once the other is known. Our work on pipe friction has indicated the general form of the relationship between friction and flow width, and thus provides a starting point for the calculation.

Table 6.1 gives the relationships suggested by the pipe-friction laws: laminar, from Table 3.2; smooth-wall turbulent, from the Blasius law (4.2.22); and rough-wall turbulent, approximately as in equation (4.3.17). These calculations implicitly assume similar profiles, and $\delta_2 \propto \delta$, any other measure of boundary-layer thickness. The laws are presented in terms of

$$\mathrm{Re}_\delta = \frac{U_1 \delta}{\nu} \tag{6.2.36}$$

the Reynolds number based on the local thickness, and

$$\mathrm{Re}_x = \frac{U_1 x}{\nu} \tag{6.2.36}$$

that based on the effective development length.

Energy Integrals

To consider the flux of enthalpy, we introduce

$$S \rightarrow H \quad \text{and} \quad J_0 \rightarrow \dot{q}_\mathrm{w}$$

On taking H_1 constant, and neglecting the dissipation-diffusion term on the right of equation (6.2.27), and also the streamwise transfer $\overline{su}$, we find that the 'source' integral of equation (6.2.32) vanishes. The *thermal energy integral* is then

$$\frac{\mathrm{d}}{\mathrm{d}x}\int_0^{y_1} U(H - H_1)\,\mathrm{d}y = \frac{\dot{q}_\mathrm{w}}{\rho} \tag{6.2.37}$$

For the case of constant wall temperature, we can use the constant $H_\mathrm{w} - H_1$ as a scale for the enthalpy variation, and

$$\frac{\mathrm{d}\delta_T}{\mathrm{d}x} = \frac{\dot{q}_\mathrm{w}}{\rho U_1(H_\mathrm{w} - H_1)} = \mathrm{St}(x) \tag{6.2.38}$$

with St(x) the local Stanton number, and

$$\delta_T = \int_0^{y_1} \frac{U}{U_1}\frac{H - H_1}{H_\mathrm{w} - H_1}\,\mathrm{d}y \tag{6.2.38}$$

a *thermal or convection thickness* for the flow. For free turbulence, we have simply

$$\delta_T = \text{constant along the flow} \tag{6.2.39}$$

in the absence of internal sources.

Integration of the turbulence energy equation (6.2.24) across the flow gives generalizations of equations (3.3.28, 30), but these will not be considered here. Much the same information is contained in the *mean-flow energy integral* obtained from equation (6.2.25), and this is also of interest because it is often used (see Section 8.2.3) as an auxiliary equation in boundary-layer calculations. The required result cannot be derived from equation (6.2.32), owing to the prior multiplication by the mean velocity U, but similar methods lead to

$$\frac{\mathrm{d}}{\mathrm{d}x}\int_0^{y_1} U(\tfrac{1}{2}U_1^2 - \tfrac{1}{2}U^2)\,\mathrm{d}y = \frac{1}{\rho}\int_0^{y_1} \tau\frac{\partial U}{\partial y}\,\mathrm{d}y + \int_0^{y_1} U\frac{\partial(\overline{u^2} - \overline{v^2})}{\partial x}\,\mathrm{d}y \tag{6.2.40}$$

The relationships (6.2.31) and (6.1.15) have been used, together with the conditions $\tau(y_1) = 0$, and either $U(0) = 0$ or $\tau(0) = 0$. When the normal stresses are omitted

$$\frac{\mathrm{d}(\frac{1}{2}\rho U_1^3 \delta_3)}{\mathrm{d}x} = \int_0^{y_1} \tau\frac{\partial U}{\partial y}\,\mathrm{d}y \tag{6.2.41}$$

with δ_3 the energy thickness of equation (6.1.7).

Referring to the energy balance (6.2.23), we see that the work integral of equations (6.2.40, 41) represents energy extraction from the mean flow. In a turbulent flow, this differs from the total dissipation to 'heat' within the section, by virtue of the net streamwise convection. Nevertheless, it is often referred to as the *dissipation integral*.

6.2.5 Axisymmetric Flows

For each of the preceding results for plane flow there is an equivalent for flow with axial and radial components $U(x,r)$ and $V(x,r)$. In the derivations we must, of course, take account of the altered geometry; this introduces a factor r here and there.

We consider *nearly axial flows*, with $U \gg V$ and $\partial/\partial r \gg \partial/\partial x$. This class includes round wakes and jets, and boundary layers on and within cylinders. An order-of-magnitude analysis like that of Section 6.2.1 reduces the conservation law (3.3.3) to

$$\frac{\mathrm{D}S}{\mathrm{D}t} = U\frac{\partial S}{\partial x} + V\frac{\partial S}{\partial r} = \frac{\partial(SU)}{\partial x} + \frac{1}{r}\frac{\partial(rSV)}{\partial r} = \frac{\dot{S}_s}{\rho} - \frac{1}{r}\frac{\partial}{\partial r}\left[r\left(\overline{sv} - K\frac{\partial S}{\partial r}\right)\right] - \frac{\partial(\overline{su})}{\partial x} \tag{6.2.42}$$

For a wall layer, this reduces further to

$$\frac{\partial}{\partial r}\left[r\left(\overline{sv} - K\frac{\partial S}{\partial r}\right)\right] = \frac{r\dot{S}_s}{\rho} \tag{6.2.43}$$

Integrating the result (6.2.42) between the constant wall radius R_0 and a cylinder of uniform diameter in the outer flow, $r = R_1$, we get

$$\frac{d}{dx}\int_{R_0}^{R_1}(SU + \overline{su})\,r\,dr + R_1 V_1 S_1 = \int_{R_0}^{R_1}\frac{\dot{S}_s}{\rho}\,r\,dr + \frac{R_0 J_0}{\rho} \qquad (6.2.44)$$

Here we consider $R_1 > R_0$; the other case can be treated in a similar way. The continuity equation (3.3.8) gives

$$R_1 V_1 = -\int_{R_0}^{R_1}\frac{\partial U}{\partial x}\,r\,dr \qquad (6.2.45)$$

and this can be used to rewrite the last result in the manner of equation (6.2.32).

Momentum Equations

Introducing $S = U$ and $\dot{S}_s = -\partial P/\partial x$ into the general results (6.4.43, 44), we obtain:

(1) the differential momentum equation

$$U\frac{\partial U}{\partial x} + V\frac{\partial U}{\partial r} + \frac{\partial(\overline{u^2} - \overline{v^2})}{\partial x} = U_1\frac{dU_1}{dx} + \frac{1}{r}\frac{\partial}{\partial r}\left(v\frac{\partial U}{\partial r} - \overline{uv}\right)$$
$$= -\frac{1}{\rho}\left[\frac{dP_1}{dx} - \frac{1}{r}\frac{\partial}{\partial r}(r\tau)\right] \qquad (6.2.46)$$

(2) the integral momentum equation

$$\frac{d}{dx}\int_{R_0}^{R_1}[U(U_1 - U) - \overline{u^2} + \overline{v^2}]\,r\,dr + \frac{dU_1}{dx}\int_{R_0}^{R_1}(U_1 - U)\,r\,dr = \frac{R_0\tau_w}{\rho} \qquad (6.2.47)$$

For *free turbulence*, $R_0 = 0$, and it is convenient to introduce radii to represent displacement and momentum loss:

$$r_1^2 = 2\int_0^{R_1}\left(1 - \frac{U}{U_1}\right)r\,dr$$

and (6.2.48)

$$r_2^2 = 2\int_0^{R_1}\frac{U}{U_1}\left(1 - \frac{U}{U_1}\right)r\,dr$$

In this case $\tau_w = 0$, and the integral (6.2.47), with normal stresses neglected, becomes

$$\frac{dr_2^2}{dx} + \frac{H' + 2}{U_1}\frac{dU_1}{dx}r_2^2 = 0 \qquad (6.2.48)$$

with $H' = (r_1/r_2)^2$ an appropriate shape factor. For self-preserving flows, H' is constant, and

$$r_2\,U_1^{\frac{1}{2}H'+1} = \text{constant} \qquad (6.2.49)$$

relates the radius and the varying external velocity.

Energy Equations

As an example, we consider the enthalpy integral analogous to (6.2.37). With $S = H$, and taking

$$H_1 = \text{constant}, \quad \dot{S}_s = 0 \quad \text{and} \quad \partial(\overline{uh})/\partial x = 0$$

we obtain from equations (6.2.44, 45) the simple result

$$\frac{d}{dx}\int_{R_0}^{R_1} U(H - H_1)\, r\, dr = \frac{R_0 \dot{q}_w}{\rho} \tag{6.2.50}$$

The enthalpy-flux integral can be related to the heat input from an initial state with $H(r) = H_1$:

$$2\pi\rho \int_{R_0}^{R_1} U(H - H_1)\, r\, dr = 2\pi R_0 \int_0^L \dot{q}_w(x)\, dx = \dot{Q}(L) \tag{6.2.51}$$

In free turbulence, the enthalpy flux is uniform along the flow.

6.2.6 Buoyant Flows

Figure 6.6 indicates the kind of motion we shall consider:

(1) The flow is nearly vertical, with $U \gg V$; it is thin in the horizontal direction.

(2) The motion may be *natural convection*, driven entirely by heat supplied (or removed) at a wall or at a thermal source, or

(3) The motion may combine *forced and free convection*, being influenced by the outer stream ($U_1 \neq 0$) or by momentum introduced through a jet.

(4) Alternatively, the buoyancy (positive or negative) may arise from a difference in the fluid species of the shear layer and outer flow, or from matter dissolved or suspended in small drops or bubbles.

For simplicity, we shall take

$$U_1 = \text{constant} \quad \text{and} \quad \frac{dP_1}{dx} = -\rho_1 g, \quad \text{a constant} \tag{6.2.52}$$

throughout the outer flow. Assuming that this pressure variation is impressed across the thin flow, we obtain the equation of motion:

$$\begin{aligned} \rho_1 \frac{DU}{Dt} &= -\frac{dP_1}{dx} - \rho g + \frac{\partial \tau}{\partial y} \\ &= g(\rho_1 - \rho) + \frac{\partial \tau}{\partial y} \end{aligned} \tag{6.2.53}$$

The quantity $g(\rho_1 - \rho)$ is called the *buoyancy* of the fluid. We can neglect the effect of small density changes on the inertia and continuity relations for the fluid, because they introduce second-order changes. For example, $\rho DU/Dt$

and $\rho_1 \mathrm{D}U/\mathrm{D}t$ differ only in a term $\sim O(\Delta\rho/\rho_1)^2$. To this approximation, the integrals (6.2.30, 31, 32) and (6.2.44, 45) are still applicable.

For compactness, we adopt conventions which allow the results for plane and round flows to be combined; thus

$$\frac{\mathrm{D}U}{\mathrm{D}t} = g\left(1 - \frac{\rho}{\rho_1}\right) + \frac{1}{y^j}\frac{\partial}{\partial y}\left[y^j\left(\nu\frac{\partial U}{\partial y} - \overline{uv}\right)\right] \tag{6.2.54}$$

with $j = 0$ for plane flow, and $j = 1$ for axisymmetric flow.

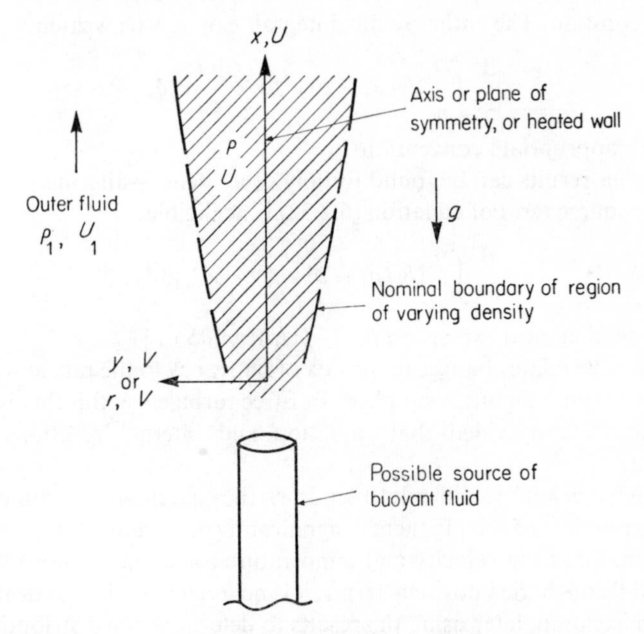

Figure 6.6 Schematic representation of flows of varying density

Integral Results

A comparison of equation (6.2.54) with the general result (6.2.11) reveals that the source for momentum is now $\dot{S}_s = g(\rho_1 - \rho)$: buoyancy simply replaces the pressure gradient. Hence the *momentum integrals* are

$$\frac{\mathrm{d}}{\mathrm{d}x}\int_{y_0}^{y_1} \rho_1 U(U - U_1) y^j \,\mathrm{d}y = \int_{y_0}^{y_1} g(\rho_1 - \rho) y^j \,\mathrm{d}y - y_0^j \tau_w \tag{6.2.55}$$

with appropriate conventions. These results show that the momentum change is balanced by the change in the integrated buoyancy and by the wall stress, if any.

Let us restrict the discussion to thermal buoyancy for the time being. Since natural convection does not involve high velocities, we may adopt the energy integrals (6.2.37, 50), with their neglect of dissipation. We accept also the restriction to H_1 constant, and to negligible internal generation of heat. While the fundamental forms of the differential and integral energy equations are not altered by buoyancy, it is convenient to rewrite them by introducing

$$H - H_1 = c_p(T - T_1) = \frac{c_p}{\beta}\left(1 - \frac{\rho}{\rho_1}\right) \tag{6.2.56}$$

with $\beta = -(1/\rho)(\partial\rho/\partial T)_p$ the coefficient of cubical expansion of the fluid, assumed constant. The enthalpy-flux integrals can now be written

$$\frac{\mathrm{d}}{\mathrm{d}x}\int_{y_0}^{y_1} Ug(\rho_1 - \rho)\, y^j \,\mathrm{d}y = \frac{g\beta}{c_p}\, y_0^j \dot{q}_w \tag{6.2.57}$$

again with appropriate conventions.

Analogous results can be found for any other density-altering property for which the source term of equation (6.2.32) is negligible:

$$\frac{\mathrm{d}}{\mathrm{d}x}\int_{y_0}^{y_1} Ug\,(\rho_1 - \rho)\, y^j \,\mathrm{d}y = g\beta_s y_0^j J_0 \tag{6.2.58}$$

with a general cubical expansion $\beta_s = -(1/\rho)(\partial\rho/\partial S)_p$.

These results relate changes in the flux of buoyancy to the rate at which heat (or another contaminant) is supplied. For free turbulence, this flux is uniform along the flow, provided that radiation and internal reactions are unimportant.

Although it is not too difficult to set down the equations governing buoyant flows, the problem of solving them is significantly more difficult than for forced convection. Since the velocity and temperature (or concentration) variations are linked through the buoyancy terms, it is no longer possible to deal with the former in isolation, later using the results to determine the distribution of the passive contaminant.

6.3 METHODS OF ANALYSIS

We now have lots of equations; what can be done with them? Our position with regard to thin developing flows is like that attained in Section 3.4 for channel flows. Time-averaged conservation laws have been established, and we have a superficial understanding of the processes they describe, with a fairly realistic picture of the variations of their terms within some simple flows.

Let us first see how we stand with respect to *closure*. The continuity, momentum and turbulence-energy equations for fully turbulent plane flow are

$$\frac{\partial U}{\partial x} + \frac{\partial V}{\partial y} = 0 \tag{6.3.1}$$

$$U\frac{\partial U}{\partial x}+V\frac{\partial U}{\partial y}=-\frac{1}{\rho}\frac{\mathrm{d}P_1}{\mathrm{d}x}-\frac{\partial \overline{uv}}{\partial y} \tag{6.3.2}$$

$$U\frac{\partial(\frac{1}{2}\overline{q^2})}{\partial x}+V\frac{\partial(\frac{1}{2}\overline{q^2})}{\partial y}+\overline{uv}\frac{\partial U}{\partial y}+\frac{\partial}{\partial y}\left[\overline{v\left(\frac{p}{\rho}+\frac{1}{2}q^2\right)}\right]=-\varepsilon_t \tag{6.3.3}$$

Even though the normal stresses $\overline{u^2}$ and $\overline{v^2}$ have been omitted, these contain two more time-mean quantities (V and $\frac{1}{2}\overline{q^2}$) than do the parallel-flow equations (3.4.1, 2), while only one additional equation (continuity) is available. Moreover, the streamwise component U plays a more complex role here. To obtain explicit predictions, we must not only specify the turbulent transfer across the flow, but must also define the role of mean-flow convection along the developing flow.

The mean-flow convection terms acknowledge the dependence of the turbulent activity at each point on the statistical history of the fluid passing that point. Hence the motion can no longer be considered to be 'locally determined' as in a wall layer where mean-flow convection is unimportant. In a broader sense, the history of the fluid is important even in a wall layer; the empirical coefficients for the wall layer take account of this.

In Section 1.4 it was pointed out that undisturbed thin flows, those spreading into still air or into a uniform stream, display streamwise similarity: their time-averaged profiles have much the same shape from section to section. Figure 6.1 illustrates this behaviour. The existence of similar profiles indicates that the mean flow and turbulence settle into a self-preserving pattern in which the larger elements develop in harmony. The fact (or assumption) of similarity implies that the motion at any section is related in a simple way to that upstream; thus it accounts in a general way for the history of the developing flow. We are still left, however, with the problem of describing cross-stream transfers.

Nearly every analysis of developing flow can be put into one of four categories of increasing complexity:

I. *Similarity arguments* based on momentum integrals, followed by the use of the gradient-diffusion models of Section 3.4, often with either eddy viscosity or mixing length assumed constant.

II. Consideration of the consequences of *self-preservation* for the differential and integral equations of motion, followed by a gradient-diffusion model.

III. *Streamwise integration* of ordinary differential equations derived from momentum integrals or from modified momentum integrals. The latter may be *strip integrals*, obtained by integrating part way across the flow, or *moment-of-momentum integrals*, with integrands weighted with factors y^m or U^m. Reduction of the integrals is achieved by introducing cross-stream profiles based on similarity and gradient diffusion, for each of the distant layers into which the flow is divided.

IV. The use of partial differential *transport equations* for certain attributes of the turbulence, usually including the turbulence kinetic energy. These account for both streamwise and cross-stream 'history'. A closed set of equations, simple enough for rapid numerical integration, is obtained by either

(a) postulating plausible forms for the transport equations, including constants which can be chosen to match measurements, or
(b) truncating exact transport equations developed from the equations of motion, on the basis of expediency and/or statistical arguments.

Moreover, it is expedient to shorten the calculations by assuming the flow near a wall to be locally determined.

The applicability of approaches I to III is severely limited:

(1) They depend on the 'thinness' of the flow and require either plane or axisymmetric mean flow.
(2) They use similarity (within individual layers, at least) to account for mean-flow convection and development.
(3) They incorporate a gradient-diffusion model of turbulent mixing, or something equally simplistic.

Even when the geometric restrictions (1) and (2) are realistic, gradient diffusion may fail to provide a reasonable (let alone accurate) representation of transfers across the flow. In Section 3.4.6 we noted some limitations of gradient models for channel flows. Analogous problems arise for free turbulence and developing wall flows:

(1) If the eddy viscosity is assumed uniform across a wake or jet, the mixing length (see equations (3.4.60)) must be infinite at the axis.
(2) If the mixing length is assumed uniform, the eddy viscosity is zero at the axis, implying that no transfer can occur from side to side of the flow.
(3) For a wall jet, a variety of ridiculous predictions follow when $\partial U/\partial y = 0$ and $\tau = 0$ do not coincide. The example of Figure 3.12 illustrates this.

To apply the simpler methods of analysis (I to III above), we must not only carefully select our flow, but must confine our attention to those aspects that are broadly independent of the details of the transfer processes. In practice, this means that we can do little more than predict distributions of mean velocity, temperature and concentration. We cannot predict with any confidence finer detail such as: points of maximum velocity; profiles of shear stress, energy flux and turbulence intensities; or mean profiles near the spreading boundary. For such details of thin flows, and for virtually any information about more complex flow geometries, we must turn to the more comprehensive transport models of category IV.

In the following two chapters we shall see what can be learned from the simpler analytical techniques: similarity arguments (I and II) tell us a good deal about free turbulence, and streamwise integration (III) comes into its own for boundary layers. In Chapter 9 we shall look briefly at the transport models which have been developed in recent years.

FURTHER READING

Rouse, H. Reference 10: Chapter VII, Sections A and B
Schlichting, H. Reference 11: Chapters II, VII and VIII
Townsend, A. A. Reference 13: Chapter 7

SPECIFIC REFERENCES

Coles, D. 'Interfaces and intermittency in turbulent shear flow', in Reference 14

Phillips, O. M. 'The irrotational motion outside a free turbulent boundary', *Proc. Camb. Phil. Soc.*, **51**, pp. 220–229 (1955)

Phillips, O. M. 'The entrainment interface', *J. Fluid Mech.*, **51**, pp. 97–118 (1972)

Townsend, A. A. 'Entrainment and the structure of turbulent flow', *J. Fluid Mech.*, **41**, pp. 13–46 (1970)

EXAMPLES

6.1 Calculate the thicknesses (6.1.5, 6, 7) for a plane wake and jet, taking the velocity variation to be that of equation (6.1.4). How do these results differ from those for a 'top-hat' profile, with U constant across the shear layer and changing abruptly at a hypothetical edge?

6.2 For a plane jet spreading into still fluid, it is found that $l_0 \propto x$ and $U_0 \propto x^{-\frac{1}{2}}$ specify the variations of width and velocity scales.

(a) Demonstrate that the first result implies that the absolute mean advance of the turbulence interface is given by $V_1 + V_i \propto U_0$.

(b) Show that the mass flow along the jet varies as $x^{\frac{1}{2}}$, and hence that the lateral inflow is given by $V_1 \propto x^{-\frac{1}{2}}$.

(c) Finally, obtain the relationships

$$V_1 \propto V_i \propto V_1 + V_i \propto x^{-\frac{1}{2}}$$

showing that the three 'entrainment' velocities remain in proportion during development. What are their signs and relative magnitudes?

6.3 Extend the transport equation (6.2.11) to include thin flows near a porous wall. Show that the effect of lateral convection on the first of the momentum equations (6.2.33) is to add a term $V_w U_1$ on the right.

6.4 (a) For free turbulence, show that the relationship corresponding to equations (6.2.14) is

$$\frac{l_0}{L} \sim 1 \quad \text{or} \quad \frac{U_0}{U_1}$$

(b) What does this tell you about a wake and about a jet?

(c) Using momentum arguments, show that $l_0 U_0$ is constant for a plane wake, and that $l_0^2 U_0$ is constant for a round wake, and hence that $l_0 \propto L^{\frac{1}{2}}$ and $L^{\frac{1}{3}}$ for the two.

6.5 Explain the trends in the mean-flow convection terms of Figures 6.5(a) and (c) in terms of the mean flow patterns within these flows.

6.6 Derive the mean-flow energy equation (6.2.25) from the general transport equation (6.2.11). Start by showing that $S + s = \frac{1}{2}U^2 + uU$.

6.7 Integrate the turbulence energy equation (6.2.24) across a thin flow to obtain a generalization of equation (3.3.28). How is the result related to equation (6.2.41)?

6.8 (a) Show that equation (6.2.31) implies that

$$\frac{V_1}{U_1} = \frac{d\delta_1}{dx} \quad \text{for} \quad \frac{dU_1}{dx} = 0$$

What does this imply for the streamlines of the outer flow?

(b) What is the corresponding result when the outer velocity varies?

(c) What is the result when there is a lateral flow, as through a porous wall? How does V vary across the layer for the patricular case of asymptotic suction layer?

6.9 Write the first of equations (6.2.33) in a form suitable for the calculation, from velocity measurements, of the friction beneath a wall jet, taking the outer flow to be at rest. How would you define a friction coefficient for this flow? (See Example 4.10.)

6.10 Using the wake data of Figure 6.1, determine the order of the error introduced into the momentum integral equation (6.2.33) by neglect of the normal-stress terms, for $x/d = 100$ and 1000. Are these results representative of other turbulent shear flows?

6.11 It is proposed that a self-preserving shear flow will be achieved when c_f is constant and $U_1 \propto x^m$. How will P_1 and δ_1 vary in these circumstances?

6.12 When the local friction coefficient varies as Re_x^n, show that it is related to the average coefficient (6.2.35) by

$$(n + 1)\, C_f = c_f$$

Evaluate for the layers of Table 6.1.

6.13 Assuming that Reynolds analogy relates heat transfer and friction in boundary layers, find how the heat transfer varies along constant-pressure laminar and turbulent layers.

6.14 Consider momentum integrals for plane flow between walls whose separation varies slowly along the flow. Take the density to be uniform.

(a) Integrate the general transport equation (6.2.11) from the wall $y = 0$ to a plane of symmetry $y = y_1(x)$ to show that equation (6.2.30) still applies. The derivative with respect to the upper limit must now be taken into account: differentiation and integration can no longer be freely interchanged.

(b) Show that the momentum integral equation is now

$$\frac{d(\rho U_1^2 \delta_2)}{dx} = \tau_w + \frac{dP_1}{dx}\delta_1 + \frac{dP_0}{dx}(y_1 - \delta_1)$$

(the normal stresses being neglected), with δ_1 and δ_2 found by integrating to y_1, where the total pressure is $P_0 = P_1 + \frac{1}{2}\rho U_1^2$. Does this result apply when the flow separates from the wall?

(c) What special results apply when:

(1) the flow has a potential core,
(2) the pressure is uniform along the channel,
(3) the flow is on the point of separating (but is still symmetrical), and
(4) the channel has parallel walls.

(d) Generalize the result (b) to any cross-section. Does the result apply when separation occurs?

6.15 Show that the thermal thickness at the end of a boundary layer is related to the overall transfer factor for the layer (U of equations (5.2.9)) by $\delta_T/L = U/(\rho c_p U_1)$, where U_1 is the free-stream velocity.

6.16 Run through the derivation of the mean-flow energy integral (6.2.40) to see how the limits are handled for a boundary layer and for a symmetrical free-turbulent flow.

6.17 Rewrite equation (6.2.47) in terms of displacement and momentum areas, related to the radii of equations (6.2.48) by $A_1 = \pi r_1^2$ and $A_2 = \pi r_2^2$. Compare with the corresponding plane-flow result.

6.18 Show that the shape factors of equations (6.2.34, 48) are 1 and 0 for wakes and for jets in still air. For the flow width to vary linearly along the stream (this will be seen, in Section 7.2, to be necessary for exact self-preservation), how must the external velocity vary?

6.19 Write equations analogous to (6.2.45, 42, 43) for axisymmetric, non-swirling flow near a plane, that is, nearly radial flow. Take U as the radial component and V as that normal to the plane. What is the form of the equations for radial flow between a pair of flat plates?

6.20 The energy equation for a 'thin' strongly swirling turbulent flow with fluctuating density $\rho + \rho'$ is

$$\rho\frac{DH_0}{Dt} + \overline{\rho' v}\frac{\partial H_0}{\partial r} + \frac{1}{r}\frac{\partial}{\partial r}[r\rho(\overline{vh} + W\overline{vw} + U\overline{vu})] = 0$$

with U, V and W the axial, radial and swirl components, and H_0 the total enthalpy (3.3.36). What is the significance of each term? These energy flows are the source of the Ranque (or Hilsch) effect, by which a uniform stream of gas passing through a vortex tube is separated into heated and cooled fractions, the former extracted near the periphery, the latter from the vortex core.

6.21 Combine the definitions (6.1.5, 6) with the first pair of equations (6.2.48) using the plane-and-round conventions of equation (6.2.54).

6.22 Consider laminar free convection between parallel vertical planes ($y = 0, b$), between which a temperature differential Δ is maintained. Take all transport properties to be uniform.

(a) Show that the temperature and density variations are

$$T = \bar{T} - \tfrac{1}{2}\Delta\frac{y}{b} \quad \text{and} \quad \rho = \bar{\rho}[1 - \beta(T - \bar{T})]$$

(b) Find the velocity distribution

$$\frac{Ub}{\nu} = \frac{1}{12}\mathrm{Gr}_b\frac{y}{b}\left(\frac{y^2}{b^2} - 1\right)$$

with Gr_b a Grashof number. Why is the lateral transfer rate independent of the fluid motion in this case? The plates must be of finite length; what happens at the top and bottom?

7

Developing Flows II: Free Turbulence

This chapter deals for the most part with five simply developing shear flows: the plane mixing layer, and plane and round wakes and jets. Their development is 'simple' because the surrounding fluid provides only uniform convection and a source of entrained fluid. Later in the chapter, we shall consider more complex patterns of development that are distorted by the outer flow or by the presence of solid boundaries. Although these motions are less well understood than simply developing flows, they are of greater practical importance: what happens in the middle of an extensive body of fluid is of less interest than events at its boundaries. Results for the simple flows are not without practical application: cooling and exhaust systems discharge into large bodies of still fluid; aircraft fly through the wakes of other aircraft; and the detection of a submarine's wake provides a way of tracing the path of the vessel. Nevertheless, the simple flows are primarily important as precursors of more general free-turbulent motions. Viewed in this way, the similarity methods which will be used extensively in this chapter appear somewhat sterile, at best severely limited, for they aim to predict the gross behaviour of a flow while saying (and revealing) very little about the turbulent activity within.

We shall look at free-turbulent flows in the following ways:

(1) using dimensional analysis to find patterns of development consistent with the momentum integrals and the general nature of the differential equations of motion;

(2) examining the differential equations to discover self-preserving patterns of development and their implications for entrainment;

(3) predicting profiles of mean velocity on the basis of constant eddy viscosity or constant mixing length;

(4) applying similarity arguments and simple transfer models to transport processes in self-preserving flows, including some driven by buoyancy; and

(5) briefly considering the interaction of a developing shear layer with the outer flow and its boundaries.

The manipulations used in predicting cross-stream profiles on the basis of a constant mixing length or eddy viscosity are not of fundamental interest, nor are the results accurate in detail. We shall concentrate on the simplest case of plane wake flow. The pattern of analysis is much the same for other simply developing flows, and the round jet is treated in examples at the end of the chapter.

7.1 SIMILARITY ARGUMENTS

The development of the five flows to be considered—the plane mixing layer, and plane and round wakes and jets—is determined by a few parameters which remain invariant along the flow. They define:

(1) the *environment* of the flow, to be specific, whether it is a jet advancing into still, unbounded fluid, a wake uniformly convected with velocity U_1, or a mixing layer between regions with uniform velocities U_1 and U_2; and

(2) the *momentum transferring attributes* of the fluid, the density ρ and, if the flow is laminar, the viscosity ν.

For four of the flows, the wakes and jets, a further parameter is relevant:

(3) a *momentum invariant*, the momentum flux M, which remains constant in the specified environments.

We conclude that the length and velocity scales, those shown in Figure 6.4, will be given by laws of the form:

$$l_0,\ U_0 = f(x, U_1, U_2, M, \rho, \nu) \tag{7.1.1}$$

For particular cases, certain of the parameters will be omitted. We can account for these omissions, and for events near the start of the flow, by measuring the coordinate x from a *virtual origin*, which will depend on the Reynolds number and on the geometry of the generating region. The role of the virtual origin parallels that of the 'slip constant' in the law of the wall, which accounts for viscosity and perhaps for the roughness geometry.

Momentum Integrals

We fix our attention first on the momentum invariant M. For a thin flow spreading into an unbounded fluid with constant external pressure P_1, the momentum integrals (6.2.33, 47) give:

(1) for plane flows

$$\rho \int_0^{y_1} [U(U_1 - U) + \overline{v^2} - \overline{u^2}]\, \mathrm{d}y = M', \quad \text{a constant} \tag{7.1.2}$$

(2) for round flows

$$2\pi\rho \int_0^{R_1} [U(U_1 - U) + \overline{v^2} - \overline{u^2}]\, r\, \mathrm{d}r = M, \quad \text{a constant} \tag{7.1.3}$$

For *jets* in still fluid, $U_1 = 0$, and these can be written:

$$\int_0^{y_1} (U^2 + \overline{u^2} - \overline{v^2})\,\mathrm{d}y = \frac{F'}{2\rho} \tag{7.1.4}$$

with F' the momentum flux per unit breadth of a plane jet, and

$$\int_0^{R_1} (U^2 + \overline{u^2} - \overline{v^2})\,r\,\mathrm{d}r = \frac{F}{2\pi\rho} \tag{7.1.5}$$

with F the total momentum flux in a round jet.

For small-deficit *wakes*, $U \simeq U_1$, giving the approximate forms:

$$U_1 \int_0^{y_1} (U_1 - U)\,\mathrm{d}y = \frac{D'}{2\rho} \tag{7.1.6}$$

with D' the drag or momentum deficit per unit breadth, and

$$U_1 \int_0^{R_1} (U_1 - U)\,r\,\mathrm{d}r = \frac{D}{2\pi\rho} \tag{7.1.7}$$

with D the total drag on the wake-generating body. Approximating the mean-flow contribution and neglecting the normal stresses introduces errors of order U_0/U_1.

These results can be used in a variety of ways. First, they suggest that the flow-defining parameters do not appear independently, but in the combinations F/ρ and $D/\rho U_1$. More specific results are obtained by restricting attention to self-similar velocity profiles, with

$$\frac{U - U_1}{U_0} = f\left(\frac{y}{l_0}\right) \quad \text{or} \quad f\left(\frac{r}{l_0}\right) \tag{7.1.8}$$

The momentum integrals can then be written in forms such as

$$U_1 U_0 l_0 \int_0^{y_1/l_0} f\left(\frac{y}{l_0}\right) \mathrm{d}\left(\frac{y}{l_0}\right) = -\frac{D'}{2\rho} \tag{7.1.9}$$

The integral is an absolute constant, characteristic of the particular similarity profile. We conclude that:

$$\left.\begin{aligned}
&\text{(1) for a plane jet,} && \frac{F'}{\rho U_0^2 l_0} = \text{constant}\\
&\text{(2) for a round jet,} && \frac{F}{\rho U_0^2 l_0^2} = \text{constant}\\
&\text{(3) for a plane wake,} && \frac{D'}{\rho U_1 U_0 l_0} = \text{constant}\\
&\text{(4) for a round wake,} && \frac{D}{\rho U_1 U_0 l_0^2} = \text{constant}
\end{aligned}\right\} \tag{7.1.10}$$

with each constant applying to all flows of the class. These results can be used to find one of the scales when the other is known.

Reynolds-number Variations

The relationships (7.1.10) indicate how the Reynolds number

$$\mathrm{Re}_0 = \frac{U_0 l_0}{\nu} \tag{7.1.11}$$

varies along the flow. Adding the obvious result for a mixing layer, we have

$$\begin{aligned} \mathrm{Re}_0 &\propto l_0 \quad \text{for a mixing layer} \\ &\propto l_0^{\frac{1}{2}} \quad \text{for a plane jet} \\ &= \text{constant,} \quad \text{for a round jet or plane wake} \\ &\propto l_0^{-1} \quad \text{for a round wake} \end{aligned} \tag{7.1.12}$$

This local mean-flow Reynolds number is a measure of the tendency of the flow to instability, as are Re_d for pipe flow and Re_δ for a boundary layer. Since all five of the flows considered spread and decelerate, we conclude that the plane mixing layer and jet become progressively less stable (or more strongly turbulent), that the round wake becomes more stable (or the turbulence ultimately vanishes), and that the stability of the other two flows remains unchanged along the flow.

For a boundary layer, the criterion for *transition* to turbulence can be inferred from the pipe-flow value: roughly speaking, $\mathrm{Re}_\delta = U_1 \delta_{99}/\nu \sim 2000$ at transition. Free shear layers, lacking the wall constraint, commonly become unstable when $\mathrm{Re}_0 = \Delta U \Delta y/\nu < 100$. Since the Reynolds number of a simple shear layer increases rapidly (see equations (7.1.12)), transition commonly occurs within such layers, before they coalesce to form a wake or jet. The nature of this breakdown makes it difficult to give a precise criterion for transition: fairly orderly waves appear, tighten into concentrated vortices, and only then break down into disordered turbulent motion.

The results (7.1.12) also tell us how the eddy viscosity varies along the flow. If the contributing scales of motion are independent of viscosity, we may assume that the eddy viscosity (taken to be constant across the flow) is specified by

$$\frac{\epsilon_c}{U_0 l_0} = \frac{1}{R_0} \tag{7.1.13}$$

with R_0 an absolute constant characteristic of a particular flow species. It has been termed the *flow constant* or the turbulent Reynolds number (yet another!). This parameter will be considered in more detail later; for the moment, we note only that

$$\frac{\epsilon_c}{\nu} = \frac{U_0 l_0}{\nu R_0} = \frac{\mathrm{Re}_0}{R_0} \tag{7.1.14}$$

varies in the same way as the local Reynolds number. This treatment of the eddy viscosity parallels that leading to equations (3.3.22) for the core of a channel flow.

Mass-flux Variations and the Outer Flow

The relationships (7.1.12) also define the variation in mass flux within a developing flow. For a plane flow, the additional mass flux is

$$\dot{m}' = 2\rho \int_0^{y_1} (U - U_1)\,\mathrm{d}y \tag{7.1.15}$$

and for a round flow

$$\dot{m} = 2\pi\rho \int_0^{R_1} (U - U_1)\,r\,\mathrm{d}r \tag{7.1.16}$$

Assuming similar profiles, we find

$$\dot{m}' \propto \rho U_0 l_0 \tag{7.1.17}$$

and

$$\dot{m} \propto \rho U_0 l_0^2 \tag{7.1.18}$$

using the arguments which led to equations (7.1.10). Comparison with those equations indicates that

$$\begin{aligned} \dot{m}', \dot{m} &\propto l_0^{\frac{1}{2}}, l_0 && \text{for the jets} \\ &= \text{constant}, && \text{for the wakes} \end{aligned} \tag{7.1.19}$$

The constancy of the mass-flux deficit for small-deficit wakes is already evident in equations (7.1.6, 7), which show it to be proportional to the momentum-flux deficit. Well downstream of the generating body, the flow outside the wake is exactly parallel. Jets, both laminar and turbulent, require a continued inflow from the otherwise still ambient fluid, in order to maintain a constant momentum flux while the velocity falls. In boundary layers, on the other hand, the momentum of the outer fluid is reduced, and the induced flow is outwards from the shear layer.

The lateral velocity in the outer flow can be calculated from continuity requirements. For U_1 constant, equations (6.2.31, 45) give:

(1) for plane flows

$$V_1 = \frac{\mathrm{d}}{\mathrm{d}x}\int_0^{y_1} (U_1 - U)\,\mathrm{d}y = -\frac{1}{2\rho}\frac{\mathrm{d}\dot{m}'}{\mathrm{d}x} \tag{7.1.20}$$

(2) for round flows

$$rV = R_1 V_1 = \frac{\mathrm{d}}{\mathrm{d}x}\int_0^{R_1} (U_1 - U)\,r\,\mathrm{d}r = -\frac{1}{2\pi\rho}\frac{\mathrm{d}\dot{m}}{\mathrm{d}x} \tag{7.1.21}$$

with r a point beyond the region of significant velocity variation.

As anticipated:

$$\begin{aligned} V_1 &< 0 \quad \text{for jets} \\ &= 0 \quad \text{for wakes} \\ &> 0 \quad \text{for boundary layers} \end{aligned} \tag{7.1.23}$$

So far as the outer flow is concerned, the shear layer acts as a pattern of material sources ($V_1 > 0$) or sinks ($V_1 < 0$), distributed along the plane or axis near which the flow is concentrated.

Let us extend this argument to the mixing layer shown in Figure 7.1, the extreme case for which $U_2 = 0$. The lack of symmetry complicates the discussion: at the upper edge the layer has the character of a wake or boundary layer, and we expect $V_1 > 0$ or $V_1 \simeq 0$; at the lower edge the flow is more like

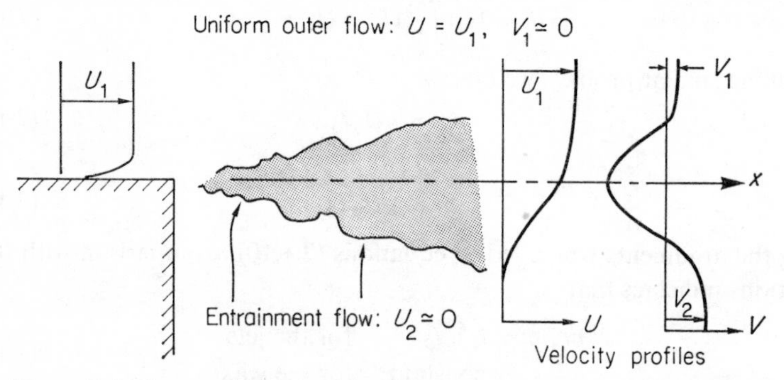

Figure 7.1 Plane turbulent mixing layer at an abrupt wall deflection. The profiles of mean streamwise and cross-stream velocities are not to the same scale

a jet, and we expect $V_2 > 0$, that is, motion towards the shear layer. The spreading of the region in which the velocity falls implies that $V < 0$ *within* the layer; this lateral flow also acts to convect momentum from the fast-moving to the slow-moving side.

Although we do not know the inflow and outflow for the two sides of the layer, we can calculate the net excess flow:

$$\dot{m}' = \rho \int_{y_2}^{y_1} (U - U_2)\, dy \tag{7.1.23}$$

with $U_2 \neq 0$ introduced for generality. Assuming similar velocity profiles, we have

$$\dot{m}' \propto \rho (U_1 - U_2)\, l_0 \tag{7.1.24}$$

which has the form (7.1.17), since $U_0 = U_1 - U_2$ in this case. Finally, we have

$$V_1 - V_2 = -\frac{1}{\rho}\frac{d\dot{m}'}{dx} < 0 \tag{7.1.25}$$

as for a jet.

Results of Dimensional Analysis

The implications of the invariance of the momentum flux have delayed our attack on the problem posed in equation (7.1.1), the prediction of variations of length and velocity scales. We noted earlier that the form of the momentum invariants implies that the combinations F/ρ and $D/\rho U_1$ must appear. For a turbulent jet, this observation (which is in fact necessary for dimensional homogeneity) reduces the relationship (7.1.1) to

$$l_0,\ U_0 = f\left(\frac{F}{\rho}, x\right) \tag{7.1.26}$$

For the round jet, we find immediately

$$l_0 \propto x \quad \text{and} \quad U_0 \propto \frac{(F/\rho)^{\frac{1}{2}}}{x} \tag{7.1.27}$$

To obtain such explicit results for wakes, we must further restrict the parameters. We argue that the coordinate x is important only as a measure of the time x/U_1 available for the convected fluid to spread. Thus

$$l_0,\ U_0 = f\left(\frac{D}{\rho U_1}, \frac{x}{U_1}\right) \tag{7.1.28}$$

which has a small enough number of variables. This grouping can also be deduced from the equation for small-deficit flow:

$$U_1 \frac{\partial U}{\partial x} = \frac{1}{\rho}\frac{\partial \tau}{\partial y} \tag{7.1.29}$$

The constant U_1 can be incorporated into x/U_1, and this combination will occur in any solution.

For laminar flows, equations (7.1.1) include an additional parameter, the viscosity ν. The way in which it is related to the other defining quantities can be inferred from the momentum balance

$$U\frac{\partial U}{\partial x} \sim \nu \frac{\partial^2 U}{\partial y^2} \tag{7.1.30}$$

We see that the constant ν can be incorporated in νx, and that, for laminar wakes, $\nu x/U_1$ is the relevant combination.

Table 7.1 is a systematic presentation of these arguments; it deals first with the selection of controlling parameters, then with the development laws for turbulent flow, and finally with the corresponding results for laminar motion. The trends in Reynolds number, eddy viscosity and mass flux are consistent with the arguments based on the momentum invariant. The laminar and turbulent growth laws are the same for the two cases in which ϵ_c is constant along the flow. However, even there, the constants of proportionality have quite different structures for turbulent and laminar flow.

The only scales which are independent of the defining parameters are the widths of the turbulent jets: all plane jets and all round jets have the same angles of expansion (not necessarily equal), provided of course that the outer fluid is at rest and that the density is uniform. Moreover, the rate of spread of turbulent mixing layers is the same for any one value of the velocity ratio U_1/U_2. It is in these flows that the implications of Reynolds-number similarity are most obvious.

Table 7.1. Similarity in developing flows

I. Controlling parameters (bracketed quantities for laminar flow)

Flow type	Basic parameters	Momentum invariant	Grouped parameters
Mixing layer	$U_1, U_2, (\nu)$	—	$U_1/U_2, x$ (or $\nu x/U_1$)
Plane wake	$U_1, D', \rho, (\nu)$	$D'/\rho U_1 U_0 l_0$	$D'/\rho U_1, x/U_1$ (or $\nu x/U_1$)
Round wake	$U_1, D, \rho, (\nu)$	$D/\rho U_1 U_0 l_0^2$	$D/\rho U_1, x/U_1$ (or $\nu x/U_1$)
Plane jet	$F', \rho, (\nu)$	$F'/\rho U_0^2 l_0$	$F'/\rho, x$ (or νx)
Round jet	$F, \rho, (\nu)$	$F/\rho U_0^2 l_0^2$	$F/\rho, x$ (or νx)

II. Free-turbulent flows

Flow type	l_0	U_0	$Re_0 \propto \epsilon_c/\nu$	$\dot{m}'$ or $\dot{m}$
Mixing layer	$f(U_1/U_2)\, x$	$U_1 - U_2$	$\propto x$	$\propto x$
Plane wake	$(D'/\rho U_1^2)^{\frac{1}{2}} x^{\frac{1}{2}}$	$(D'/\rho)^{\frac{1}{2}}/x^{\frac{1}{2}}$	$D'/\rho U_1$	$-D'/U_1$
Round wake	$(D/\rho U_1^2)^{\frac{1}{3}} x^{\frac{1}{3}}$	$(DU_1/\rho)^{\frac{1}{3}}/x^{\frac{2}{3}}$	$\propto x^{-\frac{1}{3}}$	$-D/U_1$
Plane jet	x	$(F'/\rho)^{\frac{1}{2}}/x^{\frac{1}{2}}$	$\propto x^{\frac{1}{2}}$	$\propto x^{\frac{1}{2}}$
Round jet	x	$(F/\rho)^{\frac{1}{2}}/x$	$(F/\mu\nu)^{\frac{1}{2}}$	$\propto x$

III. Laminar flows

Flow type	l_0	U_0	Re_0	$\dot{m}'$ or $\dot{m}$
Mixing layer	$(\nu/U_1)^{\frac{1}{2}} x^{\frac{1}{2}}$	$U_1 - U_2$	$\propto x^{\frac{1}{2}}$	$\propto x$
Plane wake	$(\nu/U_1)^{\frac{1}{2}} x^{\frac{1}{2}}$	$(D'^2/\rho\mu U_1)^{\frac{1}{2}}/x^{\frac{1}{2}}$	$D'/\mu U_1$	$-D'/U_1$
Round wake	$(\nu/U_1)^{\frac{1}{2}} x^{\frac{1}{2}}$	$(D/\mu)/x$	$\propto x^{-\frac{1}{2}}$	$-D/U_1$
Plane jet	$(\mu\nu/F')^{\frac{1}{3}} x^{\frac{2}{3}}$	$(F'^2/\rho\mu)^{\frac{1}{3}}/x^{\frac{1}{3}}$	$\propto x^{\frac{1}{3}}$	$\propto x^{\frac{1}{3}}$
Round jet	$(\mu\nu/F)^{\frac{1}{2}} x$	$(F/\mu)/x$	$(F/\mu\nu)^{\frac{1}{2}}$	$\propto x$

7.2 SELF-PRESERVING FLOWS

Another way of studying the development of free-turbulent flows is to examine the consequences of similarity for the differential equations of motion. This provides a way of accounting for non-uniformity in the convecting outer flow, and thus reveals a greater variety of self-similar solutions. Since the continued balancing of the various transfer processes is explicitly recognized, the name 'self-preserving' is often applied to turbulent flows revealed by this analysis.

Kinematic Conditions

We shall consider in detail only plane flow, taking

$$U = U_1 + U_0 f(\eta) \quad \text{with} \quad \eta = y/l_0 \tag{7.2.1}$$

for the mean-velocity variation, and

$$\overline{u_i u_j} = U_0^2 g_{ij}(\eta) \quad \text{with} \quad i, j = 1, 2, 3 \tag{7.2.2}$$

for the components of the mixing-stress tensor (for example, $\overline{uv} = \overline{u_1 u_2} = U_0^2 g_{12}$). These postulates express variations along the flow in terms of the scales $U_1(x)$, $U_0(x)$ and $l_0(x)$, and introduce Reynolds-number similarity for the motions contributing to the mixing stresses.

The corresponding variation of the cross-stream component can be found by integrating the continuity equation:

$$\begin{aligned} V(y) &= -\frac{\mathrm{d}}{\mathrm{d}x}\int_0^y [U_1 + U_0 f(\eta)]\,\mathrm{d}y \\ &= -\frac{\mathrm{d}U_1}{\mathrm{d}x} y - \frac{\mathrm{d}U_0}{\mathrm{d}x}\int_0^y f\,\mathrm{d}y + \frac{U_0}{l_0}\frac{\mathrm{d}l_0}{\mathrm{d}x}\int_0^y y\,\mathrm{d}f \\ &= -l_0\frac{\mathrm{d}U_1}{\mathrm{d}x}[\eta] + U_0\frac{\mathrm{d}l_0}{\mathrm{d}x}[\eta f] - \frac{\mathrm{d}(U_0 l_0)}{\mathrm{d}x}\left[\int_0^\eta f\,\mathrm{d}\eta\right] \end{aligned} \tag{7.2.3}$$

Here it has been assumed that $V(0) = 0$. The last line follows from integration by parts. In it, factors representing streamwise and cross-stream variations are separated out; the quantities in square brackets do not vary along the flow.

Some of the conditions for self-preservation are implied by the results (7.2.3). For the cross-stream profile of the lateral velocity to be invariant, it is necessary that each term change in the same way along the flow. From the second line of equations (7.2.3):

$$\frac{\mathrm{d}U_1}{\mathrm{d}x} \propto \frac{\mathrm{d}U_0}{\mathrm{d}x} \propto \frac{U_0}{l_0}\frac{\mathrm{d}l_0}{\mathrm{d}x} \tag{7.2.4}$$

Hence the requirements for exact self-preservation are

$$U_1^m \propto U_0^m \propto l_0 \tag{7.2.5}$$

when all of the scales vary. When one scale is constant, one of the relationships (7.2.4) must be omitted. For example

$$U_0^m \propto l_0 \quad \text{for } U_1 \text{ constant}$$

When two scales are constant, we learn nothing more from equations (7.2.3).

Momentum Conditions

Seeking further constraints, we turn to the momentum equation (6.2.21). This has a complicated structure when the similarity forms (7.2.1, 2, 3) are introduced. However, equations (7.2.3) indicate that

$$V = U_0 \frac{dl_0}{dx} F(\eta) \quad \text{and} \quad \frac{\partial U}{\partial x} = \frac{dU_0}{dx} G(\eta) \tag{7.2.6}$$

when the conditions of kinematic similarity are satisfied. The momentum condition can now be written

$$\begin{aligned} U_1 \frac{dU_0}{dx}[G] + U_0 \frac{dU_0}{dx}[fG] + \frac{U_0^2}{l_0}\frac{dl_0}{dx}[f'F] \\ + 2U_0 \frac{dU_0}{dx}[g_{11} - g_{22}] - \frac{U_0^2}{l_0}\frac{dl_0}{dx}[\eta(g_{11}' - g_{22}')] \\ = U_1 \frac{dU_1}{dx} - \frac{U_0^2}{l_0}[g_{12}'] + \frac{\nu U_0}{l_0^2}[f''] \end{aligned} \tag{7.2.7}$$

The primes indicate differentiation with respect to η.

On examining equation (7.2.7), we note that

$$\frac{dl_0}{dx} = \text{constant} \tag{7.2.8}$$

ensures that the mixing stress, the penultimate term, varies in the same way as two of the other terms. Moreover, when combined with the kinematic requirements (7.2.5), this single dynamic condition implies that all of the coefficients (the final, viscous-stress term excepted) are proportional. Hence exact self-preservation requires that

$$U_1^m \propto U_0^m \propto l_0 \propto x \tag{7.2.9}$$

when all of the scales vary. For U_1 constant, we have

$$U_0^m \propto l_0 \propto x \tag{7.2.10}$$

This gives the length scale of a simple jet; the velocity scale is as yet undetermined. When U_1 and U_0 are constant, we obtain

$$l_0 \propto x \tag{7.2.11}$$

the pattern of development of a simple mixing layer. Special results apply for l_0 or U_0 constant, but they are not of great practical interest.

Nothing has been learned about the plane wake, simply because the necessary approximation $U_1 \gg U_0$ has not been introduced. When this is done, most of the terms of equation (7.2.7) vanish, and the dynamic condition for approximate self-preservation is

$$U_1 \frac{dU_0}{dx} \propto \frac{U_0^2}{l_0} \tag{7.2.12}$$

replacing equation (7.2.8). The kinematic conditions (7.2.5) must be re-assessed, for some of them have also become irrelevant.

We direct our attention first to uniform convection, U_1 constant. To complete the discussion of the plane jet and wake, we use the integral-momentum conditions (7.1.10):

$$\frac{F'}{\rho U_0^2 l_0} = \text{constant}, \quad \text{for the jet} \tag{7.2.13}$$

$$\frac{D'}{\rho U_1 U_0 l_0} = \text{constant}, \quad \text{for the wake} \tag{7.2.14}$$

From the first, and equations (7.2.10), there follows immediately

$$U_0 \propto \left(\frac{F'}{\rho x}\right)^{\frac{1}{2}} \tag{7.2.15}$$

For the wake, the condition (7.2.12) gives

$$\frac{\mathrm{d}U_0}{\mathrm{d}x} \propto \frac{\rho}{D'} U_0^3 \quad \text{whence} \quad U_0 \propto \left(\frac{D'}{\rho x}\right)^{\frac{1}{2}} \tag{7.2.16}$$

The results (7.2.15, 16) correspond to those in Table 7.1, II.

Turning to non-uniform convection, U_1 varying, we find some new results, thus justifying the more complicated analysis employed here. The integral momentum condition for a freely developing flow, laminar or turbulent, in a stream of varying velocity is given in the second of equations (6.2.34). For an exactly self-preserving flow, $U_0 \propto U_1$ and all lateral scales vary in proportion; hence

$$l_0 U_0^{H+2} = \text{constant} \tag{7.2.17}$$

with $H = \delta_1/\delta_2$ the constant shape factor for a particular velocity profile. The condition (7.2.8) now gives

$$U_0 \propto x^{-1/(H+2)} \tag{7.2.18}$$

for this class of motions, exactly self-preserving save that the normal stresses were omitted in obtaining equation (7.2.17).

For jet-like flows, with $U_0/U_1 =$ a large constant:

$$H \to 0 \quad \text{and} \quad U_0 \propto x^{-\frac{1}{2}} \tag{7.2.18}$$

as for the jet in quiescent fluid. For wake-like flows, with $U_0/U_1 =$ a small constant:

$$H \to 1 \quad \text{and} \quad U_0 \propto x^{-\frac{1}{3}} \tag{7.2.18}$$

This result for an exactly self-preserving wake is not quite the same as that (7.2.16) for the approximately self-preserving wake in a uniform stream.

Other Flow Geometries

The discussion can be extended to round wakes and jets by examining the consequences of self-preservation for the axisymmetric continuity and momentum equations (6.2.45, 46). It turns out that the conditions (7.2.4, 8) still apply, since the basic structure of the equations is the same. Adding the round-flow momentum conditions (7.1.10) and (6.2.49), we find

(1) for the round jet

$$U_0 \propto l_0^{-1} \propto x^{-1} \tag{7.2.19}$$

(2) for the round wake

$$U_0 \propto l_0^{-2} \propto x^{-1} \tag{7.2.20}$$

(3) for round self-preserving flows in a varying stream

$$U_0 \propto x^{-2/(H'+2)} \tag{7.2.21}$$

The first results are equivalent to those of Table 7.1, II.

The study of self-preserving free-turbulent flows can be extended in a variety of ways:

(a) to approximately self-preserving plane and axisymmetric motions with $U_0(x) \ll U_1(x)$; these generalize the approximately self-preserving wake;

(b) to other simple forms of mean-flow straining, for example, $U_1 =$ constant, $V = -ay$, $W = az$, with a constant;

(c) to other species of axisymmetric flows, either basically radial, or swirling, or both; for swirling flows, an angular-momentum invariant must be considered; and

(d) wakes from self-propelled bodies; the momentum integral is zero and is not helpful, but certain integrals obtained by multiplying the momentum equation by a power of y or r are finite and invariant.

Some of the flows (a) and (b) are of little practical interest, an artificial environment having been selected to make self-preservation possible. However, such flows can be established in specially contrived experiments, and some are of interest in displaying, in isolation, processes which play a part in more general turbulent motions.

The consequences of self-preservation can be investigated by examining other equations governing the turbulent motion. In particular, the turbulence energy equation may be required to have self-preserving form, in addition to the continuity and momentum equations. It is then possible to relax the requirement that the velocity scales for mean flow and turbulence vary in the same way along the flow: the added constraint makes possible the determination of the extra scale q_0, for the turbulence intensity.

Entrainment

The conditions for self-preservation allow us to investigate the processes associated with the word 'entrainment'—the advance of the turbulence interface into the outer fluid, influenced by lateral mean convection arising within the spreading shear layer or imposed by the convecting stream.

The definitions preceding equation (6.1.11) imply that the velocity of interface advance relative to the mean fluid motion can be expressed as

$$V_i = (V_i + V_1) - V_1 \tag{7.2.22}$$

with $V_i + V_1$ the apparent or absolute rate of expansion of the boundary. The mean fluid velocity at the interface can be calculated from equations (7.2.3), which contain terms representing the acceleration of the convecting flow and the spreading of the shear layer. For a self-preserving development, we can write

$$V_1 = -\frac{d(U_0 l_0)}{dx} H(\eta_1) = -U_0 \frac{dl_0}{dx} I(\eta_1) \tag{7.2.23}$$

with H and I numbers characteristic of a particular flow species, that is, dependent on U_0/U_1.

The apparent lateral velocity of the interface may be expressed as

$$V_i + V_1 \propto U_1 \frac{dl_0}{dx} \tag{7.2.24}$$

The constant of proportionality depends on the definition of l_0, but is in essence the ratio y_1/l_0. Alternatively, using the kinematic condition

$$U_0 \frac{dl_0}{dx} \propto \frac{d(U_0 l_0)}{dx}$$

implied by equations (7.2.3), we have

$$V_i + V_1 \propto \frac{U_1}{U_0} \frac{d(U_0 l_0)}{dx} \tag{7.2.24}$$

For a simple jet, U_0 will replace U_1 in equations (7.2.24), on the assumption that the interface is carried along the stream at a speed keyed to the velocity scale.

Introducing the last results into equation (7.2.22), we have

$$V_i = (C_1 U_1 + C_2 U_0) \frac{dl_0}{dx}$$

or (7.2.25)

$$V_i = \left(C_3 \frac{U_1}{U_0} + C_4\right) \frac{d(U_0 l_0)}{dx}$$

with the coefficients C_1 to C_4 dependent on the flow type. It will be remembered that one of the conditions for similarity (7.2.5) is $U_1/U_0 = \text{constant}$; it follows that the terms of these equations are proportional. Finally, we remember that dl_0/dx is constant in these self-preserving flows, and conclude that

$$V_1 \propto V_i \propto V_i + V_1 \propto U_0 \tag{7.2.26}$$

(In retrospect, these results are hardly surprising for a self-preserving flow.) Hence equations of the form (7.2.25) can be written for any of these entrainment velocities.

The second of equations (7.2.25) shows that the various lateral velocities are proportional to the rate of change of volume flow along the stream. This interpretation is the basis of the *entrainment method* of boundary-layer prediction.

The preceding arguments apply to the mixing layer and jet in still air, and to the variety of exactly self-preserving jets and wakes. For the uniformly convected wake, the condition (7.2.12) implies that

$$U_0 \propto U_1 \frac{l_0}{U_0} \frac{dU_0}{dx} \propto U_1 \frac{dl_0}{dx}$$

since $U_0 l_0$ is constant. Hence $V_i \propto U_0$ as in equations (7.2.26).

In the light of these results, equations (7.2.25) are often written in the forms

$$\beta U_0 = (U_1 + \alpha U_0)\frac{dl_0}{dx}$$

and (7.2.27)

$$\beta_1 U_0 = \frac{d(U_0 l_0)}{dx}$$

The parameters β and β_1 are called *entrainment constants*. In the first of these entrainment equations, the contributions of V_1 and $V_i + V_1$ are kept separate, the coefficient α representing their relative importance. We may expect: $\alpha = 0$ for a simple wake; $\alpha > 0$ for a jet or mixing layer; and $\alpha < 0$ for a boundary layer. In practice, the value $\alpha = \frac{1}{2}$ is often adopted, this being more representative of jets and mixing layers, and the entrainment coefficient is defined as

$$\beta = \left(\frac{U_1}{U_0} + \frac{1}{2}\right)\frac{dl_0}{dx} \tag{7.2.28}$$

The alternative coefficient β_1 in the second of equations (7.2.27) incorporates all of the effects of changes in U_1/U_0; perhaps this is more realistic in our present state of understanding of entrainment processes.

7.3 MEAN VELOCITY PROFILES

7.3.1 The Nature of the Problem

The similarity conditions, by ensuring that the coefficients of the momentum equation (7.2.7) are proportional, reduce it to an *ordinary* differential equation, albeit of complicated form and involving more than one dependent variable. However, the convective acceleration, which involves the velocities U and V, save in small-deficit flow, can be expressed in terms of a single dependent variable by introducing an appropriate stream function; this technique is illustrated elsewhere, in Section 8.1.1, for instance. The assumption of similarity has accounted for the streamwise development; for turbulent flow, we are left with the problem of relating the mean velocity (or its stream function) and the mixing stress. Here we shall investigate the lateral variation of velocity using the simplest models of turbulent mixing—either an eddy viscosity or a mixing length which is constant across the flow, but not necessarily along it.

We shall consider only one flow in detail—the plane wake, for which both acceleration and stress terms have particularly simple forms. The structure of the results for this one flow reveals the general nature of the solutions for other developing flows. Although the mathematical details for the various flows are of some interest, they do not tell us much about the turbulence; they do not even give very accurate predictions of the mean velocity profiles.

This latter failing is not difficult to explain. In Figure 6.3(a) we see that the assumption $\epsilon_m = \epsilon_c$ is fairly realistic in the nearly-always-turbulent region; however, the smaller outer values of ϵ_m imply that the actual velocity profile drops off more rapidly than does the constant-viscosity model. This is partly attributable to the intermittency of the outer flow. Note that the adoption of a constant eddy viscosity will lead to velocity profiles identical to those for the corresponding laminar flows, even though the patterns of development may not be the same.

The length scales of a free-turbulent flow are also fairly constant in the central region, but they rise in the region of intermittency, since the scales in the non-vortical fluid are large. Hence it is to be expected that the constant-mixing-length model will underestimate the outer velocities. Some distortion must also be anticipated near an axis of symmetry, since $\epsilon_m = l_m^2|\partial U/\partial y|$ vanishes there.

7.3.2 The Plane Wake

For this flow the momentum equation has the simple form

$$U_1 \frac{\partial U}{\partial x} = -\frac{\partial \overline{uv}}{\partial y} + \nu \frac{\partial^2 U}{\partial y^2} \tag{7.3.1}$$

with U_1 constant. Introducing the similarity forms (7.2.1, 2), we have

$$U_1\left(\frac{l_0}{U_0^2}\frac{dU_0}{dx}f-\frac{1}{U_0}\frac{dl_0}{dx}\eta f'\right)=-g_{12}'+\frac{\nu}{U_0 l_0}f'' \quad \text{with } f'=\frac{df}{d\eta} \tag{7.3.2}$$

In this flow $\mathrm{Re}_0 = U_0 l_0/\nu$ is constant, by virtue of the momentum constraint $U_0 l_0 =$ constant. This implies also that the remaining coefficients are equal, but have opposite signs. Hence the differential equation can be written

$$C_1(f+\eta f') = -g_{12}' + \frac{f''}{\mathrm{Re}_0}$$

with the constant

$$C_1 = -\frac{U_1}{U_0}\frac{dl_0}{dx} \tag{7.3.3}$$

Integration gives

$$C_1 \eta f = -g_{12} + \frac{f'}{\mathrm{Re}_0} \tag{7.3.4}$$

for the boundary conditions

$$\frac{\partial U}{\partial y},\ f'=0 \quad \text{and} \quad \tau,\ g_{12}=0 \quad \text{at } \eta=0 \tag{7.3.5}$$

For the flow considered here, the left-hand, acceleration element of equation (7.3.4) is linear in the dependent variable; non-linearity can enter only through the mixing stress. In general, the acceleration terms will not be linear, and the equation to be solved is non-linear even for laminar flow. This is illustrated by a round jet, treated in examples at the end of this chapter.

Constant Eddy Viscosity

We take the mixing stress to be

$$\overline{uv} = -\epsilon_c \frac{\partial U}{\partial y} = -\frac{U_0^2}{R_0} f' \tag{7.3.6}$$

giving the scaled stress

$$g_{12} = \frac{\overline{uv}}{U_0^2} = -\frac{f'}{R_0} \tag{7.3.6}$$

The flow constant $R_0 = U_0 l_0/\epsilon_c$ is uniform along this flow, but this is not so in general. Equation (7.3.4) can now be written (with the viscous term omitted):

$$-2k\eta f = f' \tag{7.3.7}$$

With the help of equation (7.3.3), we obtain

$$k = -\tfrac{1}{2} C_1 R_0 = \frac{1}{4}\frac{U_1}{\epsilon_c}\frac{dl_0^2}{dx}$$

and see that the wake growth is specified by

$$l_0^2 = \frac{4\epsilon_c k}{U_1} x \tag{7.3.8}$$

Finally, integration of equation (7.3.7) gives

$$f = -\exp(-k\eta^2) \tag{7.3.9}$$

and

$$U = U_1 - U_0 \exp\left[-k\left(\frac{y}{l_0}\right)^2\right] \tag{7.3.9}$$

on requiring that $U = U_1 - U_0$ for $y = 0$.

To relate the constants of this profile to the drag associated with the wake, we introduce the profile into the momentum integral (7.1.6):

$$\begin{aligned} D' &= 2\rho U_1 U_0 l_0 \int_0^\infty e^{-k\eta^2}\, d\eta \\ &= \left(\frac{\pi}{k}\right)^{\frac{1}{2}} \rho U_1 U_0 l_0 = \left(\frac{\pi}{k}\right)^{\frac{1}{2}} \rho U_1 R_0 \epsilon_c \end{aligned} \tag{7.3.10}$$

when the value $\frac{1}{2}(\pi/k)^{\frac{1}{2}}$ is introduced for the standard integral. The eddy viscosity can now be eliminated from the width variation (7.3.8):

$$l_0^2 = \frac{4}{R_0}\left(\frac{k^3}{\pi}\right)^{\frac{1}{2}} \frac{D'}{\rho U_1^2} x \tag{7.3.11}$$

The velocity variation follows:

$$U_0^2 = \frac{R_0}{4}(k\pi)^{-\frac{1}{2}} \frac{D'/\rho}{x} \tag{7.3.12}$$

These results are consistent with those of Table 7.1, II, but now we have expressed the constants of proportionality in terms of the flow constant R_0. It may seem that the variations depend also on the constant k, but this is not really so. Examination of the solution (7.3.9) shows that k can be selected arbitrarily to specify the point on the profile to which l_0 is measured. The quantity

$$\frac{R_0}{k^{\frac{1}{2}}} = \frac{U_0 l_0 / k^{\frac{1}{2}}}{\epsilon_c}$$

is independent of k, and the velocity scale (7.3.12) is, as expected, independent of the way in which the scale l_0 is defined.

The values of R_0 for two choices of k are related by

$$\frac{R_{0_1}}{R_{0_2}} = \left(\frac{k_1}{k_2}\right)^{\frac{1}{2}} \tag{7.3.13}$$

Choices that are often made are:

$$k_1 = \tfrac{1}{2} \quad \text{giving} \quad -f = e^{-\frac{1}{4}} \quad \text{for } y = l_{0_1}, \text{ and}$$
$$k_2 = \ln 2 \quad \text{giving} \quad -f = \tfrac{1}{2} \quad \text{for } y = l_{0_2}$$

Then $R_{0_1}/R_{0_2} = (0{\cdot}5/0{\cdot}693)^{\frac{1}{2}} = 0{\cdot}85$ relates the flow constants.

Constant Mixing Length

Taking the mixing stress to be

$$\overline{uv} = l_c^2 \left(\frac{\partial U}{\partial y}\right)^2 = \left(\frac{l_c}{l_0}\right)^2 U_0^2 f'^2 \tag{7.3.14}$$

we find

$$g_{12} = \frac{\overline{uv}}{U_0^2} = \left(\frac{l_c}{l_0}\right)^2 f'^2 \tag{7.3.14}$$

To satisfy the similarity hypothesis $g_{12} = g_{12}(\eta)$, it is necessary that

$$\frac{l_c}{l_0} = \text{constant} \quad \text{along the flow} \tag{7.3.15}$$

Equation (7.3.4) now gives (with the viscous term omitted once again):

$$C_2 \eta f = f'^2 \tag{7.3.16}$$

with the constant

$$C_2 = -\left(\frac{l_0}{l_c}\right)^2 C_1 = \frac{U_1}{U_0}\left(\frac{l_0}{l_c}\right)^2 \frac{dl_0}{dx} \tag{7.3.17}$$

Integration gives

$$f^{\frac{1}{2}} = \tfrac{1}{3} C_2^{\frac{1}{2}} \eta^{\frac{3}{2}} - A$$

The flow has a finite width, since $f \to 0$ as η approaches a particular value. We fix the constant of integration by taking $f = 0$ for $\eta = 1$, implying that the lateral scale is the finite width $l_1(x)$. Hence

$$f = -(1 - \eta^{\frac{3}{2}})^2 \tag{7.3.18}$$

on requiring that $C_2 = 9$, so that $f = -1$ for $\eta = 0$. (The treatment of signs is rather odd, in consequence of the neglect of the absolute value associated with the definition of the mixing length; see equations (3.4.37).)

Turning now to the momentum integral, we have

$$D' = 2\rho U_1 U_0 l_0 \int_0^1 (1 - \eta^{\frac{3}{2}})^2 \, d\eta = \tfrac{9}{10} \rho U_1 U_0 l_0 \tag{7.3.19}$$

From equation (7.3.17)

$$C_2 = 9 = \frac{9}{10} \frac{\rho U_1^2}{D'} \left(\frac{l_0}{l_c}\right)^2 l_0 \frac{dl_0}{dx}$$

Integration gives

$$l_1^2 = 20\left(\frac{l_c}{l_1}\right)^2 \frac{D'}{\rho U_1^2}x \tag{7.3.20}$$

followed by

$$U_0^2 = \frac{5}{81}\left(\frac{l_1}{l_c}\right)^2 \frac{D'/\rho}{x} \tag{7.3.21}$$

These results parallel (7.3.11, 12), but the constants are now expressed in terms of the constant ratio l_c/l_1 of the mixing length to the local flow width. Other length scales can be obtained by introducing suitable factors into equation (7.3.20); for example, if l_0 is measured to $f=-\frac{1}{2}$, we have $l_0/l_1 = 0{\cdot}441$.

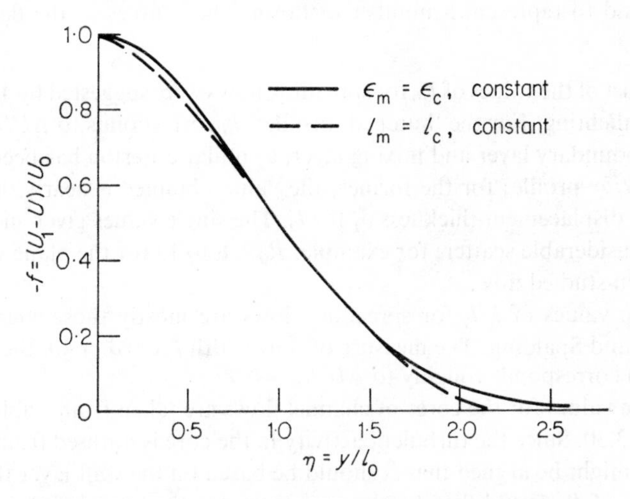

Figure 7.2 Velocity profiles for a plane turbulent wake, on the basis of constant eddy viscosity and constant mixing length

Comparing the velocity variations (7.3.21, 12), we see that the parameters R_0 and l_c/l_1 are related by

$$\left(\frac{l_c}{l_1}\right)^2 R_0 = \tfrac{20}{81}(k\pi)^{\frac{1}{2}} \tag{7.3.22}$$

Values around $R_0 = 11$ are found to represent turbulent wakes, when the scale l_0 is defined by $k=\frac{1}{2}$. This corresponds to $l_c/l_1 = 0{\cdot}17$, typical of the values introduced into the mixing-length model to make it work.

Figure 7.2 shows the two scaled profiles (7.3.9, 18), made to correspond when $f=-\frac{1}{2}$. The expected differences near the plane of symmetry and at the outer edge are evident; measured values follow the Gaussian curve for constant eddy viscosity near the centre, and lie between the two predictions at the edge. These results are consistent with Figure 6.1(a) in showing the mean profile to be insensitive to the nature of the turbulence. The sharp edge of the constant-mixing-length model is convenient in giving a definite width to the flow. Oddly enough, though this prediction is wrong in the mean, it is crudely representative of the instantaneous situation.

7.3.3 Empirical Data for Free Turbulence and Core Flows

When other free-turbulent flows are analysed, the results have the general form of those for a plane wake: the development is specified either by a value of $R_0 = U_0 l_0/\epsilon_c$, or by a value of l_c/l_1. Table 7.2 gives the values which have been found to represent a number of flows. The sources of the data are as follows:

(1) Most of the values of R_0 for spreading flows were suggested by Townsend or by Schlichting. For the 'humped' profiles, l_0 corresponds to $\Delta U/U_0 = e^{-\frac{1}{2}}$. For the boundary layer and mixing layer, a similar criterion has been applied to the $\partial U/\partial y$ profile; for the former, the value obtained is nearly that found using the displacement thickness δ_1 for l_0. The single values given in the table hide a considerable scatter; for example, $R_0 = 8$ to 13 for the plane wake, the most often studied flow.

(2) The values of l_c/l_1 for spreading flows are mostly those suggested by Launder and Spalding. The measure of flow width l_1 ($= \delta$ or $\frac{1}{2}\delta$, the width or half of it) corresponds roughly to $\Delta U/U_0 = 0{\cdot}99$.

(3) The values for the cores of channel flows are taken from Table 3.3 and Example 3.30. Since the turbulent activity in the core is diffused from the wall layers, it might be argued that l_0 should be based on the wall-layer thickness; the values of R_0 would then be closer to those for the boundary layer.

In Table 7.2 there is a steady progression in R_0 from high values for the channel flows and boundary layer, through intermediate values for wall jets and jets, and finally to the smallest values for wakes. The first part of this progression corresponds to the removal of the wall constraint; subsequent reductions in R_0 result from progressive changes in the interaction between the mean flow and the largest elements of the turbulence. Somewhat surprisingly, this interaction is strongly influenced by the relatively small cross-stream flows, whose nature is quite different for wakes and for jets. The values for axisymmetric flows follow the same trends, but are a little higher than those for the corresponding plane flows.

Although the flow constants depend on the way in which the scales l_0 and U_0 are defined, they are of the same order as the local Reynolds numbers Re_0 at which transition occurs in free shear layers. Townsend has argued that this is no coincidence, and that the largest elements of the turbulence are instabilities continually appearing in a fluid whose effective viscosity is generated by the smaller scales of motion. Viewed in this way, the alternative name for R_0, the turbulent Reynolds number, has added significance.

The ratio l_c/l_1 varies only by a factor of two or so for the flows of Table 7.2. (It should be noted, however, that a change in this parameter has a larger effect than does a change in R_0; compare equations (7.3.11, 20).) This suggests that the scale of the momentum-transferring structures within these flows is,

Table 7.2. Values of the flow constant $R_0 = U_0 l_0/\epsilon_c$ and the mixing-length ratio l_c/l_1

Flow type	R_0	l_0	l_c/l_1	l_1	$R_0(l_c/l_1)^2$
I. Core and outer turbulence of wall flows					
Plane pressure	85	$\frac{1}{2}b$	0·10	$\frac{1}{2}b$	0·85
Plane shearing	55	b	0·135	b	1·0
Pipe	91	R	0·10	R	0·91
Boundary layer: U_1 constant	55	$[\partial U/\partial y]$	0·09	δ	0·46
	60	δ_1			
Wall jet: $U_1 = 0$	50	$[U]$	0·075	δ	0·28
$U_1 = U_0$	35	$[U]$	0·075	δ	0·20
II. Plane free-turbulent flows					
Mixing layer	31	$[\partial U/\partial y]$	0·07	δ	0·15
Jet: $U_1 = 0$	27	$[U]$	0·09	$\frac{1}{2}\delta$	0·22
Convected jet: $U_1 = 2U_0$	21	$[U]$	?	–	–
$U_1 = 4U_0$	17	$[U]$	?	–	–
Wake: $U_1 > 10U_0$	11	$[U]$	0·16	$\frac{1}{2}\delta$	0·28
III. Round free-turbulent flows					
Jet: $U_1 = 0$	35	$[U]$	0·075	$\frac{1}{2}\delta$	0·20
Wake	14	$[U]$	?	–	–
Plume	14	$[U]$	?	–	–

very roughly, a constant fraction of the flow width. This scale is not in fact a very small fraction of the distance over which changes in mean velocity occur, in contradiction to the gradient-diffusion argument.

Generally speaking, large values of R_0 are associated with small values of l_c/l_1. It is not quite true to say that $(l_c/l_1)^2 \propto 1/R_0$, as suggested by the wake result (7.3.22), but the final column of Table 7.2 shows that

$$\left(\frac{l_c}{l_1}\right)^2 R_0 = 0{\cdot}25 \qquad (7.3.23)$$

is fairly representative of a variety of spreading flows. This may be compared with a conclusion reached by Gartshore concerning the width σ of the zone of intermittency:

$$\left(\frac{\sigma}{l_0}\right)^2 R_0 = 5{\cdot}4 \tag{7.3.24}$$

Taking $l_0 = l_1/2$ to $l_1/3$, we have

$$\frac{l_c}{\sigma} = 0{\cdot}4 \text{ to } 0{\cdot}6 \tag{7.3.24}$$

linking the momentum-transferring scales to those which distort the turbulence interface.

Other Empirical Coefficients

We have seen that only one empirical constant is required to specify the development of a simple self-preserving flow, provided that the momentum integral is used to relate the length and velocity scales. This statement assumes that we know the location of the virtual origin of the spreading flow, and that we have an adequate cross-stream profile to introduce into the momentum integral. The latter is not a very stringent requirement; $e^{-k\eta^2}$ is usually adequate, and even a 'top-hat' profile is useful for rough calculations. There are, however, many ways of specifying the single bit of empirical information; we now consider some alternatives to the parameters R_0 and l_c/l_1 which were used above.

An empirical specification closely related to R_0 was adopted by Prandtl:

$$\epsilon_c = KU_0 l_1 \quad (\text{or } KU_0 l_0) \tag{7.3.25}$$

Evidently, the *exchange coefficient*

$$K = \frac{l_0/l_1}{R_0} \quad \left(\text{or } \frac{1}{R_0}\right) \tag{7.3.25}$$

and lies in the range 0·035 to 0·007 (or 0·09 to 0·02). For a wall flow, the alternative formulation

$$\epsilon_c = K_f u_f l_1 \tag{7.3.26}$$

is sometimes used. The constant K_f is the reciprocal of the parameter R_f introduced in equations (3.3.22) and used in Table 3.3 and elsewhere.

For flows which grow linearly—the simple jet, the mixing layer and the self-preserving convected flows of Section 7.2—it is convenient to use the *rate of spread*:

$$\begin{aligned} \frac{dl_0}{dx} &= 0{\cdot}104 \quad \text{for a plane jet with } U_1 = 0 \\ &= 0{\cdot}085 \quad \text{for a round jet with } U_1 = 0 \\ &= 0{\cdot}098 \quad \text{for a mixing layer with } U_2 = 0 \end{aligned} \tag{7.3.27}$$

For the first two, l_0 is measured to $\Delta U/U_0 = \frac{1}{2}$ ($k = \ln 2$); for the last, l_0 is measured over $\Delta U/U_0 = 0{\cdot}1$ to $0{\cdot}9$.

For linearly growing flows, the scaled coordinate $\eta = y/l_0$ is often written $\sigma y/x$, with σ a *growth parameter*. Since $\eta = y/(\mathrm{d}l_0/\mathrm{d}x)x$ in such a flow, we have

$$\sigma = \left(\frac{\mathrm{d}l_0}{\mathrm{d}x}\right)^{-1} \tag{7.3.28}$$

and see that $\sigma \simeq 10$ for the values (7.3.27).

Yet another parameter linked to the rate of spread is the *entrainment coefficient* of equation (7.2.28). Its value, with the ratio U_1/U_0, is sufficient to fix the development laws.

Finally, we consider the use of length and velocity scales external to the developing flow. The growth of the wake behind a circular cylinder of diameter d may be specified by $l_0 = \text{constant} \times (dx)^{\frac{1}{2}}$, or its eddy viscosity by $\epsilon_c = \text{constant} \times (U_1 d)$. Similarly, the properties of a jet may be specified in terms of d and U_j, the diameter and velocity at the generating nozzle.

To see the limitations of such *external scales*, consider a plane wake behind a circular cylinder. The drag is specified by the coefficient

$$C_D = \frac{D'}{\frac{1}{2}\rho U_1^2 d} \tag{7.3.29}$$

and equations (7.3.11, 12) can be written

$$l_0^2 = f_1(C_D) \times (dx) \quad \text{and} \quad \frac{U_0^2}{U_1^2} = f_2(C_D) \times \frac{d}{x} \tag{7.3.29}$$

Evidently, scaling with the cylinder diameter gives a law valid for a single drag coefficient, in contrast to scaling using the momentum invariant, which applies to all plane turbulent wakes. Nor do the laws (7.3.29) apply to non-circular cylinders. Of course, the drag coefficient for a given section may be constant over a wide range of Reynolds numbers, especially if the section has sharp corners, and the scaling (7.3.29) will be adequate in that range. Figure 6.1(a) illustrates this.

Similar arguments apply for jets: although the momentum flux at the nozzle is often not very different from that in the developed flow, it is safer to use the latter, rather than $\rho U_j^2 d^2$.

7.4 TRANSPORT PROCESSES

7.4.1 The Nature of the Problem

We shall follow our usual practice of treating explicitly the transfer of heat by convection and turbulent and molecular diffusion. The results can be used to describe the transfer of other passive entities which move with the fluid, provided that the fluid properties remain sensibly constant.

For the transfer of thermal energy within a turbulent flow with a free boundary, a vital factor is the large magnitude of the diffusivity within the turbulence interface compared with that outside. Typically, the concentration of a transferred quantity will be fairly uniform within the turbulent region, and will change abruptly at the interface. This reasoning does not apply for Pr, Sc $\ll 1$, when the large molecular diffusivity is able to 'cut through' the interface. However, for a broad range of transfers with Pr, Sc $> 0{\cdot}03$, say, the mean profile of the transferred entity will approximate to that of the intermittency factor. More precisely, the temperature or energy profile of a hot shear layer, for example, may be expected to lie somewhat within the intermittency profile, since cool fluid is absorbed at the turbulence interface.

We have seen, in Figure 6.2, that the profile of mean velocity is narrower than that of intermittency. Hence the distribution of temperature relative to mean velocity must be as indicated in Figure 7.3; any other passively transferred quantity will have much the same distribution in the outer turbulent region. The role of the interface is equivalent to an *apparent Prandtl or Schmidt number* whose value is defined by the statistical geometry of the interface. Table 5.1, III shows that $\mathrm{Pr_a}$, $\mathrm{Sc_a} = 0{\cdot}5$ to $0{\cdot}7$ is usual. Note that the lateral scale for the transferred quantity is only a little greater than that of the velocity variation, unless Pr, Sc $\ll 1$. In this respect turbulent shear layers behave more simply than do laminar flows, whose approximate character is indicated in equation (6.2.15).

In the almost-always-turbulent region near a wall or axis of symmetry, the transfer is controlled by a Prandtl number $\mathrm{Pr_t} = f(\mathrm{Pr, Re})$ peculiar to the particular fluid and flow condition; its variation is suggested in Figure 5.2(b). However, for diffusion across a free-turbulent layer, the mean profile is nearly flat in this inner region and, in consequence, the transfer characteristic of the turbulent fluid has little effect on the shape of the profile. Moreover, for many of the situations which have been studied experimentally, and which are of practical interest, $\mathrm{Pr_t}$, $\mathrm{Sc_t} = 0{\cdot}5$ to $1{\cdot}0$ within the turbulent fluid; this is so in particular for gases ($\mathrm{Pr} \simeq 0{\cdot}7$ and $\mathrm{Sc} = 0{\cdot}6$ to $1{\cdot}0$ for diffusion in air) at the low local Reynolds numbers ($\mathrm{Re_0} \sim 10^3$) typical of free turbulence. Under these circumstances, and especially when a constant eddy viscosity or mixing length has already been assumed, the transfer can be adequately represented using a single value of $\mathrm{Pr_t}$ for the entire shear layer. This value will be nearly the apparent value $\mathrm{Pr_a}$ for the region of intermittency.

At the end of Section 5.1.3, it was suggested that the low values of $\mathrm{Pr_t}$ given for free turbulence in Table 5.1, III are the consequences of a particular combination of Re and Pr. Though this idea may be correct, it is now seen to be largely irrelevant. The apparent Prandtl number is primarily a consequence of the convolutions of the free boundary, and depends only weakly on Re and Pr.

We now consider the various classes of thin free-turbulent flows for which the variation of mean temperature may be required. Our analysis will utilize the requirement that the energy flux is invariant along the flow. Hence the results are applicable to other quantities whose integrated flux is conserved. In order of increasing complexity, we have:

(1) Pure thermal wakes. These result when heat is supplied at a steady rate at a fixed point in a uniform flow (or at a point steadily moving through still fluid), without a change in the momentum flux. An example is the wake of a self-propelled vehicle. The spreading of the contaminant is ultimately accomplished by the background diffusivity of the convecting medium.

(2) Heated shear flows. Here energy and momentum are added at the same time, and dispersion is accomplished by the mean convection and turbulence of the resulting shear flow.

(3) Transmitting shear flows. A mixing layer acts to transmit heat between the initially uniform regions on its two sides.

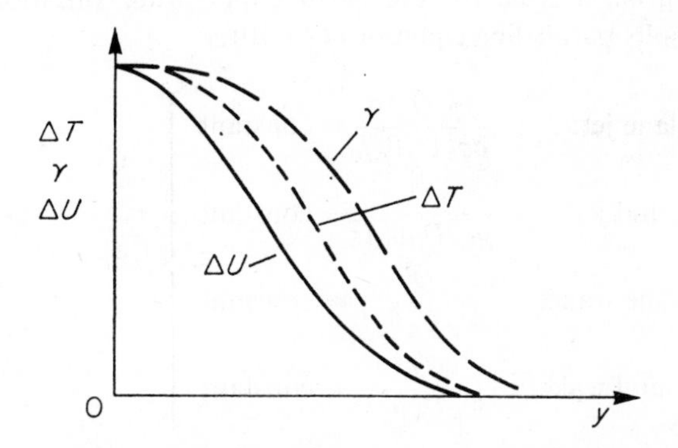

Figure 7.3 Typical relationships between variations across a free-turbulent flow in the mean velocity, the intermittency factor γ, and the mean value of a passively transferred quantity, the last exemplified by the temperature distribution

(4) Buoyancy-driven flows. The mechanisms of dispersion are essentially those of (2), but the motion is sustained by density differences in the heated fluid.

(5) Combined mechanisms. A variety of flows intermediate to those considered above can be envisaged; they will not be considered here. We might also extend the discussion to shear layers in more varied environments, for example, plumes ejected into a cross wind or into a stratified atmosphere.

7.4.2 Similarity Arguments

Energy Invariants

We begin by using similarity methods to investigate the pattern of thermal development for some of the flows identified above. For a thin, free flow, the energy integrals (6.2.37, 50) imply the existence of the energy invariants:

(1) for plane flow

$$2\rho \int_0^{y_1} U(H - H_1)\,\mathrm{d}y = \dot{Q}' \tag{7.4.1}$$

(2) for round flow

$$2\pi\rho \int_0^{R_1} U(H - H_1)\,r\,\mathrm{d}r = \dot{Q} \tag{7.4.2}$$

with $\dot{Q}'$ and $\dot{Q}$ the rates of heat addition per unit breadth and for the whole flow.

Introducing θ_0, a scale for the mean temperature variation, we obtain particular results paralleling equations (7.1.10):

$$\left.\begin{aligned}
&\text{(1) for a plane jet,} && \frac{\dot{Q}'}{\rho c_p U_0 \theta_0 l_0} = \text{constant}\\
&\text{(2) for a round jet,} && \frac{\dot{Q}}{\rho c_p U_0 \theta_0 l_0^2} = \text{constant}\\
&\text{(3) for a plane wake,} && \frac{\dot{Q}'}{\rho c_p U_1 \theta_0 l_0} = \text{constant}\\
&\text{(4) for a round wake,} && \frac{\dot{Q}}{\rho c_p U_1 \theta_0 l_0^2} = \text{constant}
\end{aligned}\right\} \tag{7.4.3}$$

These suggest the combined parameters:

$$\left.\begin{aligned}
&\frac{\dot{Q}'}{\rho c_p} \quad \text{and} \quad \frac{\dot{Q}}{\rho c_p} \quad \text{for jets}\\
&\frac{\dot{Q}'}{\rho c_p U_1} \quad \text{and} \quad \frac{\dot{Q}}{\rho c_p U_1} \quad \text{for wakes}
\end{aligned}\right. \tag{7.4.4}$$

Any other convected entity S can be treated in this way, provided that the 'source' terms of the general integral (6.2.32) are negligible within the developing flow. The scale S_0 replaces $c_p\theta_0$, and $\dot{Q}$ is replaced by the rate of injection of S into the flow.

We have tacitly assumed that the scale of lateral diffusion is linked to the velocity variation. While this is realistic for most turbulent flows, where the transferred entity is confined within the turbulence interface, the assumption limits application to those laminar cases for which Pr, Sc $\simeq 1$. There is no reason to expect that the dependence of the profile width on Prandtl or Schmidt number is complicated, but it is not apparent from simple similarity arguments.

Pure Thermal Wakes

In the absence of a maintained mean velocity variation, turbulence generation within the hot region soon ceases; thereafter, spreading is accomplished by the basic diffusivity of the convecting stream, perhaps the atmospheric wind. The diffusivity we take to be uniform throughout the fluid: $\epsilon_e = \epsilon_h + \kappa$, composed of turbulent and molecular elements. We argue, as for momentum wakes in Table 7.1, that it is the combination $\epsilon_e x/U_1$ which is relevant to the lateral transfer. Hence

$$l_0, \theta_0 = f\left(\frac{\dot{Q}'}{\rho c_p U_1}, \frac{\epsilon_e x}{U_1}\right) \tag{7.4.5}$$

will govern the scales of a plane thermal wake. Dimensional analysis gives

$$l_0 \propto \left(\frac{\epsilon_e x}{U_1}\right)^{\frac{1}{2}} \quad \text{and} \quad \theta_0 \propto \frac{\dot{Q}'/\rho c_p}{(\epsilon_e U_1 x)^{\frac{1}{2}}} \tag{7.4.5}$$

Similarly, for a round thermal wake

$$l_0 \propto \left(\frac{\epsilon_e x}{U_1}\right)^{\frac{1}{2}} \quad \text{and} \quad \theta_0 \propto \frac{\dot{Q}/\rho c_p \epsilon_e}{x} \tag{7.4.6}$$

Heated Shear Flows

The length and velocity scales for these momentum-driven flows are given in Table 7.1. The corresponding changes in temperature scale follow from equations (7.4.3):

$$\left.\begin{aligned} &\text{(1) for a plane jet,} && \theta_0 \propto \frac{\dot{Q}'/\rho c_p}{U_0 l_0} \\ &\text{(2) for a round jet,} && \theta_0 \propto \frac{\dot{Q}/\rho c_p}{U_0 l_0^2} \\ &\text{(3) for a plane wake,} && \theta_0 \propto \frac{\dot{Q}'/\rho c_p U_1}{l_0} \\ &\text{(4) for a round wake,} && \theta_0 \propto \frac{\dot{Q}/\rho c_p U_1}{l_0^2} \end{aligned}\right\} \tag{7.4.7}$$

The introduction of the length and velocity scales of Tables 7.1, II and III gives the temperature variations for free-turbulent flows, and for the corresponding laminar flows with Pr $\simeq$ 1. In each case the combination of energy and momentum invariants (7.1.10) and (7.4.3) implies that

$$\theta_0 \propto U_0 \tag{7.4.8}$$

However, the constants of proportionality depend in different ways on the defining parameters.

The plane mixing layer requires special treatment. If the temperatures in the bounding flows are uniform, we have simply $\theta_0 = T_1 - T_2$, analogous to $U_0 = U_1 - U_2$. On the other hand, if the two outer temperatures are the same, and heat is introduced at the start of the flow, the energy invariant gives

$$\theta_0 \propto \frac{\dot{Q}'/\rho c_p(U_1 - U_2)}{l_0} \tag{7.4.9}$$

Once again, length scales from Table 7.1 can be introduced.

Buoyant Flows

We consider plumes rising into otherwise still fluid; the plane flow occurs above a heated horizontal pipe, while the round flow models the 'thermals' generated in the atmosphere when the ground is strongly heated.

The basis of our dimensional analysis must now be reconsidered: while the energy invariants still apply, the momentum flux varies along the flow, by virtue of changes in the buoyancy integral of equation (6.2.55). It is convenient to work in terms of the scale ρ_0 for the density variation, rather than the temperature scale related to it by $\theta_0 = -\rho_0/(\rho_1 \beta)$, where β is the coefficient of cubical expansion.

The energy invariants follow from equation (6.2.57):

$$2\int_0^{y_1} Ug(\rho_1 - \rho)\,\mathrm{d}y = \frac{g\beta}{c_p}\dot{Q}'$$

and (7.4.10)

$$2\pi\int_0^{R_1} Ug(\rho_1 - \rho)\,r\,\mathrm{d}r = \frac{g\beta}{c_p}\dot{Q}$$

These lead to

$$\left.\begin{aligned} &\text{(1) for a plane plume,} \quad \frac{\beta\dot{Q}'}{c_p U_0 \rho_0 l_0} = \text{constant} \\ &\text{(2) for a round plume,} \quad \frac{\beta\dot{Q}}{c_p U_0 \rho_0 l_0^2} = \text{constant} \end{aligned}\right\} \tag{7.4.11}$$

We conclude that the defining parameters occur in the combinations $\beta\dot{Q}'/c_p$ and $\beta\dot{Q}/c_p$.

The momentum conditions follow from equation (6.2.55), with $U_1, \tau_w = 0$:

$$\frac{\mathrm{d}}{\mathrm{d}x}(\rho_1 U_0^2 l_0^{j+1}) \sim g\rho_0 l_0^{j+1} \tag{7.4.12}$$

with $j=0$, 1 for plane and round flows. This suggests that the parameters ρ_1 and g appear in the combination ρ_1/g. Finally, we conclude that

$$l_0,\, U_0,\, \rho_0 = f\left(\frac{\beta \dot{Q}}{c_p}, \frac{\rho_1}{g}, x\right) \tag{7.4.13}$$

with a similar result for plane flow.

Dimensional analysis gives:

(1) for both plane and round plumes, $\quad l_0 \propto x \qquad$ (7.4.14)

(2) for a plane plume, $\quad U_0 = \text{constant} \quad \text{and} \quad \rho_0,\, \theta_0 \propto x^{-1} \qquad$ (7.4.15)

(3) for a round plume, $\quad U_0 \propto x^{-\frac{1}{3}} \quad \text{and} \quad \rho_0,\, \theta_0 \propto x^{-\frac{5}{3}} \qquad$ (7.4.16)

The determination of the scaling constants is straightforward, and will be left as an exercise.

7.4.3 Cross-stream Profiles

Again we concentrate, for brevity, on the simple case of a plane wake. From equations (6.2.29) we obtain

$$U_1 \frac{\partial H}{\partial x} = -\frac{\partial(\overline{vh})}{\partial y} + \kappa \frac{\partial^2 H}{\partial y^2} \tag{7.4.17}$$

exactly paralleling the momentum equation (7.3.1). Following the pattern of equations (7.2.1, 2), we adopt the similarity forms

$$\begin{aligned} H &= H_1 + c_p \theta_0\, j\left(\frac{y}{l_0}\right) \\ \overline{vh} &= c_p \theta_0 U_0\, k_2\left(\frac{y}{l_0}\right) \end{aligned} \tag{7.4.18}$$

When these are introduced and account is taken of the condition $\theta_0 l_0 = \text{constant}$, equation (7.4.17) becomes

$$C_1(j + \eta j') = -k_2' + \frac{j''}{\text{Pe}_0} \quad \text{with } j' = \frac{\mathrm{d}j}{\mathrm{d}\eta}$$

with $\text{Pe}_0 = U_0 l_0/\kappa$ the local Péclet number, and C_1 the constant of equations (7.3.3, 4). Integration gives

$$C_1 \eta j = -k_2 + \frac{j'}{\text{Pe}_0} \tag{7.4.19}$$

on taking $j', k_2 = 0$ at $\eta = 0$, conditions appropriate to a temperature distribution which is symmetrical about the axis.

Omitting the molecular-transfer terms of equations (7.3.4) and (7.4.19), we have $C_1\eta f = -g_{12}$ and $C_1\eta j = -k_2$, whence

$$\frac{f}{j} = \frac{g_{12}}{k_2} \tag{7.4.20}$$

This remarkably simple result relates the scaled fluxes of momentum and enthalpy to the scaled velocity and temperature. It holds for other constant-pressure free-turbulent flows, as may be seen by noting that the relationship (7.4.8), namely, $U_0 \propto \theta_0$, implies that the convective operator $U\partial/\partial x + V\partial/\partial y$ will generate the same structure when applied to H as it does when applied to U.

Let us see what can be said about the right-hand side of equation (7.4.20). The ratio of the two transfers can be expressed using equations (3.4.37):

$$\frac{\overline{uv}}{\overline{vh}} = \frac{\epsilon_m}{\epsilon_h}\frac{\partial U/\partial y}{\partial H/\partial y} = \frac{l_m^2}{l_h^2}\frac{\partial U/\partial y}{\partial H/\partial y} \tag{7.4.21}$$

or

$$\frac{g_{12}}{k_2} = \mathrm{Pr_t}\frac{f'}{j'} = \left(\frac{l_m}{l_h}\right)^2\frac{f'}{j'} \tag{7.4.21}$$

These results involve only the formal definitions of the diffusivities and mixing lengths; no assumption is made regarding the species of the (thin) flow, or regarding the actual mechanism of transfer. Whatever these may be, we have

$$\mathrm{Pr_t} = \left(\frac{l_m}{l_h}\right)^2 \tag{7.4.22}$$

For $\mathrm{Pr_t} \simeq 0{\cdot}5$, a value typical of free turbulence (see Table 5.1), we have $2l_m^2 \simeq l_h^2$. This is close to the relationship (3.4.52) between l_m and l_ω, the mixing length for vorticity on the Taylor model. Hence it appears that l_ω and l_h are not very different in free-turbulent flow.

Using the results (7.4.22), we can convert equation (7.4.20) formally to

$$\frac{f}{j} = \mathrm{Pr_t}\frac{f'}{j'} \tag{7.4.23}$$

This provides a means of determining the variation of $\mathrm{Pr_t}$ from measured velocity and temperature profiles. Alternatively, on adopting a constant Prandtl number, we have

$$j = (-f)^{\mathrm{Pr_t}} \tag{7.4.24}$$

(Remember that $f < 0$ for a wake.) In terms of the velocities and temperatures themselves

$$\frac{H - H_1}{c_p\theta_0} = \frac{T - T_1}{\theta_0} = \left(\frac{U_1 - U}{U_0}\right)^{\mathrm{Pr_t}} \tag{7.4.24}$$

Since these results apply to any model of the transfers across a wake, we can introduce either of the velocity profiles (7.3.9, 18) for constant eddy viscosity and constant mixing length. The results (7.4.24) also apply to a laminar wake, with the molecular Pr replacing Pr_t. Once a temperature distribution is obtained, the energy integral (7.4.1) can be used to specify the constant of the development law whose basic structure is indicated in equations (7.4.7).

The results (7.4.22, 23, 24) apply to other species of turbulent self-preserving flow, since they follow from the result (7.4.20) and the formal definitions (7.4.21).

7.5 THE INFLUENCE OF THE OUTER FLOW AND BOUNDARIES

In the preceding discussion of nearly parallel shear flows within a large volume of fluid at rest or in uniform motion, we have considered the outer fluid simply as a passive supplier of boundary conditions. We now extend the discussion to include three more positive aspects of the outer flow, dealing in turn with situations in which it:

(1) is prescribed by the shear layer;
(2) prescribes the development of the shear layer; and
(3) interacts with the shear flow, so that identification of cause and effect is impossible.

We shall sometimes broaden the interpretation of 'outer flow' to include its boundaries or a boundary which replaces it. Thus the discussion blends into the following chapter on boundary layers.

Motion Induced in the Outer Fluid

At several points (Sections 6.1.4, 7.1 and 7.2) we have considered the lateral flows required to maintain overall mass and momentum balances within thin shear layers. Then we were interested primarily in the effect of these flows on the shear layers, and assumed that the outer-flow boundaries were such that the required flow could develop spontaneously. The way in which an induced flow can be put to use is illustrated in Figure 7.4. A jet blown backwards along the surface of an aerofoil draws in the outer fluid, preventing separation and

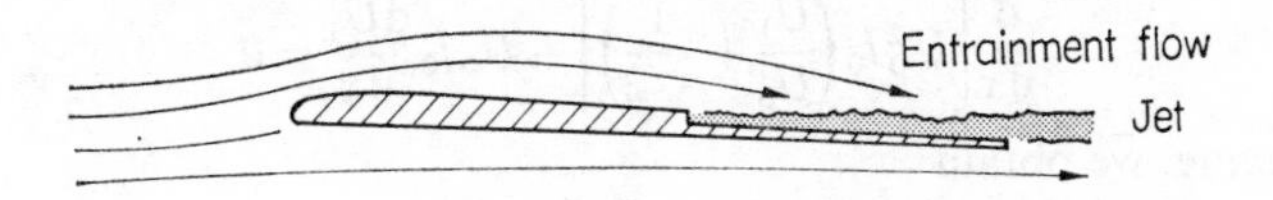

Figure 7.4 Application of entrainment in controlling the flow over an aerofoil

increasing the circulation around the lifting surface. Not only does the jet provide some thrust, but it acts in the manner of a flap to augment the lift. This artifice is often referred to as a *jet flap*.

Convection and Pressure Gradient

We now reverse our point of view, and ask what effect the linked variations in outer pressure and convective velocity have on an embedded shear layer. For definiteness, we consider the particular case shown in Figure 7.5: a plane jet advancing through a stream distorted symmetrically by a surrounding duct. More realistic examples arise in aeronautical and marine engineering.

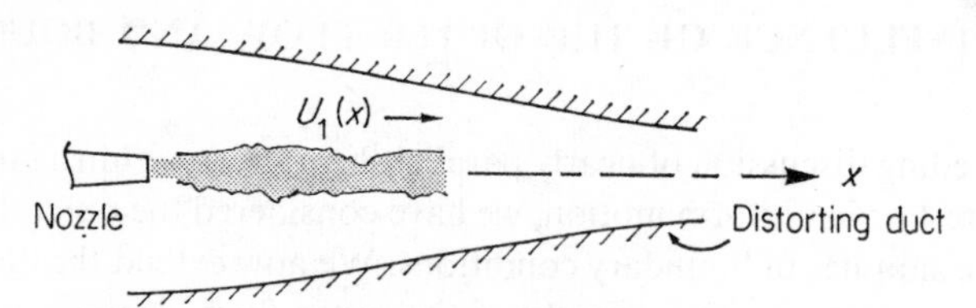

Figure 7.5 An example of jet development with varying outer flow

To answer the question posed in the preceding paragraph, we turn to the integral momentum equation (6.2.33). With normal stresses omitted:

$$\frac{\mathrm{d}}{\mathrm{d}x}\int_0^{y_1} U(U_1-U)\,\mathrm{d}y+\frac{\mathrm{d}U_1}{\mathrm{d}x}\int_0^{y_1}(U_1-U)\,\mathrm{d}y=0 \tag{7.5.1}$$

Assuming that the profile shape is invariant once the 'potential' core near the nozzle is left behind, we write this as

$$\frac{\mathrm{d}}{\mathrm{d}x}\left[U_0U_1l_0\int_0^\infty f\,\mathrm{d}\eta+U_0^2l_0\int_0^\infty f^2\,\mathrm{d}\eta\right]+U_0l_0\frac{\mathrm{d}U_1}{\mathrm{d}x}\int_0^\infty f\,\mathrm{d}\eta=0 \tag{7.5.2}$$

where the integrals are constant along the flow. Adopting the Gaussian profile $f=\mathrm{e}^{-k\eta^2}$, as predicted in equation (7.3.9) and found more widely to be an acceptable approximation, we have

$$\int_0^\infty f\,\mathrm{d}\eta=\sqrt{2}\int_0^\infty f^2\,\mathrm{d}\eta=\frac{1}{2}\left(\frac{\pi}{k}\right)^{\frac{1}{2}} \tag{7.5.3}$$

giving the momentum equation

$$\frac{\mathrm{d}}{\mathrm{d}x}\left[U_0^2l_0\left(\frac{U_1}{U_0}+\frac{1}{\sqrt{2}}\right)\right]+U_0l_0\frac{\mathrm{d}U_1}{\mathrm{d}x}=0 \tag{7.5.4}$$

Re-arranging, we obtain

$$\left(1+\frac{U_0}{\sqrt{2}U_1}\right)\frac{1}{l_0}\frac{\mathrm{d}l_0}{\mathrm{d}x}+\left(\sqrt{2}+\frac{U_1}{U_0}\right)\frac{\mathrm{d}(U_0/U_1)}{\mathrm{d}x}+\left(3+\frac{\sqrt{2}U_0}{U_1}\right)\frac{1}{U_1}\frac{\mathrm{d}U_1}{\mathrm{d}x}=0 \tag{7.5.5}$$

Some simple cases are contained within this last result:

(1) For U_1 constant and $U_0 \ll U_1$, we recover the plane-wake result $U_0 l_0 = \text{constant}$.

(2) For U_0/U_1 constant, as required for exact self-preservation in equations (7.2.5), we obtain

$$l_0 U_1^{H+2} = \text{constant}$$

as in equations (6.2.34), since the shape factor is

$$H = \left(1 + \frac{U_0}{\sqrt{2}U_1}\right)^{-1} \tag{7.5.6}$$

for the Gaussian profile.

To determine the development for a particular $U_1(x)$, we must know something more about the relationship between U_0 and l_0. Abramovich has suggested that this additional information be introduced through the entrainment constant of equation (7.2.28):

$$\frac{dl_0}{dx} = \frac{\beta}{U_1/U_0 + \frac{1}{2}} \tag{7.5.7}$$

As was pointed out earlier, there is no very good reason (save expediency) for assuming that a constant value of β will realistically account for the changes in the roles of streamwise and lateral convection as U_1/U_0 varies. In other words, we would expect that $\beta = f(U_0/U_1)$, as the alternative parameter R_0 is seen to be in Table 7.2. Nevertheless, satisfactory predictions can be obtained for 'strong' jets (U_1/U_0 not too large) using the value for $U_1/U_0 = 0$. Introducing the growth rate from equations (7.3.27), we have

$$\beta = \beta_0 = \frac{1}{2}\left(\frac{dl_0}{dx}\right)_0 = 0{\cdot}052 \tag{7.5.8}$$

Equation (3.5.7) now has a definite form and, once initial values of l_0 and U_0 are specified, it can be solved with the momentum equation (7.5.5) to find the variations of these scales in a prescribed outer stream $U_1(x)$. Although the form of the governing equations was suggested by the properties of self-preserving solutions, we now suppose that the equations apply for any variation of the outer velocity.

Curvature

We consider yet another way in which the outer flow can influence a shear layer within it—curvature. Generally speaking, the rate of spread of a free jet is increased at the convex boundary, and is decreased on the concave. For a boundary layer or wall jet, these trends are reversed. A simple explanation is often offered: when high-velocity fluid moves radially outwards, it finds itself in a region where the centripetal pressure gradient is defined by slower-moving fluid, and hence tends to move even further; in a like manner, the lateral motion of low-velocity fluid is damped by the centripetal gradients. However, this reasoning is too simple. The rate of spread will also be influenced by changes in the shear-stress profile, by the rotation of the mean strains which feed the expansion-producing 'eddies', and by the simple increase in the scale of motion on the convex side of the flow.

For a jet, the overall effect of curvature is to increase the growth rate. For a particular class of curved jets to be considered shortly (Figure 7.6(a)), Schwartzbach found

$$\frac{\mathrm{d}l_0}{\mathrm{d}x} = 0{\cdot}186 \tag{7.5.9}$$

with l_0 measured to $\Delta U/U_0 = \frac{1}{2}$ and x measured along the jet axis.

Effects of Boundaries: *Interactions*

By limiting the ability of the outer flow to respond to the demands of the shear layer, solid boundaries near a shear flow give rise to complex interactions between outer flow and the developing layer. Figure 7.6(a) shows such a situation—a plane jet emitted parallel to a solid wall. On one side of the jet, fluid can be entrained without limit; on the other, the jet must feed on itself by forming a recirculating 'outer' flow. Such recirculating flows also develop when a channel expands abruptly, and when a uniform stream passes over a backwards-facing step in a wall. In these cases, a boundary layer develops downstream, rather than a wall jet.

The jet of Figure 7.6(a) is 'attracted' to the wall by the necessity of entrainment. We may expect that a strong wake or a boundary layer will be 'repelled' from a solid boundary, and that a weak wake will merely be convected downstream. When attraction or repulsion occurs, the interaction of shear layer and outer flow must be taken into account.

Within flows like that of Figure 7.6(a) occurs the rapid dissipation typical of transitions and fittings in channels; see Section 4.5.4. The processes within the characteristic dissipative mechanism of open-channel flow—the hydraulic jump shown in Plate II (facing p. 47)—are essentially similar. These flows contain free shear layers where the dissipation rate is very much higher than in a boundary layer, since the turbulence is not inhibited by an adjacent solid wall.

The zone of reattachment shown in Figure 7.6(a) can be moved upstream or downstream by extracting fluid from or admitting fluid to the separation bubble. In this way we attain a degree of control over the jet, and this can be put to use in *fluidic switching* devices, such as that sketched in Figure 7.6(b).

The jet can be moved from one wall to the other by changes in the control flows $\dot{m}_1$ and $\dot{m}_2$; the supply flow (or the pressure it produces) is then delivered to one or other of the two ports downstream. Once the jet is attached to one wall (by a pulse in $\dot{m}_2$, say), it may be stable to considerable changes in the downstream conditions (p_1 and p_2) and to small changes in the control flows. Alternatively, for other boundary geometries, the output may vary smoothly with $\dot{m}_1$ and $\dot{m}_2$; this gives proportional control, the basis of *fluidic amplification.*

The entrainment-attraction of a jet to a nearby wall is the basis of the

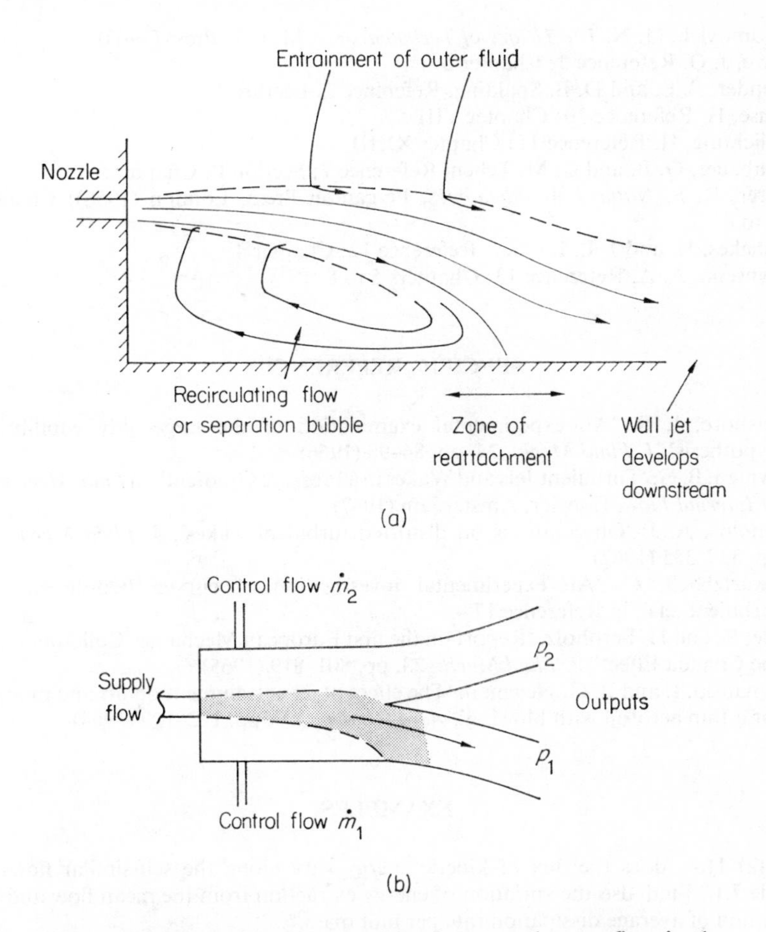

Figure 7.6 Interaction between shear layers and outer flow in the presence of solid boundaries. (a) Attachment of a plane jet to a wall: solid lines represent mean-flow streamlines; dashed lines denote nominal edges of the jet. (b) A simple fluidic control device

Coanda effect, the adherence of a wall jet to a convex surface. This can be observed quite easily by bringing the side of a finger up to a stream of water flowing from a tap. The Coanda effect is utilized in a variety of fluidic controls, and in jet-flap and slotted devices with aeronautical and marine applications. As noted above, the rate of spread of the wall jet is influenced by the curvature of the surface.

FURTHER READING

Abramovich, G. N. *The Theory of Turbulent Jets*, M. I. T. Press (1963)
Hinze, J. O. Reference 5: Chapter 6
Launder, B. E. and D. B. Spalding. Reference 2: Lecture 2
Rouse, H. Reference 10: Chapter VIII
Schlichting, H. Reference 11: Chapter XXIII
Schubauer, G. B. and C. M. Tchen, Reference 7: Section B, Chapter 5
Scorer, R. S. *Natural Aerodynamics*, Pergamon Press, London (1958): Chapters 5 to 8
Tennekes, H. and J. L. Lumley. Reference 12: Chapter 4
Townsend, A. A. Reference 13: Chapters 5 to 8

SPECIFIC REFERENCES

Gartshore, I. S. 'An experimental examination of the large-eddy equilibrium hypothesis', *J. Fluid Mech.*, **24**, pp. 84–98 (1966)
Newman, B. G. 'Turbulent Jets and Wakes in a Pressure Gradient', in *Fluid Mechanics of Internal Flow*, Elsevier, Amsterdam (1967)
Reynolds, A. J. 'Observations on distorted turbulent wakes', *J. Fluid Mech.*, **13**, pp. 335–355 (1962)
Schwartzbach, C. 'An experimental investigation of curved two-dimensional turbulent jets', in Reference 17
Wille, R. and H. Fernholz. 'Report on the first European Mechanics Colloquium, on the Coanda Effect', *J. Fluid Mech.*, **23**, pp. 801–819 (1965)
Wygnanski, I. and B. G. Newman. 'The effect of jet entrainment on lift and moment for a thin aerofoil with blowing', *Aero. Quart.*, **XV**, pp. 122–150 (1964)

EXAMPLES

7.1 (a) How does the flux of kinetic energy vary along the self-similar flows of Table 7.1? Find also the variation of energy extraction from the mean flow and the variation of average dissipation rate per unit mass.

(b) What distributions of sources are required to represent entrainment into these flows?

7.2 For laminar flows, what conditions are implied by equation (7.2.7) for: (1) exactly self-similar development, and (2) approximately self-similar wakes? Combine with the conditions (7.2.5) and (7.1.10) to find the variations of length and velocity scales, and compare with Table 7.1, III.

7.3 For the self-preserving wake of a self-propelled vehicle, the condition (7.2.12) is still applicable, but the momentum flux is no longer invariant. A suitable invariant can be derived from the momentum equation, on the assumption that the momentum transfer can be represented by a constant eddy viscosity.

(a) Multiply the equation of motion

$$U_1 \frac{\partial U}{\partial x} = \frac{\epsilon_m}{r} \frac{\partial}{\partial r}\left(r \frac{\partial U}{\partial r}\right)$$

by r^{m+1} and integrate by parts to obtain

$$U_1 \frac{d}{dx} \int_0^{R_1} r^{m+1}(U_1 - U)\,dr = \epsilon_m m^2 \int_0^{R_1} r^{m-1}(U_1 - U)\,dr$$

(b) Show that the right-hand side is zero for $m = 2$. How do U_0, l_0 and Re_0 vary along the flow? Does this wake remain turbulent? If the wake has a small but finite momentum, how will it ultimately develop?

7.4 Show that equation (7.3.4) follows from equation (7.3.1) on adopting the form $U = x^{-\frac{1}{2}} f(y/x^{\frac{1}{2}})$, which is suggested by the results of Table 7.1.

7.5 Write equations (7.3.11, 12) in forms appropriate to laminar flow.

7.6 Derive laws $l_0/\delta_2 = \text{constant} \times (x/\delta_2)^{\frac{1}{2}}$ and $U_0/U_1 = \text{constant} \times (\delta_2/x)^{\frac{1}{2}}$ for a plane turbulent wake, with δ_2 the momentum thickness of equation (6.1.6). State clearly how l_0 is defined.

7.7 (a) Develop formulae paralleling (7.3.11, 12) for a round turbulent wake.

(b) For the wake behind a sphere of diameter d, how are Re_d, Re_0 and C_D related?

(c) Develop formulae in which d is used as a length scale for the wake of a sphere.

7.8 (a) Show that the entrainment coefficients for plane wake, jet, and mixing layer are given by

$$\frac{\beta}{dl_0/dx} = \frac{U_1}{U_0}, \frac{1}{2}, \frac{3}{2}$$

What values has β for these three flows?

(b) For the wake, show that $\beta = 2k/R_0$, and find how it is related to the other empirical constant l_c/l_1.

7.9 (a) Adopting the profile (6.1.4), show that

$$F = \frac{\pi}{2k} \rho U_0^2 l_0^2$$

for a round jet.

(b) Using the results (7.3.27), obtain results of the form $U_0 = \text{constant} \times (F/\rho)^{\frac{1}{2}}/x$, $U_0/U_J = \text{constant} \times (d/x)$ and $\epsilon_c = \text{constant} \times (U_J d)$, on the assumption that F is equal to the momentum at a nozzle of diameter d.

(c) Is the final assumption correct: (1) for a jet emerging from a long pipe, and (2) for a jet from an orifice in a plane wall?

7.10 To analyse the velocity profile of a jet, we introduce a stream function to represent the mean flow pattern. The form

$$\psi = U_0 l_0^2 F\left(\frac{r}{l_0}\right) \quad \text{with} \quad rU = \frac{\partial \psi}{\partial r} \quad \text{and} \quad rV = -\frac{\partial \psi}{\partial x}$$

is suitable for a round, self-preserving flow.

(a) Show that continuity is satisfied by the velocity components defined in this way, and that

$$U = U_0\left[\frac{F'}{\eta}\right] \quad \text{and} \quad V = U_0 \frac{\mathrm{d}l_0}{\mathrm{d}x}\left[F' - \frac{F}{\eta}\right]$$

(b) Using the momentum conditions (7.1.10), show that the equation of motion (with viscous and normal stresses omitted) reduces to

$$\frac{\mathrm{d}l_0}{\mathrm{d}x}\left[\frac{F'^2}{\eta} + F\frac{\mathrm{d}(F'/\eta)}{\mathrm{d}\eta}\right] = \frac{\mathrm{d}(\eta g_{12})}{\mathrm{d}\eta}$$

(c) Show that the conditions

$$\frac{F}{\eta} = F' = g_{12} = 0 \quad \text{at} \quad \eta = 0$$

must be satisfied, and hence that

$$\frac{\mathrm{d}l_0}{\mathrm{d}x} FF' = \eta^2 g_{12}$$

(d) For a constant eddy viscosity, show that

$$F = \frac{\frac{1}{2}\eta^2}{1 + \frac{1}{4}k\eta^2} \quad \text{with} \quad k = R_0 \frac{\mathrm{d}l_0}{\mathrm{d}x}$$

is a solution satisfying appropriate boundary conditions.

(e) Find U and V. For what value of k does l_0 correspond to $\Delta U/U_0 = \frac{1}{2}$? What is the significance of the condition $V = 0$? Sketch the flow pattern. What confidence have you in the predictions for the outer flow?

7.11 The hot gases ejected vertically from a chimney have a finite initial momentum. How are the development laws (7.4.13) modified? How long does the flow remain a jet before turning into a plume?

7.12 (a) For the constant-eddy-viscosity model of wake flow, show that equations (7.4.24) imply that corresponding points on velocity and temperature profiles are related by $l_U/l_T = \mathrm{Pr}_t^{\frac{1}{2}}$, as in the laminar result (6.2.15).

(b) In what way does the constant-mixing-length model differ? Sketch the variations of j and f for the two models, taking $\mathrm{Pr}_t = \frac{1}{2}$.

7.13 Adopting the velocity variation (7.3.9), show that the energy integral for a plane wake becomes

$$\left(\frac{\pi}{k\,\mathrm{Pr}}\right)^{\frac{1}{2}} \rho c_p U_1 \theta_0 l_0 = \dot{Q}'$$

and hence find the variation of temperature in the wake. Considering a wake in atmospheric air at sea-level conditions, find the variation of the temperature scale as a function of $\dot{Q}'$ and D' only.

7.14 During static tests on the ground, a jet engine generates a thrust of 7000 lbf through a nozzle of effective (that is, referred to ambient density) area 5 ft². If the exhaust temperature is 500 °K, estimate the highest temperature achieved on a set of deflector vanes located 75 ft behind the nozzle. The nozzle axis is 8 ft above the runway surface. What temperature will the surface achieve below the vanes?

7.15 A power station rejects heat at the rate of 500 MW by raising the temperature of cooling water from 12 °C to 32 °C. The cooling water is ejected along the surface of a lake through a duct 1 m high and 2 m wide. Estimate the areas on the surface over which the mean temperature is higher than 25 °C and higher than 13 °C. Should the jet be taken to be round or plane?

7.16 To account for the variations in the outer velocity $U_1(x)$, the stream function for a round flow (see Example 7.10) can be modified to

$$\psi = \tfrac{1}{2}r^2 U_1 + U_0 l_0^2 F(\eta)$$

with U_1/U_0 constant for self-preservation.

(a) Show that the equation of motion reduces to

$$\frac{1}{2}\frac{dl_0}{dx}\frac{d}{d\eta}\left[F^2 + \frac{U_1}{U_0}\eta^2 F\right] = \eta^2 g_{12}$$

(b) Derive the simpler results for pure jet and pure wake flow. Do you expect that it will be possible to find a simple formula incorporating the two limiting velocity profiles for constant eddy viscosity?

7.17 (a) How is equation (7.5.4) modified if a 'top-hat' profile is used instead of the Gaussian form? How does this change alter the pattern of self-preserving development?

(b) Write a computer program to carry out the streamwise integration of equations (7.5.4, 7, 8). Compare with one of the standard Runge–Kutta routines.

7.18 What happens to the point of reattachment of Figure 7.6(a) when fluid is injected into or is extracted from the separation bubble? Why is it more meaningful to think of reattachment as occurring in a zone rather than at a single point? How is the location of this zone influenced by injection and extraction?

7.19 The dissipation in a rapidly expanding duct flow is determined by the basic geometry of the flow, independent of the details of the turbulence. This fact imposes a limit on the length of the separated shear layer in which the dissipation is accomplished. Consider constant-density flow through a duct of uniform rectangular section, partially blocked by a gate which leaves open the fraction f and reduces the minimum flow area to the fraction r of the duct area.

(a) How do you expect the ratio r/f to vary with f?

(b) Show that the total dissipation (per unit of channel width) in the region of rapid expansion is given by $\tfrac{1}{2}\rho b U_a(U_m - U_a)^2$, with b the channel height, U_a the average velocity of the undisturbed flow, and $U_m = U_a/r$ the maximum 'contracted' velocity.

(c) The dissipation per unit of plan area of the shear layer may be estimated as $\rho\overline{uv}\,\Delta u \simeq U_m^3/100$ where $\overline{uv}$ is a typical mixing stress in the flow, and ΔU is the velocity change across the layer. Justify this estimate. What is the corresponding result for a boundary layer on the duct wall? Why is the dissipation much more rapid in the shear layer?

(d) Show that the length of the separation bubble must be of order $L/b = 50r(1-r)^2$. Over what range is this result likely to be valid? Plot the likely variation of L/b *vs* f, the degree of opening. What part of the channel wall should be armoured against violent pressure fluctuations?

8

Developing Flows III: Boundary Layers

The turbulent flow bounded by a fixed wall and by a steady stream combines a wall layer, essentially like that of a channel flow, and an outer turbulent layer possessing the advancing interface which is characteristic of free turbulence. When the outer stream is uniform, the wall layer dominates the flow, and certain gross features can be predicted using results for channel flows. When the outer velocity falls and the pressure rises along the stream, the velocity change in the outer region is a larger fraction of the total; then inner and outer elements must be matched to determine the behaviour of the whole layer. Finally, there are situations in which the local wall layer is no longer influential—flows near separation or reattachment. As was noted earlier, such flows cannot be considered as 'thin', in the sense of the boundary-layer approximations.

Only under rather special circumstances are the separate demands of the wall constraint and outer flow consistent, allowing a self-preserving pattern of development. Accordingly, the emphasis in this chapter differs markedly from that in the preceding discussion of free turbulence. Here our approach is more like that for channel flows: we take the characteristics of the wall layer as the starting point, and say as little as we can about the outer flow. The following pattern will be followed:

(1) the friction and growth of constant-pressure boundary layers will be discussed on the basis of pipe-flow results;

(2) layers with non-uniform outer flow will be analysed using a 'two-layer' model to derive a local friction law, and consideration will be given to the information required (in addition to the momentum integral) to specify the development;

(3) transport processes within constant-pressure layers will be discussed, including the buoyancy-driven boundary layer; and

(4) brief consideration will be given to the two cataclysmic events in the life of a boundary layer—transition and separation—and to some other species of wall-bounded turbulent flows.

At several points, laminar boundary layers will be discussed briefly, partly for their intrinsic interest, and partly to provide a standard of comparison for turbulent layers.

8.1 THE CONSTANT-PRESSURE LAYER

8.1.1 Laminar Flow: the Blasius Solution

We begin our study of wall-bounded developing flows with the simplest, that formed when an initially uniform, parallel, and non-turbulent stream flows along a flat plate. There is in fact an even simpler boundary layer—the asymptotic suction layer—but this does not develop, since outwards diffusion is balanced by mean convection towards the porous wall.

We look first at the initial portion of the boundary layer, where the flow is laminar, and assume also that the properties of the fluid are uniform throughout the flow. Save near the nose of the plate, where the motion does not satisfy the thin-flow conditions, the laminar flow near the surface is adequately described by the Prandtl boundary-layer equations (6.2.22). Following Blasius, a pupil of Prandtl's who first examined this problem in detail, around 1908, we introduce a *stream function* related to the velocity components by

$$U = \frac{\partial \psi}{\partial y} \quad \text{and} \quad V = -\frac{\partial \psi}{\partial x} \tag{8.1.1}$$

This way of defining the velocity field automatically satisfies the continuity equation (6.2.22), and allows the momentum equation (6.2.22) to be written as

$$\frac{\partial \psi}{\partial y}\frac{\partial^2 \psi}{\partial x \partial y} - \frac{\partial \psi}{\partial x}\frac{\partial^2 \psi}{\partial y^2} = \nu \frac{\partial^3 \psi}{\partial y^3} \tag{8.1.2}$$

We have omitted the pressure gradient and the local acceleration with respect to time, since the outer flow is required to be uniform and steady.

To reduce this to an ordinary differential equation, we seek a similarity solution; this step can be justified by dimensional analysis or other similarity arguments, or by noting that successive measured profiles do have the same shape. The form of the cross-stream scale is apparent from equation (6.2.14) and again in Table 6.1: $l_0 \propto (\nu x/U_1)^{\frac{1}{2}}$. Hence we expect

$$U = U_1 f(\eta) \quad \text{with} \quad \eta = \frac{y}{(\nu x/U_1)^{\frac{1}{2}}} \tag{8.1.3}$$

and obtain it by taking

$$\psi = (\nu U_1 x)^{\frac{1}{2}} F(\eta) \tag{8.1.3}$$

When the velocity components

$$U = U_1 F' \quad \text{and} \quad V = \frac{1}{2}\left(\frac{\nu U_1}{x}\right)^{\frac{1}{2}} (\eta F' - F) \quad \text{with} \quad F' = \frac{\mathrm{d}F}{\mathrm{d}\eta} \tag{8.1.3}$$

are introduced, the momentum equation (8.1.2) becomes

$$FF'' + 2F''' = 0 \tag{8.1.4}$$

This non-linear, but ordinary equation is known as *Blasius equation*. It is to be solved subject to the boundary conditions:

$$F, F' = 0 \quad \text{for } \eta = 0 \quad \text{and} \quad F' = 1 \quad \text{for } \eta \to \infty \tag{8.1.5}$$

The general nature of the solution is indicated in Figure 5.3. The numerical methods used in finding it are not of interest here, but we shall require a number of characteristics of the layer which can be calculated from $F(\eta)$:

(1) the friction is specified by the local coefficient

$$c_f = \frac{\tau_w}{\frac{1}{2}\rho U_1^2} = 0{\cdot}664\ \text{Re}_x^{-\frac{1}{2}} \tag{8.1.6}$$

and by the average coefficient (see equations (6.2.35))

$$C_f = 1{\cdot}328\ \text{Re}_L^{-\frac{1}{2}} \tag{8.1.6}$$

(2) the 'thickness' by

$$\frac{\delta}{x} \simeq 5\ \text{Re}_x^{-\frac{1}{2}} \quad \text{to} \quad U = 0{\cdot}99\ U_1 \tag{8.1.6}$$

(3) the displacement and momentum thicknesses by

$$\frac{\delta_1}{x} = 1{\cdot}721\ \text{Re}_x^{-\frac{1}{2}} \quad \text{and} \quad \frac{\delta_2}{x} = 0{\cdot}664\ \text{Re}_x^{-\frac{1}{2}} \tag{8.1.6}$$

(4) the shape factor by

$$H = \frac{1{\cdot}721}{0{\cdot}664} = 2{\cdot}59 \tag{8.1.6}$$

a constant for these self-similar profiles.

The skeletal forms of these development laws were given in Table 6.1.

8.1.2 Velocity and Friction in the Turbulent Layer

The turbulent counterpart of the Blasius flow can be generated either by allowing the laminar layer to develop until it undergoes spontaneous transition (that is, transition initiated by arbitrarily small disturbances), or by forcing transition with a trip wire or similar artifice. Our immediate aim is to obtain results paralleling those given above for laminar flow, but two difficulties prevent us from adopting a parallel method of analysis. The momentum equation (6.2.21) contains an additional dependent variable, the mixing stress. Moreover, we cannot expect to find a similarity profile applicable from wall to free stream: the roles of viscosity and density are very different in the viscous layer and in the turbulent region; their joint effect cannot be lumped into a single parameter, ν, as was possible for laminar flow.

For a constant-pressure layer, these obstacles are of little consequence, for we can obtain many useful results by introducing another kind of simplifying assumption. In turbulent pipe flow, most of the velocity change occurs within the wall layer; the same is true for this particular boundary layer. Hence it is acceptable to extend the wall-layer profile to the free stream, compensating for this artificiality by modifying the empirical constants of the resulting formulae.

Logarithmic Laws

We consider first the most realistic of the wall-layer velocity variations—the logarithmic profile

$$\frac{U}{u_f} = A \ln\left(\frac{y}{y_f}\right) + B \tag{4.1.3}$$

Integrating as indicated in equations (6.1.5, 6), we find the displacement and momentum thicknesses:

$$\begin{aligned} \frac{\delta_1}{\delta} &= A\frac{u_f}{U_1} = A(\tfrac{1}{2}c_f)^{\frac{1}{2}} \\ \frac{\delta_2}{\delta} &= A(\tfrac{1}{2}c_f)^{\frac{1}{2}}[1 - 2A(\tfrac{1}{2}c_f)^{\frac{1}{2}}] \end{aligned} \tag{8.1.7}$$

The form of the resulting shape factor

$$H = \frac{\delta_1}{\delta_2} = \frac{1}{1 - A(2c_f)^{\frac{1}{2}}} \tag{8.1.7}$$

is consistent with our conclusion that self-preserving development is impossible, save in the limit of very high Reynolds numbers, when c_f is very small. More typically, H falls slowly as the layer develops and c_f decreases. This is not in fact a very accurate estimate of the shape factor, but it does show that the variation with c_f is considerable: $H = 1{\cdot}55$ to $1{\cdot}2$ for the range of c_f indicated in Figure 4.3(a). Note that these results apply for both rough and smooth walls, since they are independent of the value of the slip constant B.

Applying the logarithmic law at the 'edge' of the layer, we obtain

$$\frac{U_1}{u_f} = \left(\frac{2}{c_f}\right)^{\frac{1}{2}} = A \ln\left(\frac{\delta}{y_f}\right) + B \tag{8.1.8}$$

von Kármán, who first adopted this approach, recommended

$$\frac{1}{\sqrt{c_f}} = 1{\cdot}8 \ln(\mathrm{Re}_\delta \sqrt{c_f}) + 3{\cdot}6 \tag{8.1.9}$$

which corresponds to the effective values

$$A_e = 2{\cdot}55 \quad \text{and} \quad B_e = 6{\cdot}0 \tag{8.1.9}$$

As anticipated, these are close to the values obtained directly from the velocity profile; see Table 4.2.

The information contained in von Kármán's result can be expressed in other ways which prove convenient in applications:

(1) A form much used by naval architects is that attributed to Schoenherr:

$$\frac{1}{\sqrt{C_f}} = 1{\cdot}8 \ln (\mathrm{Re}_L C_f) \tag{8.1.10}$$

with $\mathrm{Re}_L = U_1 L/\nu$ and C_f the average coefficient defined in equations (6.2.35). This result assumes that the layer is turbulent over almost all of its length, a realistic assumption for a ship's hull.

(2) Schlichting proposed the convenient results

$$C_f = \frac{0{\cdot}455}{(\log_{10} \mathrm{Re}_L)^{2{\cdot}58}} \tag{8.1.11}$$

and

$$c_f = (2 \log_{10} \mathrm{Re}_x - 0{\cdot}65)^{-2{\cdot}3} \tag{8.1.11}$$

These too apply when the layer is turbulent nearly from the beginning; they are within a few per cent of measured values for $10^5 < \mathrm{Re}_L < 10^9$.

(3) Squire and Young adopted the form

$$\frac{1}{\sqrt{c_f}} = 1{\cdot}81 \ln \mathrm{Re}_2 + 2{\cdot}54 \quad \text{with} \quad \mathrm{Re}_2 = \frac{U_1 \delta_2}{\nu} \tag{8.1.12}$$

and Rotta the form

$$\frac{1}{\sqrt{c_f}} = 1{\cdot}77 \ln \mathrm{Re}_1 + 2{\cdot}62 \quad \text{with} \quad \mathrm{Re}_1 = \frac{U_1 \delta_1}{\nu} \tag{8.1.13}$$

Their advantage over equation (8.1.9) is the use of the more clearly defined thicknesses δ_2 and δ_1. Since they use a local Reynolds number, these laws are less dependent on upstream conditions.

The problem of a 'non-standard' upstream history is illustrated in Figure 8.1, which shows a layer with an extensive laminar section. Assuming that the friction is given everywhere by the limiting formulae for laminar and turbulent flow (this neglects irregularities at initiation and in the region of transition), we find the total drag to be

$$\begin{aligned} D(L) &= \rho U_1^2 \delta_2(L) \\ &= D_t(L - x') + D_l(x_c) - D_t(x_c - x') = D_t(L - x') \end{aligned} \tag{8.1.14}$$

Here $D_l(x)/\rho U_1^2 x = C_f(x)$ from equations (8.1.6), and $D_t(x)/\rho U_1^2 x = C_f(x)$ from equations (8.1.10, 11). Equations (6.2.35) have been used to relate drag, friction coefficient and momentum thickness. Note that this analysis assumes that the friction is locally determined (that is, linked to the local thickness and outer velocity), once we decide whether the local flow is laminar or turbulent.

The results (8.1.14) make possible the calculation of:

(1) the laminar friction $D_l(x_c)$ up to the specified point of transition;

(2) the effective origin x' for the turbulent layer, that having the same friction and momentum thickness;

(3) the net drag of the laminar and turbulent layers $D(L)$, equal to the drag of the effective turbulent layer; and

(4) the final momentum thickness $\delta_2(L)$ and the average friction coefficient $C_f(L)$ over the entire length L.

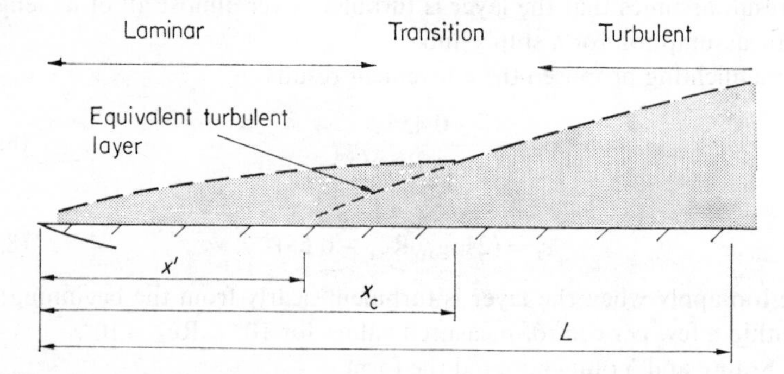

Figure 8.1 Turbulent boundary layer with an initial laminar section. The vertical scale is very greatly expanded

Similar arguments can be devised for more varied upstream conditions, for example, a blunt leading edge or an initially rough wall. Provided that we are told the drag upstream of some point, or the momentum thickness at that point, we can determine an effective origin for the subsequent 'standard' development, laminar or turbulent.

Power Laws

Proceeding as in Section 4.2.3, we consider power-law approximations to the velocity profile and friction characteristic:

$$\frac{U}{U_1} = \left(\frac{y}{\delta}\right)^{1/n} \tag{8.1.15}$$

and

$$c_f = C\,\mathrm{Re}^{-1/p} \tag{8.1.16}$$

These give explicit relationships and simplify calculations, while providing accuracy adequate for many purposes. The similarity arguments leading to the relationship (4.2.20) are still applicable; hence

$$n = 2p - 1 \tag{8.1.17}$$

Equations (6.1.5, 6) give the profile characteristics

$$\frac{\delta_1}{\delta} = \frac{1}{1+n} \quad \text{and} \quad \frac{\delta_2}{\delta} = \frac{n}{(1+n)(2+n)} \tag{8.1.18}$$

whence

$$H = 1 + \frac{2}{n} \tag{8.1.18}$$

Comparison with the displacement thickness for the logarithmic law (8.1.7) gives

$$\frac{1}{1+n} = A(\tfrac{1}{2}c_f)^{\frac{1}{2}} \tag{8.1.19}$$

showing that $n = n(x)$ increases along the layer. The ranges $n = 5$ to 11, $p = 3$ to 6, and $H = 1{\cdot}4$ to $1{\cdot}2$ are obtained for c_f varying over the range 0·01 to 0·0025. Note that the profile (8.1.15) is not exactly a similarity postulate, although it may appear to be at first sight.

The result (8.1.19) is very like that (4.2.17) for pipe flow; slightly different relationships are obtained by equating other characteristics of logarithmic power laws, for instance, their shape factors.

The most used power laws are those for $n = 7$ and $p = 4$, corresponding to the pipe-flow results (4.2.21, 22). The latter

$$c_f = \frac{\tau_w}{\frac{1}{2}\rho U_a^2} = 0{\cdot}079\left(\frac{\nu}{U_a d}\right)^{\frac{1}{4}}$$

can be converted to a form appropriate to a boundary layer by introducing $2\delta = d$ and $U_1 = 1{\cdot}25\ U_a$, say. Then

$$c_f = \frac{\tau_w}{\frac{1}{2}\rho U_1^2} = 0{\cdot}045\left(\frac{\nu}{U_1 \delta}\right)^{\frac{1}{4}} \tag{8.1.20}$$

This form was introduced into the momentum integral

$$\frac{d\delta_2}{dx} = \tfrac{1}{2}c_f \tag{6.2.35}$$

to find the pattern of development given in Table 6.1. The neglect of the variation in the indices p and n is justified by the slow change in c_f in any one layer.

Using equations (6.2.35) and (8.1.18), we can convert equation (8.1.20) into formulae describing other aspects of the developing layer. In view of the rough estimate made for the constant of proportionality, we need not be surprised that these results do not fit the measured values exactly. The best constants for the range $5 \times 10^5 < \mathrm{Re}_L < 10^7$ are those in

$$c_f = 0.059\left(\frac{\nu}{U_1 x}\right)^{\frac{1}{5}} = 0.0256\left(\frac{\nu}{U_1 \delta_2}\right)^{\frac{1}{4}} = 0.046\left(\frac{\nu}{U_1 \delta}\right)^{\frac{1}{4}}$$

and (8.1.21)

$$C_f = 0.074\left(\frac{\nu}{U_1 L}\right)^{\frac{1}{5}}$$

In their range of applicability, these results can be used in the same way as the more widely valid logarithmic formulae (8.1.9) to (8.1.13).

8.1.3 The Role of Roughness

There is not a great deal to add to the discussion of roughness in Section 4.3. It is the outer part of a boundary layer which distinguishes it from a channel flow, and the motion there is influenced by roughness only through the friction stress at the overlap with the wall layer. The pattern of development is slightly modified when the wall is rough, as noted in Table 6.1, since the dependence of friction on thickness is different in the absence of a progressively deepening viscous layer.

As a boundary layer develops, the potential viscous layer thickens, and the flow normally moves towards the smooth-wall state. In many cases this is achieved in the initial laminar section, and the effects of roughness can then be lumped with those of other early irregularities. To minimize the effect of roughness on the drag of a slender body, the front part must be made smooth, in particular, because roughness promotes transition to turbulence. Of course, for a large body, such as the hull of a ship, the areas of laminar and rough-wall flow may be relatively small, and the drag cannot be appreciably reduced in this way.

It is a simple matter to determine the *admissible roughness*, that is, the roughness height beyond which the friction is modified. From Figure 4.2, we find that

$$\frac{k_s}{y_f} = \frac{u_f}{U_1}\frac{U_1 k_s}{\nu} = (\tfrac{1}{2}c_f)^{\frac{1}{2}}\frac{U_1 k_s}{\nu} < 5 \tag{8.1.22}$$

specifies effectively smooth-wall conditions, with k_s the equivalent sand-grain scale. In Figure 4.3(a) c_f varies over the range 0·01 to 0·0025; hence

$$\frac{U_1 k_s}{\nu} < 70 \text{ to } 140 \tag{8.1.22}$$

The lower value applies near the beginning of the layer where the friction is highest. However, we may accept a small region in which roughness plays some part, since the layer thickens rapidly there and thus soon achieves the smooth-wall condition. The admissible-roughness criterion

$$k_s < 100\frac{\nu}{U_1} \tag{8.1.22}$$

is often adopted in practice.

Since a boundary layer on a rough wall spontaneously acts to generate effectively smooth conditions, it is often necessary to consider layers that are engaged in the conversion ($5 < k_s/y_f < 50$ in Figure 4.2) between fully rough and effectively smooth-wall flow. This process is not easy to analyse, since the 'transition' is peculiar to a particular roughness geometry, and cannot be specified by a single sand-grain scale. These considerations suggest that the fully rough state is of less interest for boundary layers than for channel flows, which cannot expand to 'escape' the roughness.

A general friction law is given by the logarithmic profile (4.3.5):

$$\frac{U_1}{u_f} = \left(\frac{2}{c_f}\right)^{\frac{1}{2}} = A\ln\left(\frac{\delta}{k}\right) + g\left(\frac{k}{y_f}\right) \tag{8.1.23}$$

with k some roughness scale, and $g(k^+)$ an empirical function specifying the rough-to-smooth transition. For fully rough conditions, equation (4.3.20) gives the friction in terms of the sand-grain scale:

$$\frac{U_1}{u_f} = \left(\frac{2}{c_f}\right)^{\frac{1}{2}} = A\ln\left(\frac{\delta}{k_s}\right) + B_s \tag{8.1.24}$$

Schlichting devised alternative formulae for the fully rough state:

$$c_f = \left[2{\cdot}87 - 1{\cdot}58\log_{10}\left(\frac{x}{k_s}\right)\right]^{-2\cdot5}$$

and

$$C_f = \left[1{\cdot}89 - 1{\cdot}62\log_{10}\left(\frac{L}{k_s}\right)\right]^{-2\cdot5} \tag{8.1.25}$$

for the range $10^2 < L/k_s < 10^6$.

Alternatively, we can adapt the power law (4.3.17) using the methods which led to equation (8.1.20). The introduction of $2\delta = d$ and $U_1 = 1{\cdot}25\ U_a$ leads to

$$c_f = 0{\cdot}021\left(\frac{k_s}{\delta}\right)^{0\cdot31} \tag{8.1.26}$$

for $10 < \delta/k_s < 1000$. We can obtain results paralleling the smooth-wall formulae (8.1.21) by introducing (8.1.26) into the integral momentum relation (6.2.35); the choice of the index n to relate δ and δ_2 should be guided by equation (8.1.19).

Diagrams generally similar to the Moody diagram for pipe flow can be prepared for boundary-layer friction. The directly analogous form,

$c_f = f(U_1\delta/\nu,\ k_s/\delta)$, is not convenient, since both parameters vary along the flow, and δ is not clearly defined. The alternative $c_f = f(U_1\delta_2/\nu,\ U_1k_s/\nu)$ is more easily used, but another parameter is needed if the variation along the layer is to be found directly: $c_f = f(U_1x/\nu,\ U_1k_s/\nu,\ x/k_s)$. We shall see shortly that another parameter (at least) is required to represent the pressure gradient when there is one. These complications reduce the utility of friction charts, and it is usual to determine boundary-layer friction from formulae like those developed above.

8.2 LAYERS WITH NON-UNIFORM OUTER FLOW

8.2.1 Laminar Flow: the Falkner–Skan Results

Our investigation of the linked effects of changes in pressure and free-stream velocity begins with self-similar laminar solutions which generalize Blasius's constant-pressure flow. The scales for these flows can be deduced from the conditions for self-preservation found in Section 7.2. For a boundary layer, the scales U_1 and U_0 are identical, and equations (7.2.5) give only

$$U_1^m \propto l_0 \tag{8.2.1}$$

Turning to the momentum equation (7.2.7), and neglecting the turbulence contributions, we obtain the dynamic condition

$$\frac{l_0^2}{\nu}\frac{\mathrm{d}U_1}{\mathrm{d}x} = n, \quad \text{a constant} \tag{8.2.2}$$

These two conditions ensure that all of the terms have the same streamwise variation.

Combining the conditions (8.2.1, 2), we obtain

$$l_0^2 \propto \frac{x}{U_1} \quad \text{and} \quad U_1 \propto x^n \tag{8.2.3}$$

The second result gives the velocity variation found (in ideal, frictionless flow) on the surface of an upstream-pointing wedge with semi-angle

$$\theta = \frac{\pi n}{n+1} \tag{8.2.4}$$

The presence of a thin boundary layer on the surface will not significantly alter this outer flow, save near the apex of the wedge. The values $n = 0$ and 1 correspond, respectively, to the Blasius, constant-pressure solution and to plane stagnation-point flow, for which $l_0 =$ constant. For the more interesting cases with $n < 0$, the wedge-flow interpretation is rather forced, but appropriate outer-velocity variations can be produced in suitably distorting channels.

Note that the boundary-layer development is formally dependent only on the external pressure gradient, and not on the wall curvature, so long as this is small; see equations (6.2.18).

Generalizing the constant-pressure forms (8.1.3) we take

$$\psi = U_1 l_0 F(\eta) \quad \text{with} \quad \eta = \frac{y}{l_0} = \frac{y}{(\nu x/U_1)^{\frac{1}{2}}} \tag{8.2.5}$$

The equation of motion (8.1.2), now with the pressure-gradient term $U_1 \mathrm{d}U_1/\mathrm{d}x$ included, reduces to

$$F''' + \tfrac{1}{2}(n+1)FF'' - nF'^2 + n = 0 \tag{8.2.6}$$

This *Falkner–Skan equation* generalizes the Blasius result; its solution must satisfy the same boundary conditions (8.1.5). The final term represents the pressure gradient; the other modifications arise from changes in the convection pattern.

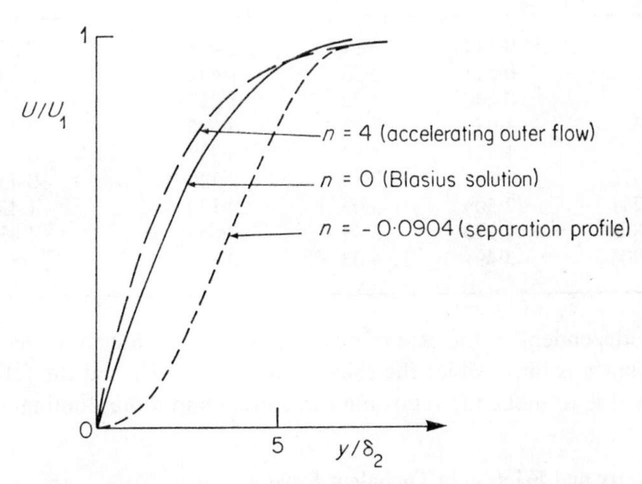

Figure 8.2 Self-similar velocity profiles for Falkner–Skan laminar boundary layers subjected to positive and negative pressure gradients

Figure 8.2 indicates the general nature of the solutions. The shape of the profile is little altered by acceleration of the outer flow ($n > 0$, $\mathrm{d}P_1/\mathrm{d}x < 0$), but is profoundly changed by an adverse pressure gradient in a decelerating outer stream ($n < 0$, $\mathrm{d}P_1/\mathrm{d}x > 0$). These trends are also apparent in Table 8.1, which gives some leading characteristics of the solutions. The rise in δ_1/l_0 shows the thickening of the layer arising from the deceleration imposed near the wall by the pressure gradient. The shape factor falls only a little in accelerating flows ($n > 0$), but rises considerably in decelerating conditions ($n < 0$). Only a small rate of deceleration is required to produce a flow continuously close to separation, with $\tau_w \simeq 0$. This indicates that viscous stresses can balance only small adverse pressure forces, and shows why stirring of a laminar boundary layer is effective in inhibiting separation.

Members of this family of boundary layers can also be distinguished using the parameter

$$N = \frac{\delta_1}{\tau_w}\frac{dP_1}{dx} = -\frac{\delta_1}{l_0}\frac{n}{F''(0)} \tag{8.2.7}$$

This is the ratio of the pressure-gradient and wall-stress terms of the integral momentum equation (6.2.33). When the outer flow accelerates, the layer becomes thinner, the wall stress rises, and the two forces fall into a balance

Table 8.1. Characteristics of Falkner–Skan boundary layers

$n = (l_0^2/\nu)\mathrm{d}U_1/\mathrm{d}x$	δ_1/l_0	$H = \delta_1/\delta_2$	$F''(0) = l_0\tau_w/\mu U_1$	$N = (\delta_1/\tau_w)\mathrm{d}P_1/\mathrm{d}x$
4	0·344	2·17	2·405	−0·572
1·5	0·543	2·20	1·494	−0·545
1	0·648	2·22	1·233	−0·525
0·25	1·079	2·33	0·675	−0·400
0	1·721	2·59	0·332	0
−0·0476	2·092	2·80	0·220	0·453
−0·0741	2·509	3·09	0·130	1·437
−0·0868	2·970	3·48	0·058	4·435
−0·0904	3·497	4·03	0	∞

broadly independent of the rate of change. When the outer flow decelerates, such a balance is impossible: the thickening of the layer and the fall in wall stress combine to make the retarding force more and more dominant.

8.2.2 Velocity and Friction in Turbulent Layers

Figure 8.3 shows profiles of mean velocity and shear stress typical of accelerating and decelerating turbulent boundary layers. As in laminar flow, the more dramatic effects occur in decelerating layers. The velocity profiles of Figure 8.3(a) are typical of layers advancing into an increasing pressure gradient; such profiles are adopted one after another at successive stations along the flow. In the decelerated layers, the velocity rise is distributed more uniformly across the turbulent region; this is mirrored in the stress distributions, which show the maximum stress near the middle of the boundary layer. Thus these retarded flows have something of the character of free-turbulent flows, which they do in fact become, following separation.

In strongly retarded boundary layers, the energy extraction ($\tau\,\partial U/\partial y$) is not concentrated in the viscous and logarithmic layers, but is disseminated through the flow. The energy-extracting turbulence is relatively larger and is to some extent freed from the constraint of the wall. The net effect of the enlargement of the region of dissipation and the relaxation of the wall constraint is a large increase in the dissipation per unit length of the layer, moderated a little by the reduction in the velocity gradient. The dissipation in the wall layer itself may rise as a result of scouring by the hyper-active outer turbulence.

This added dissipation determines the efficiency of a diffuser in converting kinetic energy into pressure or enthalpy. Even in the absence of separation, the dissipation rate in a decelerating duct is far greater than in a comparable

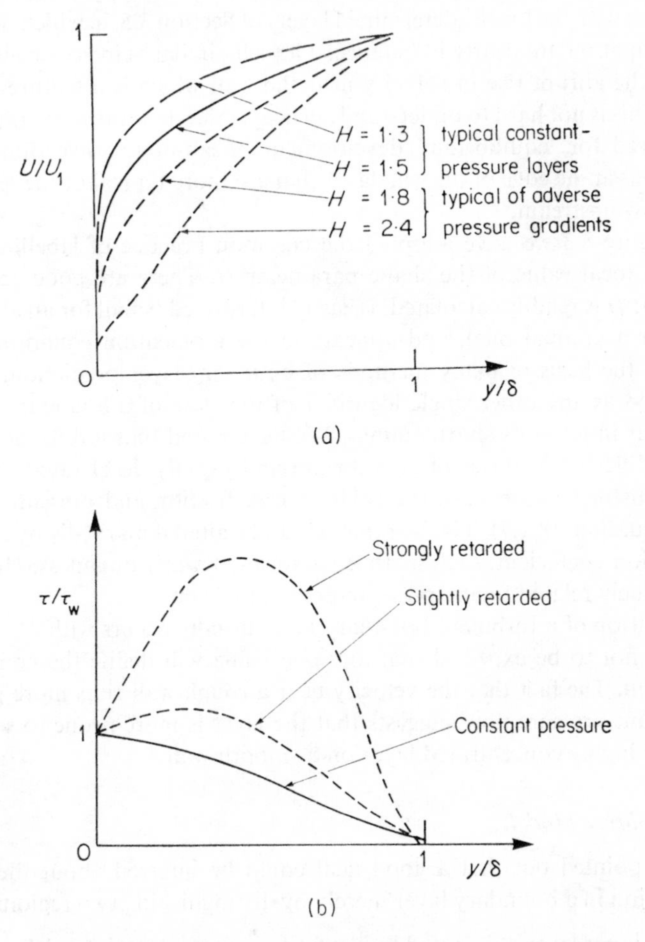

Figure 8.3 Typical characteristics of turbulent boundary layers subjected to pressure gradients. (a) Velocity profiles. (b) Shear-stress profiles

flow in a uniform channel. It is apparent also that the performance of a diffuser will depend strongly on the initial velocity distribution: the effective retarding force, measured by $N = (\delta_1/\tau_w)\mathrm{d}P_1/\mathrm{d}x$, has a head start if the entering flow already has thick boundary layers.

The situation in which the profiles of Figure 8.3(a) are generated differs in important ways from that postulated in the Falkner–Skan analysis. There, the adverse pressure gradient decreases steadily along the flow: $\mathrm{d}P_1/\mathrm{d}x \propto x^{2n-1}$ for $n \neq 0$. The cases of most practical interest are those in which the flow advances against an increasing pressure gradient. It is possible to produce turbulent analogues of the Falkner–Skan layers; these self-preserving flows are often called *equilibrium layers*. This term is not a happy one, as it leads to confusion with the locally determined layers of Section 3.5, in which production and dissipation are nearly in balance. The self-similar velocity profiles do not display the abrupt rise in velocity near the wall which is a feature of Figure 8.3(a). This is not hard to understand: in a layer decelerating more rapidly than is required for 'equilibrium' development, mean-flow convection provides conditions at the edge of the wall layer that are more typical of the less decelerated flow upstream.

In Figure 8.3 we have adopted the common practice of labelling profiles with the local value of the shape parameter H. There are good reasons for doing so: H is readily calculated, is easily interpreted (small for an abrupt rise, large for a gradual one), and appears in the momentum equation (6.2.34), which is the basis of many methods of boundary-layer prediction. While H is as good as any other single identifier of the state of a boundary layer, we must bear in mind its shortcomings. We have noted that it does not account for the differing histories of equilibrium and rapidly decelerated flows. Nor does it distinguish between the roles of wall friction and pressure gradient. From equations (8.1.7), it is clear that H can be altered markedly by changes in the friction coefficient, such as those associated with roughness. Hence H is not uniquely related to any pressure-gradient parameter.

Separation of a turbulent boundary layer usually occurs with $H = 2{\cdot}5$ to 3, but it is not to be expected that any one value will define the condition of separation. The fact that the velocity near a rough wall rises more gradually in constant-pressure flow suggests that the layer is more prone to separation than is a highly concentrated layer on a smooth wall.

The Two-layer Model

Coles pointed out that a good deal could be inferred about the velocity distribution in a boundary layer merely by distinguishing two regions:

(1) an inner layer dominated by the friction and constraints of the wall, with the velocity (in the fully turbulent part) specified by the logarithmic law or one of its generalizations; and

(2) an outer turbulent layer, more akin to free turbulence, but with the friction velocity providing the velocity scale, since the motion is controlled by the stress at the junction between the two layers.

This led Coles to represent the velocity variation as

$$\frac{U}{u_f} = A \ln\left(\frac{y}{y_f}\right) + B + A\,\Pi(x)\,w\left(\frac{y}{\delta}\right) \quad \text{for } y < \delta \tag{8.2.8}$$

outside the viscous or roughness layer. In the final term, the difference between the full profile and the logarithmic component is described in the manner of a self-preserving free-turbulent flow. The term 'two-layer' is something of a misnomer, since the two elements overlap, save near the wall, where $w \to 0$, and near the outer edge, where the logarithmic term provides nearly uniform convection.

Coles referred to the component $w(y/\delta)$ as the *law of the wake*, since the boundary conditions which it satisfies are like those for a wake, and since his thinking was influenced by the concept of a wake as a convector of the upstream effects of friction. He did not use this parallel to find the shape of the profile, but analysed a great deal of boundary-layer data to isolate it, at the same time establishing that this element does indeed have much the same shape for the varied flows. For most purposes, analytical approximations are adequate; two of the simpler are:

(1) the simple profile used by Hinze and others

$$w = 1 - \cos\left(\frac{\pi y}{\delta}\right) \tag{8.2.9}$$

(2) the form used by Coles himself

$$w = 2\sin^2\left(\frac{\frac{1}{2}\pi y}{\delta}\right) \tag{8.2.10}$$

These are least satisfactory near the 'edge' of the layer and near the wall, where in any case the viscous layer must be taken into account.

It is possible, of course, to predict the profile using mixing-length and eddy-viscosity models. However, their inadequacies near the outer edge (in their simplest forms at least; see Figure 6.3(a)) suggest that they are unlikely to improve upon Coles' brand of empiricism.

The amplitude factor $\Pi(x)$ represents, together with $c_f(x)$ and $\delta(x)$, the streamwise development of the flow. Its variation cannot be specified initially, but must be determined as part of the procedure of boundary-layer prediction. Nevertheless, some leading characteristics may be noted:

(1) Π must be small for a constant-pressure layer, since the logarithmic law gives reasonable results for this case. Experimental values of $\Pi = 0{\cdot}6 \pm 10$ per cent are found for a smooth wall; there is no reason to expect a unique value.

(2) Values of $\Pi < 0{\cdot}1$ are seldom found, even in strongly accelerated flow. Seemingly, the logarithmic law is the effective limit for such flows.

(3) $\Pi \to \infty$ near separation or reattachment. The reversal of flow near the wall, following separation, can be formally represented by letting Π take on negative values and changing the sign of U. This can be done because no use is made of the boundary-layer equations.

The Local Friction Law

At the edge of the layer the general profile (8.2.8) gives

$$\frac{U_1}{u_f} = \left(\frac{2}{c_f}\right)^{\frac{1}{2}} = A \ln\left(\frac{\delta}{y_f}\right) + B + 2A\,\Pi \tag{8.2.11}$$

since it is arranged that $w = 2$ there, as in the forms (8.2.9, 11). This result may be used to calculate Π. Alternatively, if $\Pi(x)$ is presumed known, or to be specified in terms of other parameters, the result may be interpreted as a generalization of the constant-pressure friction law (8.1.8). Implicitly, we have

$$c_f = f(\mathrm{Re}_\delta, \Pi) \tag{8.2.12}$$

specifying the friction in terms of local values of $U_1(x)$, $\Pi(x)$ and $\delta(x)$.

Our next task is to express this friction law in the parameters of equation (6.2.34), the integral momentum equation, which can be written

$$\frac{d\delta_2}{dx} + \frac{H+2}{U_1}\frac{dU_1}{dx}\delta_2 = \tfrac{1}{2}c_f$$

or

$$\frac{d\delta_2}{dx} = \tfrac{1}{2}c_f\left[1 + \frac{H+2}{H}N\right] \quad \text{with} \quad N = \frac{\delta_1}{\tau_w}\frac{dP_1}{dx} \tag{8.2.13}$$

The relationships hinge upon the near viscosity-independence of the scaled velocity defect, given by equations (8.2.8, 11) as

$$\frac{U_1 - U}{u_f} = -A \ln\left(\frac{y}{\delta}\right) + A\Pi\left[2 - w\left(\frac{y}{\delta}\right)\right] \quad \text{for } y < \delta \tag{8.2.14}$$

Introducing one of (8.2.9, 10), or any other suitably normalized 'wake' function, we obtain

$$\delta_1 = \frac{\delta}{U_1}\int_0^1 (U_1 - U)\,d\left(\frac{y}{\delta}\right) = \frac{A\,\delta\,u_f}{U_1}(1 + \Pi) \tag{8.2.15}$$

Note also that

$$\delta_1 - \delta_2 = \frac{\delta}{U_1^2}\int_0^1 (U_1 - U)^2\,d\left(\frac{y}{\delta}\right)$$

$$= \delta\left(\frac{Au_f}{U_1}\right)^2 F(\Pi) \tag{8.2.16}$$

with $F(\Pi)$ a quadratic function. Combining these results, we obtain

$$1 - \frac{1}{H} = \frac{\delta_1 - \delta_2}{\delta_1}$$

$$= A \frac{u_f}{U_1} \frac{F(\Pi)}{1 + \Pi} = (\tfrac{1}{2}c_f)^{\frac{1}{2}} G(\Pi) \tag{8.2.17}$$

These calculations assume that the contribution of the viscous layer is small; it is not difficult to introduce corrections if greater accuracy is required.

The results given above are sometimes expressed in terms of the *defect and defect-squared thicknesses*:

$$\Delta_1 = \int_0^\delta \left[\frac{U_1 - U}{u_f}\right] dy = \left(\frac{2}{c_f}\right)^{\frac{1}{2}} \delta_1$$

$$\Delta_2 = \int_0^\delta \left[\frac{U_1 - U}{u_f}\right]^2 dy = \frac{2}{c_f}(\delta_1 - \delta_2) \tag{8.2.18}$$

The relationship (8.2.17) is then simply

$$\frac{\Delta_2}{\Delta_1} = G(\Pi) \tag{8.2.18}$$

Equation (8.2.17) and equation (8.2.15) written as

$$\frac{\delta_2}{\delta} = A(1 + \Pi)\frac{(\tfrac{1}{2}c_f)^{\frac{1}{2}}}{H} \tag{8.2.19}$$

show that the local friction law (8.2.12) is implicitly

$$c_f = f(\mathrm{Re}_2, H) \tag{8.2.20}$$

now in terms of the parameters of the momentum equation (8.2.13). Ludwieg and Tillmann developed a semi-empirical law of this form:

$$c_f = 0{\cdot}246\left(\frac{U_1 \delta_2}{\nu}\right)^{-0\cdot268} \times 10^{-0\cdot678H} \tag{8.2.21}$$

A law of such simple structure inevitably fails to represent some aspects of the more complex relationships indicated above. Rotta has shown that this result is consistent with Cole's formulation under some circumstances (for example, $H = 1{\cdot}5$, $\mathrm{Re}_2 = 10^5$ and $H = 2{\cdot}5$, $\mathrm{Re}_2 = 10^6$), but that it overestimates c_f when H is large and Re_2 small, and underestimates c_f when H is small and Re_2 large; errors up to 40 per cent are possible.

Equilibrium Layers

It was noted earlier that the pressure gradient can be tailored to maintain a uniform velocity profile during development. Townsend has shown that self-preserving turbulent boundary layers can be achieved in three ways:

(1) exactly, for $U_1 \propto x^{-1}$, as in the flow between converging planes;

(2) approximately, at very (impractically) high Reynolds numbers, as noted with respect to the shape factor (8.1.7); and

(3) approximately, when $U_1 \propto x^a$ with $a > -\frac{1}{3}$, and for some allied accelerating flows.

Flows of the last type have been studied experimentally by Clauser and others, by keeping $N = (\delta_1/\tau_w)\mathrm{d}P_1/\mathrm{d}x$ constant along the flow, as for the Falkner–Skan profiles of Table 8.1.

For a self-preserving or 'equilibrium' boundary layer, the wake parameter Π is constant. Hence, for this class of layers the function $G(\Pi)$ can be specified as a function of N only. Nash proposed

$$G = 6{\cdot}1(N + 1{\cdot}81)^{\frac{1}{2}} - 1{\cdot}7 \qquad (8.2.22)$$

for this function. The parameter G varies little for constant-pressure and accelerated layers. This suggests that Π does not change much in these flows (as we noted earlier); moreover, the contribution of the logarithmic element is dominant when Π is small.

At one time it was hoped that the relationship $G = G(N)$ for equilibrium layers would be more widely valid. This hope was not fulfilled, although equation (8.2.22) does provide a useful indication of trends. The nature of the most radical departures from it is instructive: when the pressure-gradient parameter N is reduced from a peak level, the value of G does not follow, but stays constant or slowly increases. This demonstrates the need for a somewhat more dynamic relationship than is provided by a single algebraic equation.

8.2.3 Auxiliary Equations

At this point our orderly progress is abruptly arrested, as though firm ground beneath our feet had been replaced by a quagmire. We have two points of support—the momentum integral (8.2.13) and the local friction law, either an empirical form like (8.2.21), or that implicit in Coles' similarity laws. To proceed further, we need another relationship among the local values of H, c_f, Re_2 and N, or perhaps including the related parameters G and Π. Our problem is not in finding this relationship, but in choosing among the many which have been devised over the past forty years. This plethora of auxiliary equations testifies to the practical importance of predictions of skin friction and separation, particularly for lifting surfaces, where the pressure varies of necessity. The great variety in auxiliary equations arises because there are very many ways of reintroducing the information which was rejected on integrating with respect to time and across the layer.

Rotta and Bradshaw have pointed out that most of the older closing relationships are of the form

$$\delta_2 \frac{\mathrm{d}H}{\mathrm{d}x} = \frac{\delta_2}{U_1}\frac{\mathrm{d}U_1}{\mathrm{d}x} f_1 + f_2 \qquad (8.2.23)$$

with $f_1, f_2 = f(H, \mathrm{Re}_2, c_f)$. These are indeed somewhat more 'dynamic' than the algebraic relationship (8.2.22) in that they take some account of the mode of development. Since the algebraic equation went some way towards predicting patterns of development, the form (8.2.23) can hardly fail to describe a considerable range of boundary-layer behaviour, provided that the coefficients are consistent with the empirical facts concerning that range. However, the single derivative appearing in equation (8.2.23) is not adequate for the prediction of every possible mode of development.

The auxiliary equations used in practice were not obtained simply by seeking the functions f_1 and f_2 which gave the best predictions. In retrospect, this procedure appears attractive, but this was not so before the introduction of high-speed computers. Most auxiliary equations were derived by manipulating the boundary-layer equations to give results into which empirical information could be introduced conveniently. We shall not examine in detail the arguments which were used; this would take too long; and, in any case, more comprehensive methods of representing turbulent flows have recently been developed. These methods, which will be introduced in the following chapter, take more realistic account of upstream conditions and of the cross-flow transfer processes whose role has been suppressed in the integral formulation.

W. C. Reynolds has reviewed the arguments and analytical methods from which auxiliary equations have been derived. (His 'morphology' was part of the Stanford conference which compared the predictions given by twenty-eight computational techniques for a specified group of boundary layers.) Setting aside the 'differential methods' to be considered in the next chapter, he divided the 'integral models' into three classes:

(1) dissipation-integral methods based on the mean-flow energy equation (6.2.41);

(2) entrainment methods, which in effect introduce an assumption concerning the entrainment laws (7.2.27); and

(3) a more heterogeneous group, with less obvious physical bases; some use moment-of-momentum equations or strip integrals for parts of the flow; others are little more than techniques of numerical analysis.

The *dissipation-integral methods*, associated with the names of Rotta and Walz, among others, are more typical of the broad range of integral prediction techniques, being based on an integral of the boundary-layer equations, in this case

$$\frac{\mathrm{d}(U_1^3 \delta_3)}{\mathrm{d}x} = \frac{2\mathscr{D}}{\rho} \tag{8.2.24}$$

where $\mathscr{D}$ is the integral of equation (6.2.41), termed the dissipation integral or, more properly, the shear-work integral. To link this result to the other development equations, we introduce an empirical expression for $\mathscr{D}$, the simplest being Rotta and Truckenbrodt's

$$\frac{\mathscr{D}}{\rho U_1^3} = 0{\cdot}0056\,\mathrm{Re}_2^{-\frac{1}{6}} \tag{8.2.25}$$

It is also necessary to relate the energy thickness δ_3 to δ_2 and H; this can be done using outer-layer similarity arguments and empirical data.

The *entrainment arguments*, first used by Head, parallel those which led to the entrainment laws (7.2.27) for free turbulence. Head postulated that the change in the flux of turbulent fluid (equivalent to $V_\mathrm{i} + V_1$, the absolute velocity of 'entrainment') could be specified by

$$\frac{\mathrm{d}}{\mathrm{d}x}\int_0^\delta U\,\mathrm{d}y = \frac{\mathrm{d}}{\mathrm{d}x}[U_1(\delta - \delta_1)] = \beta_2(H) \times U_1 \tag{8.2.26}$$

This is similar in structure to equations (7.2.27), but now the entrainment constant is conceived as a function of the shape factor. Head developed a rather complicated power law for $\beta_2(H)$.

The entrainment method has been extended in a number of ways, notably by Head and Patel, who introduced into the entrainment constant some dependence on the pressure gradient and shear stress (N and c_f). Without this, the entrainment law does not distinguish between cases in which a large value of H is associated with pressure variation and with a large wall stress.

8.3 TRANSPORT PROCESSES

In extending our treatment of boundary layers to include the transport of a passive entity (heat will as usual be taken as the exemplar), we restrict the discussion again to constant-pressure layers. We assume further that the wall flux and temperature vary only slowly along the flow. The analysis of more general boundary layers is at a rather primitive level; simple analogies between local wall stress and flux are often used, even though the thermal history of the flow is known to be important. However, the more specific transport models of the next chapter are potentially capable of predicting transfer rates within arbitrary boundary layers.

Further important restrictions to be adopted, for brevity rather than necessity, are the requirements that the density and transport properties of the fluid are constant across the boundary layer. This removes from our field of study the numerous investigations of 'kinetic heating' in boundary layers, stimulated by the problems of high-speed motion through the atmosphere. These effects were touched on in Section 5.2.2, and their consequences for skin friction are dealt with in Example 8.19.

The Laminar Thermal Layer

For an introduction to the relationship between the fields of velocity and of a convected and diffused entity, we look at heat transfer in the Blasius constant-pressure boundary layer. We assume that all of the fluid properties are uniform throughout the flow. The thermal energy equation, a particular case of equation (6.2.29), is

$$U\frac{\partial H}{\partial x}+V\frac{\partial H}{\partial y}=\kappa\frac{\partial^2 H}{\partial y^2} \tag{8.3.1}$$

We seek a similarity solution, representing the velocity and enthalpy fields by

$$\psi=(\nu U_1 x)^{\frac{1}{2}}F(\eta) \quad \text{and} \quad H=H_w-(H_w-H_1)\,j(\eta) \tag{8.3.2}$$

with $\eta=y/(\nu x/U_1)^{\frac{1}{2}}$, as in equations (7.4.18) and (8.1.3). We have restricted attention to a uniform enthalpy differential between wall and outer flow, and have arranged that the boundary conditions for j (representing ΔH) follow the pattern of those for F', given in equations (8.1.5):

$$j=0 \quad \text{for} \quad \eta=0 \qquad \text{and} \qquad j=1 \quad \text{for} \quad \eta\to\infty \tag{8.3.3}$$

Equation (8.3.1) now becomes

$$j''+\tfrac{1}{2}\mathrm{Pr}\,F\,j'=0 \tag{8.3.4}$$

Because the velocity distribution has been assumed independent of temperature, this is a linear equation for the latter.

We use our knowledge of the Blasius flow in calculating the transport within it. For $\mathrm{Pr}=1$, the equations and boundary conditions for the two profiles match exactly. Hence

$$j'=F'' \tag{8.3.5}$$

throughout the flow. By virtue of the similarity forms (8.3.2)

$$\frac{j'}{F''}=\frac{\mathrm{Pr}\,(\dot{q}_h/\tau)\,U_1}{H_w-H_1} \tag{8.3.6}$$

for any Prandtl number. Evaluating at the wall, we have

$$\frac{\mathrm{St}}{\frac{1}{2}c_f}=\frac{1}{\mathrm{Pr}}\frac{j'(0)}{F''(0)} \tag{8.3.6}$$

and see that equation (8.3.5) implies the simple Reynolds analogy:

$$\mathrm{St}=\tfrac{1}{2}c_f \quad \text{for} \quad \mathrm{Pr}=1 \tag{8.3.7}$$

For the general case with $\mathrm{Pr}\neq 1$, we have the same value of $F''(0)$—Table 8.1 gives 0·332—but now $j'(0)=f(\mathrm{Pr})$ must be found from equation (8.3.4). Integrating twice, and using the boundary conditions (8.3.3), we find

$$\frac{1}{j'(0)}=\int_0^\infty \exp\left[-\tfrac{1}{2}\mathrm{Pr}\int_0^\eta F\,d\eta\right]d\eta \tag{8.3.8}$$

Schlichting has approximated the rather complicated function (8.3.8) by

$$j'(0)=0{\cdot}332\,\mathrm{Pr}^{0\cdot343} \tag{8.3.9}$$

for $\mathrm{Pr} > 0{\cdot}5$. It is often further approximated to

$$j'(0) = 0{\cdot}332\,\mathrm{Pr}^{\frac{1}{3}} \tag{8.3.10}$$

for $0{\cdot}6 < \mathrm{Pr} < 15$, and this gives

$$\frac{\mathrm{St}}{\frac{1}{2}c_f} = \frac{\mathrm{Nu}}{\frac{1}{2}c_f\,\mathrm{Re}\,\mathrm{Pr}} = \mathrm{Pr}^{-\frac{2}{3}} \tag{8.3.11}$$

Finally, introducing the local and average coefficients for Blasius flow (8.1.6), we obtain

$$\begin{aligned} \mathrm{Nu}_x &= 0{\cdot}332\,\mathrm{Pr}^{\frac{1}{3}}\,\mathrm{Re}_x^{\frac{1}{2}} \\ \mathrm{Nu}_L &= 0{\cdot}664\,\mathrm{Pr}^{\frac{1}{3}}\,\mathrm{Re}_L^{\frac{1}{2}} \end{aligned} \tag{8.3.12}$$

The only approximation (other than the basic laminar boundary-layer assumptions) is in the factor $\mathrm{Pr}^{\frac{1}{3}}$; better representations are indicated above (8.3.8, 9).

Turbulent Flow: the Prandtl–Taylor Model

Much of the discussion of Section 5.3 is applicable to transfer across a turbulent boundary layer. Indeed, the Prandtl–Taylor analysis, and its various extensions, are more directly applicable here than to transfer within a pipe, since they relate both friction and heat transfer to conditions at a hypothetical boundary for the transfer processes. A much used transfer law for a constant-pressure layer is that obtained from the simple Prandtl–Taylor result (5.3.7), using equation (5.3.9) to define the role of the thin molecular-diffusion layers, and adopting the simple friction laws (8.1.21):

$$\mathrm{St}_x = \frac{0{\cdot}0295\,\mathrm{Re}_x^{-\frac{1}{5}}}{1 + 1{\cdot}29\,\mathrm{Re}_x^{-\frac{1}{10}}\,\mathrm{Pr}^{-\frac{1}{6}}\,(\mathrm{Pr}-1)} \tag{8.3.13}$$

The assumptions behind this result are made apparent by the more general form (5.3.6). The restriction to moderate Prandtl numbers, $0{\cdot}5 < \mathrm{Pr} < 50$, say, is not too serious, since calculations for developing layers usually relate to air or water. Another source of inaccuracy is the use of $\mathrm{Pr}_t = 1$. We have seen that the effective value for the region of intermittency is around 0·5 to 0·7; in this respect the Prandtl–Taylor model is less appropriate for a boundary layer than for pipe flow. However, the discrepancy is not serious at high Reynolds numbers, when the velocity and temperature changes occur mostly in the wall layer.

The Colburn Analogy

We now consider the use of power laws to represent the transfer characteristic of a boundary layer. From equation (5.3.23) for pipe flow, we have

$$\mathrm{St} = \frac{\mathrm{Nu}}{\mathrm{Re}_d\,\mathrm{Pr}} = 0{\cdot}023\,\mathrm{Re}_d^{-0{\cdot}2}\,\mathrm{Pr}^{m-1} \tag{8.3.14}$$

with the index $m = 0{\cdot}3$ to $0{\cdot}4$. To see what this implies about the relationship between St and c_f, we utilize the friction law (4.2.23), which has the same Reynolds-number dependence. Thus

$$\frac{\mathrm{St}}{\frac{1}{2}c_f} = \frac{\mathrm{Nu}}{\frac{1}{2}c_f\,\mathrm{Re}\,\mathrm{Pr}} = \mathrm{Pr}^{m-1} \tag{8.3.15}$$

The fact that the constant of proportionality is unity follows from the selection of 0·023 in equations (4.2.23) and (5.3.23) and (8.3.14); slightly different choices could be made with equal justification, and these would alter equation (8.3.15) a little.

The result (8.3.15), with $m = \frac{1}{3}$, is referred to as the Colburn analogy. For this index, the turbulent and laminar results (8.3.11, 15) are identical. This unity could also have been achieved with $m = 0{\cdot}343$, giving the structure of the more accurate laminar result (8.3.9). The formulae for turbulent and laminar flows do not actually have the same form, but reasonable accuracy is obtained by nudging them gently to produce a convenient parallel structure. With this proviso, we have in equation (8.3.15), a relationship valid for laminar and turbulent boundary layers and for turbulent pipe flow; it is not valid for laminar pipe flow, as is evident in the results of Example 3.18.

Introducing the friction laws for a turbulent boundary layer on a smooth wall, and taking $m = \frac{1}{3}$, we obtain

$$\begin{aligned} \mathrm{Nu}_x &= 0{\cdot}0295\,\mathrm{Pr}^{\frac{1}{3}}\,\mathrm{Re}_x^{\frac{4}{5}} \\ \mathrm{Nu}_L &= 0{\cdot}037\,\mathrm{Pr}^{\frac{1}{3}}\,\mathrm{Re}_L^{\frac{4}{5}} \end{aligned} \tag{8.3.16}$$

These apply when the layer is entirely turbulent, and the latter assumes also that the temperature is uniform over the whole of the surface on which the layer develops.

A widely recommended, but not very convincing method of allowing for an initial laminar section is to calculate the transfer following transition as though the boundary layer had always been turbulent. Thus

$$\mathrm{Nu}_L = 0{\cdot}037\,\mathrm{Pr}^{\frac{1}{3}}\left[\mathrm{Re}_L^{0\cdot8} - \mathrm{Re}_c^{0\cdot5}\left(\mathrm{Re}_c^{0\cdot3} - \frac{0{\cdot}664}{0{\cdot}037}\right)\right] \tag{8.3.17}$$

with $\mathrm{Re}_c = \mathrm{Re}_x$ for transition. A somewhat more realistic estimate can be made using procedures paralleling the friction calculation outlined with reference to Figure 8.1.

The Buoyancy-driven Boundary Layer

Finally, we consider a boundary layer of the kind sketched in Figure 6.6, a motion generated in otherwise still fluid by heating along a vertical plane surface. We shall not attempt a detailed solution using the coupled differential momentum and energy equations (6.2.54, 29), but use similarity methods to determine only the structure of the solution. It is plausible to expect a similarity solution for this flow (unlike the forced-convection turbulent layer), since the velocity field develops spontaneously to match the density and wall-stress distributions.

We introduce the similarity forms

$$U = U_0(x)\, f(\eta) \quad \text{and} \quad H - H_1 = (H_w - H_1)\, j(\eta) \tag{8.3.18}$$

with $H_w - H_1$ constant along the layer. The energy and momentum integrals (6.2.37, 55) become

$$\begin{aligned} \frac{d}{dx}\left[U_0 l_0 \int_0^\infty f j \, d\eta\right] &= \frac{\dot{q}_w}{\rho_1 (H_w - H_1)} \\ \frac{d}{dx}\left[U_0^2 l_0 \int_0^\infty f^2 \, d\eta\right] &= \frac{\beta g}{c_p}(H_w - H_1)\, l_0 \int_0^\infty j \, d\eta - \frac{\tau_w}{\rho_1} \end{aligned} \tag{8.3.19}$$

Arguing that the local transfer rates are related to the local velocity and length scales as in a forced boundary layer (this implies that most of the temperature change occurs in the wall layer), we take

$$\begin{aligned} \frac{\tau_w}{\rho_1} &\propto U_0^2 \left(\frac{\nu}{U_0 l_0}\right)^{\frac{1}{4}} \\ \frac{\dot{q}_w}{\rho_1 (H_w - H_1)} &\propto U_0 \left(\frac{\nu}{U_0 l_0}\right)^{\frac{1}{4}} \mathrm{Pr}^{-\frac{2}{3}} \end{aligned} \tag{8.3.20}$$

guided by the turbulent laws (8.1.21) and (8.3.15). The integral equations now have the forms:

$$\begin{aligned} \frac{d(U_0 l_0)}{dx} &= C_1 U_0 \left(\frac{\nu}{U_0 l_0}\right)^{\frac{1}{4}} \mathrm{Pr}^{-\frac{2}{3}} \\ \frac{d(U_0^2 l_0)}{dx} &= C_2 \beta g (T_w - T_1)\, l_0 - C_3 U_0^2 \left(\frac{\nu}{U_0 l_0}\right)^{\frac{1}{4}} \end{aligned} \tag{8.3.21}$$

Assuming scales of the form

$$U_0 = A x^m \quad \text{and} \quad l_0 = B x^n \tag{8.3.22}$$

we find the values

$$m = 0{\cdot}5 \quad \text{and} \quad n = 0{\cdot}7 \tag{8.3.22}$$

from equations (8.3.21). The three relationships between m and n provided by these equations are consistent; this is in agreement with the hypothesis of self-preservation. The 'amplitudes' A and B are also related by equations (8.3.21):

$$\begin{aligned} AB &= D_1 \mathrm{Pr}^{-\frac{2}{3}} A \left(\frac{\nu}{AB}\right)^{\frac{1}{4}} \\ A^2 B &= D_2 \beta g (T_w - T_1)\, B - D_3 A^2 \left(\frac{\nu}{AB}\right)^{\frac{1}{4}} \end{aligned} \tag{8.3.23}$$

(The Ds differ from the Cs of equations (8.3.21) only by factors $m+n$ and $2m+n$ introduced by differentiation.) Solving these equations, we find

$$U_0 \propto \left[\frac{\mathrm{Gr}_x}{1+(D_3/D_1)\,\mathrm{Pr}^{\frac{2}{3}}}\right]^{\frac{1}{2}} \frac{\nu}{x}$$
$$l_0 \propto \left[\frac{1+(D_3/D_1)\,\mathrm{Pr}^{\frac{2}{3}}}{\mathrm{Gr}_x}\right]^{\frac{1}{10}} \mathrm{Pr}^{-\frac{8}{15}}\, x \tag{8.3.24}$$

with $\mathrm{Gr}_x = \beta g(T_w - T_1)x^3/\nu^2$ a Grashof number.

Finally, returning to the second of equations (8.3.20), we calculate the local Nusselt number:

$$\mathrm{Nu}_x \propto \mathrm{Gr}_x^{\frac{2}{5}}\,\mathrm{Pr}^{\frac{7}{15}}\left[1+\frac{D_3}{D_1}\mathrm{Pr}^{\frac{2}{3}}\right]^{-\frac{2}{5}} \tag{8.3.25}$$

The dimensionless constants can be determined experimentally, or by introducing reasonable profiles f and j, and realistic constants for the transfer laws (8.3.20). Eckert and Jackson, who first worked through this calculation, suggested 0·0295 for the constant of proportionality in equation (8.3.25) and $D_3/D_1 = 0{\cdot}494$. These calculated values are consistent with experimental results.

Empirical laws of the form

$$\mathrm{Nu}_L \propto (\mathrm{Gr}_L\,\mathrm{Pr})^q = \mathrm{Ra}^q \tag{8.3.26}$$

with Ra a Rayleigh number, are often used to specify free-convection heat-transfer. It will be clear from (8.3.25) that they are unlikely to be valid over a wide range of Pr.

8.4 SOME LOOSE ENDS

As is so often necessary, we end this chapter by briefly considering some interesting and practically important processes whose detailed study is prohibited by lack of space. Here they fall into two classes:

(1) the natural disasters to which boundary layers are prone—transition and separation; and

(2) more varied wall-bounded flows which generalize the plane boundary layers examined earlier.

Transition

There exists an extensive and elegant body of analysis concerning hydrodynamic stability, and a great deal of supporting experimental information. Here we shall consider only the simplest empirical facts, drawing on a survey made by Dryden. Knowing that transition in pipe flow occurs when $\mathrm{Re}_d =$

2400 or more, we can easily make a rough estimate of the position within a constant-pressure laminar layer at which transition is likely. Taking $U_1\delta/\nu = U_a d/\nu = 2400$, we find from the third of equations (8.1.6):

$$\mathrm{Re}_c = \frac{U_1 x_c}{\nu} = \left(\frac{U_1 \delta}{5\nu}\right)^2 = 2 \times 10^5 \quad \text{or more}$$

Too much confidence should not be invested in this prediction; we now consider some of the factors which alter the point of transition.

(1) Free-stream turbulence. For a flat plate, it has been found that $9 \times 10^4 < \mathrm{Re}_c < 2{\cdot}8 \times 10^6$ for $0{\cdot}025 > (\overline{q^2}/3)^{\frac{1}{2}}/U_1 > 0{\cdot}001$; outside this range, Re_c is only weakly dependent on changes in the turbulence level.

(2) Pressure gradient. An adverse pressure gradient encourages transition by thickening the boundary layer and by generating a less stable profile; a halving of Re_c from its constant-pressure value is easily achieved. On the other hand, values as high as $\mathrm{Re}_c = 14 \times 10^6$ have been attained on carefully designed laminar-flow aerofoils, where the pressure steadily increases over much of the surface.

(3) Curvature. Stabilization and destabilization by centripetal accelerations prove to be significant for:

$$\frac{\delta_1}{R} > 0{\cdot}0026 \quad \text{on a convex, stabilizing surface}$$

$$\frac{\delta_1}{R} > 0{\cdot}00013 \quad \text{on a concave, destabilizing surface}$$

Here R is the radius of curvature along the stream.

(4) Heat transfer. Cooling of a surface stabilizes the flow on it, while heating promotes transition; variations of Re_c by factors of 2 and $\frac{1}{2}$ can be achieved. In high-speed aeronautics, recovery effects or 'kinetic heating' also play a part.

(5) Vibration. The imposition of periodic fluctuations with a frequency close to that of the least stable perturbations can profoundly reduce the threshold of stability. 'Untuned' vibrations act in much the same way as free-stream turbulence.

(6) Roughness. The transition point can be $\mathrm{Re}_c = 10^5$ or lower on a rough surface. It is found that transition is produced by a single element whose height $k > \delta_1$, approximately. This is a useful guide in selecting a trip wire to promote transition. Behind an isolated element, there forms a wedge-shaped turbulent region with an angle of spread around 16°. The spontaneous breakdown to turbulence on an effectively smooth surface displays rather similar features—intermittent, unsteady turbulent 'streaks' and 'spots' which spread to cover the whole of the shear flow.

Reverse transition or relaminarization can be achieved in a variety of ways:

by rapid acceleration of the outer flow, by deceleration to subcritical values of the local Reynolds number, by stabilization through centripetal forces, and by other alterations in the factors listed above.

Separation

About this complex subject we shall have even less to say than about its partner, transition. The partnership is sometimes close: transition often follows even a local separation, while separation can be prevented by transition. The latter effect may cause marked differences between model and prototype behaviour, unless transition is forced at an appropriate point on the model. While transition is usually a once-for-all event, repeated separation is often found. On a lifting surface, there may be a brief laminar separation, followed by reattachment to form a separation bubble, and finally a separation of the turbulent flow. On a wavy surface, periodic separation is encountered, with a region of recirculation behind each wave crest.

The analysis of separation, and of the allied problem of reattachment, is considerably more difficult than that of transition: the flow may be either laminar or turbulent; the boundary-layer equations are no longer applicable; large-scale unsteadiness often occurs; and, for three-dimensional flows, it is difficult even to visualize the kinematics of the process.

Some of the prediction procedures for two-dimensional boundary layers give realistic estimates of separation points, but none of them can extend right up to or beyond separation. For turbulent layers, no simple 'criterion' for separation can be given, other than the failure of the prediction procedure: separation depends on the nature of the surface and on the upstream history of the flow, and is not defined by any single parameter, such as the shape factor.

Other Wall-bounded Flows

Here we take note of some flows which differ in only one or two particulars from the plane boundary layers considered earlier.

(1) Boundary layers on axisymmetric bodies. These behave like plane layers when thin, but have their own peculiar properties when of a thickness comparable to the body diameter.

(2) Boundary layers with an outer flow which varies in time. Such flows are produced by water waves, by breathing and by the intermittent exhaust from an engine.

(3) Three-dimensional boundary layers, including swirling motions. Here even the faithful law of the wall breaks down.

(4) Boundary layers, such as that in the atmosphere, in which stratification and buoyancy are important.

(5) The wall flow in a diffuser fed with a more-or-less well developed turbulent flow. This differs from the standard boundary layer in that the 'free stream' is itself turbulent, and the pattern of development is different from that in a layer whose spreading is controlled by the turbulence interface.

(6) The wall jet. This is like the boundary layer in combining an inner, wall-dominated region and a spreading outer layer more akin to free turbulence. But for a boundary layer, the velocity change across the outer layer may be comparatively small, and some important features can be estimated by considering only the wall layer. This is never possible for a wall jet, where the two velocity changes are of the same order.

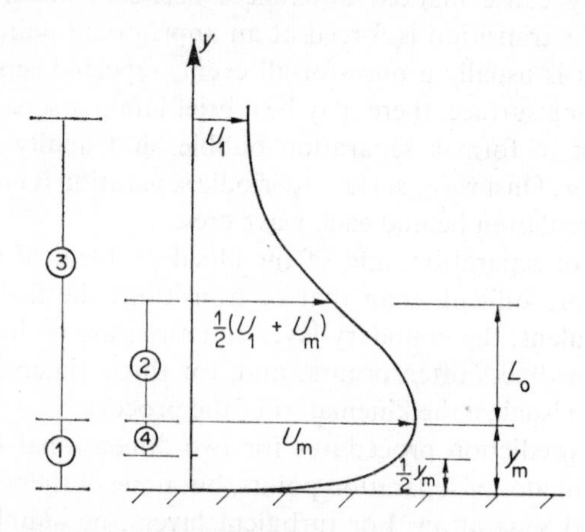

Figure 8.4 Selection of integration strips for analysis of a wall jet in streaming flow, following Gartshore and Newman

Wall jets have often been analysed using strip-integration methods, in which the momentum equation is integrated across parts of the flow. Figure 8.4 illustrates an extreme (but necessary) development of the technique by Gartshore and Newman, who set up integrals for the four strips shown. When suitable assumptions were made for the shear stresses at the patching points, there resulted four equations for U_m, L_0, y_m and n (the index of the velocity power law assumed for the inner layer), in terms of the prescribed outer velocity variation $U_1(x)$. In their calculation, the flow constant R_0 for the outer flow was made to vary along the flow, as is seen to be necessary in Table 7.2.

FURTHER READING

Chang, P. A. *Separation of Flow*, Pergamon Press, Oxford (1970)
van Driest, E. R. Reference 7: Section F
Dryden, H. L. Reference 7: Section A

Hinze, J. O. Reference 5: Chapter 7
Kutateladze, S. S. and A. I. Leont'ev (trans. D. B. Spalding). *Turbulent Boundary Layers in Compressible Gases*, Arnold, London (1964)
Patankar, S. V. and D. B. Spalding. *Heat and Mass Transfer in Boundary Layers*, Morgan-Grampian, London (1967)
Rohsenow, W. M. and H. Y. Choi. Reference 9: Chapters 7, 8 and 11
Schlichting, H. Reference 11: Chapters XIV to XVII, XXI, XXII
Schubauer, G. B. and C. M. Tchen. Reference 7: Section B
Tennekes, H. and J. L. Lumley. Reference 12: Chapter 5
Townsend, A. A. Reference 13: Chapters 10 and 11
Walz, A. (trans. H. J. Oser). *Boundary Layers of Flow and Temperature*, M. I. T. Press (1969)

SPECIFIC REFERENCES

Bradshaw, P. 'The turbulence structure of equilibrium boundary layers', *J. Fluid Mech.*, **29**, pp. 625–645 (1967)
Coles, D. 'The law of the wake in the turbulent boundary layer', *J. Fluid Mech.*, **1**, pp. 191–226 (1956)
Gartshore, I. S. and B. G. Newman. 'The turbulent wall jet in an arbitrary pressure gradient', *Aero. Quart.*, **XX**, pp. 25–56 (1969)
Head, M. R. and V. C. Patel. 'Improved entrainment method for calculating turbulent boundary layer development', *Rep. and Memo.* **3643**, Aero. Res. Council, London (1968)
Nash, J. F. 'Turbulent boundary layer behaviour and the auxiliary equation', in Reference 15
Reynolds, W. C. 'A morphology of prediction methods', in Reference 16
Rotta, J. C. 'Critical review of existing methods for calculating the development of turbulent boundary layers', in *Fluid Mechanics of Internal Flow*, Elsevier, Amsterdam (1967)

EXAMPLES

8.1 The simplest boundary layer is the laminar asymptotic suction layer with V_w constant.

(a) For this flow, show that $U = U_1[1 - \exp(V_w y/\nu)]$ and $\tau_w = -\rho V_w U_1$, and that the 'capture' drag is specified by $c_d = -2V_w/U_1$.

(b) Find the ratios of δ_1, δ_2, δ_{99} for this layer.

(c) Theoretical predictions of the critical condition for transition in this layer give $U_1 \delta_1/\nu = 70{,}000$. What are the corresponding values of the suction coefficient $-V_w/U_1$ and capture drag coefficient. Compare with the values of Re_1 and c_f in a suction-free laminar layer, for transition at $\mathrm{Re}_x = 10^6$.

(d) It is desired to minimize the drag of a wing section by applying suction. Is it appropriate to maintain the critical value of suction over the whole surface? How would you set about choosing the optimum distribution of V_w?

8.2 An aeroplane with a wing of chord 8 ft and thickness 10 per cent (maximum thickness to chord) flies at 200 miles per hour near sea level.

(a) Neglecting separation and the extreme effects of pressure gradients, compare the displacement thickness of the boundary layers at the trailing edge with the wing thickness. Assume transition at the quarter-chord point on the upper surface, and at half-chord on the lower.

(b) Will boundary-layer growth significantly alter the pressure distribution over the wing? What is the friction-drag coefficient for the profile, based on the chord length? What are the laws of development for the wake behind the wing?

8.3 A train on the Japanese high-speed Tokaido line runs at 170 km/hr. It has ten cars, each 25 m long, 3·4 m wide, and 4·5 m high. The bottom clears the road-bed by 0·5 m.

(a) Estimate the friction on the train, assuming a constant-pressure, smooth-wall boundary layer over the entire surface. Is the pressure differential on the ends of the train significant?

(b) Over what portion of the bottom of the train will a Couette flow be developed? Estimate the added drag there.

(c) Estimate the aerodynamic drag on the train when passing through a tunnel of diameter 5·5 m; assume the pressure to be uniform along the train. In which direction will the pressure gradient act in practice? What effect will this have on the drag?

8.4 (a) Justify the calculation of the drag of a ship of waterline length L, travelling at speed U_1, from the formula

$$D = \tfrac{1}{2}\rho U_1^2 L^2\left[C_p + C_w\left(\frac{U_1}{\sqrt{(gL)}}\right) + \frac{P}{L} C_f\left(\frac{U_1 L}{\nu}\right)\right]$$

where C_p is a form-drag coefficient, dependent only on the hull form; C_w is a wave-drag coefficient, a function of the Froude number $U_1/\sqrt{(gL)}$ for a particular hull form; and C_f is the average friction coefficient over the hull whose mean wetted perimeter is P; it is a function of Reynolds number.

(b) In testing ship models, it is usual to maintain the Froude number constant, giving a wave pattern similar to that around the prototype, and thus establishing the same value of C_w. In tests for a tanker of length 300 m and design speed 32 km/hr, a model of length 3 m is operated in this way. Where would boundary-layer transition normally occur on this model? Where should it be forced to occur?

(c) The total drag on the model of (b) is found to be 6 kgf when the loading is such that $P = 50$ cm, and when transition occurs at the appropriate point. Taking 0·5 for the form-drag coefficient based on the submerged frontal area of 250 cm², estimate the three drag contributions for this test. Calculate them also for the prototype, and estimate the engine power needed.

8.5 (a) For a constant-pressure boundary layer represented by $U/U_1 = (y/\delta)^{1/n}$ show that the momentum integral equation gives

$$U_1^2 \frac{d\delta}{dx} = (1+n)\left(1 + \frac{2}{n}\right)\frac{\tau_w}{\rho}$$

(b) Show that the stream function

$$\psi = \frac{n}{n+1} U_1 \left(\frac{y}{\delta}\right)^{1/n} y$$

gives the profile of (a), and hence that the accelerations in the layer are specified by

$$\frac{DU}{Dt} = -\frac{U_1^2}{n+1}\frac{1}{\delta}\frac{d\delta}{dx}\left(\frac{y}{\delta}\right)^{2/n} = \frac{1}{\rho}\frac{\partial\tau}{\partial y}$$

(c) Finally, show that the stress distribution is

$$\frac{\tau}{\tau_w} = 1 - \left(\frac{y}{\delta}\right)^H = 1 - \left(\frac{U}{U_1}\right)^{nH}$$

with H the shape factor (8.1.18). Under what conditions is the variation nearly linear?

8.6 When the boundary layer on a body of revolution is thick enough, the laws for a plane layer lose their validity.

(a) For a body of uniform diameter, save for a pointed nose, show that $L/d < \mathrm{Re}_L^{\frac{1}{5}}$ specifies conditions in which the plane-flow friction laws are appropriate, for an entirely turbulent boundary layer.

(b) Derive the parallel result for a wholly laminar layer. Plot the two variations of limiting L/d, indicate where each will be valid, and sketch the transition between them. Show also the logarithmic modification of the result (a) which is necessary for high Re_L.

(c) Mark on the ranges of L/d and Re_L for ship hulls, aircraft fuselages and trains. Consider also the lower-Reynolds-number cases of canoes and arrows, and show that plane-flow results will be valid for nearly all of these.

8.7 (a) For an axisymmetric boundary layer on a body of uniform radius R, show that $r\tau = \tau_w R$ is a plausible replacement for the plane wall-layer condition $\tau =$ constant.

(b) Show that $U^+ = R^+ \ln(r/R)$ in the sublayer, and that $U^+ = A \ln\{4R^+[(r/R)^{\frac{1}{2}} - 1]/[(r/R)^{\frac{1}{2}} + 1]\} + B$ in the turbulent part of the wall layer.

(c) Determine the constant B using the technique of Problem 5.29(c), namely, arguing that the two laws of (b) always meet at the same value of ϵ_m/ν. Under what conditions does B have its plane-flow value?

(d) In what way is the axisymmetric layer like a round wake? Will the flow remain turbulent as it develops along a long rod?

8.8 Rewrite the Colebrook–White friction formula (4.3.18) in a form appropriate to boundary layers. Sketch the development (c_f *vs* x) for flow over a flat plate with uniform roughness scale k_s, showing its relationship to the ultimate, effectively smooth-wall flow. Sketch also the development at the entry to a large pipe with uniformly rough wall.

8.9 A boundary layer in air develops on a surface whose uniform roughness is specified by $k_s = 0{\cdot}1$ mm; the outer velocity is 250 m/sec. Assuming the flow to be turbulent from the beginning, predict the variation in wall stress using numerical integration. Assume the sand-grain characteristics of Figure 4.2, and continue the integration to the point where the surface is effectively smooth. Find the average friction coefficient for the region considered.

8.10 Show that, according to Coles' two-layer model of a boundary layer, $U/U_1 = \frac{1}{2}w(y/\delta)$ at separation. Sketch the profiles just upstream and downstream of this point.

8.11 For a self-preserving or equilibrium boundary layer, show that c_f is constant and that $\delta \propto x$.

8.12 (a) Show that the stress distribution in the wall layer, where $U^+ = f(y^+)$, is given by

$$\tau = \tau_w + y_f \frac{dP_1}{dx} y^+ + \mu \frac{du_f}{dx} \int_0^{y^+} f^2 \, dy^+$$

(Suggestion: integrate continuity first to find V.)

(b) For an adverse pressure gradient, $dP_1/dx > 0$, show that the final term is negative. Sketch the resulting stress variation, and show that the assumption of a constant-stress layer is more widely valid than would be supposed from a consideration of the first two terms only.

8.13 For boundary layers developing on the walls of a channel, the pressure gradient is prescribed by the continuity requirement for the entire flow. For example, in a plane channel of breadth b, this is $U_1(b - 2\delta_1) = \dot{v}$, the constant flow per unit width, with U_1 the centre-line velocity.

(a) For the initial part of a parallel-sided channel where the developing flow possesses a non-turbulent core, show that the integral momentum equation (see Example 6.14) gives

$$\frac{d\delta_2}{dx} = \frac{\frac{1}{2}c_f(b - 2H\delta_2)}{b + 2H(H+1)\delta_2}$$

on the assumption that H is constant. What values of H might be adopted for the laminar and turbulent sections?

(b) How is the result of (a) modified when the two boundary layers meet?

(c) Generalize the result of (a) to include thin boundary layers in diffusers and nozzles, taking into account variations in b and H. In which case are the latter more important?

(d) What additional information is required to calculate these patterns of boundary-layer development?

8.14 (a) Assuming uniform wall and free-stream temperatures, find the temperature distribution in the laminar asymptotic layer, and the heat-transfer rate at the wall.

(b) How are the temperature and velocity profiles related? How are the thicknesses of velocity and thermal layers related?

8.15 Use the Colburn analogy to find the heat-transfer characteristic, $\mathrm{Nu} = f(\mathrm{Pr}, \mathrm{Re})$, for a turbulent boundary layer heated only over the length L_1 to L_2. Under what conditions would you have confidence in this prediction? How can the heat transfer associated with an arbitrarily varying temperature distribution be calculated using this analogy?

8.16 Sketch plausible variations of Pr_t across boundary layers in air and water.

8.17 Find the average friction coefficient for an initially laminar, then turbulent boundary layer, using arguments like those leading to equation (8.3.17).

8.18 In the analysis of the buoyancy-driven layer given at the end of Section 8.3, the friction laws for a constant-pressure flow were used to specify the transfer characteristics, even though a vertical pressure gradient necessarily exists. Justify this step.

8.19 For an insulated wall, the dependence of skin friction on Mach number is primarily through the alteration of the temperatures near the wall by recovery effects.

(a) Taking $\mu_1/\mu_R = (T_1/T_R)^m$, show that $\mathrm{Re}_R = \mathrm{Re}_i(T_1/T_R)^{1+m}$ relates the Reynolds number based on the reference temperature T_R to that based on free-stream properties.

(b) Assuming that $c_f = f(\mathrm{Re}_R)$ is the same function of Reynolds number as in isothermal conditions, show that

$$\frac{c_f}{c_{f_1}} = \frac{T_1}{T_R}\frac{f(\mathrm{Re}_R)}{f(\mathrm{Re}_i)}$$

(c) Assuming that $C_f \propto (\ln \mathrm{Re})^{-n}$ as in the Schoenherr formula (8.1.10), show that

$$\frac{C_f}{C_{f_1}} = \frac{T_1}{T_R}\left[1 + \frac{(1+m)\ln(T_1/T_R)}{\ln \mathrm{Re}_1}\right]^{-n}$$

(d) Taking $T_R = \frac{1}{2}(T_1 + T_w)$, show that the reference temperature is related to the Mach number by $T_R/T_1 = 1 + \frac{1}{4}r(\gamma - 1)M_1^2$, with r the recovery factor, and $\gamma = c_p/c_v$ the ratio of specific heats.

(e) For air, the values $r = 1$ and $m = 0{\cdot}75$ are found to be appropriate. Using the constant n of the Schoenherr formula, plot C_f/C_{f_1} *vs* $M_1(=0$ to $8)$ for $\mathrm{Re}_1 = \mathrm{Re}_L = 10^5$ and 10^8. Compare the influences of M_1 and Re_L.

8.20 A boundary layer on a smooth flat plate develops beneath an essentially non-turbulent uniform air flow with $U_1 = 30$ m/sec.

(a) Plot the variations of δ_1 and δ_2 for $x = 0$ to 3 m, using an exaggerated y-scale.

(b) Plot the variation of δ_{99} for $x = 0$ to 3 m, maintaining a correct proportion of x and y coordinates.

(c) Plot the variations of c_f, C_f, y_f/δ_{99} and n, the most appropriate power-law index.

9

More Complex Flows

This title gives a rather too narrow impression of the developments of this chapter. The methods to be discussed are indeed applicable to flows with more complex boundary geometry and internal structure, for example, regions of recirculation and mixing streams. But these methods can also describe details of flows whose simpler features were considered earlier, for example, the distribution of fluctuation intensity across the flow, the secondary currents in channel flows, the part of a wake near the generating body, and the boundary layer near separation.

The results of the preceding chapters were, for the most part, based on integral equations for thin flows, those described adequately by the boundary-layer equations. For flows that are not thin in this sense, and for many details of thin flows, this approach is inadequate. Hence we turn away from the integral formulation, though it remains in the background to provide checks on more detailed calculations, and to answer sufficiently simple questions. Here we work directly with the differential equations governing the motion and, in principle at least, need not adopt the thin-flow approximations.

The methods of the preceding chapters have been seen to be deficient in two respects: they fail to account adequately for the history of a developing flow (except when the pattern is particularly simple); nor do they describe cross-flow transfers realistically. We can combine these criticisms into a single theme: we require a better representation of transport processes within the turbulent flow, both turbulent diffusion and mean-flow convection. We shall attack this problem by converting the differential equations of motion into transport equations for certain vital entities, though their identity is not immediately obvious. The analysis has two major elements:

(1) The derivation of exact general equations for a statistically steady turbulent motion with constant density. These are derived by averaging, over time, the statements of conservation of mass, momentum and energy for the instantaneous motion. They involve the mean velocity and mixing stresses, the mean vorticity and vorticity fluctuations, and other fluxes within the turbulent fluid.

(2) The reduction of the general equations, using similarity arguments and empirical data, to systems simple enough for numerical solution on high-speed computers. Here we shall, for simplicity, restrict consideration to thin flows.

The general equations have been known for decades and are, indeed, the basis of the extensive statistical theory of turbulence. The ideas behind their simplification—and considerable reduction is necessary, even with the computers now available—have been dormant in the literature for decades. Not until rapid and convenient computing facilities became available, within the last decade, has it been possible to utilize these ideas directly in engineering calculations.

Here we shall indicate only the general line of development of methods of numerical prediction. Techniques are improving rapidly, and a textbook is inevitably some years out of date, as are the references on which it is based. With this warning, the reader who requires a more detailed discussion of these 'differential' or 'turbulent field' methods of prediction may refer to the reviews of W. C. Reynolds (cited in Chapter 8) and of Rotta, Bradshaw, and Launder and Spalding (listed at the end of this chapter).

9.1 GENERAL TRANSFER EQUATIONS

9.1.1 Tensor Notation

In order to obtain a compact and easily assimilable description of a general turbulent motion, we adopt the tensor notation widely used in the literature of turbulence and other branches of fluid mechanics. The quantities considered here are *Cartesian tensors*, the species whose components are referred to straight, orthogonal axes. These are a restricted form of the more general covariant and contravariant tensors. We begin by using Cartesian tensor notation to express some of the quantities with which we shall have to deal later.

Vectors and Tensors

When discussing correlations in Section 2.6 and spectra in Section 2.7, we found it convenient to represent multicomponent entities using a compact suffix notation. Thus the components of the velocity vector are

$$u_i \quad \text{with } i = 1, 2, 3 \text{ understood}$$

and the *gradient vector* of the pressure field is

$$\frac{\partial p}{\partial x_j} \quad \text{with } j = 1, 2, 3 \text{ understood}$$

These vectors have three components (more are possible in a hypothetical, multi-dimensioned space); a single suffix denotes the direction of the component. *Scalar* quantities, such as temperature, pressure and energy, do not depend on the direction of the coordinate axes, and have no directional suffix.

We also expressed the double-velocity correlation (2.6.5, 6) as

$$\overline{u_i u_j} \quad \text{with } i, j = 1, 2, 3 \text{ understood}$$

and interpreted these as the components of a *second-order tensor*, with two independent suffices and nine components. Another second-order tensor is

$$e_{ij} = \frac{\partial U_i}{\partial x_j} + \frac{\partial U_j}{\partial x_i} \tag{9.1.1}$$

giving the *rates of strain* associated with the mean motion; a similar expression can be written for the fluctuations. Only six of the components (9.1.1) are distinct; check that this is so. For $i = j$, we have the three direct strains; for $i \neq j$, we have three shearing strains. The related *vorticity* tensor

$$\zeta_{ij} = \frac{\partial U_j}{\partial x_i} - \frac{\partial U_i}{\partial x_j} \tag{9.1.2}$$

is described as the *curl* or *rotation* of the velocity vector. Since there are only three distinct components (check this), the entity can be treated as a vector for some purposes, its components related to the tensor components by

$$\Omega_k = \zeta_{ij} \tag{9.1.2}$$

Each vector component is normal to the plane of the rotation which it represents.

The manipulations which follow will generate higher-order tensors, for example, $\overline{u_i u_j u_k}$, with more than two independent suffices.

The Einstein Summation Convention

This is one of the great conveniences of tensor notation: in the absence of a specific note to the contrary, repetition of a suffix implies summation over all its possible values. Thus

$$\tfrac{1}{2}u_i^2 = \tfrac{1}{2}u_1^2 + \tfrac{1}{2}u_2^2 + \tfrac{1}{2}u_3^2$$

represents the instantaneous kinetic energy per unit mass of fluid. This is a scalar, since summation removes the directional dependence which is characteristic of vectors and tensors; in other words, there is no 'free' suffix. Another example of summation is

$$\frac{\partial J_j}{\partial x_j} = \frac{\partial J_1}{\partial x_1} + \frac{\partial J_2}{\partial x_2} + \frac{\partial J_3}{\partial x_3} \tag{9.1.3}$$

This is termed the *divergence* of the vector J_j. Reference to equation (3.3.1) shows that this scalar quantity represents the net outflow of the entity whose flux is denoted by the vector J_j.

A quantity such as $\frac{1}{2}\overline{u_i u_j^2}$, with three suffices but two alike, is a vector—in this case, the vector giving the net flux of kinetic energy. A further example of this kind is

$$\frac{DA_i}{Dt} = U_j \frac{\partial A_i}{\partial x_j} \tag{9.1.4}$$

the *convective derivative* of the component A_i; compare with equations (3.3.9). We shall use the notation D/Dt, as above, to indicate differentiation following the *mean* motion.

The summation convention can be used in conjunction with the *Kronecker delta* (or substitution tensor), a notational second-order tensor with components:

$$\begin{aligned} \delta_{ij} &= 1 \quad \text{for } i = j \\ &= 0 \quad \text{for } i \neq j \end{aligned} \tag{9.1.5}$$

Thus $\delta_{1j}u_j = u_1$, and $\delta_{ik}u_k = u_i$.

9.1.2 The Momentum Equation and its Associates

Hitherto we have used the momentum equation of a flowing fluid in its time-averaged forms, the most general being the thin-flow approximations (6.2.21, 22). To extract more information from the momentum principle, we return to the more fundamental result which holds at each instant. For constant density and viscosity, the *Navier–Stokes equations* for the three coordinate directions are, in tensor notation

$$\frac{\partial}{\partial t}(U_i + u_i) + (U_k + u_k)\frac{\partial}{\partial x_k}(U_i + u_i) = -\frac{1}{\rho}\frac{\partial}{\partial x_i}(P + p) + \nu\frac{\partial^2}{\partial x_k^2}(U_i + u_i) \tag{9.1.6}$$

We have represented the velocity components and the pressure as sums of time-mean values and fluctuations about those values, anticipating application to a turbulent flow which is steady in the mean. These equations state that the instantaneous total acceleration, the sum of local and convective elements, is a consequence of the pressure gradient and the viscous stresses. They can be derived by introducing the instantaneous momentum-flux components, like that from which equation (3.2.1) was derived, into the three-dimensional form of the conservation law (3.3.1). As an equivalent derivation is given in many textbooks on fluid mechanics, it need not concern us here.

Averaging equation (9.1.6) over time, and taking the flow to be basically steady, we obtain the three *Reynolds equations*:

$$\frac{\mathrm{D}U_i}{\mathrm{D}t} = U_k \frac{\partial U_i}{\partial x_k} = -\frac{1}{\rho}\frac{\partial P}{\partial x_i} + \nu \frac{\partial^2 U_i}{\partial x_k^2} - \overline{u_k \frac{\partial u_i}{\partial x_k}}$$
$$= -\frac{1}{\rho}\frac{\partial P}{\partial x_i} + \nu \frac{\partial^2 U_i}{\partial x_k^2} - \frac{\partial \overline{u_i u_k}}{\partial x_k} \tag{9.1.7}$$

The thin-flow results (6.2.16, 21) are special cases. The final rewriting of the mixing-stress term makes use of the *continuity equation*, whose instantaneous form for source-free flow at constant density is

$$\frac{\partial (U_j + u_j)}{\partial x_j} = 0 \tag{9.1.8}$$

Averaging gives

$$\frac{\partial U_j}{\partial x_j} = \frac{\partial u_j}{\partial x_j} = 0 \tag{9.1.9}$$

In words, the divergence of the velocity is zero, for both mean and fluctuating fields. It follows that

$$\frac{\partial (u_i u_k)}{\partial x_k} = u_k \frac{\partial u_i}{\partial x_k} + u_i \frac{\partial u_k}{\partial x_k} = u_k \frac{\partial u_i}{\partial x_k} \tag{9.1.10}$$

Such manipulations often prove helpful in recasting these equations.

The transition between the Navier–Stokes and Reynolds equations displays the problem of *closure*: we start with four fluctuating quantities, and end up with ten time-mean values (U_i, P, and six distinct components $\overline{u_i u_j}$). It is possible to derive equations for the mixing stresses which have been added to our list of dependent variables. The procedure is as follows: multiply through the instantaneous equations (9.1.6) by u_j; rewrite with subscripts i and j interchanged; add the two equivalent forms to obtain a symmetrical result; and, finally, average and carry out some operations like (9.1.10). Thus the *transport equations for the Reynolds stresses* are found:

$$\frac{\mathrm{D}\overline{u_i u_j}}{\mathrm{D}t} = U_k \frac{\partial \overline{u_i u_j}}{\partial x_k} = -\overline{u_j u_k}\frac{\partial U_i}{\partial x_k} - \overline{u_i u_k}\frac{\partial U_j}{\partial x_k}$$
$$-\frac{1}{\rho}\left[\overline{u_i \frac{\partial p}{\partial x_j}} + \overline{u_j \frac{\partial p}{\partial x_i}}\right] - \frac{\partial \overline{u_i u_j u_k}}{\partial x_k} \tag{9.1.11}$$
$$+\nu\left[\overline{u_i \frac{\partial^2 u_j}{\partial x_k^2}} + \overline{u_j \frac{\partial^2 u_i}{\partial x_k^2}}\right]$$

We have six new equations, but they contain a host of new time-mean quantities. Evidently, such multiplying-and-averaging procedures can produce an indefinitely large number of equations, but always governing an even larger number of time-mean quantities.

Before looking further at the complicated result (9.1.11), we examine the form obtained by setting $i=j$; this step automatically introduces summation, and contracts the several transport equations to the single result:

$$\frac{\mathrm{D}(\tfrac{1}{2}\overline{u_i^2})}{\mathrm{D}t} = -\overline{u_i u_k}\frac{\partial U_i}{\partial x_k} - \frac{\partial}{\partial x_k}\left[\overline{u_k\left(\frac{p}{\rho} + \tfrac{1}{2}u_i^2\right)}\right] + \nu\overline{u_i\frac{\partial^2 u_i}{\partial x_k^2}} \tag{9.1.12}$$

This equation for the *turbulence kinetic energy* extends equations (3.3.26) and (6.2.24). From our discussion of those restricted forms, we see that the terms on the right represent, in turn: production or generation; turbulent 'diffusion'; and viscous diffusion and, more important, dissipation of turbulence energy. The dual role of the viscous terms is revealed by writing

$$\nu\overline{u_i\frac{\partial^2 u_i}{\partial x_k^2}} = \nu\left[\frac{\partial^2(\tfrac{1}{2}\overline{u_i^2})}{\partial x_k^2} + \frac{\partial^2 \overline{u_i u_k}}{\partial x_i\,\partial x_k}\right] - \tfrac{1}{2}\nu\overline{\left(\frac{\partial u_i}{\partial x_k} + \frac{\partial u_k}{\partial x_i}\right)^2} \tag{9.1.13}$$

The second, always-negative term gives the dissipation; it is not difficult to see how this reduces to the form (2.6.27) when the dissipating scales are isotropic.

Returning to the set (9.1.11) for the individual stress components, we can form most of the terms into groups generalizing those of the energy equation (9.1.12), and representing aspects of production, diffusion and dissipation. One group of terms is left over:

$$\frac{1}{\rho}\overline{p(\partial u_i/\partial x_j + \partial u_j/\partial x_i)}$$

These sum to zero when $i=j$, by virtue of the continuity results (9.1.9). They represent the working of the pressure fluctuations against the fluctuating rates of strain (9.1.1), and may be thought of as describing redistribution of 'energy' amongst the several components. Some of the $\overline{u_i u_j u_k}$ components may also be interpreted as carrying out this function.

Among the many other transport equations which can be developed by the multiplying-and-averaging procedure are:

(1) Mean-flow and total energy equations obtained after multiplication by U_j and by $U_j + u_j$. The former contains the 'production' term with reversed sign, since the turbulence energy has its source here. The latter does not contain a production term, since the contributions of mean flow and fluctuations cancel; it does contain both mean-flow and turbulence dissipation contributions, in the manner of equation (3.3.26).

(2) The vorticity equation relating the mean components Ω_i to the fluctuating elements ω_i:

$$\frac{\mathrm{D}\Omega_i}{\mathrm{D}t} = \Omega_k\frac{\partial U_i}{\partial x_k} + \nu\frac{\partial^2 \Omega_i}{\partial x_k^2} + \frac{\partial\overline{u_i\omega_k}}{\partial x_k} - \frac{\partial\overline{u_k\omega_i}}{\partial x_k} \tag{9.1.14}$$

This is found by taking the curl (9.1.2) before averaging; note that the pressure is eliminated when this is done.

(3) The vorticity-intensity equation

$$\frac{D(\frac{1}{2}\overline{\omega_i^2})}{Dt} - \nu\,\overline{\omega_i \frac{\partial^2 \omega_i}{\partial x_k^2}} = -\overline{\omega_i u_k \frac{\partial \omega_i}{\partial x_k}} - \overline{\omega_i u_k}\frac{\partial \Omega_i}{\partial x_k} + \text{similar terms} \tag{9.1.15}$$

This is obtained by multiplying the instantaneous vorticity equation by ω_i before averaging. It describes the vortex-stretching process which was discussed in a simple way in Section 1.4. Note that the production terms on the right are all of third order in the vorticity, while the dissipation is of the second order. This is consistent with the abrupt intermittency of turbulence: until the vorticity reaches a certain intensity, it is damped more rapidly than it can be generated.

9.1.3 Other Transport Equations

The equation governing the instantaneous transfer of a property $S+s$ has a structure similar to that of the Navier–Stokes equations:

$$\frac{\partial(S+s)}{\partial t} + (U_k + u_k)\frac{\partial(S+s)}{\partial x_k} = \frac{\dot{S}_s}{\rho} + K\frac{\partial^2(S+s)}{\partial x_k^2} \tag{9.1.16}$$

Immediate averaging gives an equation for the transport of S:

$$\frac{DS}{Dt} = U_k\frac{\partial S}{\partial x_k} = \frac{\overline{\dot{S}_s}}{\rho} + K\frac{\partial^2 S}{\partial x_k^2} - \frac{\partial \overline{u_k s}}{\partial x_k} \tag{9.1.17}$$

This generalizes equations (6.2.2), and also the Reynolds equations (9.1.7), provided that a suitable interpretation is made for $\dot{S}_s$.

Prior multiplication by s gives a transport equation for the fluctuation intensity:

$$\frac{D(\frac{1}{2}\overline{s^2})}{Dt} = \frac{\overline{s\dot{S}_s}}{\rho} - \overline{su_k}\frac{\partial S}{\partial x_k} - \overline{su_k\frac{\partial s}{\partial x_k}} + K\overline{s\frac{\partial^2 s}{\partial x_k^2}} \tag{9.1.18}$$

Comparison with the turbulence energy equation (9.1.12) shows that the last three terms can be interpreted as representing: production through interaction with the mean gradients; turbulent 'diffusion'; and erasure and diffusion by molecular mixing. The significance of the first term on the right depends on the entity S. It may represent the generation (or destruction) of fluctuation energy through interaction with the basic source $\dot{S}_s$, or it may correspond to turbulent diffusion, as in the pressure terms of equation (9.1.12). Alternatively, if the entity is passively convected, without influencing the mean or turbulent motion, the source term can be negligible.

The boundary-layer approximations to equations (9.1.17, 18) are obtained, for a flow which is 'thin' in the y-direction, by setting $k = 2$ and $u_k = u_2 = v$. The enthalpy equation (6.2.29) is one such.

Transport equations for the individual components $\overline{su_i}$ can be obtained by multiplying equation (9.1.16) by u_i before averaging. They are rather like equations (9.1.11) in structure, although they have 'production' and 'dissipation' terms of two kinds.

9.2 MODELS OF TURBULENT FLOW

9.2.1 The Energy Equation

The transport equation for the turbulence kinetic energy (9.1.12) was the basis of the first attempts to provide a more rational representation of transport processes than is afforded by gradient-diffusion models. This equation remains an important element in many of the more comprehensive systems developed subsequently. Its central position is easily justified: the equation is easily derived; it is readily interpretable in terms of essential activities; and its terms have been measured for a variety of flows.

Here we consider the form of the energy equation applicable to a plane, thin flow; with these restrictions, and with molecular diffusion neglected, equations (9.1.12, 13) give

$$\begin{aligned}\frac{\mathrm{D}k}{\mathrm{D}t} &= U\frac{\partial k}{\partial x} + V\frac{\partial k}{\partial y} \\ &= -\overline{uv}\frac{\partial U}{\partial y} - \frac{\partial}{\partial y}\left[\overline{v\left(\frac{p}{\rho} + k\right)}\right] - \varepsilon \end{aligned} \tag{9.2.1}$$

For compactness, the kinetic energy is given by

$$k = \tfrac{1}{2}\overline{u_i^2} = \tfrac{1}{2}\overline{q^2} \tag{9.2.1}$$

and the dissipation rate per unit mass by

$$\varepsilon = \tfrac{1}{2}\nu\overline{\left(\frac{\partial u_i}{\partial x_k} + \frac{\partial u_k}{\partial x_i}\right)^2} \tag{9.2.1}$$

Strictly, this is the dissipation within the turbulence alone, denoted by ε_t in Chapter 3.

We shall examine two ways of converting this energy equation to a form compatible with the continuity and momentum equations

$$\frac{\partial U}{\partial x} + \frac{\partial V}{\partial y} = 0 \tag{9.2.2}$$

and

$$\frac{\mathrm{D}U}{\mathrm{D}t} = U\frac{\partial U}{\partial x} + V\frac{\partial U}{\partial y} = U_1\frac{\mathrm{d}U_1}{\mathrm{d}x} + \frac{\partial \tau}{\partial y} \tag{9.2.3}$$

(The latter is equation (6.2.21), with normal stresses omitted.) First we use the shear stress τ as a scale for all of the terms of (9.2.1), then the turbulence kinetic energy k itself.

Stress Scaling

In Section 3.5 we used the local shear stress as a scale for every measure of turbulent activity in a wall layer; the flow thus specified was called a locally determined layer. Bradshaw and his coworkers carried these arguments to their logical conclusion, adopting this scaling for every term of the energy equation for a boundary layer, and applying it not only in the wall layer, but across the outer turbulent region. The turbulence energy itself is specified by

$$\frac{\tau}{\rho q^2} = \frac{\tau}{2\rho k} = a, \quad \text{a constant} \tag{9.2.4}$$

This assumption has considerable experimental support, with $a \simeq 0{\cdot}15$. As in equations (3.5.5, 6), the dissipation is taken as

$$\varepsilon = \frac{(\tau/\rho)^{\frac{3}{2}}}{L_\varepsilon} = \frac{(\tau/\rho)^{\frac{3}{2}}}{\delta f(y/\delta)} \tag{9.2.5}$$

with $f(y/\delta)$ an empirical function and L_ε a dissipation length parameter. Diffusion is represented, partly guided by analytical convenience, as

$$\overline{v\left(\frac{p}{\rho} + k\right)} = \frac{\tau_m}{\rho U_1}\frac{\tau}{\rho}\, g\left(\frac{y}{\delta}\right) \tag{9.2.6}$$

with $g(y/\delta)$ another empirical function, and τ_m the maximum value of the stress within the layer, representing the intensity of the large 'eddies' which play an important part in cross-stream diffusion.

When these assumed forms are introduced, the energy equation becomes an equation for the shear stress:

$$\frac{D(\tau/2a)}{Dt} - \tau\frac{\partial U}{\partial y} + \frac{\tau_m}{\rho U_1}\frac{\partial(g\tau)}{\partial y} + \frac{\tau(\tau/\rho)^{\frac{1}{2}}}{\delta f} = 0 \tag{9.2.7}$$

With the constant a and the functions f and g specified, equations (9.2.2, 3, 7) can be solved, numerically, for the velocity and stress fields—U, V and τ. Integration proceeds in the direction of the flow, as in the integral formulations of the preceding chapters, and also across the flow at each stage.

This model, a development of the mixing-length hypothesis and of the concept of locally determined turbulence, retains some of their limitations: the turbulence 'collapses' when $\tau \to 0$, leading to an unrealistic description near a maximum in velocity. Hence the use of the stress-scaled equations is restricted to flows where $\partial U/\partial y$ and τ have the same sign everywhere and, of course, to flows for which f and g are known with sufficient accuracy.

Energy Scaling

A somewhat artificial feature of the stress scaling used above is the conversion of the energy equation into an equation for the shear stress. This can be justified when the stress distribution has a simple, determinate form, in particular, for the linear variation of a pipe or channel flow, and for the nearly linear variation of a constant-pressure boundary layer. In other circumstances, we are led to ask: why not use the energy as the scale instead? This idea is consistent with suggestions made by Prandtl and by Kolmogoroff in the 1940s, years before the development of computers made them useful in practice. These workers sought to free the eddy diffusivity from its rigid links with shear stress and velocity gradient. In equations (3.4.56) it is evident that $(\tau/\rho)^{\frac{1}{2}}$ and $l_m \partial U/\partial y$ act as velocity scales for the eddy viscosity; these choices lead to anomalous behaviour when $\partial U/\partial y$, $\tau = 0$. Prandtl suggested the more widely applicable

$$\epsilon_m = k^{\frac{1}{2}} l \tag{9.2.8}$$

with l a length scale of the turbulence. About the same time, Kolmogoroff made an equivalent proposal:

$$\epsilon_m = \frac{k}{f} \tag{9.2.9}$$

with f a characteristic frequency for the turbulence. Either form allows the diffusivity to remain finite when stress and velocity gradient vanish.

At first sight, the proposals (9.2.8, 9) appear to be unhelpful for a wall layer and free turbulent flow, because they take no account of the simple forms of τ, ϵ_m or l_m in these situations. However, they offer hope when these special features are absent, and the possibility of a comprehensive method of prediction encompassing such disparate cases.

We proceed to scale the terms of the energy equation (9.2.1) using the energy k and the length l, introduced in equation (9.2.8). The shear stress is represented as

$$\frac{\tau}{\rho} = -\overline{uv} = k^{\frac{1}{2}} l \frac{\partial U}{\partial y} = \epsilon_m \frac{\partial U}{\partial y} \tag{9.2.10}$$

and the dissipation by

$$\varepsilon = \frac{c_D k^{\frac{3}{2}}}{l} \tag{9.2.11}$$

We argue, as in obtaining equations (3.5.3, 5), that the dissipation rate is set by energy extraction in the larger scales of the turbulence. Since the length l is defined by equation (9.2.8), we must introduce a constant c_D into equation (9.2.11); Launder and Spalding suggest that $c_D \simeq 0{\cdot}08$ for wall flows. In this respect, equation (9.2.11) differs from equation (3.5.5), which was used to *define* the alternative length scale L_ε.

Finally, we express the diffusion term as a consequence of diffusion down the gradient of k, with a diffusivity defined by k and l:

$$-\overline{v\left(\frac{p}{\rho}+k\right)}=\frac{k^{\frac{1}{2}}l}{\sigma_k}\frac{\partial k}{\partial y}=\frac{\epsilon_m}{\sigma_k}\frac{\partial k}{\partial y} \tag{9.2.12}$$

with $\sigma_k = \epsilon_m/\epsilon_k$ the diffusivity ratio relating transfers of momentum and turbulence energy. Values of $\sigma_k \simeq 1$ give satisfactory predictions in practice.

With these forms introduced, the energy equation is

$$\frac{Dk}{Dt}=\frac{\partial}{\partial y}\left[\frac{\epsilon_m}{\sigma_k}\frac{\partial k}{\partial y}\right]+\epsilon_m\left(\frac{\partial U}{\partial y}\right)^2-\frac{c_D k^{\frac{3}{2}}}{l} \tag{9.2.13}$$

Following specification of the constants c_D and σ_k, and of the function l $(=l(y/\delta)$, say), this equation can be solved with equations (9.2.2, 3), the latter with the stress represented, through equations (9.2.10), in terms of ϵ_m, or k and l. The constants can be estimated by considering simple cases, such as the wall layer. In view of the rough-and-ready way in which equation (9.3.13) was generated, it is more realistic to fix the constants by '*computer optimization*', that is, by carrying out calculations for a variety of constants, to find the combinations giving optimum results.

The model defined by equations (9.3.10) to (9.3.13) deals successfully with the problem which prompted its development, the representation of transfer across a plane where $\tau = 0$. However, its range of application is not wide: the stress, dissipation and diffusion have been related in arbitrary ways, and it is still necessary to provide a length-scale variation appropriate (hopefully) to a particular class of flows.

9.2.2 Other Dynamic Transport Equations

We shall examine the consequences of adding first one equation, then more than one, to our model of turbulent flow. But first we round off our ideas on one-equation models of turbulence. In addition to the two we have considered, both based on the energy equation, there is a model developed by Nee and Kovasznay, incorporating a postulated transport equation for the eddy viscosity. This approach is attractive since, as is evident above, the diffusivity has a central role in the relationships among the attributes of turbulence. In practice, this model suffers from limitations like those of the scaled energy equation; in particular, a length scale must be specified. We conclude that no single transport equation can provide a major extension of our ability to predict turbulent flows.

Two-equation Models

Our immediate problem appears to be that of finding the variation of the length scale l. Although this can be specified fairly realistically for boundary layers (it is rather like the mixing length), it varies in a complicated way in irregular channel sections, and in reattaching, recirculating and mixing flows.

It is possible to develop a transport equation for l, and this path has been followed by some workers. However, it is more fruitful to think of the length in terms of the relationship $l \propto k^{\frac{3}{2}}/\varepsilon$. This reveals that an equation for ε determines the length scale, in conjunction with an equation for k. Indeed, by using a transport equation for ε, we are able to eliminate the hypothetical l from the problem.

The elimination of the scale l can be achieved using any quantity $Z = f(l,k)$. The form of the transport equation for such a quantity is suggested by equation (9.1.17):

$$\frac{\mathrm{D}Z}{\mathrm{D}t} - \frac{\partial}{\partial y}\left[\frac{\epsilon_{\mathrm{m}}}{\sigma_z}\frac{\partial Z}{\partial y}\right] = \frac{\dot{S}_z}{\rho} \tag{9.2.14}$$

Turbulent diffusion has been related to the cross-stream gradient of Z; the diffusivity ratio $\sigma_z \simeq 1$ for most quantities which might be considered. Since the equations for various functions Z are obtained by manipulating the Navier–Stokes equation, the source term $\dot{S}_z$ has a structure somewhat like that equation (9.2.12).

Launder and Spalding have pointed out that

$$\frac{\dot{S}_z}{\rho} = \frac{Z}{k}\left[C_1\,\epsilon_{\mathrm{m}}\left(\frac{\partial U}{\partial y}\right)^2 - C_2\frac{k^2}{\epsilon_{\mathrm{m}}}\right] + \dot{s}_z \tag{9.2.15}$$

represents the source terms for such quantities as:

(1) $Z = l$, the length scale itself;
(2) $Z = k^{\frac{3}{2}}/l$, the dissipation; and
(3) $Z = k/l^2$, the mean-square of the vorticity fluctuations, related to the frequency of Kolmogoroff's proposal (9.2.9).

The final term $\dot{s}_z$ of equation (9.2.15) represents elements which cannot be fitted into the pattern. The constants C_1 and C_2 depend on the form of Z, and may be estimated by applying the transport equation to decaying grid turbulence and to wall-layer turbulence. Save for $Z = l$, $C_1, C_2 > 0$; in general, $|C_1| \sim 0{\cdot}1$ and $|C_2| \sim 1$. Working values are best determined by computer optimization.

Although no one function Z has emerged with a clear-cut advantage—success depends to a large extent on the effort and ingenuity of the practitioner—the dissipation ε is currently the favourite choice for the second equation of two-equation models.

Multi-equation Models

The models discussed above give some hope of representing recirculating and mixing flows, because they generate a length scale (or its equivalent) when one is not apparent initially. But they lack the normal stresses needed if secondary flows are to be calculated; see Section 4.5.1. Moreover, the dissipation and shear stress are still related in an arbitrary way. A solution to these difficulties is offered by equations (9.1.11), transport equations for the individual Reynolds stresses. When these are adopted, it is unnecessary to introduce ϵ_m and l to relate the velocity, shear stress and dissipation. We pay the price of admitting higher-order correlations, and come face to face with the problem of closure. This can be resolved by resolute 'pruning' of the higher-order terms, and by scaling the remnants using the basic transported entities. Both procedures are guided by the near-isotropy of the smaller scales of motion.

As an example, we consider the transport equation for the shear-stress correlation $\overline{uv}$. From equations (9.1.11) we obtain the thin-flow approximation

$$\begin{aligned} \frac{\mathrm{D}\overline{uv}}{\mathrm{D}t}+\overline{v^2}\frac{\partial U}{\partial y}+\varepsilon_{12} &= -\frac{\partial \overline{uv^2}}{\partial y}-\frac{1}{\rho}\left(\overline{u\frac{\partial p}{\partial y}+v\frac{\partial p}{\partial x}}\right) \\ &= -\frac{\partial}{\partial y}\left[\overline{v\left(uv+\frac{p}{\rho}\right)}\right]+\frac{1}{\rho}\overline{p\left(\frac{\partial u}{\partial y}+\frac{\partial v}{\partial x}\right)} \end{aligned} \tag{9.2.16}$$

Here ε_{12} is the contribution of molecular mixing to changes in $\overline{uv}$. In the last line, the 'redistribution' term is separated out, a streamwise diffusion term being omitted.

We shall consider two kinds of scaling, using U, $\overline{uv}$ and k, and one of l and ε; the choice between these two depends on the other equations of the model. The production term of equation (9.2.16) we take simply as

$$\overline{v^2}\frac{\partial U}{\partial y} \propto k\frac{\partial U}{\partial y} \tag{9.2.17}$$

The molecular term may be neglected for a flow at high Reynolds number (save in the viscous layer), since the smaller scales are then nearly isotropic:

$$\varepsilon_{12}=0 \tag{9.2.18}$$

The diffusion is represented by

$$\overline{v\left(uv+\frac{p}{\rho}\right)} \propto k^{\frac{1}{2}}\,l\,\frac{\partial\overline{uv}}{\partial y} \quad \text{or} \quad \frac{k^2}{\varepsilon}\frac{\partial \overline{uv}}{\partial y} \tag{9.2.19}$$

Finally, the redistribution is given by

$$\overline{p(\partial u/\partial y+\partial v/\partial x)} \propto \left(\frac{k^{\frac{1}{2}}}{l}\right)\overline{uv} \quad \text{or} \quad \frac{\varepsilon}{k}\overline{uv} \tag{9.2.20}$$

Other forms of scaling can be devised using the parameters available. For example, redistribution might be taken $\propto k\partial U/\partial y$, or to be the sum of elements like this and like equation (9.2.20). This opens up further variety in the construction of skeletal transport equations.

Introducing these forms into equation (9.2.16), we have

$$\frac{D\overline{uv}}{Dt} = C_1 \frac{\partial}{\partial y}\left[\frac{k^2}{\varepsilon}\frac{\partial \overline{uv}}{\partial y}\right] - C_2 \frac{\varepsilon}{k}\overline{uv} - C_3 k \frac{\partial U}{\partial y} \tag{9.2.21}$$

An equation of this kind was used by Hanjalić and Launder, together with equations for ε and k, and the continuity and momentum equations (9.2.2, 3). One of the flows to which this model was applied is that illustrated in Figure 4.6; not only was the mean-velocity profile predicted, but also the variations of turbulence kinetic energy, and the main elements of the energy balance within the flow. The dissipation equation used

$$\frac{D\varepsilon}{Dt} = C_4 \frac{\partial}{\partial y}\left[\frac{k^2}{\varepsilon}\frac{\partial \varepsilon}{\partial y}\right] - C_5 \frac{\varepsilon^2}{k} + C_6 \frac{\overline{uv}}{k}\frac{\partial U}{\partial y} \tag{9.2.22}$$

has a form like that of the stress equation (9.2.21) and energy equation (9.2.13).

This line of development is fed by a never-exhausted fount of transport equations which can be derived from the Navier–Stokes equations. Provided that a corresponding never-ceasing supply of empirical data is available, there seems to be no barrier to the construction of more-and-more elaborate models, capable of representing ever more varied turbulent flows. The performance of these models, and the discipline required for their construction, will undoubtedly lead to a better understanding of the nature of turbulent transport, and this will lead in turn to a better organization of the empirical data, in other words, to more efficient models. An attractive feature of this line of attack is the possibility of matching the complexity of the model to the flow considered and to the level of empirical information currently available. We shall soon be able to choose among a variety of prediction techniques, ranging from integral models, adequate for a few simply developing flows, to very flexible multi-equation systems, whose use will be limited by the cost of programming and computing.

A note of caution must be added. With the level of realism attained in the turbulence models now working, meaningful predictions are possible only for those entities for which transport equations are provided, and only when the equations have been 'tuned' by careful optimization of their empirical constants. Lacking a fundamental understanding of the transfer processes, we have succeeded only in simulating their effects using equations which crudely represent the gross features of the turbulent activity.

9.2.3 Passive Transport Processes

So far we have considered the transport of entities which contribute to the turbulent activity and thus influence the mean flow as well. Now we discuss the simpler problem (simpler, once the velocity field is known) of calculating transfers which do not modify the fluid motion, for example, heat and mass transfers with small changes in fluid properties. The variation of a mean property S is given by equation (9.1.17); for a thin plane flow:

$$U\frac{\partial S}{\partial x}+V\frac{\partial S}{\partial y}=K\frac{\partial^2 S}{\partial y^2}=\frac{\partial}{\partial y}\left[\frac{\epsilon_m}{\sigma_s}\frac{\partial S}{\partial y}\right] \tag{9.2.23}$$

with $\epsilon_m = k^{\frac{1}{2}}l \propto k^2/\varepsilon$, on the assumption that there are no sources of S in the field. The distribution $S(x,y)$ can be found if the diffusivity ratio σ_s is specified and the flow field is known, that is, U, V, k, and l or ε.

It is also possible to calculate the distribution of fluctuation intensity; equation (9.1.18) gives

$$\frac{D(\frac{1}{2}\overline{s^2})}{Dt}=-\frac{\partial}{\partial y}(\tfrac{1}{2}\overline{vs^2})-\overline{sv}\frac{\partial S}{\partial y}+K\overline{s\frac{\partial^2 s}{\partial x_k^2}} \tag{9.2.24}$$

This is similar in structure to the transport equations considered earlier, in particular, to the energy equation (9.2.1), and can be scaled in a broadly similar manner. The empirical constants which appear in the skeletal equation can once again be selected to match measured values.

FURTHER READING

Jeffreys, H. *Cartesian Tensors*, Cambridge U. P. (1931)
Launder, B. E. and D. B. Spalding. Reference 6
Tennekes, H. and J. L. Lumley. Reference 12: Chapters 2 and 3
Townsend, A. A. Reference 13: Chapter 2

SPECIFIC REFERENCES

Bradshaw, P. 'The understanding and prediction of turbulent flow', *Aero. Journal*, **76**, pp. 403–418 (1972)

Bradshaw, P., D. H. Ferriss and N. P. Atwell. 'Calculation of boundary-layer development using the turbulent energy equation', *J. Fluid Mech.*, **28**, pp. 593–616 (1967)

Hanjalić, K. and B. E. Launder. 'A Reynolds stress model of turbulence and its application to thin shear flows', *J. Fluid Mech.*, **52**, pp. 609–638 (1972)

Rotta, J. C. 'Recent attempts to develop a generally applicable calculation method for turbulent shear flow layers', in Reference 17

Nee, V. W. and L. S. G. Kovasznay. 'Simple phenomenological theory of turbulent shear flows', *Phys. Fluids*, **12**, pp. 473–484 (1969)

EXAMPLES

9.1 Check that the terms of equations (9.1.7) and (9.1.11) have uniform tensor character.

9.2 For turbulence which is isotropic in the smaller scales, show that the viscous terms of equations (9.1.11) can be expressed

$$2\nu \overline{\frac{\partial u_i}{\partial x_k} \cdot \frac{\partial u_j}{\partial x_k}} = \tfrac{2}{3}\,\delta_{ij}\,\varepsilon$$

with δ_{ij} the Kronecker delta.

9.3 (a) For two-dimensional motion, show that the transport of vorticity obeys the kind of law which governs heat transfer. Why is this?

(b) Show that one of the Reynolds momentum equations can be written

$$\frac{\mathrm{D}U_1}{\mathrm{D}t} = -\frac{\partial(P/\rho + \frac{1}{2}\overline{u_k^2})}{\partial x_1} + \nu \frac{\partial^2 U_1}{\partial x_k^2} + \overline{u_2\,\omega_3} - \overline{u_3\,\omega_2}$$

displaying the mixing-stress term explicitly as an interaction between the fluctuating velocity and vorticity fields.

(c) Derive the complete result (9.1.15).

9.4 Why were the postulates (9.2.8, 9) not helpful until rapid computing facilities became available?

9.5 (a) Apply equation (9.2.13) to the wall layer (where convection and diffusion are negligible) and show that

$$\epsilon_{\mathrm{m}} \frac{\partial U}{\partial y} = \frac{\tau}{\rho} = c_D^{\frac{1}{2}} k$$

(b) By introducing the constant a of equations (9.2.4), show that $c_D \simeq 0{\cdot}09$ in the wall layer.

9.6 (a) By applying equations (9.2.13, 14, 15) to decaying grid turbulence, show that

$$\frac{\mathrm{d}Z}{\mathrm{d}k} = \frac{C_2 Z}{c_D\, k} \quad \text{whence} \quad Z \propto k^{C_2/c_D}$$

(b) Show also that $\mathrm{d}l/\mathrm{d}x \propto k^{\frac{1}{2}}$ and, taking $Z = k^m l^n$, that

$$q\frac{C_2}{c_D} = mq - n(1 - \tfrac{1}{2}q)$$

when the decay behind the grid follows the law $k \propto x^{-q}$.

(c) Taking $q = 1$, evaluate C_2 for the three choices of Z following equation (9.2.15).

9.7 (a) For the fully turbulent part of a wall layer, show that the assumption (9.2.4) implies that: k is constant; $l = c_D^{\frac{1}{4}} K y$; and $\epsilon_{\mathrm{m}} = c_D^{\frac{1}{4}} k^{\frac{1}{2}} K y$.

(b) Show that equations (9.2.14, 15) reduce (for $Z \propto l^n$) to

$$C_1 c_D = C_2 - \frac{c_D^{\frac{1}{2}} n^2 K^2}{\sigma_z}$$

with the further assumption that σ_z is constant.

(c) Evaluate the constant C_1 for the three choices of Z following equation (9.2.15). The diffusion of k is assumed negligible in these developments. Is this true of the diffusion of Z, represented by the last term of the result (b)?

9.8 (a) Starting from equation (9.1.16), develop transport equations for the transfer correlations $\overline{u_i s}$.

(b) Specialize to an equation for $\overline{vs}$ appropriate to thin flows.

(c) Find the particular result for the enthalpy flux $\overline{vh}$ in a thin flow, and interpret the several terms. The nature of the source term can be inferred from one of equations (9.1.6), multiplied by h.

Appendix: Molecular Transport Properties of Fluids

These data were extracted from the following sources:

Tables A.1 and A.2 from Reference 3, with permission of the McGraw-Hill Book Co.;

Table A.3 from *Liquid Metals Handbook* of the U. S. Atomic Energy Commission, by courtesy of the U. S. Department of the Navy;

Table A.4 from Reference 9, with permission of Prentice-Hall, Inc.;

Table A.5 from *Convective Heat Transfer* by D. B. Spalding, with permission of Edward Arnold (Publishers) and Professor D. B. Spalding.

Table A.1. Viscous and thermal properties of gases at atmospheric pressure

Gas	T °K	ρ kg/m³	c_p kJ/kg °K	μ μNsec/m²	ν mm²/sec	k mW/m °K	κ mm²/sec	Pr –
Air	100	3·601	1·027	6·92	1·923	9·25	2·50	0·770
	200	1·768	1·006	13·29	7·490	18·09	10·17	0·739
	300	1·177	1·006	19·83	15·68	26·24	22·16	0·708
	400	0·883	1·014	22·86	25·90	33·65	37·60	0·689
	500	0·705	1·030	26·71	37·90	40·38	55·64	0·680
	1000	0·352	1·142	41·52	117·8	67·52	167·8	0·702
	1500	0·236	1·230	54·0	229·1	94·6	326·2	0·705
	2000	0·176	1·338	65·0	369·0	124	526·0	0·702
	2500	0·139	1·688	75·7	543·5	175	744·1	0·730
Ammonia	273	0·793	2·177	9·35	11·8	22·0	13·08	0·90
	473	0·441	2·395	16·49	37·4	46·7	44·21	0·84
Carbon dioxide	250	2·166	0·804	12·59	5·813	12·88	7·401	0·793
	450	1·192	0·980	21·34	17·90	28·97	24·813	0·721
Carbon monoxide	250	0·841	1·043	15·40	11·28	21·44	15·06	0·750
	450	0·758	1·055	24·18	31·88	43·6	44·39	0·718
Helium	255	0·1906	5·200	18·17	95·50	135·7	136·75	0·70
	477	0·1020	5·200	27·50	269·3	197	371·6	0·72
Hydrogen	250	0·0982	14·06	7·919	80·64	146·1	113·0	0·713
	450	0·0546	14·50	11·779	215·6	251	316·4	0·682
Oxygen	250	1·562	0·916	17·87	11·45	22·59	15·79	0·725
	450	0·868	0·957	27·77	31·99	38·28	46·09	0·694
Water vapour	450	0·490	1·980	15·25	31·1	29·9	30·7	1·010
	650	0·338	2·056	22·47	66·4	46·4	66·6	0·995

Table A.2. Viscous and thermal properties of liquids in saturation states

Liquid	T °C	ρ kg/m³	c_p kJ/kg °K	ν mm²/sec	k mW/m °K	κ mm²/sec	Pr –	β 1/°K
Water	0	1002	4·218	1·778	552	0·1308	13·6	
	20	1001	4·182	1·006	597	0·1430	7·02	0·00018
	40	995	4·178	0·658	628	0·1512	4·34	
	60	985	4·184	0·478	651	0·1554	3·02	
	80	974	4·196	0·364	668	0·1636	2·22	
	100	961	4·216	0·294	680	0·1680	1·74	
	200	867	4·505	0·160	665	0·1706	0·937	
	300	714	5·728	0·135	540	0·1324	1·019	
Ammonia	−50	704	4·463	0·435	547	0·1742	2·60	
	20	612	4·798	0·359	521	0·1775	2·02	0·00245
	50	564	5·116	0·330	476	0·1654	1·99	
Carbon dioxide	−50	1156	1·84	0·119	85·5	0·04021	2·96	
	20	773	5·0	0·091	87·2	0·02219	4·10	0·0140
	30	598	36·4	0·080	70·3	0·00279	28·7	
Freon-12 ($C\,Cl_2F_2$)	−50	1547	0·875	0·310	67	0·0501	6·2	0·00263
	20	1330	0·966	0·198	73	0·0560	3·5	
	50	1216	1·022	0·190	67	0·0545	3·5	
Sulphur dioxide	−50	1561	1·360	0·484	242	0·1141	4·24	
	20	1386	1·365	0·210	199	0·1050	2·00	0·00194
	50	1299	1·368	0·162	177	0·0999	1·61	
Calcium chloride solution (29·9%)	−50	1320	2·608	36·35	402	0·1166	312·0	
	20	1287	2·788	2·72	498	0·1394	19·6	
	50	1273	2·868	1·65	535	0·1468	11·3	
Ethylene glycol	0	1131	2·29	57·53	242	0·0934	615	
	20	1117	2·38	19·18	249	0·0939	204	0·00065
	100	1059	2·74	2·03	263	0·0908	22·4	
Glycerin	0	1276	2·26	8310	282	0·0983	84700	
	20	1264	2·39	1180	286	0·0947	12500	0·00050
	50	1245	2·58	150	287	0·0893	1630	
Lubricating oil	0	899	1·796	4280	147	0·0911	47100	
	20	888	1·880	900	145	0·0872	10400	0·00070
	100	840	2·219	20·3	137	0·0738	276	
	160	806	2·483	5·6	132	0·0663	84	

Table A.3. Viscous and thermal properties of liquid metals

Metal	T °C	ρ kg/m³	c_p J/kg °K	ν mm²/sec	k W/m °K	κ mm²/sec	Pr –
Bismuth	316	10011	144·4	0·1617	16·4	11·38	0·0142
	538	9739	154·5	0·1133	15·6	10·35	0·0110
	760	9467	164·5	0·0834	15·6	10·01	0·0083
Lead	371	10540	159	0·2276	16·1	10·84	0·024
	482	10412	155	0·1849	15·6	12·23	0·017
Mercury	0	13628	140·3	0·124	8·20	4·299	0·0288
	100	13385	137·3	0·0928	10·51	5·716	0·0162
	200	13145	157·0	0·0802	12·34	6·908	0·0116
Potassium	149	807	800	0·4608	45·0	69·9	0·0066
	427	742	750	0·2397	39·5	70·7	0·0034
	704	674	750	0·1905	33·1	65·5	0·0029
Lead/bismuth (44½/55½)	371	10236	147	0·1496	11·86	7·90	0·0189

Table A.4. Diffusive properties of gases at 1 atm

Medium	Dilute diffusing substance	T °C	D mm²/sec	Sc –
Air	Ammonia	0	21·65	0·634
	Carbon dioxide	0	11·98	1·14
	Carbon tetrachloride	0	6·19	2·13
	Chlorine	0	9·28	1·42
	Hydrogen	0	54·72	0·25
	Methane	0	15·73	0·84
	Naphthalene	0	5·16	2·57
	Oxygen	0	15·33	0·90
	Toluene	0	6·96	1·86
	Water	8	20·62	0·615
		16	28·15	0·488
Nitrogen	Hydrogen	12·5	73·76	0·187
	Mercury	18·4	3250	0·00424
	Oxygen	11·8	20·25	0·681

Table A.5. Diffusive properties of liquids at 20°C

Medium	Dilute diffusing substance	Sc –
Water	Acetic acid	1140
	Ammonia	570
	Carbon dioxide	559
	Chlorine	824
	Common salt	745
	Glycerol	1400
	Hydrogen	196
	Lactose	2340
	Nitrogen	613
	Oxygen	558
	Phenol	1200
	Sucrose	2230
Ethyl alcohol	Carbon dioxide	445
	Chloroform	1230
	Phenol	1900

Index

This is primarily a subject index, but it also gives the names of workers mentioned in more than one connection in the text.